T0212468

Lecture Notes in Computer Science   10335

Commenced Publication in 1973
Founding and Former Series Editors:
Gerhard Goos, Juris Hartmanis, and Jan van Leeuwen

More information about this series at http://www.springer.com/series/7407

Domenico Salvagnin · Michele Lombardi (Eds.)

Integration of AI and OR Techniques in Constraint Programming

14th International Conference, CPAIOR 2017
Padua, Italy, June 5–8, 2017
Proceedings

 Springer

Editors
Domenico Salvagnin
University of Padua
Padua
Italy

Michele Lombardi ⓘ
University of Bologna
Bologna
Italy

ISSN 0302-9743   ISSN 1611-3349 (electronic)
Lecture Notes in Computer Science
ISBN 978-3-319-59775-1   ISBN 978-3-319-59776-8 (eBook)
DOI 10.1007/978-3-319-59776-8

Library of Congress Control Number: 2017943007

LNCS Sublibrary: SL1 – Theoretical Computer Science and General Issues

Printed on acid-free paper

This Springer imprint is published by Springer Nature
The registered company is Springer International Publishing AG
The registered company address is: Gewerbestrasse 11, 6330 Cham, Switzerland

Preface

This volume is a compilation of the research program of the 14th International Conference on the Integration of Artificial Intelligence and Operations Research Techniques in Constraint Programming (CPAIOR 2017), held in Padova, Italy, during June 5–8, 2017.

After a successful series of five CPAIOR international workshops in Ferrara (Italy), Paderborn (Germany), Ashford (UK), Le Croisic (France), and Montreal (Canada), in 2004 CPAIOR evolved into a conference. More than 100 participants attended the first meeting held in Nice (France). In the subsequent years, CPAIOR was held in Prague (Czech Republic), Cork (Ireland), Brussels (Belgium), Paris (France), Pittsburgh (USA), Bologna (Italy), Berlin (Germany), Nantes (France), Yorktown Heights (USA), Cork (Ireland), Barcelona (Spain), and Banff (Canada), in 2017 CPAIOR returned to Italy.

The aim of the CPAIOR conference series is to bring together researchers from constraint programming (CP), artificial intelligence (AI), and operations research (OR) to present new techniques or applications in the intersection of these fields, as well as to provide an opportunity for researchers in one area to learn about techniques in the others. A key objective of the conference is to demonstrate how the integration of techniques from different fields can lead to highly novel and effective new methods for large and complex problems. Therefore, papers that actively combine, integrate, or contrast approaches from more than one of the areas were especially welcome. Application papers showcasing CP/AI/OR techniques on innovative and challenging applications or experience reports on such applications were also strongly encouraged.

In all, 73 long and short papers were submitted to the conference. The papers underwent a rigorous peer-reviewing process, with each submission receiving at least three reviews. Overall, 36 papers were selected by the international Program Committee. Four of the accepted papers were then selected for a fasttrack publication process in the *Constraints* journal: only their abstracts appear in these proceedings.

The technical program of the conference was preceded by a master-class on "Computational Techniques for Combinatorial Optimization," with lectures from Tobias Achterberg, Laurent Michel, Pierre Schaus, and Pascal Van Hentenryck. The main program also enjoyed two invited talks, one from Andrea Lodi on "On the Role of (Machine) Learning in (Mathematical) Optimization" and one from Kristian Kersting on "Relational Quadratic Programming: Exploiting Symmetries for Modelling and Solving Quadratic Programs." The conference hosted one of the instantiations the DSO Workshop 2017, organized by the EURO Working Group on Data Science Meets Optimization.

Putting together a conference requires help from many sources. We want to thank the Program Committee and all other reviewers, who worked very hard in a busy period of the year: Their effort to provide detailed reviews and discuss all papers in depth after the author feedback to come up with a strong technical program is greatly appreciated. We are also deeply grateful to the staff from the Department of Information Engineering of the University of Padova for their great support in the event organization, to the chairs from past CPAIOR editions for their advice, and finally to the CPAIOR Steering Committee for giving us the opportunity to contribute to the conference series.

The cost of holding an event like CPAIOR would be much higher without the help of generous sponsors. We received outstanding support from Decision Brain, IBM Research, and Data61. We also thank Gurobi Optimization, GAMS, AIMMS, and COSLING. A final acknowledgment goes to EasyChair and Springer, who allowed us to put together these proceedings.

April 2017   Michele Lombardi
                                                                Domenico Salvagnin

Organization

Program Chairs

Domenico Salvagnin   University of Padova and IBM Italy, Italy
Michele Lombardi   University of Bologna, Italy

Conference Chair

Domenico Salvagnin   University of Padova and IBM Italy, Italy

Steering Committee

Nicolas Beldiceanu   Ecole des Mines de Nantes, France
Natashia Boland   Georgia Institute of Technology, USA
Bernard Gendron   Université de Montréal, Canada
Willem-Jan van Hoeve   Carnegie Mellon University, USA
John Hooker   Carnegie Mellon University, USA
Jeff Linderoth   University of Wisconsin-Madison, USA
Michela Milano   Università di Bologna, Italy
George Nemhauser   Georgia Institute of Technology, USA
Gilles Pesant   Ecole Polytechnique de Montréal, Canada
Jean-Charles Régin   Université de Nice-Sophia Antipolis, France
Louis-Martin Rousseau   Ecole Polytechnique de Montréal, Canada
Michael Trick   Carnegie Mellon University, USA
Pascal Van Hentenryck   University of Michigan, USA
Mark Wallace   Monash University, Australia

Program Committee

Chris Beck   University of Toronto, Canada
David Bergman   University of Connecticut, USA
Timo Berthold   Fair Isaac Germany, Germany
Pierre Bonami   IBM Spain, Spain
Hadrien Cambazard   Grenoble INP, France
Andre A. Cire   University of Toronto, Canada
Matteo Fischetti   University of Padova, Italy
Bernard Gendron   Université de Montréal, Canada
Ambros Gleixner   Zuse Institute Berlin – ZIB, Germany
Carla Gomes   Cornell University, USA
Tias Guns   Vrije Universiteit Brussel, Belgium
John Hooker   Tepper School of Business and Carnegie Mellon
                                University, USA

Matti Järvisalo	University of Helsinki, Finland
Serdar Kadioglu	Oracle Corporation, USA
Philip Kilby	NICTA and Australian National University, Australia
Joris Kinable	Carnegie Mellon University, USA
Jeff Linderoth	University of Wisconsin-Madison, USA
Andrea Lodi	Ecole Polytechnique de Montréal, Canada
Ines Lynce	Instituto Superior Tecnico – Lisboa, Portugal
Laurent Michel	University of Connecticut, USA
Michele Monaci	University of Bologna, Italy
Siegfried Nijssen	Leiden University, The Netherlands
Barry O'Sullivan	University College Cork and Insight Center, Ireland
Claude-Guy Quimper	Université Laval, Canada
Jean-Charles Régin	Université de Nice-Sophia Antipolis, France
Louis-Martin Rousseau	Ecole Polytechnique de Montréal, Canada
Ashish Sabharwal	Allen Institute for Artificial Intelligence, USA
Scott Sanner	University of Toronto, Canada
Pierre Schaus	UC Louvain, Belgium
Christian Schulte	KTH Royal Institute of Technology, Sweden
Helmut Simonis	University College Cork, Ireland
Christine Solnon	LIRIS CNRS UMR 5205, France
Peter-J. Stuckey	University of Melbourne, Australia
Michael Trick	Carnegie Mellon University, USA
Pascal Van-Hentenryck	University of Michigan, USA
Willem-Jan Van-Hoeve	Tepper School of Business and Carnegie Mellon University, USA
Sicco Verwer	Delft University of Technology, The Netherlands
Toby Walsh	UNSW and Data61, Australia
Alessandro Zanarini	ABB CRC, Switzerland
Yingqian Zhang	TU Eindhoven, The Netherlands

The CPAIOR 2017 conference was sponsored by:

Invited Talks (Abstracts)

On The Role of (Machine) Learning in (Mathematical) Optimization

Andrea Lodi

Polytechnique Montréal, Montreal, Canada
andrea.lodi@polymtl.ca

Abstract. In this talk, I try to explain my point of view as a Mathematical Optimizer – especially concerned with discrete (integer) decisions – on Big Data. I advocate a tight integration of Machine Learning and Mathematical Optimization (among others) to deal with the challenges of decision-making in Data Science. For such an integration I concentrate on three questions: (1) what can optimization do for machine learning? (2) what can machine learning do for optimization? (3) which new applications can be solved by the combination of machine learning and optimization? Finally, I will discuss in details two areas in which machine learning techniques have been (successfully) applied in the area of mixed-integer programming.

Relational Quadratic Programming: Exploiting Symmetries for Modelling and Solving Quadratic Programs

Kristian Kersting

TU Dortmund, Dortmund, Germany
kristian.kersting@cs.tu-dortmund.de

Abstract. Symmetry is the essential element of lifted inference that has recently demonstrated the possibility to perform very efficient inference in highly-connected, but symmetric probabilistic models models aka. relational probabilistic models. This raises the question, whether this holds for optimization problems in general. In this talk I shall demonstrate that for a large class of mathematical programs this is actually the case. More precisely, I shall introduce the concept of fractional symmetries of linear and convex quadratic programs (QPs), which lie at the heart of many machine learning approaches, and exploit it to lift, i.e., to compress them. These lifted QPs can then be tackled with the usual optimization toolbox (off-the-shelf solvers, cutting plane algorithms, stochastic gradients etc.): If the original QP exhibits symmetry, then the lifted one will generally be more compact, and hence their optimization is likely to be more efficient. This talk is based on joint works with Martin Mladenov, Martin Grohe, Leonard Kleinhans, Pavel Tokmakov, Babak Ahmadi, Amir Globerson, and many others.

Fast Track Papers
for the "Constraints" Journal

Auto-tabling for Subproblem Presolving
in MiniZinc (Summary)

Jip J. Dekker[1], Gustav Björdal[1], Mats Carlsson[2], Pierre Flener[1],
and Jean-Noël Monette[3]

[1] Uppsala University, Uppsala, Sweden
Pierre.Flener@it.uu.se
[2] SICS, Kista, Sweden
Mats.Carlsson@sics.se
[3] Tacton Systems AB, Stockholm, Sweden
Jean-Noel.Monette@tacton.com

If poor propagation is achieved within part of a constraint programming (CP) model, then a common piece of advice is to table that part. *Tabling* amounts to computing all solutions to that part and replacing it by an extensional constraint requiring the variables of that part to form one of these solutions. If there are not too many solutions, then the hope is that the increased propagation leads to faster solving. While powerful, the tabling reformulation is however often not tried, because it is tedious and error-prone to perform; further, it obfuscates the original model if it is actually performed.

In [1], in order to encourage modellers to try tabling, we extend the MiniZinc toolchain to perform the automatic tabling of suitably annotated constraint predicates, without requiring any changes to solvers, thereby eliminating both the tedium and the obfuscation. We introduce a `presolve(autotable)` annotation and apply it on four case studies: the black-hole patience problem, the block party metacube problem, the JP-encoding problem, and the handball tournament scheduling problem.

We see three advantages to our extension. First, the modeller is more likely to experiment with the well-known and powerful tabling reformulation if its tedium is eliminated by automation. Second, the original model is not obfuscated by tables, at the often negligible cost of computing them on-the-fly. Our experiments show that automated tabling yields the same tables as manual tabling. Third, tabling is technology-neutral: our experiments show that tabling can be beneficial for CP backends, with or without lazy clause generation, constraint-based local search (CBLS) backends, Boolean satisfiability (SAT) backends, SAT modulo theory (SMT) backends, and hybrid backends; we have no evidence yet that a mixed-integer programming (MIP) backend to MiniZinc can benefit from tabling.

Modelling is an art that is difficult to master: finding a model that is solved efficiently on at least one solver is often difficult. We argue that our contribution lowers the skill level required for designing a good model: the way an auto-tabled predicate is formulated matters very little, as the *same* table will be generated.

Reference

1. Dekker, J.J., Björdal, G., Carlsson, M., Flener, P., Monette, J.N.: Auto-tabling for subproblem presolving in MiniZinc. In: Constraints Journal Fast Track of CPAIOR 2017 (2017)

Cumulative Scheduling with Variable Task Profiles and Concave Piecewise Linear Processing Rate Functions (Abstract)

Margaux Nattaf, Christian Artigues, and Pierre Lopez

LAAS-CNRS, Université de Toulouse, INSA, CNRS, Toulouse, France
{nattaf,artigues,lopez}@laas.fr

We consider a cumulative scheduling problem where a task duration and resource consumption are not fixed. The consumption profile of the task, which can vary continuously over time, is a decision variable of the problem to be determined and a task is completed as soon as the integration over its time window of a non-decreasing and continuous processing rate or efficiency function of the consumption profile has reached a predefined amount of energy. This model is well suited to represent energy consuming tasks, for which the instantaneous power can be modulated at any time while a global amount energy has to be received the task through a non linear efficiency function. This is for example the case of melting operation in foundries.

The goal is to find a feasible schedule, which is an NP-complete problem. Previous studies considered identity and linear processing rate functions [1]. In this paper we study the case where processing rate functions are concave and piecewise linear. This is motivated by the fact that there exist many real-world concave processing rate functions. Besides, approximating a function by a concave piecewise linear one is always at least as good as an approximation by a linear function. We present two propagation algorithms. The first one is the adaptation to concave functions of the variant of the energetic reasoning previously established for linear functions. Furthermore, a full characterization of relevant intervals for time-window adjustments is provided. The second algorithm combines a flow-based checker with time-bound adjustments derived from the time-table disjunctive reasoning for the cumulative constraint. Complementarity of the algorithms is assessed via their integration in a hybrid branch-and-bound and computational experiments on a set of randomly generated instances. This abstract refers to the full paper [2].

References

1. Nattaf, M., Artigues, C., Lopez, P., Rivreau, D.: Energetic reasoning and mixed-integer linear programming for scheduling with a continuous resource and linear efficiency functions. OR Spectrum 1–34 (2015)
2. Nattaf, M., Artigues, C., Lopez, P.: Cumulative scheduling with variable task profiles and concave piecewise linear processing rate functions constraints (2017, to appear)

Efficient Filtering for the Resource-Cost AllDifferent Constraint

Sascha Van Cauwelaert and Pierre Schaus

UCLouvain, Louvain-la-Neuve, Belgium
{sascha.vancauwelaert,pierre.schaus}@uclouvain.be

Abstract. High energy consuming industries are increasingly concerned about the energy prices when optimizing their production schedule. This is by far due to today politics promoting the use of renewable energies. A consequence of renewable energies is the high fluctuation of the prices. Today one can forecast accurately the energy prices for the next few hours and up to one day in advance. In this work, we consider a family of problems where a set of items, each requiring a possibly different amount of energy, must be produced at different time slots. The goal is to schedule them such that the overall energy bill is minimized. In Constraint Programming, one can model this cost in two ways: (a) with a sum of ELEMENT constraints; (b) with a MINIMUMASSIGNMENT constraint. In the latter case, the cost of an edge linking a variable (i.e., an item) to a value (i.e., a time slot) is the product of the item consumption with the energy price of the time slot. Unfortunately, both approaches have limitations. Modeling with a sum of ELEMENT constraints does not take into account the fact that the items must all be assigned to different time slots. The MINIMUMASSIGNMENT incorporates the ALLDIFFERENT constraint, but the algorithm has a quite high time complexity of $\mathcal{O}(n^3)$, where n is the number of items. This work proposes a third approach by introducing the RESOURCECOSTALLDIFFERENT constraint [1] and an associated scalable filtering algorithm, running in $\mathcal{O}(n.m)$, where m is the maximum domain size. Its purpose is to compute the total cost by dealing with the fact that all assignments must be different in a scalable manner. The constraint RESOURCECOSTALLDIFFERENT(X, C, P, T) ensures that

$$\wedge \begin{cases} \text{ALLDIFFERENT}(X) \\ T = \sum_{i=1}^{|X|} C(X_i) * P(X_i) \end{cases}$$

where X is a sequence of integer variables; C is a sequence of $|X|$ integer constants; P is a sequence of H integer constants ($H \geq |X|$); and T is an integer variable that is the total resource cost. We first evaluate the efficiency of the new filtering on a real industrial problem and then on the Product Matrix Travelling Salesman Problem, a special case of the Asymmetric Travelling Salesman Problem. The study shows experimentally that our approach generally outperforms the decomposition and the MINIMUMASSIGNMENT ones.

Reference

1. Van Cauwelaert, S., Schaus, P.: Efficient filtering for the resource-cost all different constraint. Constraints (2017)

Mining Time-Constrained Sequential Patterns with Constraint Programming

John O.R. Aoga[1] (iD), Tias Guns[2,3] and Pierre Schaus[1]

[1] UCLouvain, ICTEAM, Louvain-la-Neuve, Belgium
{john.aoga,pierre.schaus}@uclouvain.be
[2] VUB Brussels, Brussels, Belgium
tias.guns@vub.ac.be
[3] KU Leuven, Leuven, Belgium
tias.guns@cs.kuleuven.be

This abstract is related to the original paper in the Constraint Journal [1].

Constraint Programming has proven to be an effective platform for constraint-based sequential pattern mining. Previous work has focussed on standard sequential pattern mining, as well as sequential pattern mining with a maximum 'gap' between two matching events in a sequence. The main challenge in the latter is that this constraint can not be imposed independently of the omnipresent frequency constraint. In this work, we go beyond that and investigate the integration of timed events, the gap constraint as well as the span constraint that constrains the time between the first and last matching event. We show how the three are interrelated, and what the required changes to the frequency constraint are.

Our contributions can be summarized as follows: (1) we have the backtracking-aware datastructure to store all possible occurences of the pattern in a sequence, including the first matching symbol to support *span* constraints; (2) we avoid scanning a sequence for a symbol beyond the (precomputed) last occurence of that symbol in the sequence; (3) we introduce the concept of *extension window* of an embedding and avoid to scan overlapping windows multiple times; (4) we avoid scanning for the start of an extension window, which is specific to the gap constraint, by precomputing these in advance; and finally (5) we experimentally demonstrate that the proposed approach outperforms both specialised and CP-based approaches in most cases, and that the difference becomes increasingly large for low frequency thresholds. Furthermore, this time-aware global constraint can be combined with other constraints such as regular expression constraints, item inclusion/exclusion constraints or pattern length constraints.

Our proposal, called Prefix Projection Incremental Counting with time restrictions Propagator (*PPICt*), is implemented in CP-Solver Oscar[1].

Reference

1. Aoga, J.O.R., Guns, T., Schaus, P.: Mining time-constrained sequential patterns with constraint programming. In: Salvagnin, D., Lombardi, M. (eds.) CPAIOR 2017, LNCS, vol. 10335, p. XX. Springer, Switzerland (2017)

---

[1] The data and code are available at http://sites.uclouvain.be/cp4dm/spm/ppict/.

Contents

Technical Papers

Sharpening Constraint Programming Approaches for Bit-Vector Theory

Zakaria Chihani$^{(\boxtimes)}$, Bruno Marre, François Bobot, and Sébastien Bardin

CEA, LIST, Software Security Lab, Gif-sur-Yvette, France
{zakaria.chihani,bruno.marre,francois.bobot,sebastien.bardin}@cea.fr

Abstract. We address the challenge of developing efficient Constraint Programming-based approaches for solving formulas over the quantifier-free fragment of the theory of bitvectors (BV), which is of paramount importance in software verification. We propose CP(BV), a highly efficient BV resolution technique built on carefully chosen anterior results sharpened with key original features such as thorough domain combination or dedicated labeling. Extensive experimental evaluations demonstrate that CP(BV) is much more efficient than previous similar attempts from the CP community, that it is indeed able to solve the majority of the standard verification benchmarks for bitvectors, and that it already complements the standard SMT approaches on several crucial (and industry-relevant) aspects, notably in terms of scalability w.r.t. bit-width, theory combination or intricate mix of non-linear arithmetic and bitwise operators. This work paves the way toward building competitive CP-based verification-oriented solvers.

1 Introduction

Context. Not so long ago, program verification was such an ambitious goal that even brilliant minds decided it was "bound to fail" [37]. At the time, the authors concluded their controversial paper saying that if, despite all their reasons, "verification still seems an avenue worth exploring, so be it". And so it was. Today, software verification is a well established field of research, and industrial adoption has been achieved is some key areas, such as safety-critical systems.

Since the early 2000's, there is a significant trend in the research community toward reducing verification problems to satisfiability problems of first-order logical formulas over well-chosen theories (e.g. bitvectors, arrays or floating-point arithmetic), leveraging the advances of modern powerful SAT and SMT solvers [6,30,38,43]. Besides weakest-precondition calculi dating back to the 1970's [19], most major recent verification approaches follow this idea [15,25,29,35].

The problem. While SMT and SAT are the *de facto* standard in verification, a few teams explore how Constraint Programming (CP) techniques can be used in

Work partially funded by ANR under grants ANR-14-CE28-0020 and ANR-12-INSE-0002. The CP solver COLIBRI is generously sponsored by IRSN, the French nuclear safety agency.

© Springer International Publishing AG 2017
D. Salvagnin and M. Lombardi (Eds.): CPAIOR 2017, LNCS 10335, pp. 3–20, 2017.
DOI: 10.1007/978-3-319-59776-8_1

that setting [16, 26, 27, 33, 42]. Indeed, CP could in principle improve over some of the well-known weaknesses of SMT approaches, such as non-native handling of finite domains theories (encoded in the Boolean part of the formula, losing the high-level structure) or very restricted theory combinations [39].

Yet, currently, there is no good CP-based resolution technique for (the quantifier-free fragment of) the theory of bitvectors [30], i.e. fixed-size arrays of bits equipped with standard low-level machine instructions, which is of paramount importance in verification since it allows to encode most of the basic datatypes found in any programming language.

Goal and challenge. We address the challenge of developing efficient Constraint Programming-based approaches for solving formulas over the quantifier-free fragment of the theory of bitvectors (BV). Our goal is to be able to solve many practical problems arising from verification (with the SMTCOMP challenge[1] as a benchmark) and to be at least complementary to current best SMT approaches. The very few anterior results were still quite far from these objectives [44], even if preliminary work by some of the present authors was promising on conjunctive-only formulas [1].

Proposal and contributions. We propose CP(BV), a highly efficient BV resolution technique built on carefully chosen anterior results [1, 36] sharpened with key original features such as thorough domain combination or dedicated labeling. Extensive experimental evaluations demonstrate that CP(BV) is much more efficient than previous similar attempts from the CP community, that it is indeed able to solve the majority of the standard verification benchmarks for bitvectors, and that it already complements the standard SMT approaches on several crucial (and industry-relevant) aspects, notably in terms of scalability w.r.t. bitwidth, theory combination or intricate mix of non-linear arithmetic and bitwise operators. Our main contributions are the following:

- We present CP(BV), an original framework for CP-based resolution of BV problems, which built on anterior results and extend them with several key new features in terms of thorough domain combination or dedicated labeling. This results in a competitive CP-based solver, excelling in key aspects such as scalability w.r.t. bitwidth and combination of theories. A comparison of CP(BV) with previous work is presented in Table 1.
- We perform a systematic and extensive evaluation of the effect of our different CP improvements on the efficiency of our implementation. This compares the options at our disposal and justifies those we retained, establishing a firm ground onto which future improvements can be made. It also shows the advantage of our approach relative to the other CP approaches applied to BV, and that our approach is able to solve a very substantial part of problems from the SMTCOMP challenge.
- Finally, we perform an extensive comparison against the best SMT solvers for BV problems, namely: Z3 [8], Yices [20], MathSAT [14], CVC4 [4] and

---

[1] smtcomp.sourceforge.net.

Table 1. Overview of our method

	Bardin *et al.* [1]	Michel *et al.* [36]	CP(BV)
Bitvector domain	+	++	++ [36]
Arithmetic domain	++	+	++ [1]
Domain combination	+	+	++
Simplifications	+	x	++
BV-aware labeling	x	x	++
Implemented	Yes	No	Yes
Benchmark	Conjunctive formulas $\approx$ 200 formulas		Arbitrary formulas $\approx$ 30,000 formula

Boolector [11]. This comparison exhibits the strenghts of CP over SMT approaches on particular instances. Specifically, our implementation surpasses several (and sometimes *all*) solvers on some examples involving large bit-vectors and/or combination with floating-point arithmetic.

Discussion. Considering that our current CP(BV) approach is far from optimal compared to existing SMT solvers (implemented in Prolog, no learning), we consider this work as an important landmark toward building competitive CP-based verification-oriented solvers. Moreover, our approach clearly challenges the well-accepted belief that bitvector solving is better done through bitblasting, opening the way for a new generation of word-level solvers.

2 Motivation

The standard (SMT) approach for solving bit-vector problem, called *bit-blasting* [7], relies on a boolean encoding of the initial bitvector problem, one boolean variable being associated to each bit of a bitvector. This *low-level encoding* allows for a direct reuse of the very mature and ever-evolving tools of the SAT community, especially DPLL-style SAT solvers [38,43]. Yet, crucial high-level structural information may be lost during bitblasting, leading to potentially poor reasoning abilities on certain kinds of problems, typically those involving many arithmetic operations [12] and large-size bitvectors. Following anterior work [1,36], we propose a *high-level encoding* of bitvector problems, seen as Constraint Satisfaction Problems (CSP) over finite (but potentially huge) domains. Each bitvector variable of the original BV problem is now considered as a (CSP) bounded-arithmetic variable, with dedicated domains and propagators.

Now illustrating with concrete examples, we show the kind of problems where our approach can surpass existing SMT solvers. Consider the three following formulas:

$$x \times y = (x \mathbin{\&} y) \times (x \mid y) + (x \mathbin{\&} \bar{y}) \times (\bar{x} \mathbin{\&} y) \tag{A}$$

$$x_1 < x_2 < \cdots < x_n < x_1 \tag{B}$$

$$(\bigwedge_{i=1}^{n-2} x_i < x_{i+1} \mathbin{\&} x_{i+2}) \wedge (x_{n-1} < x_n \mathbin{\&} x_1) \wedge (x_n < x_1 \mathbin{\&} x_2) \tag{C}$$

where $\bar{\cdot}$ (resp. $\cdot$ & $\cdot$, $\cdot$ | $\cdot$) is the bit-wise negation (resp. conjunction, disjunction), $\wedge$ is the logical conjunction, $<$ is an unsigned comparison operator, n was chosen to be 7. As an example, the SMT-LIB language [5] encoding of formula A is:

```
(assert (= (bvmul X Y) (bvadd (bvmul (bvand X Y) (bvor X Y))
(bvmul (bvand X (bvnot Y)) (bvand (bvnot X) Y)))))
```

Table 2 shows the time in seconds according to bit-vector size, both for the satisfiability proof of the *valid* formula A and the unsatisfiablity of B and C. CP(BV) is the name of our technique, and TO means that the solver was halted after a 60-second timeout.

Table 2. Comparison of performance (time in sec.) for different solvers

Formula	size(bits)	Z3	Yices	MathSAT	CVC4	Boolector	CP(BV)
A	512	TO	1.60	6.04	17.28	20.55	0.24
	1024	TO	7.25	26.72	TO	TO	0.23
	2048	TO	31.83	TO	TO	TO	0.23
B	512	0.53	0.82	1.37	0	2.75	0.26
	1024	1.75	4.89	4.23	0	7.39	0.22
	2048	5.73	16.15	22.76	0	16.81	0.21
C	512	0.15	0.85	1.55	0.76	3.15	0.25
	1024	0.33	1.25	4.53	3.49	3.81	0.22
	2048	0.70	5.55	19.57	8.82	14.73	0.25

On these examples, CP(BV) clearly surpasses SMT solvers, reporting no TO and a very low (size-independent) solving time. In light of this, we wish to emphasize the following advantages of our CP(BV) high-level encoding for bitvector solving:

- Each variable is attached to *several and complementary domain representations*, in our case: *intervals plus congruence, bitlist* [1] (i.e. constraints on the possible values of specific bits of the bitvector) and *global difference constraint* (delta). Each domain representation comes with its own *constraint propagation* algorithms and deals with different aspects of the formula to solve. We will call *integer domains* or *arithmetic domains* those domains dealing mainly with high-level arithmetic properties (here: intervals, congruence, deltas), and *BV domains* or *bitvector domains* those domains dealing with low-level aspects (here: bitlist).
- These domains can *freely communicate between each other*, each one refining the other through a *reduced product* (or *channeling*) mechanism [41]. For example, in formula B, adding the difference constraint to the delta domain does allow CP(BV) to conclude *unsat* directly at propagation. Having multiple domains also allows to search for a solution in the smallest of them (in terms of the cardinality of the concretization of the domain abstraction).

- In the case of formula C, a reduced product is not enough to conclude at propagation. Here, a BV constraint *itself* refines an arithmetic domain: indeed, with the simple observation that, if a & $b = c$ then $c \leqslant a$ and $c \leqslant b$, the bit-vector part of CP(BV) not only acts on the bit-vector representation of the variables, but also "informs" the global difference constraints of a link it could not have found on its own.

3 Background

This section lays down the ground on which our research was carried out, both the theoretical foundations, anterior works and the COLIBRI CP solver [33].

3.1 BV Theory

We recall that BV is the quantifier-free theory of bitvectors [30], i.e. a theory where variables are interpreted over fixed-size arrays (or vectors) of bits along with their basic operations: logical operators (conjunction " & ", disjunction "|" and xor "$\oplus$", *etc.*), modular arithmetic (addition $+$, multiplication $\times$, *etc.*) and other structural manipulations (concatenation $::$, extraction $\lfloor . \rfloor_{i.j}$, etc.).

3.2 CP for Finite-Domain Arithmetic

A *Constraint Satisfaction Problem* [18] (CSP) consists in a finite set of variables ranging over some domains, together with a set of constraints over these variables – each constraint defining its own set of solutions (valuations of the variables that satisfy the constraint). Solving a CSP consists in finding a solution meeting all the constraints of the CSP, or proving that no such solution exists. We are interested here only in the case where domains are finite. *Constraint Programming* [18] (CP) consists in solving a CSP through a combination of *propagation* and *search*. Propagation consists mainly in reducing the potential domains of the CSP variables by deducing that some values cannot be part of any solution. Once no more propagation is possible, search consists in assigning a value to a variable (taken from its reduced domain) and continue the exploration, with backtrack if required.

The CP discipline gets its strenght from global constraints and capabilities for dense interreductions between domain representations, along with constraint solving machinery. We present in this section standard domains and constraints for bounded arithmetic.

Union of intervals. A simple interval $[a; d]$, where $a, d \in N$ represents the fact that a given variable can take values only between a and d. A natural extension of this notion is the *union of intervals* (Is). As a shortened notation, if $x \in \{a\} \uplus [b; c] \uplus \{d\}$, one writes $[x] = [a, b{\cdot}{\cdot}c, d]$.

Congruence [31]. If the division remainder of a variable x by a divisor d is r (*i.e.*, x satisfies the equation $x\%d = r$), then $\langle x \rangle = r[d]$ is a *congruence* and represents all values that variable x can take, *e.g.*, $\langle x \rangle = 5[8]$ means $x \in \{5, 13, 21, \ldots\}$.

Global difference constraint (Delta) [23]. A global difference constraint is a set of linear constraints of the form $x - y \diamond k$ where $\diamond \in \{=, \neq, <, >, \leqslant, \geqslant\}$. Tracking such sets of constraints allows for better domain propagation and early infeasibility detection, thanks to a global view of the problem compared with the previous (local) domains.

3.3   The COLIBRI Solver for FD Arithmetic

The COLIBRI CP solver [33] was initially developped to assist CEA verification tools [2,9,17,47]. COLIBRI supports bounded integers (both standard and modular arithmetic [28]), floating-points [34] and reals. Considering arithmetic, COLIBRI already provides all the domains described in Sect. 3.2, together with standard propagation techniques and strong interreductions between domains. Search relies mostly on a standard fail-first heuristics. COLIBRI is implemented in ECLiPSe Prolog, yielding a significant performance penalty (compared with compiled imperative languages such as C or C++) but allowing to quickly prototype new ideas.

3.4   Former CP(BV) Approaches

Two papers must be credited with supplying the inspiration for this work, written by Bardin et al. [1] and by Michel and Van Hentenryck [36]. Put together, these two papers had good ideas, which we adopted, unsatisfactory ideas which were disgarded, and finally ideas that were not advanced enough which we extended.

The first paper [1] introduces the bitlist domain, i.e. lists of four-valued items ranging over $\{0, 1, ?, \perp\}$ – indicating that the ith bit of a bitvector must be 0, 1, any of these two values (?), or that a contradiction has been found ($\perp$) – together with its propagators for BV operators. Moreover, the authors also explain how arithmetic domains (union of intervals and congruence) can be used for BV, and describe first interreduction mechanisms between bitlists and arithmetic domains.

The second paper [36] introduces a very optimized implementation of bitlists, using two BVs $\langle {}^1x, {}^0x \rangle$ to represent the bitlist of x, where ${}^1x$ (resp. ${}^0x$) represents bits known to be set (resp. cleared) in x. The efficiency comes from the use of machine-level operations for performing domain operations, yielding constant time propagation algorithms for reasonable bitvector sizes.

Basically, we improve the technique described in [1] by: borrowing the optimized bit-vector domain representation from [36] (with a few very slight improvements), significantly improving inter-domain reduction, and designing a BV-dedicated search labeling strategy.

As improving inter-domain reduction is one of our key contributions, we present hereafter the reductions between BV domains and arithmetic domains described in [1]:

With congruence: the BV domain interacts according to the longest sequence of known least significant bits. For example, a BV domain $[\![10?00?101]\!]$ of a variable b indicates that b satisfies the equation $b[8] = 5$, which therefore constrains the congruence domain using $5[8]$. Conversely a known congruence of some power of 2 fixes the least significant bits.

With _Is_: for a variable x, the union of intervals can refine the most significant bits of the BV domain by clearing bits according to the power of two that is immediately greater than the maximum extremum of the _Is_. And the BV domain influences _Is_ by (only) refining the extrema through the maximum and minimum bit-vectors allowed, and by removing _singletons_ that do not conform to the BV domain.

4 Boosting CP(BV) for Efficient Handling of Bit-Vectors

In this section, we delve into the specificities of our approach, leaving complex details to a technical report[2]. We first start by presenting a significantly improved inter-domain reduction, followed by new reductions from BV constraints to other domains, then we show some simplifications and factorisations at the constraint level, and we finish by presenting our BV-dedicated labeling strategy.

4.1 Better Interreduction Between BV- and Arithmetic- Domains

We present here several significant improvements to the inter-domain reduction techniques proposed in [1] between bitlists and unions of intervals. Our implementation also borrows the inter-reduction between bitlists and congruence from [1].

_Is_ to BVs. Let m and M be respectively the minimal and the maximal value of a union of intervals. Then, the longest sequence of most-significant bits on which they "agree" can also be fixed in the bit-vector domain. For example, $m = 48$ and $M = 52$ (00110000 and 00110100 in binary) share their five most-significant bits, denoted $[\![00110???]\!]$. Therefore, a bit-vector $bl = [\![0??1???0]\!]$ can be refined into $[\![00110??0]\!]$. For comparison, the technique in [1] only reduces bl to $[\![00?1???0]\!]$.

BV to _Is_. Consider a variable b with a _Is_ domain $[b] = [153, 155, 158 \cdot\cdot 206, 209]$, and a bit-vector domain $(\!|b|\!) = [\![1??1??01]\!] = \{\cdots, 153, 157, 177, 181, 185, 189, 209, \cdots\}$, as illustrated in Fig. 1. The inter-domain reduction from [1] can refine the extremum of the _Is_ (here: nothing to do, since 153 and 209 both conforms to $(\!|b|\!)$) and removes the singletons that are not in $(\!|b|\!)$ (here: 155), yielding $[b] = [153, 158 \cdot\cdot 206, 209]$. We propose to go a step further by refining each bound inside a _Is_, such that after reduction each bound of $[b]$ conforms to $(\!|b|\!)$. Here, 158 (resp. 206) is not allowed and should be replaced by its closest upper (resp. lower) value in $(\!|b|\!)$, i.e. 177 (resp. 189), yielding $[b] = [153, 177 \cdot\cdot 189, 209]$.

We have designed such a correct and optimal reduction algorithm from bitlist to _Is_. Since we work on the $\langle {}^1x, {}^0x \rangle$ representation of bitlists, the algorithm relies

---

[2] sites.google.com/site/zakchihani/cpaior.

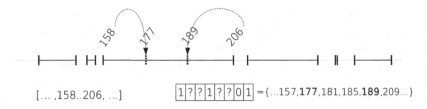

Fig. 1. Enlarging the gaps in the *Is* according to the BV domain

on machine-level operations and is linear in the size of the bitvector (cf. technical report). We describe the procedure for increasing a lower bound in Algorithm 1; decreasing the upper bound (symmetrically) follows the same principle (cf. technical report). In order to calculate a new bound r accepted by the bit-vector domain $(\!(b)\!)$, we start by imposing on the lower bound l what we already know, i.e., set what is set in ^1b and clear what is cleared in ^0b (line 1 of Algorithm 1). Then flag the bits that were changed by this operation, going from cleared to set and from set to cleared.

To refine the lower bound, we must raise it as much as necessary but not one bit higher, i.e., we should look for the smallest amount to add to the lower bound in order to make it part of the concretisation of $(\!(b)\!)$. This entails two things: a cleared bit can only become set if all bits of lower significance get cleared. For example, to increase the binary represented integer **0**10 exactly until the leftmost bit gets set, we will pass by 011 and *stop* at **1**00 : going to 101 would increase more than strictly necessary. Similarly, the *smallest* increase that clears a set bit i is one where the *first cleared bit on the left* of i can be set (line 12 of Algorithm 1, function `left-cl-can-set-of`). For example, to clear the third most significant bit

Algorithm 1. Increasing the lower bound l according to $(\!(b)\!)$

1: $r := {^1b} \mid l \;\&\; {^0b}$
2: $set2cl := l \;\&\; \bar{r}$
3: $cl2set := \bar{l} \;\&\; r$
4: **if** $cl2set > set2cl$ **then**
5:      $size := log_2\,(cl2set)$
6:      $mask0 := -1 \ll size$
7:      $can\text{-}cl := mask0 \mid {^1b}$
8:      $r := r \;\&\; can\text{-}cl$
9: **else**
10:     $size := log_2\,(set2cl)$
11:     $cl\text{-}can\text{-}set := \bar{r} \;\&\; {^0b}$
12:     $next\text{-}to\text{-}set := left\text{-}cl\text{-}can\text{-}set\text{-}of(size,$ $cl\text{-}can\text{-}set)$
13:     $r := set(r, next\text{-}to\text{-}set)$
14:     $mask0 := -1 \ll next\text{-}to\text{-}set$
15:     $can\text{-}cl := mask0 \mid {^1b}$
16:     $r := r \;\&\; can\text{-}cl$
17: **end if**

(in bold) in 01**1**011, one needs to increase to 011100, 011101, 011110, 011111 then reach 100000. Doing so clears not only the target bit i but all the bits of lower significance.

Drilling the *Is* according to BV. If the *Is* contains only one interval, then our technique does not improve over [1], and is only slighly superior to the channeling method of [36]. For this reason, we force the bit-vector domain to create at least one gap in the union of intervals. Consider for example a domain $bl = [\![0?10?1?]\!]$.

When observing the concretisation $\{18, 19, 22, 23, 50, 51, 54, 55\}$, the largest gap is between 23 and 50, *i.e.*, 0010111 and 0110010, obtained by fixing the *most significant unknown bit (msub)*. More generally, for a variable x the largest gap is created by intersecting $[x]$ with $[{}^1x \cdot \cdot a, b \cdot \cdot {}^0x]$, where a is obtained by clearing the *msub* and setting all other unknown bits, and b is obtained by setting the *msub* and clearing all other unknown bits. One can of course enforce more gaps, but there is a tradeoff between their propagation cost (as *Is*) and their benefits. In this work, using one gap was satisfactory.

4.2 BV Constraints Reducing Arithmetic Domains

Our CP(BV) approach strives to keep each of its domains as aware as possible of the other domains. We now show how non-BV domains can be reduced through BV constraints. In the following, we recall that $[x]$ denotes the union of intervals attached to variable x.

4.2.1 BV Constraints on *Is*

It turns out that most BV constraints can influence unions of intervals.

Bitwise binary operands: a disjunction $x \mid y = z$ can refine $[z]$ in more than one way, but experimentation showed a notable effect only when $[x]$ and $[y]$ contain only singletons, at which case $[z]$ can be refined by the pairwise disjunction of those singletons. Similar refinements can occure through the bitwise conjunction and exclusive disjunction. For the latter, one can also refine in the same manner the *Is* of the operands, since $x \oplus y = z$ implies the same constraint for all permutation of x, y, z.

Negation: from a negation constraint $x = \overline{y}$, one can refine the *Is* of one variable from that of the other. By mapping each singleton $\{c\} \in [x]$ and interval $a \cdot \cdot b \in [x]$ to $\{\overline{c}\}$ and $\overline{b} \cdot \cdot \overline{a}$, we build a *Is* to populate $[y]$. The symmetric construction populates $[x]$.

Shifts: from a right shift $x \gg y = z$, which is equivalent to a natural division (*i.e.*, $x/2^y = z$), one can refine $[z]$ simply by right-shifting all elements of $[x]$ (singletons and bounds of internal intervals) by y. The left-shift constraint is treated mostly in the same way but requires extra care as it is a modular multiplication and it can overflow.

Sign-Extension: when extending the sign of x by i positions to obtain z, the method consists in splitting the $[x]$ by the integers that are interpreted as negative, most significant bit is 1, and the one interpreted as positive, most significant bit is 0 and to apply the sign extention separately, disjunction with $2^i - 1 \ll i$ for the firsts and identity for the seconds.

Extractions: when extracting from the left-most to any position, it's the same as a right logical shift. The more general case is tricky. Take $\lfloor x \rfloor_{i.j} = y$ to mean the extraction from bit i to j of x to obtain y (with $\|x\| > i \geqslant j \geqslant 0$), then a singleton in for an interval $x_a \cdot \cdot x_b$,

- If $(x_b \gg j) - (x_a \gg j) > 2^{i-j}$, then the interval necessarily went through all integer coded on i bits, so the integer domain cannot be refined.
- else, if $(x_b \oplus x_a)$ & $2^i = 0$, then no power of 2 was traversed, the bounds can simply be truncated and stay in that order: $(\lfloor x_a \rfloor_{i.j}) \cdots (\lfloor x_b \rfloor_{i.j})$
- else, 2^i was traversed, then the Is is $[0 \cdots (\lfloor x_b \rfloor_{i.j}), (\lfloor x_a \rfloor_{i.j}) \cdots (2^{(i-j)} - 1)]$

For example, using the binary representation for the integer bounds of a union of intervals, an extraction of the 3 rightmost bits of a variable whose union of intervals contains $01110 \cdots 10011$ would not produce the invalid interval $110 \cdots 011$ because its lower bound is greater than its upper bound. This falls in the third case above, and would generate the two intervals $000 \cdots 011$ and $110 \cdots 111$.

Concatenation: for a concatenation $x :: y = z$, the inner structure of $[z]$ can be refined from $[x]$ and $[y]$. Let $v^{\ll x}$ be $v \mid (x \ll \|y\|)$ and $(a \cdots b)^{\ll x}$ be $a^{\ll x} \cdots b^{\ll x}$. For example, if $[x] = [x_a \cdots x_b]$ and $[y] = [y_1, y_2 \cdots y_3, y_4 \cdots y_5]$, then $[z]$ can be refined by $[(y_1)^{\ll x_a}, (y_2 \cdots y_3)^{\ll x_a}, (y_4^{\ll x_a}) \cdots (y_1^{\ll x_b}), (y_2 \cdots y_3)^{\ll x_b}, (y_4 \cdots y_5)^{\ll x_b}]$. The algorithm is described in the technical report. One can also refine $[x]$ and $[y]$ from $[z]$.

4.2.2   BV Constraints on Deltas
As seen in the motivation section, keeping the deltas informed of the relationship between different variables can be an important factor for efficient reasoning.

Bitwise operations: a constraint $x \mid y = c$ implies that that $(c - y \leqslant x \leqslant c) \wedge (c - x \leqslant y \leqslant c)$. Symmetric information can be derived for conjunction. Exclusive disjunction, however, does not derive knowledge regarding deltas. The bitwise negation has limited effect and can only impose that its argument and its result be different.

Extraction: regardless of the indices, the result of an extraction is always less than or equal to its result. As a matter of fact, an extraction $\lfloor x \rfloor_{i.j} = (x \% 2^i)/2^j$ and can enjoy the same propagations on the non-BV domains.

More generally: many BV constraints can be mapped to an integer counterpart and propagate on non-BV domains. For example, a concatenation $x :: y$ can have the same effect as the (overflowless) integer constraint $z = x \times 2^{\|y\|} + y$ would.

4.3   Factorizations and Simplifications

In the course of solving, a constraint can be simplified or become duplicate or a subsumption of another constraint. These shortcuts can be separated in two categories.

Simplifications. Neutral and absorbing elements provide many rewriting rules which replace constraints by simpler ones. In addition to these usual simplifications one can detect more sophisticated ones, such as if $z \gg y = z$ and $y > 0$ then $z = 0$ (without restricting the value of y), and when z is x rotated i times, if the size of x is 1 or if $i \% \|x\| = 0$, then $z = x$. Furthermore, if i and $\|x\|$ are coprimes, and $x = z$, then $x = 0$ or $x = 2^{\|x\|} - 1$.

Factorizations. The more constraints are merged or reduced to simpler constraints, the closer we get to a proof. Functional factorization allows to detect instances based on equality of arguments, but some other instances can be factored as well, for example:

- from $x \oplus y = z$ and $x \oplus t = y$, we deduce that $t = z$, unifying the two variables and removing one of the constraints, now considered duplicates
- when $x \ll y_1 = z_1$ and $x \ll y_2 = z_2$ and $y_1 < y_2$, and z_1 is a constant, then one can infer the value of z_2. A similar operation can be carried out for $\gg$.
- two constraints $x \mathbin{\&} y = 0$ and $x \mid y = 2^{\|x\|} - 1$ can be replaced by $x = \overline{y}$
- a constraint $x \mathbin{\&} y = z$ (resp. $x \mid y = z$) is superfluous with the constraint $x = \overline{y}$ once z is deducted to be equal to 0 (resp. $2^{\|x\|} - 1$).
- the constraints $x = \overline{y}$ and $x \mathbin{\&} z = y$ (resp. $x \mid z = y$) can both be removed once deducted that $x = 2^{\|x\|} - 1, y = z = 0$ (resp. $x = 0, y = z = 2^{\|x\|} - 1$).

4.4   BV-Dedicated Labeling Strategies

A labeling strategy (a.k.a. search strategy) consists mainly of two heuristics: *variable selection* and *value selection*. For variable selection, we rely on the fail-first approach implemented in COLIBRI [33]. Basically, the variable to be selected is the one with the smallest domain (in terms of concretization). Adding the BV domain allows here to refine the notion of smallest. For value selection, in the event that BV is the smallest domain, our strategy is the following:

- First, we consider certain values that can simplify arithmetic constraints. In particular, we start by trying 0 (for $+, -, \times, /, \mathbin{\&}, \mid, \oplus$), 1 (for $\times, /$) and $2^s - 1$ where s is the bitvector size (for $\mathbin{\&}, \mid$);
- Second, we fix the value of several most significant and least significant unknown bits *(msub, lsub)* at once, allowing to strongly refine all domains thanks to inter-reduction, and to fix early the sign of the labeled variable (useful for signed BV operations). Currently, we fix at each labeling step one *msub* and two *lsub*, yielding 8 possible choices. We choose whether to set first or clear first in an arbitrary (but deterministic) way, using a fixed seed.

5   Experimentation

We describe in this section our implementation and experimental evaluation of CP(BV).

Implementation. We have implemented a BV support inside the COLIBRI CP solver [33] (cf. Sect. 3.3). Modular arithmetic domains and propagators are treated as blackbox, and we add the optimized bitlist domain and its associated propagators from [36], as well as all improvements discussed in Sect. 4. Building on top of COLBRI did allow us to prototype our approach very quickly, compared with starting from scratch.

Because it is written in a Prolog dialect, the software must be interpreted at each run, inducing a systematic 0.2 s starting time. This obstacle is not troubling for us because any real-world application would execute our software once and feed its queries in a chained manner through a server mode. Yet, the SMTCOMP rules impose that the software be called on command line with exactly one .smt2 file, which excludes a "server mode".

Experimental setup and caveats. We experiment CP(BV) on the 32 k BV-formulas from the SMTCOMP benchmark, the leading competition of the SMT community. These formulas are mostly taken from verification-oriented industrial case-studies, and can be very large (up to several hundreds of MB). The first set of experiments (Sect. 5.1) has been run on the StarExec server[3] provided by SMTCOMP, they are made public[4]. The second set of experiments (Sect. 5.2) is run on a Intel® Core^{TM} i7-4712HQ CPU @ 2.30 GHz with 16 GB memory. Two points must be kept in mind.

- We fix a low time-out (60 s) compared with the SMTCOMP rules (40 min), yet we argue that our results are still representative: first, such low time-outs are indeed very common in applications such as bug finding [25] or proof [19]; second, it is a common knowledge in first-order decision procedures that *"solvers either terminate instantly, or timeout"* – adding more time does not dramatically increase the number of solved formulas;
- SMT solvers such as Z3 and CVC4 are written in efficient compiled languages such as C/C++, with near-zero starting time. Hence, we have a constant-time disadvantage here – even if such a burden may not be so important in verification: since we are anyway attacking NP-hard problems, we are looking for exponential algorithmic improvements ; constant-time gains can only help marginally.

5.1   Evaluation Against State-of-the-Art Benchmark

Absolute performance. Our implementation solved 24 k formula out of 32 k (75%). While it would not have permitted us to win the SMTCOMP, it is still a *significantly more thorough comparison with the SMT community than any previous CP effort on bitvectors*, demonstrating that CP can indeed be used for verification and bitvectors.

Comparing different choices of implementation. Improvements offered by our different optimizations are very dependent on the type of formulas, and these details would be diluted if regrouping all of the benchmarks. The reader is invited to consult our detailed results on the StarExec platform. Yet, as a rule of thumb, *our extensions yield a significant improvement on some families of examples, and do not incur any overhead on the other, proving their overall utility.* For example :

- On the family of formulas named stp_samples($\simeq$400 formulas), when deactivating the reductions from BV constraints to other domains (Sect. 4.2), the

---

[3] www.starexec.org.

[4] www.starexec.org/starexec/secure/explore/spaces.jsp?id=186070.

solver is unable to solve *a quarter less* formulas that it did with the full implementation. Removing the interreduction with the *Is* (Sect. 4.1), the loss rises to *half*;

- Solving the `spear` family suffers little from deactivating BV/*Is* interreductions, but *half* (200) formulas are lost without the BV constraints reducing other domains (Sect. 4.2);
- On some other families, such as `pspace` and `dwp_formulas`, there is no tangible effect (neither positive nor negative) to the deactivation of improvements.

5.2   Comparison to State-of-the-Art SMT Solvers

We demonstrate in the following that CP(BV) is actually *complementary* to SMT approaches, especially on problems with large bitvectors or involving multitheories. As competitors, we select the five best SMT solvers for BV theory: Z3, Yices, MathSAT, CVC4 and Boolector.

SMTCOMP: large formulas. To study the effect of our method on *scalability*, we show here the results on three categories of formulas, regrouped according to the (number of digits of the) size of the their largest bit-vectors: 3-digit (from 100 to 999), 4-digit and 5-digit, respectively having 629, 298 and 132 formulas. Results in Table 3 show on the one hand the *scalability* of CP(BV) – the larger BV sizes, the greater impact the CP approach has – and on the other hand its *complementarity* with SMT. In particular, it shows the result of *duels* between CP(BV) and each of the SMT solvers : a formula is considered a win for a solver if it succeeds (TO = 60 s) while the other solver does not. We report results on a format Win/Lose (Solve), where Win and Lose are from CP(BV) point of view, and Solve indicates the number of formulas solved by the SMT solver. For example, MathSAT could solve 17 of the 132 5-digit formulas – all of which being solved by CP(BV), while CP(BV) could solve 63 formulas – 46 of which were unsolved by MathSAT. Here, *CP(BV) solves the higher number of 5-digit size formulas (equality with CVC4), and no solver but Boolector solves formulas that CP(BV) does not. On other sizes, CP(BV) solves less formulas, but it can still solve formulas that SMT solvers do not.*

SMTCOMP: hard formulas. We define *hard formulas* by separating 5 classes of difficulty, with class *i* regrouping the formulas on which *i* SMT solvers out of 5 spend more than 5 s. We compare CP(BV) to SMT solvers on these hard

Table 3. Comparing CP(BV) with five state-of-the-art solvers on large formulas

sz	#f	CP(BV) #solved	Z3 w/l (s)	Yices w/l (s)	MathSAT w/l (s)	CVC4 w/l (s)	Boolector w/l (s)
5	132	63	63/0 (0)	53/0 (10)	46/0 (17)	0/0 (63)	32/10 (41)
4	298	44	34/153 (163)	40/87 (91)	43/68 (69)	42/150 (152)	43/204 (205)
3	629	35	24/496 (507)	23/262 (274)	23/419 (431)	23/511 (523)	25/507 (517)

sz: size (#digits) - #f: # of formulas
w/l (s): #win/#lose for CP(BV), s: #formulas solved by SMT solver

Table 4. Overall comparison on *hard examples*

Category	1-fail	2-fail	3-fail	4-fail	All-fail
#benchs	1083	382	338	1075	873
CP(BV) under 5 s	139	108	10	68	61

problems. Results are presented in Table 4, where we report for each class i the number of formulas from this class that CP(BV) solves quickly. Especially, the 5th column (All-fail) shows that 61 formulas are solved only by CP(BV) in less than 5 s.

Mixing bitvectors and floats. Considering multi-theory formulas combining BV and FP arithmetics, COLIBRI has been tested on 7525 industrially-provided formulas (not publically available). It was able to solve 73% of them, standing half-way between Z3 / MathSAT and CVC4 (Table 5). Considering now the last SMTCOMP QF_BVFP category (Table 6), *even with the 0.2 s starting time, CP(BV) would have won the competition* – admittedly, there are only few formulas in this category.

Table 5. Industrial formula with bitvectors and floats

	CP(BV)	Z3	MathSAT	CVC4
#solved	5512	7225	7248	2245
Ratio	73%	96%	96%	29%

Total: 7525 formulas

Table 6. SMTCOMP, QF_BVFP category

	Z3	MathSAT	CP(BV)
int_to_float_complex_2.smt2	1.04	0.13	0.25
int_to_float_simple_2.smt2	2.17	0.22	0.21
int_to_float_complex_1.smt2	0.95	0.08	0.25
int_to_float_simple_1.smt2	0.02	0.02	0.25
nan_1.smt2	0	0	0.26
incr_by_const.smt2	8.20	30.50	0.26
int_to_float_complex_3.smt2	1.89	0.44	0.25
quake3_1.smt2	TO	TO	TO

6   Related Work

CP-based methods for BV. This work strongly stands upon the prior results of Bardin *et al.* [1] and Michel *et al.* [36]. Our respective contribution is already discussed at length in Sects. 2 and 3. Basically, while we reuse the same general ideas, we sharpen them through a careful selection of the best aspects of each of these works and the design of new mechanisms, especially in terms of domain combination and labeling strategies. As a result, experiments in Sect. 5 demonstrate that our own CP(BV) approach performs much better than previous attempts. Moreover, we perform an extensive comparison with SMT solvers on the whole SMTCOMP benchmark, while these previous efforts were either limited to conjunctive formulas or remain only theoretical. The results by Michel *et al.* have been applied to a certain extent [46] as an extension of MiniSat [21], yet with no reduced product, to a limited set of BV operations and on BV sizes no larger than 64 bits. Older word-level approaches consider straightforward translations of bit-vector problems into disjunctive or non-linear arithmetic problems [10,22,40,45,48,49] (including bitblasting-like transformation for logical bitwise operators), and then rely on standard methods from linear integer programming or CP. Experimental results reported in [1,44] demonstrate that such straightforward word-level encoding yield only very poor results on formulas coming from software verification problems.

SMT-based methods for BV. While state-of-the-art methods heavily rely on bitblasting and modern DPLL-style SAT solvers [38,43], the community is sensing the need for levels of abstraction "where structural information is not blasted to bits"[12]. Part of that need comes from the knowledge that certain areas, arithmetic for example, are not efficiently handled by bit-level reasoning tools. As a mitigation, SMT solvers typically complement optimized bitblasting [12,13,32] with word-level preprocessing [3,24]. Compared to these approaches, we lack the highly-efficient learning mechanisms from DPLL. Yet, our domains and propagations yield more advanced simplifications, deeply nested with the search mechanism.

7   Conclusion

This work addresses the challenge of developing efficient Constraint Programming-based approaches for solving formulas over (the quantifier-free fragment of) the theory of bitvectors, which is of paramount importance in software verification. While the Formal Verification community relies essentially on the paradigm of SMT solving and reduction to Boolean satisfiability, we explore an alternative, high-level resolution technique through dedicated CP principles. We build on a few such anterior results and sharpen them in order to propose a highly efficient CP(BV) resolution method. We show that CP(BV) is much more efficient than the previous attempts from the CP community and that it is indeed able to solve the majority of the standard verification benchmarks for bitvectors. Moreover CP(BV) already complements the standard SMT approach on several crucial (and industry-relevant) aspects, such as scalability w.r.t. bit-width, formulas combining bitvectors with bounded integers or floating-point arithmetic, and formulas deeply combining non-linear arithmetic and bitwise operators.

Considering that our current CP(BV) approach is far from optimal compared with existing SMT solvers, we believe this work to be an important landmark toward building competitive CP-based verification-oriented solvers. Moreover, our approach clearly challenges the well-accepted belief that bitvector solving is better done through bitblasting, opening the way for a new generation of word-level solvers.

References

1. Bardin, S., Herrmann, P., Perroud, F.: An alternative to SAT-based approaches for bit-vectors. In: Esparza, J., Majumdar, R. (eds.) TACAS 2010. LNCS, vol. 6015, pp. 84–98. Springer, Heidelberg (2010). doi:10.1007/978-3-642-12002-2_7
2. Bardin, S., Herrmann, P.: OSMOSE: automatic structural testing of executables. Softw. Test. Verification Reliab. **21**(1), 29–54 (2011)
3. Barret, C., Dill, D., Levitt, J.: A decision procedure for bit-vector arithmetic. In: DAC (1998)
4. Barrett, C., Conway, C.L., Deters, M., Hadarean, L., Jovanović, D., King, T., Reynolds, A., Tinelli, C.: CVC4. In: Gopalakrishnan, G., Qadeer, S. (eds.) CAV 2011. LNCS, vol. 6806, pp. 171–177. Springer, Heidelberg (2011). doi:10.1007/978-3-642-22110-1_14
5. Barrett, C., et al.: The SMT-LIB Standard: Version 2.0. Technical report (2010)
6. Barrett, C.W., et al.: Satisfiability modulo theories. In: Handbook of Satisfiability (2009)
7. Biere, A., Cimatti, A., Clarke, E., Zhu, Y.: Symbolic model checking without BDDs. In: Cleaveland, W.R. (ed.) TACAS 1999. LNCS, vol. 1579, pp. 193–207. Springer, Heidelberg (1999). doi:10.1007/3-540-49059-0_14
8. Bjørner, N.: Taking satisfiability to the next level with Z3. In: Gramlich, B., Miller, D., Sattler, U. (eds.) IJCAR 2012. LNCS, vol. 7364, pp. 1–8. Springer, Heidelberg (2012). doi:10.1007/978-3-642-31365-3_1
9. Blanc, B., et al.: Handling state-machines specifications with GATeL. Electron. Notes Theor. Comput. Sci. **264**(3), 3–17 (2010)
10. Brinkmann, R., Drechsler, R.: RTL-datapath verification using integer linear programming. In: 15th International Conference on VLSI Design (2002)
11. Brummayer, R., Biere, A.: Boolector: an efficient SMT solver for bit-vectors and arrays. In: Kowalewski, S., Philippou, A. (eds.) TACAS 2009. LNCS, vol. 5505, pp. 174–177. Springer, Heidelberg (2009). doi:10.1007/978-3-642-00768-2_16
12. Bruttomesso, R., Cimatti, A., Franzén, A., Griggio, A., Hanna, Z., Nadel, A., Palti, A., Sebastiani, R.: A lazy and layered SMT(BV) solver for hard industrial verification problems. In: Damm, W., Hermanns, H. (eds.) CAV 2007. LNCS, vol. 4590, pp. 547–560. Springer, Heidelberg (2007). doi:10.1007/978-3-540-73368-3_54
13. Bryant, R.E., Kroening, D., Ouaknine, J., Seshia, S.A., Strichman, O., Brady, B.: Deciding bit-vector arithmetic with abstraction. In: Grumberg, O., Huth, M. (eds.) TACAS 2007. LNCS, vol. 4424, pp. 358–372. Springer, Heidelberg (2007). doi:10.1007/978-3-540-71209-1_28
14. Cimatti, A., Griggio, A., Schaafsma, B.J., Sebastiani, R.: The MathSAT5 SMT solver. In: Piterman, N., Smolka, S.A. (eds.) TACAS 2013. LNCS, vol. 7795, pp. 93–107. Springer, Heidelberg (2013). doi:10.1007/978-3-642-36742-7_7
15. Clarke, E., Kroening, D., Lerda, F.: A tool for checking ANSI-C programs. In: Jensen, K., Podelski, A. (eds.) TACAS 2004. LNCS, vol. 2988, pp. 168–176. Springer, Heidelberg (2004). doi:10.1007/978-3-540-24730-2_15

16. Collavizza, H., Rueher, M., Hentenryck, P.: CPBPV: a constraint-programming framework for bounded program verification. In: Stuckey, P.J. (ed.) CP 2008. LNCS, vol. 5202, pp. 327–341. Springer, Heidelberg (2008). doi:10.1007/978-3-540-85958-1_22

17. David, R., et al.: BINSEC/SE: A dynamic symbolic execution toolkit for binary-level analysis. In: SANER 2016 (2016)

18. Dechter, R.: Constraint Processing. Morgan Kaufmann Publishers Inc., Massachusetts (2003)

19. Dijkstra, E.W.: A Discipline of Programming, vol. 1. Prentice-Hall Englewood Cliffs, New Jersey (1976)

20. Dutertre, B.: Yices 2.2. In: Biere, A., Bloem, R. (eds.) CAV 2014. LNCS, vol. 8559, pp. 737–744. Springer, Cham (2014). doi:10.1007/978-3-319-08867-9_49

21. Eén, N., Sörensson, N.: An extensible SAT-solver. In: Giunchiglia, E., Tacchella, A. (eds.) SAT 2003. LNCS, vol. 2919, pp. 502–518. Springer, Heidelberg (2004). doi:10.1007/978-3-540-24605-3_37

22. Ferrandi, F., Rendine, M., Sciuto, D.: Functional verification for SystemC descriptions using constraint solving. In: Design, Automation and Test in Europe (2002)

23. Feydy, T., Schutt, A., Stuckey, P.J.: Global difference constraint propagation for finite domain solvers. In: PPDP (2008)

24. Ganesh, V., Dill, D.L.: A decision procedure for bit-vectors and arrays. In: Damm, W., Hermanns, H. (eds.) CAV 2007. LNCS, vol. 4590, pp. 519–531. Springer, Heidelberg (2007). doi:10.1007/978-3-540-73368-3_52

25. Godefroid, P.: Test generation using symbolic execution. In: D'Souza, D., Kavitha, T., Radhakrishnan, J. (eds.) FSTTCS, vol. 18, pp. 24–33. Schloss Dagstuhl, Germany (2012)

26. Gotlieb, A.: TCAS software verification using constraint programming. Knowl. Eng. Rev. 27(3), 343–360 (2012)

27. Gotlieb, A., Botella, B., Rueher, M.: Automatic test data generation using constraint solving techniques. In: ISSTA (1998)

28. Gotlieb, A., Leconte, M., Marre, B.: Constraint Solving on Modular Integers (2010)

29. Henzinger, T.A., et al.: Lazy abstraction. In: POPL (2002)

30. Kroening, D., Strichman, O.: Decision Procedures: An Algorithmic Point of View, 1st edn. Springer Publishing Company Incorporated, Heidelberg (2008)

31. Leconte, M., Berstel, B.: Extending a CP solver with congruences as domains for program verification. In: Trends in Constraint Programming (2010)

32. Manolios, P., Vroon, D.: Efficient circuit to CNF conversion. In: Marques-Silva, J., Sakallah, K.A. (eds.) SAT 2007. LNCS, vol. 4501, pp. 4–9. Springer, Heidelberg (2007). doi:10.1007/978-3-540-72788-0_3

33. Marre, B., Blanc, B.: Test selection strategies for Lustre descriptions in GaTeL. Electron. Notes Theor. Comput. Sci. 111, 93–111 (2005)

34. Marre, B., Michel, C.: Improving the floating point addition and subtraction constraints. In: Cohen, D. (ed.) CP 2010. LNCS, vol. 6308, pp. 360–367. Springer, Heidelberg (2010). doi:10.1007/978-3-642-15396-9_30

35. McMillan, K.L.: Lazy abstraction with interpolants. In: Ball, T., Jones, R.B. (eds.) CAV 2006. LNCS, vol. 4144, pp. 123–136. Springer, Heidelberg (2006). doi:10.1007/11817963_14

36. Michel, L.D., Hentenryck, P.: Constraint satisfaction over bit-vectors. In: Milano, M. (ed.) CP 2012. LNCS, pp. 527–543. Springer, Heidelberg (2012). doi:10.1007/978-3-642-33558-7_39

37. Millo, R.A.D., Lipton, R.J., Perlis, A.J.: Social processes and proofs of theorems and programs. Commun. Assoc. Comput. Mach. 22(5), 271–280 (1979)

38. Moskewicz, M.W., et al.: Chaff: engineering an efficient SAT solver. In: Design Automation Conference, DAC (2001)
39. Nelson, G., Oppen, D.C.: Simplification by cooperating decision procedures. ACM Trans. Program. Lang. Syst. **1**(2), 245–257 (1979)
40. Parthasarathy, G., et al.: An efficient finite-domain constraint solver for circuits. In: 41st Design Automation Conference (2004)
41. Pelleau, M., Miné, A., Truchet, C., Benhamou, F.: A constraint solver based on abstract domains. In: Giacobazzi, R., Berdine, J., Mastroeni, I. (eds.) VMCAI 2013. LNCS, vol. 7737, pp. 434–454. Springer, Heidelberg (2013). doi:10.1007/978-3-642-35873-9_26
42. Scott, J.D., Flener, P., Pearson, J.: Bounded strings for constraint programming. In: ICTAI (2013)
43. Silva, J.P.M., Sakallah, K.A.: GRASP: a search algorithm for propositional satisfiability. IEEE Trans. Comput. **48**(5), 506–521 (1999)
44. Sülflow, A., et al.: Evaluation of SAT like proof techniques for formal verification of word level circuits. In: 8th IEEE Workshop on RTL and High Level Testing (2007)
45. Vemuri, R., Kalyanaraman, R.: Generation of design verification tests from behavioral VHDL programs using path enumeration and constraint programming. IEEE Trans. VLSI Syst. **3**(2), 201–214 (1995)
46. Wang, W., Søndergaard, H., Stuckey, P.J.: A bit-vector solver with word-level propagation. In: Quimper, C.-G. (ed.) CPAIOR 2016. LNCS, vol. 9676, pp. 374–391. Springer, Cham (2016). doi:10.1007/978-3-319-33954-2_27
47. Williams, N., Marre, B., Mouy, P.: On-the-fly generation of K-path tests for C functions. In: ASE 2004 (2004)
48. Zeng, Z., Ciesielski, M., Rouzeyre, B.: Functional test generation using constraint logic programming. In: 11th International Conference on Very Large Scale Integration of Systems-on-Chip (2001)
49. Zeng, Z., Kalla, P., Ciesielski, M.: LPSAT: a unified approach to RTL satisfiability. In: 4th Conference on Design, Automation and Test in Europe (2001)

Range-Consistent Forbidden Regions
of Allen's Relations

Nicolas Beldiceanu[1] , Mats Carlsson[2(✉)] , Alban Derrien[5] ,
Charles Prud'homme[1] , Andreas Schutt[3,4] , and Peter J. Stuckey[3,4]

[1] TASC (LS2N-CNRS), IMT Atlantique, Nantes, France
{Nicolas.Beldiceanu,Charles.Prudhomme}@imt-atlantique.fr
[2] RISE SICS, Kista, Sweden
Mats.Carlsson@sics.se
[3] Data61, CSIRO, Canberra, Australia
{Andreas.Schutt,Peter.Stuckey}@data61.csiro.au
[4] University of Melbourne, Melbourne, Australia
[5] Lab-STICC UMR 6285 – CNRS, Université de Bretagne-Sud,
Lorient, France
Alban.Derrien@univ-ubs.fr

Abstract. For all 8192 combinations of Allen's 13 relations between
one task with origin o_i and fixed length ℓ_i and another task with origin
o_j and fixed length ℓ_j, this paper shows how to systematically derive a
formula $F(\underline{o_j}, \overline{o_j}, \ell_i, \ell_j)$, where $\underline{o_j}$ and $\overline{o_j}$ respectively denote the earliest
and the latest origin of task j, evaluating to a set of integers which are
infeasible for o_i for the given combination. Such forbidden regions allow
maintaining range-consistency for an Allen constraint.

1 Introduction

More than 30 years ago Allen proposed 13 basic mutually exclusive relations [1]
to exhaustively characterise the relative position of two tasks. By considering all
potential disjunctions of these 13 basic relations one obtains 8192 general relations. While most of the work has been focussed on qualitative reasoning [5,8]
with respect to these general relations, and more specifically on the identification
and use of the table of transitive relations [11], or on logical combinators involving Allen constraints [4,10], no systematic study was done for explicitly charac-
terising the set of infeasible/feasible values of task origin/length with respect to
known consistencies. In the context of range consistency the contribution of this
paper is to derive from the structure of basic Allen Relations the *exact formulae*
for the lower and upper bounds of the intervals of infeasible values for the 8192
general relations and to synthesised a corresponding data base [2].

After recalling the definition of Basic Allen's relations, Sect. 2.1 gives the forbidden regions for these basic Allen's relation, Sect. 2.2 unveils a regular structure
on the limits of those forbidden regions, and Sect. 2.3 shows how to systematically compute a compact normal form for the forbidden regions of all the 8192
general relations.

© Springer International Publishing AG 2017
D. Salvagnin and M. Lombardi (Eds.): CPAIOR 2017, LNCS 10335, pp. 21–29, 2017.
DOI: 10.1007/978-3-319-59776-8_2

Definition 1 (Basic Allen's relations). *Given two tasks i, j respectively defined by their origin o_i, o_j and their length $\ell_i > 0$, $\ell_j > 0$, the following 13 basic Allen's relations systematically describe relationships between the two tasks, i.e. for two fixed tasks only one basic Allen's relation holds.*

- b : $o_i + \ell_i < o_j$
- m : $o_i + \ell_i = o_j$
- o : $\begin{aligned} o_i < o_j \wedge \\ o_i + \ell_i > o_j \wedge \\ o_i + \ell_i < o_j + \ell_j \end{aligned}$

- s : $\begin{aligned} o_i = o_j \wedge \\ o_i + \ell_i < o_j + \ell_j \end{aligned}$
- d : $\begin{aligned} o_j < o_i \wedge \\ o_i + \ell_i < o_j + \ell_j \end{aligned}$

- f : $\begin{aligned} o_j < o_i \wedge \\ o_i + \ell_i = o_j + \ell_j \end{aligned}$
- e : $\begin{aligned} o_i = o_j \wedge \\ o_i + \ell_i = o_j + \ell_j \end{aligned}$

The basic relations bi, mi, oi, si, di and fi are respectively derived from b, m, o, s, d and f by permuting task i and task j. The expression i r j denotes that the basic relation r holds between task i and task j.

Definition 2 (Allen's constraint). *Given two tasks i, j respectively defined by their origin o_i, o_j and their length $\ell_i > 0$, $\ell_j > 0$, and a basic relation r, the* ALLEN*$(r, o_i, \ell_i, o_j, \ell_j)$ constraint holds if and only if the condition i r j holds.*

> if $r = $ b then
>   propagate $o_i + \ell_i < o_j$
> else if $r = $ m then
>   propagate $o_i + \ell_i = o_j$
> else if $\ldots$ then
> $\ldots$
> end if

Note that o_i, o_j, ℓ_i, ℓ_j are integer variables. Similarly, the basic relation r is an integer variable r, whose initial domain is included in $\{$b, bi, m, mi, o, oi, s, si, d, di, f, fi, e$\}$. This constraint could be decomposed as shown above, but such a decomposition would propagate nothing until r has been fixed, whereas our formulae capture perfect constructive disjunction for all the 8192 general relations, e.g. for use in a range-consistency propagator.

2   Range Consistency

Given an integer variable o, $D(o)$, $\underline{o}$, $\overline{o}$ respectively denote the *set of values*, the *smallest value*, the *largest value* that can be assigned to o. The *range* of a variable o is the interval $[\underline{o}..\overline{o}]$ and is denoted by $R(o)$. A constraint CTR is *range consistent* (RC) [3] if and only if, when a variable o of CTR is assigned any value in its domain $D(o)$, there exist values in the ranges of all the other variables of CTR such that the constraint CTR holds.

2.1   Forbidden Regions Normal Form of Basic Allen's Relations

For each of the 13 basic Allen's relations column **RC** of Table 1 provides the corresponding normalised forbidden regions.

Table 1. Inconsistent values for RC for the 13 basic Allen's relations between two tasks i and j respectively defined by their origin o_i, o_j and their length ℓ_i and ℓ_j subject to Allen's relation i r j with r $\in \{b, bi, \ldots, e\}$ (for reasons of symmetry we only show the filtering of task i).

Rel	RC	
	Parameter cases	Inconsistent values
b		$o_i \notin [\overline{o_j} - \ell_i.. + \infty)$
bi		$o_i \notin (-\infty..o_j + \ell_j]$
m		$o_i \notin (-\infty..o_j - \ell_i - 1] \cup [\overline{o_j} - \ell_i + 1.. + \infty)$
mi		$o_i \notin (-\infty..o_j + \ell_j - 1] \cup [\overline{o_j} + \ell_j + 1.. + \infty)$
o	$\ell_i > 1 \wedge \ell_j > 1 \wedge \ell_i \leq \ell_j$:	$o_i \notin (-\infty..o_j - \ell_i] \cup [\overline{o_j}.. + \infty)$
	$\ell_i > 1 \wedge \ell_j > 1 \wedge \ell_i > \ell_j$:	$o_i \notin (-\infty..o_j - \ell_i] \cup [\overline{o_j} + \ell_j - \ell_i.. + \infty)$
	$\ell_i = 1 \vee \ell_j = 1$:	$o_i \notin (-\infty.. + \infty)$
oi	$\ell_i > 1 \wedge \ell_j > 1 \wedge \ell_i \leq \ell_j$:	$o_i \notin (-\infty..o_j + \ell_j - \ell_i] \cup [\overline{o_j} + \ell_j.. + \infty)$
	$\ell_i > 1 \wedge \ell_j > 1 \wedge \ell_i > \ell_j$:	$o_i \notin (-\infty..o_j] \cup [\overline{o_j} + \ell_j.. + \infty)$
	$\ell_i = 1 \vee \ell_j = 1$:	$o_i \notin (-\infty.. + \infty)$
s	$\ell_i < \ell_j$:	$o_i \notin (-\infty..o_j - 1] \cup [\overline{o_j} + 1.. + \infty)$
	$\ell_i \geq \ell_j$:	$o_i \notin (-\infty.. + \infty)$
si	$\ell_j < \ell_i$:	$o_i \notin (-\infty..o_j - 1] \cup [\overline{o_j} + 1.. + \infty)$
	$\ell_j \geq \ell_i$:	$o_i \notin (-\infty.. + \infty)$
d	$\ell_i + 1 < \ell_j$:	$o_i \notin (-\infty..o_j] \cup [\overline{o_j} + \ell_j - \ell_i.. + \infty)$
	$\ell_i + 1 \geq \ell_j$:	$o_i \notin (-\infty.. + \infty)$
di	$\ell_j + 1 < \ell_i$:	$o_i \notin (-\infty..o_j + \ell_j - \ell_i] \cup [\overline{o_j}.. + \infty)$
	$\ell_j + 1 \geq \ell_i$:	$o_i \notin (-\infty.. + \infty)$
f	$\ell_i < \ell_j$:	$o_i \notin (-\infty..o_j + \ell_j - \ell_i - 1] \cup [\overline{o_j} + \ell_j - \ell_i + 1.. + \infty)$
	$\ell_i \geq \ell_j$:	$o_i \notin (-\infty.. + \infty)$
fi	$\ell_j < \ell_i$:	$o_i \notin (-\infty..o_j + \ell_j - \ell_i - 1] \cup [\overline{o_j} + \ell_j - \ell_i + 1.. + \infty)$
	$\ell_j \geq \ell_i$:	$o_i \notin (-\infty.. + \infty)$
e	$\ell_j = \ell_i$:	$o_i \notin (-\infty..o_j - 1] \cup [\overline{o_j} + 1.. + \infty)$
	$\ell_j \neq \ell_i$:	$o_i \notin (-\infty.. + \infty)$

Lemma 1. *(a) The correct and complete forbidden region for $o_i + \ell < o_j$ is $o_i \notin [\overline{o_j} - \ell.. + \infty)$. (b) The correct and complete forbidden region for $o_j + \ell < o_i$ is $o_i \notin (-\infty..o_j + \ell]$.*

Proof. (a) Given $o_i + \ell < o_j$ then clearly $o_i < \overline{o_j} - \ell$ and hence $o_i \notin [\overline{o_j} - \ell.. + \infty)$. Given $v < \overline{o_j} - \ell$ then $o_i = v, o_j = \overline{o_j}$ is a solution of the constraint. (b) Given $o_j + \ell < o_i$ then clearly $o_i > \underline{o_j} + \ell$ and hence $o_i \notin (-\infty..\underline{o_j} + \ell]$. Given $v < \underline{o_j} + \ell$ then $o_i = v, o_j = \underline{o_j}$ is a solution of the constraint. □

Lemma 2. *Given constraint* $c \equiv c_1 \wedge c_2$ *if* $o_i \notin R_1$ *is a correct forbidden region of* c_1 *and* $o_i \notin R_2$ *is a correct forbidden region of* c_2 *then* $o_i \notin R_1 \cup R_2$ *is a correct forbidden region of* c.

Proof. Since there can be no solution of c_1 with $o_i \in R_1$ and no solution of c_2 with $o_i \in R_2$ there can be no solution of c with $o_i \in R_1 \cup R_2$. □

Theorem 1. *Forbidden intervals of consecutive values shown in column* **RC** *of* Table 1 *are correct and complete.*

Proof. We proceed by cases and omit relations bi, mi, oi, si, di, fi for which the reasoning is analogous to b, m, o, s, d, f.

b   Follows from Lemma 1(a).

m   Correctness: given i meets j then $o_i + \ell_i = o_j$ thus $o_i + \ell_i \leq o_j \wedge o_i + \ell_i \geq o_j$ thus $o_i + (\ell_i - 1) < o_j \wedge o_j + (-\ell_i - 1) < o_i$. From Lemma 1 we have that $[\overline{o_j} - \ell_i + 1 .. + \infty)$ and $(-\infty .. o_j - \ell_i - 1]$ are correct forbidden regions and by Lemma 2 correctness holds. Completeness: choose $v \in [o_j - \ell_i .. \overline{o_j} - \ell_i]$ then $o_i = v, o_j = v + \ell_i$ is a solution.

o   Correctness: given i overlaps with j we have $o_i < o_j \wedge o_i + \ell_i > o_j \wedge o_i + \ell_i < o_j + \ell_j$. Suppose $\ell_i = 1$ then this implies $o_i < o_j \wedge o_i + 1 > o_j$ contradiction, or suppose $\ell_j = 1$ then this implies $o_i + \ell_i > o_j \wedge o_i + \ell_i < o_j + 1$ contradiction hence $(-\infty .. + \infty)$ is a correct forbidden region. Lemma 1 gives us correct forbidden regions $(-\infty .. o_j - \ell_i]$, $[\overline{o_j} + \ell_j - \ell_i .. + \infty)$, $[\overline{o_j} .. + \infty)$. If $\ell_i \leq \ell_j$ this is equivalent to $(-\infty .. o_j - \ell_i] \cup [\overline{o_j} .. + \infty)$. If $\ell_i > \ell_j$ this is equivalent to $(-\infty .. o_j - \ell_i] \cup [\overline{o_j} + \ell_i - \ell_j .. + \infty)$. Completeness: when $\ell_i = 1$ or $\ell_j = 1$ then completeness follows from the contradiction. Choose $v \in [o_j - \ell_i + 1 .. \overline{o_j} + \min(0, \ell_j - \ell_i) - 1]$ then $o_i = v, o_j = v + \ell_i - 1$ is a solution.

s   Correctness: If $\ell_i \geq \ell_j$ then the constraints are unsatisfiable and $(-\infty .. + \infty)$ is a correct forbidden region. Otherwise from s we have that $o_i < o_j + 1 \wedge o_i > o_j - 1 \wedge o_i + \ell_i < o_j + \ell_j$ and Lemma 1 gives us correct forbidden regions $[\overline{o_j} + 1 .. + \infty)$, $(-\infty .. o_j - 1]$ and $[\overline{o_j} + \ell_j - \ell_i .. + \infty)$. If $\ell_i < \ell_j$ then this gives $(-\infty .. o_j - 1] \cup [\overline{o_j} + 1 .. + \infty)$. Completeness: If $\ell_i \geq \ell_j$ then completeness follows from the unsatisfiability. Otherwise choose $v \in [o_j .. \overline{o_j}]$ then $o_i = v, o_j = v$ is a solution.

d   Correctness: If $\ell_i + 1 \geq \ell_j$ then $o_i + \ell_i \geq o_i + \ell_j - 1 \geq o_j + \ell_j$ but this contradicts $o_i + \ell_i < o_j + \ell_j$ hence $(-\infty .. + \infty)$ is a correct forbidden region. Otherwise Lemma 1 gives us correct forbidden regions $(-\infty .. o_j]$ and $[\overline{o_j} + \ell_j - \ell_i .. + \infty)$ whose union is the correct forbidden region. Completeness: The contradiction proves completeness when $\ell_i + 1 \geq \ell_j$. Otherwise choose $v \in [o_j + 1 .. \overline{o_j} + \ell_j - \ell_i - 1]$ then $o_i = v, o_j = v - 1$ is a solution.

f   Correctness: Suppose $\ell_i \geq \ell_j$ then $o_j < o_i = o_j + \ell_j - \ell_i \leq o_j$, a contradiction, hence $(-\infty .. + \infty)$ is a correct forbidden region. Otherwise $o_j + \ell_j = o_i + \ell_i$ is equivalent to $o_j + \ell_j - 1 < o_i + \ell_i \wedge o_j + \ell_j + 1 > o_i + \ell_i$. From these two inequalities and from $o_j < o_i$, Lemma 1 gives us correct forbidden regions $(-\infty .. o_j + \ell_j - \ell_i - 1]$, $[\overline{o_j} + \ell_j - \ell_i + 1 .. + \infty)$ and $(-\infty .. o_j]$. Since $\ell_i < \ell_j$ the correct union is $(-\infty..o_j + \ell_j - \ell_i - 1] \cup [\overline{o_j} + \ell_j - \ell_i + 1.. + \infty)$. Completeness: If $\ell_i \geq \ell_j$ then the contradiction gives the completeness. Otherwise choose $v \in [o_j + \ell_j - \ell_i..\overline{o_j} + \ell_j - \ell_i]$ then $o_i = v, o_j = v + \ell_i - \ell_j$ is a solution.

e   Correctness: Suppose $\ell_i \neq \ell_j$ then the constraints $o_i = o_j \wedge o_i + \ell_i = o_j + \ell_j$ contradict and $(-\infty.. + \infty)$ is a correct forbidden region. When $\ell_i = \ell_j$ Lemma 1 gives us correct forbidden regions $(-\infty..o_j - 1], [\overline{o_j} + 1.. + \infty)$ from both constraints, and their union is the correct answer. Completeness: If $\ell_i \neq \ell_j$ then the contradiction proves completeness, otherwise choose $v \in [o_j..\overline{o_j}]$ then $o_i = v, o_j = v$ is a solution.   □

2.2   Structure of the Normalised Forbidden Regions

All forbidden regions of the basic Allen's relations given in Sect. 2.1 consist of one or two intervals of the form $(-\infty..up]$, $[low.. + \infty)$ or $(-\infty.. + \infty)$. Indeed, only the forbidden regions for b and bi consist of a single (nonuniversal) forbidden region. In the following, we call *upper limit* (resp. *lower limit*) the terms *up* (resp. *low*). In the case of a single universal forbidden region, $up = +\infty$ and $low = -\infty$.

We show that all upper limits (resp. lower limits) can be totally ordered provided we know the relative order between the lengths ℓ_i and ℓ_j of the corresponding tasks. This is because all upper limits (resp. lower limits) correspond to linear expressions involving $+o_j$ (resp. $+\overline{o_j}$). Figure 1 illustrates this for the case $\ell_i < \ell_j$, where each limit is a node mentioning the associated formula, the basic Allen's relation(s) from which it is generated and the restriction on the parameters. We also show that we always have that the kth upper limit is strictly less than the $k + 1^{th}$ lower limit. This is because the kth upper limit and the $k + 1^{th}$ lower limit are issued from the same basic Allen's relation. Within Fig. 1 a solid arrow from a start node to an end node indicates that the limit attached to the start node is necessarily strictly less than (resp. strictly less by one than) the limit attached to the end node.

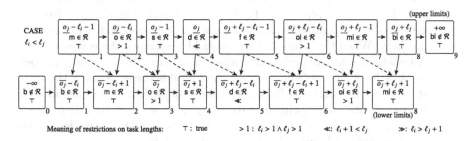

Fig. 1. Ordering the upper limits (resp. lower limits) of the forbidden regions of a general Allen relation $\mathcal{R}$ depending on the relative length of the two tasks i and j when $\ell_i < \ell_j$; a solid arrow from a limit x to a limit y represents an inequality of the form $x < y$, while a dashed arrow represents an inequality of the form $x + 1 < y$. Upper (resp. lower) limits of each of the three cases are identified by a unique identifier located on the corresponding lower rightmost corner.

2.3 Normal Form for the Forbidden Regions

Given any general Allen's relation $\mathcal{R}$ we now show how to synthesise a normalised sequence of forbidden regions for this relation under the different cases regarding the relative sizes of the two tasks to which $\mathcal{R}$ applies (i.e., $\ell_i < \ell_j$, $\ell_i = \ell_j$, $\ell_i > \ell_j$). This will lead to a data base [2] of normalised forbidden regions for the 8192 general relations. A typical entry of that data base, for instance for relation $\{b, bi, d, di, e, f, fi, m, mi, si\}$, looks like:

$$\begin{cases} [\overline{o_j} - \ell_i + 1..\underline{o_j}] \cup [\overline{o_j} + \ell_j - \ell_i + 1..\underline{o_j} + \ell_j - 1] & \text{if } \ell_i < \ell_j \wedge \ell_i > 1 \\ [\overline{o_j}..\underline{o_j}] & \text{if } \ell_i = 1 \wedge \ell_j > 1 \\ \emptyset & \text{if } \ell_j = 1 \\ [\overline{o_j} - \ell_i + 1..\underline{o_j} + \ell_j - \ell_i - 1] \cup [\overline{o_j} + 1..\underline{o_j} + \ell_j - 1] & \text{if } \ell_i \geq \ell_j \wedge \ell_i > 1 \wedge \ell_j > 1 \end{cases} \quad (1)$$

Each case consists of a normalised sequence of forbidden regions F and of a condition C involving the lengths of the tasks; such a case will be denoted as (F if C). Generating such cases is done by using the normalised forbidden regions of the 13 Allen's basic relations given in column **RC** of Table 1, as well as the strong ordering structure between the limits (see Fig. 1) of these forbidden regions we identified in Sect. 2.2 in three steps as follows.

1. EXTRACTING THE LOWER/UPPER LIMITS OF FORBIDDEN REGIONS OF BASIC ALLEN'S RELATIONS IN $\mathcal{R}$.
   (a) First, we filter from the considered general Allen's relation $\mathcal{R}$ those basic relations which are neither mentioned in the upper nor in the lower limits of the forbidden regions attached to the relevant case (i.e., $\ell_i < \ell_j$, $\ell_i = \ell_j$, $\ell_i > \ell_j$). This is because such Allen's basic relations generate one single forbidden region of the form $(-\infty.. + \infty)$ and can be therefore removed from the disjunction. For the same reason, we also filter from $\mathcal{R}$ those basic relations for which the parameter restriction does not hold.
   (b) Second, we group together the set of restrictions attached to the remaining basic Allen's relations. This leads to a set of restrictions in $\{\top, \ell_i > 1 \wedge \ell_j > 1, \ell_i + 1 < \ell_j, \ell_i > \ell_j + 1\}$. For those restrictions different from $\top$ we consider all possible combinations where each relation holds or does not hold. When the relation does not hold we remove the corresponding Allen's basic relation for the same reason as before. This gives us a number of cases for which we will generate the forbidden regions using the next steps. Since to each lower limit correspond an upper limit we remain with n lower limits low_{α_k} (with $0 \leq \alpha_1 < \alpha_2 < \cdots < \alpha_n \leq n$) and n upper limits up_{β_k} (with $1 \leq \beta_1 < \beta_2 < \cdots < \beta_n \leq n + 1$). A case for which $n = 0$ means a full forbidden region $(-\infty.. + \infty)$.
2. COMBINING THE LIMITS OF FORBIDDEN REGIONS OF BASIC ALLEN'S RELATIONS TO GET THE FORBIDDEN REGIONS OF $\mathcal{R}$.
   Second, the forbidden regions of the considered general Allen's relation $\mathcal{R}$ are given by $\bigcup_{k \in [1,n] | \alpha_k \neq \beta_k} [low_{\alpha_k}..up_{\beta_k}]$.

3. Removing empty forbidden regions of $\mathcal{R}$.

Using the following steps, we eliminate from $\bigcup_{k \in [1,n] \mid \alpha_k \neq \beta_k} [low_{\alpha_k}..up_{\beta_k}]$ the intervals that are necessarily empty when the origin of task j is fixed.

(a) if $\ell_i < \ell_j \wedge \ell_i = 1$ then eliminate $[low..up]$ such that low (resp. up) is attached to a limit associated with m (resp. s). In the following, for simplicity, we just say *eliminate* [m,s]. Similarly we eliminate [f,mi].

(b) if $\ell_i < \ell_j \wedge \ell_i + 1 = \ell_j$ then eliminate [s,f].

(c) if $\ell_i = \ell_j \wedge \ell_i = 1$ then eliminate [m,e] and [e,mi].

(d) if $\ell_i > \ell_j \wedge \ell_j = 1$ then eliminate [m,fi] and [si,mi].

(e) if $\ell_i > \ell_j \wedge \ell_j + 1 = \ell_i$ then eliminate [fi,si].

We now show that the previous three steps procedure generates a symbolic normal form for the forbidden regions of a general relation $\mathcal{R}$.

Lemma 3. *For a general relation $\mathcal{R}$ by systematically combining the three cases $\ell_i < \ell_j$, $\ell_i = \ell_j$, $\ell_i > \ell_j$ with all possible restrictions from $\{\top, \ell_i > 1 \wedge \ell_j > 1, \ell_i + 1 < \ell_j, \ell_i > \ell_j + 1\}$ we generate all possible cases for that relation $\mathcal{R}$.*

Proof. The Cartesian product of $\{\ell_i < \ell_j, \ell_i = \ell_j, \ell_i > \ell_j\} \times \{\top\} \times \{\ell_i > 1 \wedge \ell_j > 1\} \times \{\ell_i + 1 < \ell_j\} \times \{\ell_i > \ell_j + 1\}$ is considered. $\square$

Lemma 4. *For a general relation $\mathcal{R}$ consider one of its case generated in step 1 and the corresponding limits low_{α_k} and up_{β_k}. The forbidden regions of $\mathcal{R}$ are given by $\bigcup_{k \in [1,n] \mid \alpha_k \neq \beta_k} [low_{\alpha_k}..up_{\beta_k}]$.*

Proof. A forbidden region of $\mathcal{R}$ is an interval of consecutive values that are forbidden for *all* basic relations of $\mathcal{R}$. Since both the lower limits low_{α_k} and the upper limits up_{β_k} are sorted in increasing order, and since $[low_p..up_q] = \emptyset$ for all $p \geq q$ we pick up for each start of a forbidden region low_{α_k} the smallest end up_{β_k} of the forbidden region that was starting before low_{α_k}. $\square$

Lemma 5. *When o_j is fixed the intervals removed by step 3 are the only empty intervals $[low_p..up_q]$ where $p < q$.*

Proof. The other cases being similar we only show the proof for the lower limit $\overline{o_j} - \ell_i + 1$ that was generated from m when $\ell_i < \ell_j$.

– Within the case $\ell_i < \ell_j$, $\overline{o_j} - \ell_i + 1$ is the lower limit of index 2 in Fig. 1. Consequently we first look at the upper limit of index 3, namely $o_j - 1$ that was generated from s. Since we want to check when $\overline{o_j} - \ell_i + 1$ will be strictly greater than $o_j - 1$ when o_j is fixed, we get $o_j - \ell_i + 1 > o_j - 1$, which simplifies to $-\ell_i + 1 > -1$ and to $\ell_i \leq 1$, which means that we can eliminate [m,s] when $\ell_i = 1$.

– We now need to compare $\overline{o_j} - \ell_i + 1$ with the next upper limit, namely the upper limit of index 4, i.e. o_j. We get $o_j - \ell_i + 1 > o_j$, which simplifies to $\ell_i < 1$ which is never true. Consequently the interval [m, d] is not empty when o_j is fixed. This implies that the other intervals [m, f], [m, oi], [m, mi], [m, bi] are also not empty when o_j is fixed since their upper limit are located after the upper limit of index 4. $\square$

Example 1. Assuming $\ell_i < \ell_j$ we successively illustrate how to generate the normalised forbidden regions for the relation $\mathcal{R}_1 = \{b, m, mi, bi\}$ (i.e. nonoverlapping), for $\mathcal{R}_2 = \{b, m\}$, and for $\mathcal{R}_3 = \{b, s, bi\}$.

1. By keeping the limits related to the basic relations b, m, mi, bi of $\mathcal{R}_1$ we get $\alpha_1 = 1, \alpha_2 = 2, \alpha_3 = 8$ and $\beta_1 = 1, \beta_2 = 7, \beta_3 = 8$. Since $\alpha_1 = \beta_1$ and $\alpha_3 = \beta_3$ we only keep α_2 and β_2 and get the interval $[low_{\alpha_2}..up_{\beta_2}] = [low_2..up_7] = [\overline{o_j} - \ell_i + 1..o_j + \ell_j - 1]$, the expected result for a nonoverlapping constraint between two tasks.
2. By keeping the limits related to the basic relations b, m of $\mathcal{R}_2$ we get $\alpha_1 = 1$, $\alpha_2 = 2$ and $\beta_1 = 1$, $\beta_2 = 9$. Since $\alpha_1 = \beta_1$ we only keep α_2 and β_2 and get the interval $[low_{\alpha_2}..up_{\beta_2}] = [low_2..up_9] = [\overline{o_j} - \ell_i + 1.. + \infty)$.
3. By keeping the limits related to the basic relations b, s, bi of $\mathcal{R}_3$ we get $\alpha_1 = 1$, $\alpha_2 = 4$ and $\beta_1 = 3$, $\beta_2 = 8$, which leads to $[low_{\alpha_1}..up_{\beta_1}] \cup [low_{\alpha_2}..up_{\beta_2}] = [low_1..up_3] \cup [low_4..up_8] = [\overline{o_j} - \ell_i..o_j - 1] \cup [\overline{o_j} + 1..o_j + \ell_j]$.

Merging Similar Cases. For a given Allen's general relation $\mathcal{R}$, two cases $(D_1$ if $C_1)$ and $(D_2$ if $C_2)$ can be merged to a single case $(D_{12}$ if $C_{12})$ if the following conditions all hold:

– C_{12} is equivalent to $C_1 \vee C_2$ and can be expressed as a conjunction of primitive restrictions.
– D_1, D_2, and D_{12} consist of the same number of intervals.
– For every interval $[b_1, u_1] \in D_1$ there are intervals $[b_2, u_2] \in D_2$ and $[b_{12}, u_{12}] \in D_{12}$ at the same position such that:
  • $b_1 = b_{12}$ and $u_1 = u_{12}$, for any values taken by ℓ_i and ℓ_j such that C_1 holds.
  • $b_2 = b_{12}$ and $u_2 = u_{12}$, for any values taken by ℓ_i and ℓ_j such that C_2 holds.

We used a semi-automatic approach to discover such endpoint generalisation rules. For every Allen's general relation, using these rules, we identified and merged pairs and triples of cases until no more merging was possible. As the result of this process, the data base [2] consists of 32396 cases covering all the 8192 general relations. In this data base, the maximum number of intervals for a case is 5, the average number of intervals is 2.14 and the median is 2.

3 Conclusion

This work belongs to the line of work that tries to synthesise in a systematic way constraint propagators for specific classes of constraints [6,7,9]. Future work may generalise this for getting a similar normal form for other families of qualitative constraints.

Acknowledgment. The Nantes authors were partially supported both by the INRIA TASCMELB associated team and by the GRACeFUL project, which has received funding from the European Union's Horizon 2020 research and innovation programme under grant agreement No 640954.

References

1. Allen, J.F.: Maintaining knowledge about temporal intervals. Commun. ACM **26**(11), 832–843 (1983)
2. Beldiceanu, N., Carlsson, M., Derrien, A., Schutt, A., Stuckey, P.J.: Range-consistent forbidden regions of Allen's relations. Technical report T2016-2, Swedish Institute of Computer Science (2016). http://soda.swedishict.se
3. Bessière, C.: Constraint propagation. In: Rossi, F., van Beek, P., Walsh, T. (eds.) Handbook of Constraint Programming, chap. I.3, pp. 29–83. Elsevier (2006)
4. Derrien, A., Fages, J.-G., Petit, T., Prud'homme, C.: A global constraint for a tractable class of temporal optimization problems. In: Pesant, G. (ed.) CP 2015. LNCS, vol. 9255, pp. 105–120. Springer, Cham (2015). doi:10.1007/978-3-319-23219-5_8
5. Gennari, R., Mich, O.: E-Learning and deaf children: a logic-based web tool. In: Leung, H., Li, F., Lau, R., Li, Q. (eds.) ICWL 2007. LNCS, vol. 4823, pp. 312–319. Springer, Heidelberg (2008). doi:10.1007/978-3-540-78139-4_28
6. Gent, I.P., Jefferson, C., Linton, S., Miguel, I., Nightingale, P.: Generating custom propagators for arbitrary constraints. Artif. Intell. **211**, 1–33 (2014)
7. Laurière, J.-L.: Constraint propagation or automatic programming. Technical report 19, IBP-Laforia (1996). In French. https://www.lri.fr/~sebag/Slides/Lauriere/Rabbit.pdf
8. Ligozat, G.: Towards a general characterization of conceptual neighborhoods in temporal and spatial reasoning. In: AAAI 1994 Workshop on Spatial and Temporal Reasoning (1994)
9. Monette, J.-N., Flener, P., Pearson, J.: Towards solver-independent propagators. In: Milano, M. (ed.) CP 2012. LNCS, pp. 544–560. Springer, Heidelberg (2012). doi:10.1007/978-3-642-33558-7_40
10. Roy, P., Perez, G., Régin, J.-C., Papadopoulos, A., Pachet, F., Marchini, M.: Enforcing structure on temporal sequences: the allen constraint. In: Rueher, M. (ed.) CP 2016. LNCS, vol. 9892, pp. 786–801. Springer, Cham (2016). doi:10.1007/978-3-319-44953-1_49
11. van Beek, P., Manchak, D.W.: The design and experimental analysis of algorithms for temporal reasoning. J. Artif. Intell. Res. (JAIR) **4**, 1–18 (1996)

MDDs are Efficient Modeling Tools:
An Application to Some Statistical Constraints

Guillaume Perez and Jean-Charles Régin[(✉)]

I3S, CNRS, Université Nice-Sophia Antipolis, Sophia Antipolis, France
guillaume.perez06@gmail.com, jcregin@gmail.com

Abstract. We show that from well-known MDDs like the one modeling a sum, and operations between MDDs we can define efficient propagators of some complex constraints, like a weighted sum whose values satisfy a normal law. In this way, we avoid defining ad-hoc filtering algorithms. We apply this idea to different dispersion constraints and on a new statistical constraint we introduce: the Probability Mass Function constraint. We experiment out approach on a real world application. The conjunction of MDDs clearly outperforms all previous methods.

1 Introduction

Several constraints, like `spread` [16], `deviation` [21–24], `balance` [3,5] and `dispersion` [17], have mainly been defined to balance certain features of a solution. For example, the balanced academic curriculum problem [1] involves courses that have to be assigned to periods so as to balance the academic load between periods. Most of the time the mean of the variables is fixed and the goal is to minimize the standard deviation, the distance or the norm.

The `dispersion` constraint is a generalization of the `deviation` and `spread` constraints. It ensures that X, a set of variables, has a mean (i.e. $\mu = \sum_{x \in X} x$) belonging to a given interval and Δ a norm (i.e. $\sum_{x \in X} (x - \mu)^p$) belonging to another given interval. If $p = 1$ then it is a `deviation` constraint and $p = 2$ defines a `spread` constraint. Usually, the goal is to minimize the value of Δ or find a value below a given threshold.

In some problems, variables are independent from a probabilistic point of view and are associated with a distribution (e.g. a normal law) that specifies probabilities for their values. Thus, globally the values taken by the variables have to respect that law and we can define a constraint ensuring this property, either by using a `spread`, a `dispersion`, a `KolmogorovSmirnov` or a `Student's t-test` constraint [19]. However, if only a subset of variables is involved in a constraint, then the values taken by these variables should be compatible with the distribution (e.g. the normal law), but we cannot impose the distribution for a subset of values because this is a too strong constraint. Therefore, we need to consider interval of values for μ and Δ. The definition of an interval for μ can be done intuitively. For instance we can consider an error rate of 10%. Unfortunately, this is not the case for Δ. It is hard to control the relation

© Springer International Publishing AG 2017
D. Salvagnin and M. Lombardi (Eds.): CPAIOR 2017, LNCS 10335, pp. 30–40, 2017.
DOI: 10.1007/978-3-319-59776-8_3

between two values of Δ, because data are coming from measures and there are some errors and because it is difficult to apply a continuous law on finite set of values. Since we use constraint programming solvers we have to make sure that we do not forbid tuples that could be acceptable. This is why, in practice, the problem is not defined in term of μ and Δ but by the probability mass function (PMF). The probability mass function gives the probability that X_r, a discrete random variable, is exactly equal to some value. In other words, if X_r takes its values in V, then the PMF gives the probability of each value of V. The PMF is usually obtained from the histogram of the values. From f_P, a PMF, we can compute the probability of any tuple by multiplying the probability of the values it contains, because variables are independent. Then, we can avoid outliers of the statistical law but imposing that the probability of a tuple belongs to a given interval $[P_{min}, P_{max}]$. With such a constraint we can select a subset of values from a large set having a mean in a given interval while avoiding outliers of the statistical law. Roughly, the minimum probability avoids having tuples with values having only very little chance to be selected and the maximum probability avoids having tuples whose values have only the strongest probability to be selected. Thus, we propose to define the PMF constraint, a new statistical constraint from μ, f_P, P_{min} and P_{max}.

Since we define a new constraint we need to define a propagator for it. Instead of designing an ad-hoc propagator we propose to represent the constraint by an MDD and to use MDD propagators, like MDD4R [13], for establishing arc consistency of the constraint. MDDs of constraints can also be intersected in order to represent the combination of the constraints and MDD operators applied on them. Recent studies show that such combinations give excellent results in practice [20].

Pesant has proposed a specific algorithm for filtering the dispersion constraint. His propagator establishes domain consistency on the X variables. Unfortunately, it is ad-hoc and so it cannot be easily combined with other constraints. Thus, we propose to also use MDD propagators for the classical version of the dispersion constraint.

The advantage of using MDDs representing the constraints in these cases, is that these MDDs are defined on sum constraints which are well-known MDDs.

We tested all propagators on a part of a real world application mainly involving convolutions which is expressed by a knapsack constraint (i.e. $\sum \alpha_i x_i$). The results show the advantage of our generic approach.

The paper is organized as follows. First, we recall some basics about MDDs, MDD propagators and the dispersion constraint. Then, we introduce simple models using MDDs for modelling the dispersion constraint with a fixed or a variable mean, and we show how we can combine them in order to obtain only one MDD. Next, we present the PMF constraint and show how it can be represented by an MDD and filtered by MDD propagators. We give some experiments supporting our approach. At last, we conclude.

2   Preliminaries

Multi-valued decision diagram (MDD). An MDD is a data-structure representing discrete functions. It is a multiple-valued extension of BDDs [6]. An MDD, as used in CP [2,4,7,9–11,13], is a rooted directed acyclic graph (DAG) used to represent some multi-valued function $f : \{0...d - 1\}^r \to \{true, false\}$, based on a given integer d. Given the r input variables, the DAG representation is designed to contain $r + 1$ layers of nodes, such that each variable is represented at a specific layer of the graph. Each node on a given layer has at most d outgoing arcs to nodes in the next layer of the graph. Each arc is labeled by its corresponding integer. The arc (u, v, a) is from node u to node v and labeled by a. All outgoing arcs of the layer r reach the true terminal node (the false terminal node is typically omitted). There is an equivalence between $f(a_1, ..., a_r) = true$ and the existence of a path from the root node to the true terminal node whose arcs are labeled $a_1, ..., a_r$ (Fig. 1).

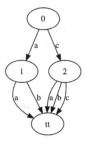

Fig. 1. An MDD of the tuple set $\{(a,a),(a,b),(c,a),(c,b),(c,c)\}$. For each tuple, there is a path from the root node (node 0) to the terminal node (node tt) whose arcs are labeled by the tuple values.

MDD of a constraint. Let C be a constraint defined on $X(C)$. The MDD associated with C, denoted by MDD(C), is an MDD which models the set of tuples satisfying C. More precisely, MDD(C) is defined on $X(C)$, such that layer i corresponds to the variable x_i and the labels of arcs of the layer i correspond to values of x_i, and a path of MDD(C) where a_i is the label of layer i corresponds to a tuple $(a_1, ..., a_r)$ on $X(C)$.

Consistency with MDD(C). A value a of the variable x is valid iff $a \in D(x)$, where $D(x)$ is the possible values of the variable x. An arc (u, v, a) at layer i is valid iff $a \in D(x_i)$. A path is valid iff all its arcs are valid.

  Let $path_{tt}^s(\text{MDD}(C))$ be the set of paths from s, the root node, to tt in MDD(C). The value $a \in D(x_i)$ is consistent with MDD(C) iff there is a valid path in $path_{tt}^s(\text{MDD}(C))$ which contains an arc at layer i labeled by a.

MDD propagator. An MDD propagator associated with a constraint C is an algorithm which removes some inconsistent values of $X(C)$. The MDD propagator establishes arc consistency of C if and only if it removes all inconsistent values with MDD(C). This means that it ensures that there is a valid path from the root to the true terminal node in MDD(C) if and only if the corresponding tuple is allowed by C and valid.

Cost-MDD. A cost-MDD is an MDD whose arcs have an additional information: the cost c of the arc. That is, an arc is a 4-uplet $e = (u, v, a, c)$, where u is the head, v the tail, a the label and c the cost. Let M be a cost-MDD and p be a path of M. The cost of p is denoted by $\gamma(p)$ and is equal to the sum of the costs of the arcs it contains.

Cost-MDD of a constraint [9]. Let C be a constraint and f_C be a function associating a cost with each value of each variable of $X(C)$. The cost-MDD of C and f_C is denoted by cost-MDD(C, f_C) and is MDD(C) whose the cost of an arc labeled by a at layer i is $f_C(x_i, a)$.

Cost-MDD propagator [8,15]. A cost-MDD propagator associated with C, f_C, a value H, and a symbol $\prec$ (which can be $\leq$ or $\geq$) is an MDD propagator on MDD(C) which ensures that for each path p of cost-MDD(C, f_C) we have $\gamma(p) \prec H$. A cost-MDD propagator establishes arc consistency of C iff each arc of cost-MDD(C) belongs to p a valid path of $path_{tt}^s($cost-MDD(C)$)$ with $\gamma(p) \prec H$.

MDD of a Generic Sum Constraint [25]. We define the generic sum constraint $\Sigma_{f,[a,b]}(X)$ which is equivalent to $a \leq \sum_{x_i \in X} f(x_i) \leq b$, where f is a non negative function. The MDD of the constraint $\sum_{x_i \in X} f(x_i)$ is defined as follows.

For the layer i, there are as many nodes as there are values of $\sum_{k=1}^{i} f(x_k)$. Each node is associated with such a value. A node n_p at layer i associated with value v_p is linked to a node n_q at layer $i + 1$ associated with value v_q if and only if $v_q = v_p + f(a_i)$ with $a_i \in D(x_i)$. Then, only values v of the layer $|X|$ with $a \leq v \leq b$ are linked to tt. The reduction operation is applied after the definition and delete invalid nodes [14]. The construction can be accelerated by removing states that are greater than b or that will not permit to reach a. For convenience, $\Sigma_{id,[\alpha,\alpha]}(X)$ is denoted by $\Sigma_\alpha(X)$.

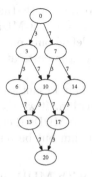

Fig. 2. MDD of the $\sum x_i = n\mu$ constraint

Figure 2 is an example of $\mathrm{MDD}(\Sigma_{20}(X))$ with $\{3,7\}$ as domains. Since f_C in non negative, the number of nodes at each layer of $\mathrm{MDD}(\Sigma_{f,[a,b]}(X))$ is bounded by b.

Dispersion Constraint [17]. Given $X = \{x_1, ..., x_n\}$, a set of finite-domain integer variables, μ and Δ, bounded-domain variables and p a natural number. The constraint DISPERSION(X, μ, Δ, p) states that the collection of values taken by the variables of X exhibits an arithmetic mean $\mu = \sum_{i=1}^{n} x_i$ and a deviation $\Delta = \sum_{i=1}^{n} |x_i - \mu|^p$.

The deviation constraint is a dispersion constraint with $p = 1$ and a spread constraint is a dispersion constraint with $p = 2$.

The main complexity of the dispersion constraint is the relation between μ and Δ variables, because μ is defined from the X variables, and Δ is defined from X and from μ. So, some information is lost when these two definitions are considered separately. However, when μ is assigned, the problem becomes simpler because we can independently consider the definitions of μ and Δ. Therefore, we propose to study some models depending on the fact that μ is fixed or not.

3 Dispersion Constraint with Fixed Mean

Arc consistency for the X variables has been established by Pesant [17], who proposed an ad-hoc dynamic programming propagator for this constraint. However, it exists a simpler method avoiding such problems of ad-hoc algorithms: we define a cost-MDD from μ and Δ and obtain a propagator having the same complexity.

3.1 MDD on μ and Δ as Cost

The mean μ is defined as a sum constraint. Since μ is fixed, we propose to use the cost-MDD of the constraint $\sum x_i = n\mu$ and the cost function defined by Δ.

The constraint $\sum x_i = n\mu$ can be represented by $\mathrm{MDD}(\Sigma_{n\mu}(X))$.

Δ ***as cost.*** We represent the dispersion constraint by cost-MDD$(\Sigma_\mu(X), \Delta)$. There are two possible ways to deal with the boundaries of Δ. Either we define two cost-MDD propagators on cost-MDD$(\Sigma_\mu(X), \Delta)$, one with a and $\geq$, and one with b and $\leq$; or we define only one cost-MDD propagator on cost-MDD$(\Sigma_\mu(X), \Delta)$ which integrates the costs at the same time as proposed by Hoda et al. [11].

These methods are simpler than Pesant's algorithm because they do not require to develop any new algorithm. If we use an efficient algorithm [15] for maintaining arc consistency for cost-MDDs then we obtain the the same worst case complexity as Pesant's algorithm but better result in practice.

3.2 MDD on μ Intersected with MDD on Δ

Since μ is fixed, then the definition of Δ corresponds to a generic sum as previously defined. Thus, the dispersion constraint can be model by defining the

MDD of $\Sigma_\mu(X)$ and the MDD of $\Sigma_{\Delta,[\underline{\Delta},\overline{\Delta}]}(X)$ and then by intersecting them. Replacing a cost-MDD by the intersection of two MDDs may strongly improve the computational results [15]. In addition, we can intersect the resulting MDD with some other MDDs in order to combine more the constraints. This method is the first method establishing arc consistency for both μ and Δ. The drawback is the possible size of the intersection.

With similar models we can also give an efficient implementation of the Student's t-test constraint and close the open question of Rossi et al. [19].

4   Dispersion Constraint with Variable Mean

In order to deal with a variable mean, we can consider all acceptable values for $n\mu$, that is the integers in $[n\lfloor\mu\rfloor, n\lceil\overline{\mu}\rceil]$, and for each value we separately apply the previous models for the fixed mean. Unfortunately, this often leads to a large number of constraints. Therefore it is difficult to use this approach in practice. In addition, note that there is no advantage in making the union of these constraints because they are independent.

Thus, we propose another model using the probability mass function.

5   Probability Mass Function (PMF) Constraint

In this section we define the PMF constraint which aims at respecting a variable mean and avoiding outliers according to a statistical law given by a probability mass function.

Given a discrete random variable X taking values in $X = \{v_1, ...v_m\}$ its probability mass function P: $X \rightarrow [0, 1]$ is defined as $P(v_i) = Pr[X = v_i]$ and satisfies the following condition: $P(v_i) \geq 0$ and $\sum_{i=1}^m P(v_i) = 1$

The PMF gives for each value v, $P(v)$ the probability that v is taken. Let f_P be a PMF and consider a set of variables independent from a probabilistic point of view and associated with f_P that specifies probabilities for their values. Since the variables are independent, we can define the probability of an assignment of all the variables (i.e. a tuple) as the product of the probabilities of the assigned values. Then, in order to avoid outliers we can constrain this probability to be in a given interval.

Definition 1. Given a set of finite-domain integer variables $X = \{x_1, x_2, ..., x_n\}$ that are independent from a probabilistic point of view, a probability mass function f_P, a bounded variable μ (not necessarily fixed), a minimum probability P_{min} and a maximum probability P_{max}. The constraint $\text{PMF}(X, f_P, \mu, P_{min}, P_{max})$ states that the probabilities of the values taken by the variables of X is specified by f_P, the collection of values taken by the variables of X exhibits an arithmetic mean μ and that $\Pi_{x_i \in X} x_i$ the probability of any allowed tuple satisfies $P_{min} \leq \Pi_{x_i \in X} f_P(x_i) \leq P_{max}$.

This constraint can be represented by cost-MDD$(\Sigma_{id,[\underline{\mu},\overline{\mu}]}(X), logP)$ where $logP$ is the logarithm of the PMF that is $logP(x) = \log(\overline{f_P}(x))$. We take the logarithm because in this way we have a sum function instead of a product function: $\log(\Pi f_P(x_i)) = \sum \log(f_P(x)) = \sum logP(x)$. Then, we define a cost-MDD propagator on cost-MDD$(\Sigma_{id,[\underline{\mu},\overline{\mu}]}(X), logP$ with $\log(P_{min})$ and $\geq$ and with $\log(P_{max})$ and $\leq$.

6   Experiments

The experiments were run on a macbook pro (2013) Intel core i7 2.3 GHz with 8 Go. The constraint solver used is or-tools. MDD4R [13] is used as MDD propagator and cost-MDD4R as cost-MDD propagator [15].

The data come from a real life application: the geomodeling of a petroleum reservoir [12]. The problem is quite complex and we consider here only a subpart. Given a seismic image we want to find the velocities. Velocities values are represented by a probability mass function (PMF) on the model space. Velocities are discrete values of variables. For each cell c_{ij} of the reservoir, the seismic image gives a value s_{ij} and the from the given seismic wavelet (α_k) we define a sum constraint $\sum_{k=1}^{22} \alpha_k log(x_{i-11+k-1}j) = s_{ij}$. Locally, that is for each sum, we have to avoid outliers w.r.t. the PMF for the velocities. Globally we can use the classical dispersion constraint. The problem is huge (millions of variables) so we consider here only a very small part.

The first experiment involves 22 variables and a constraint C_α: $\sum_{i=1}^{n} \alpha_i x_i = I$, where I is an tight interval (i.e. a value with an error variation). C_α is represented by $mdd_\alpha = \text{MDD}(\Sigma_{a_i,I}(X))$ where $a_i(x_i) = \alpha_i x_i$.

First, we impose that the variables have to be distributed with respect to a normal distribution with μ, a fixed mean.

$M_{\sigma<,\sigma>}$ represents the model of Sect. 3.2: one cost-MDD propagator on $mdd_\mu = \text{cost-MDD}(\Sigma_{n\mu}(X), \sigma)$ with σ and $\leq$ and one with σ and $\geq$. This model is similar to Pesant's model.

M_{GCC} involves a GCC constraint [18] where the cardinalities are extracted from the probability mass function.

$M_{\mu\cap\sigma}$ represents the mean constraint by $mdd_\sigma = \text{MDD}(\Sigma_{n\mu}(X))$. It represents the sigma constraint by the $\text{MDD}(\Sigma_\sigma(X))$. Then the two MDDs are intersected. An MDD propagator is used on this MDD, named $mdd_{\mu\sigma}$. See Sect. 3.2.

$M_{\mu\cap\sigma\cap\alpha}$ intersects mdd_α, the MDD of the constraint C_α, with $mdd_{\mu\sigma}$ the previous MDD to obtain mdd_{sol}. In this case, all constraints are combined.

Then, we consider a PMF constraint and that μ is variable:

M_{log}.
We define a cost-MDD propagator on $mdd_{I_\mu} = \text{cost-MDD}(\Sigma_{id,[\underline{\mu},\overline{\mu}]}(X), logP)$ with $\log(P_{min})$ and $\geq$ and with $\log(P_{max})$ and $\leq$. See Sect. 5.

$M_{log\cap\alpha}$. We define $mdd_{I_{log}} = \text{MDD}(\Sigma_{logP,I_{log}}(X))$ and we intersect it with mdd_{I_μ}. Then, we intersect it with mdd_α, the MDD of C_α, to obtain $mdd_{log\alpha}$.

Table 1 shows the result of these experiments. As we can see when the problem involves many solutions, all the methods perform well (excepted M_{GCC}). We can see that an advantage of the intersection methods is that they contain all the solutions of problem. Table 2 shows the different sizes of the MDDs.

Table 1. Comparison solving times (in ms) of models. 0 means that this is immediate. T-O indicates a time-out of 500 s.

Sat?	#sol	Fixed μ				Variable μ	
		$M_{\sigma<,\sigma>}$	M_{GCC}	$M_{\mu\cap\sigma}$	$M_{\mu\cap\sigma\cap\alpha}$	M_{log}	$M_{log\cap\alpha}$
Sat	Build	50	31	138	2,203	34	317,921
	10 sol	14	T-O	16	0	14	0
	All sol	T-O	T-O	T-O	0	T-O	0
UnSat	Build	55	28	121	151	37	133,752
	10 sol	T-O	T-O	T-O	0	T-O	0
	All sol	T-O	T-O	T-O	0	T-O	0

Table 2. Comparison of MDD sizes (in thousands) of different models. 0 means that the MDD is empty.

Sat?	N/A	Fixed μ					Variable μ		
		mdd_α	mdd_μ	mdd_σ	$mdd_{\mu\sigma}$	mdd_{sol}	mdd_{I_μ}	$mdd_{I_{log}}$	$mdd_{log\alpha}$
Sat	#nodes	3	3	5	67	521	2	18	24,062
Sat	#arcs	44	27	55	660	4,364	30	268	341,555
UnSat	#nodes	3	2	5	67	0	2	18	0
UnSat	#arcs	46	27	55	660	0	30	268	0

Random instances. The intersection methods $M_{\mu\cap\sigma}$ and $M_{\mu\cap\sigma\cap\alpha}$ have been tested on random bigger instances. Tables 3 and 4 gives some results showing how this method scales with the number of variables. In the first line, the couple is #var/#val. Times are in ms. Experiments of Table 3 set $0 < \sigma < 4n$ for having a delta depending on the number of variables like in [17], whereas experiments of Table 4 impose $100 < \sigma < 400$, these numbers come from our real world problem.

These experiments show that the $M_{\mu\cap\sigma}$ model can often be a good trade-off between space and time. Using the lower bound of the expected size of the MDD [15], we can estimate and decide if it is possible to process $M_{\mu\cap\sigma\cap\alpha}$. The last two columns of Table 3 show that it is not always possible to build such an intersection.

Table 3. Time (in ms) and size (in thousands) of the MDDs of models $M_{\mu\cap\sigma}$ and $M_{\mu\cap\sigma\cap\alpha}$. 0 means that the MDD is empty. M-O means memory-out.

$0 < \sigma < 4n$									
Method	n/d	20/20	30/20	40/30	40/40	50/40	50/50	100/40	100/100
$M_{\mu\cap\sigma}$	T(ms)	26	132	391	401	848	875	12,780	14,582
	#nodes	18	63	153	1578	306	311	2,285	2,532
	#arcs	198	808	2,308	2,427	5,196	5,354	53,757	62,057
$M_{\mu\cap\sigma\cap\alpha}$	T(ms)	561	3,084	11,864	10,789	58,092	60,513	M-O	M-O
	#nodes	163	764	0	0	0	0	M-O	M-O
	#arcs	1,788	8,416	0	0	0	0	M-O	M-O

Table 4. Time (in ms) and size (in thousands) of the MDDs of models $M_{\mu\cap\sigma}$ and $M_{\mu\cap\sigma\cap\alpha}$. $M_{\mu\cap\sigma\cap\alpha}$ is empty because there is no solution.

$100 < \sigma < 400$					
Method	n/d	20/20	30/20	40/30	40/40
$M_{\mu\cap\sigma}$	T(ms)	162	333	586	602
	#nodes	81	184	326	338
	#arcs	823	1,865	3,329	3,479
$M_{\mu\cap\sigma\cap\alpha}$	T(ms)	2,663	10,379	21,063	26,393
	#nodes	1,098	2,555	35	0
	#arcs	11,166	23,764	151	0

7   Conclusion

We have shown that modeling constraints by MDDs has several advantages in practice. It avoids to develop ad-hoc algorithms, gives competitive results and leads to efficient combination of constraints outperforming the other approaches. We have emphasized our approach on statistical constrains including the new PMF constraint we proposed.

References

1. Problem 30 of CSPLIB. www.csplib.org
2. Andersen, H.R., Hadzic, T., Hooker, J.N., Tiedemann, P.: A constraint store based on multivalued decision diagrams. In: Bessière, C. (ed.) CP 2007. LNCS, vol. 4741, pp. 118–132. Springer, Heidelberg (2007). doi:10.1007/978-3-540-74970-7_11
3. Beldiceanu, N., Carlsson, M., Demassey, S., Petit, T.: Global constraint catalog: past, present and future. Constraints **12**(1), 21–62 (2007)
4. Bergman, D., Hoeve, W.-J., Hooker, J.N.: Manipulating MDD relaxations for combinatorial optimization. In: Achterberg, T., Beck, J.C. (eds.) CPAIOR 2011. LNCS, vol. 6697, pp. 20–35. Springer, Heidelberg (2011). doi:10.1007/978-3-642-21311-3_5

5. Bessiere, C., Hebrard, E., Katsirelos, G., Kiziltan, Z., Picard-Cantin, É., Quimper, C.-G., Walsh, T.: The balance constraint family. In: O'Sullivan, B. (ed.) CP 2014. LNCS, vol. 8656, pp. 174–189. Springer, Cham (2014). doi:10.1007/978-3-319-10428-7_15

6. Bryant, R.E.: Graph-based algorithms for boolean function manipulation. IEEE Trans. Comput. C **35**(8), 677–691 (1986)

7. Cheng, K., Yap, R.: An mdd-based generalized arc consistency algorithm for positive and negative table constraints and some global constraints. Constraints **15**, 265–304 (2010)

8. Demassey, S., Pesant, G., Rousseau, L.-M.: A cost-regular based hybrid column generation approach. Constraints **11**(4), 315–333 (2006)

9. Gange, G., Stuckey, P., Szymanek, R.: MDD propagators with explanation. Constraints **16**, 407–429 (2011)

10. Hadzic, T., Hooker, J.N., O'Sullivan, B., Tiedemann, P.: Approximate compilation of constraints into multivalued decision diagrams. In: Stuckey, P.J. (ed.) CP 2008. LNCS, vol. 5202, pp. 448–462. Springer, Heidelberg (2008). doi:10.1007/978-3-540-85958-1_30

11. Hoda, S., Hoeve, W.-J., Hooker, J.N.: A systematic approach to MDD-based constraint programming. In: Cohen, D. (ed.) CP 2010. LNCS, vol. 6308, pp. 266–280. Springer, Heidelberg (2010). doi:10.1007/978-3-642-15396-9_23

12. Pennington, W.D.: Reservoir Geophys. **66**(1), 25–30 (2001)

13. Perez, G., Régin, J.-C.: Improving GAC-4 for table and MDD constraints. In: O'Sullivan, B. (ed.) CP 2014. LNCS, vol. 8656, pp. 606–621. Springer, Cham (2014). doi:10.1007/978-3-319-10428-7_44

14. Perez, G., Régin, J-C.: Efficient operations on MDDs for building constraint programming models. In: International Joint Conference on Artificial Intelligence, IJCAI 2015, Argentina, pp. 374–380 (2015)

15. Perez, G., Régin, J.-C.: Soft and cost MDD propagators. In: Proceedings of the AAAI 2017 (2017)

16. Pesant, G., Régin, J.-C.: SPREAD: a balancing constraint based on statistics. In: Beek, P. (ed.) CP 2005. LNCS, vol. 3709, pp. 460–474. Springer, Heidelberg (2005). doi:10.1007/11564751_35

17. Pesant, G.: Achieving domain consistency and counting solutions for dispersion constraints. INFORMS J. Comput. **27**(4), 690–703 (2015)

18. Régin, J.-C.: Generalized arc consistency for global cardinality constraint. In: Proceedings of the AAAI 1996, Portland, Oregon, pp. 209–215 (1996)

19. Rossi, R., Prestwich, S.D., Armagan Tarim, S.: Statistical constraints. In: ECAI 2014–21st European Conference on Artificial Intelligence, Prague, Czech Republic - Including Prestigious Applications of Intelligent Systems (PAIS 2014), 18–22 August 2014, pp. 777–782 (2014)

20. Roy, P., Perez, G., Régin, J.-C., Papadopoulos, A., Pachet, F., Marchini, M.: Enforcing structure on temporal sequences: the allen constraint. In: Rueher, M. (ed.) CP 2016. LNCS, vol. 9892, pp. 786–801. Springer, Cham (2016). doi:10.1007/978-3-319-44953-1_49

21. Schaus, P., Deville, Y., Dupont, P., Régin, J.-C.: The deviation constraint. In: Hentenryck, P., Wolsey, L. (eds.) CPAIOR 2007. LNCS, vol. 4510, pp. 260–274. Springer, Heidelberg (2007). doi:10.1007/978-3-540-72397-4_19

22. Schaus, P., Deville, Y., Dupont, P., Régin, J.-C.: Simplification and extension of the SPREAD constraint. In: Future and Trends of Constraint Programming, pp. 95–99. ISTE (2007)

23. Schaus, P., Régin, J.-C.: Bound-consistent spread constraint **2**(3) (2014)
24. Schaus, P., Deville, Y., Dupont, P.: Bound-consistent deviation constraint. In: Bessière, C. (ed.) CP 2007. LNCS, vol. 4741, pp. 620–634. Springer, Heidelberg (2007). doi:10.1007/978-3-540-74970-7_44
25. Trick, M.: A dynamic programming approach for consistency and propagation for knapsack constraints. In CPAIOR 2001 (2001)

On Finding the Optimal BDD Relaxation

David Bergman[1] and Andre Augusto Cire[2(✉)]

[1] Department of Operations and Information Management,
University of Connecticut, Mansfield, USA
`david.bergman@uconn.edu`
[2] Department of Management, University of Toronto Scarborough,
Toronto, USA
`acire@utsc.utoronto.ca`

Abstract. This paper presents an optimization model for identifying limited-width relaxed binary decision diagrams (BDDs) with tightest possible relaxation bounds. The model developed is a network design model and is used to identify which nodes and arcs should be in a relaxed BDD so that the objective function bound is as close to the optimal value as possible. The model is presented specifically for the 0–1 knapsack problem, but can be extended to other problem classes that have been investigated in the stream of research on using decision diagrams for combinatorial optimization problems. Preliminary experimental results indicate that the bounds provided by the relaxed BDDs are far superior to the bounds achieved by relaxed BDDs constructed via previously published compilation algorithms.

1 Introduction

The use of decision diagrams (DDs) for optimization has driven an accelerating stream of research in both mathematical programming and constraint programming (CP) communities [8]. The topic stemmed from the idea of approximating the feasible solution space of a problem using limited-size *relaxed* DDs. Such an approximation could either play the role of the constraint store in CP [2,18], or function as an alternative relaxation in discrete optimization problems [5,6,9]. In all these cases, the use of relaxed DDs provided strong optimization bounds that were effective in pruning large portions of the search space, thereby significantly speeding up the solution process in a variety of problems.

The success of relaxed DDs therefore hinges on the tightness of the optimization bounds they provide. This results in fundamentally different challenges than the ones that are typically addressed in the classical use of DDs in Boolean logic [1,11,12,21], where the focus is on minimizing the *size* of a DD that exactly represents a Boolean formula (see, e.g., [3,13,14,17,22,23]). In the context of optimization, the question of interest changes: Given a maximum size of a relaxed DD, what is the best possible optimization *bound* that can be achieved?

To date, this research question has remained largely unanswered. Previous papers indicate that the order of the decision variables within a DD affects

© Springer International Publishing AG 2017
D. Salvagnin and M. Lombardi (Eds.): CPAIOR 2017, LNCS 10335, pp. 41–50, 2017.
DOI: 10.1007/978-3-319-59776-8_4

the bounds [5], in that variable orderings with smaller exact DDs also lead to stronger relaxations. Relaxed DD constructions have since then primarily focused on heuristics that search for good variable orderings, and very few theoretical or computational guarantees on their quality have been provided thus far.

The contribution of this paper is to present the first methodology for obtaining the relaxed DD with the provably tightest optimization bound. Specifically, we propose a new linear integer programming (IP) model that constructs the optimal-bound relaxed DD for a given fixed size. Besides providing a new technique to build a relaxed DD (e.g., by finding any feasible solution to this IP), such a methodology can be used, for instance, to benchmark the quality of the heuristically-constructed relaxed DDs in current branch-and-bound and CP techniques [4]. This model may also allow for new theoretical results that link the quality of the bound with the underlying polyhedral structure of the IP, similar to what is investigated in approximate dynamic programming [10]. It also shares connections with existing policy construction mechanisms in Markov decision processes [24] and general approximate inference schemes [15,16].

In this work we restrict our attention to the *0–1 knapsack problem* and therefore to *binary* decision diagrams (BDDs), but later discuss how the overall concepts can be generalized. The proposed IP determines which nodes and arcs to include in the BDD so that the longest path from the root to the terminal is as small as possible (assuming a maximization problem). It shares some similarities with the problem of finding a variable ordering that minimizes the size of the exact BDD, also formulated as an IP for the knapsack problem [3]. However, it differs from previous work in that we only need to enforce that no feasible solution is lost in our BDD representation, which makes it easier to generalize.

Preliminary experimental results on small-scale problems indicates that, even when not solved to optimality, the model can identify far superior relaxed BDDs than those obtained through the standard compilation technique from the literature. This is a surprising result that may be used to directly improve existing DD-based technology and open new potential research directions, while showing that existing BDD-relaxations can be substantially improved.

The remainder of the paper is organized as follows. In Sect. 2 we formally define BDDs and later provide an example of a knapsack instance which exhibits the property that, for a fixed maximum width, current methods result in a relaxed BDD with weaker bounds than the relaxed BDD with the best possible relaxation bound. In Sect. 3 an IP formulation for determining the optimal BDD relaxation is presented. Section 4 describes the results on preliminary experimental evaluation, and a conclusion is provided in Sect. 5.

2   BDDs for Optimization

We focus the discussion on the *0–1 knapsack problem* (KP) defined by n items $\mathcal{I} = \{1, \ldots, n\}$, a *knapsack capacity* C, and where each item $i \in \mathcal{I}$ has a *profit* and *weight* $p_i, w_i \in \mathbb{Z}^+$, respectively. Each subset $I \subseteq \mathcal{I}$ has weight $w(I) = \sum_{i \in I} w_i$ and profit $p(I) = \sum_{i \in I} p_i$. The goal is to find the subset $I^* \subseteq \mathcal{I}$ with maximum profit whose weight does not exceed the knapsack capacity. Let $\mathcal{S}$ be the family of subsets of $\mathcal{I}$ that do not exceed the knapsack capacity.

A standard IP model associates a 0–1 variable x_i with each item $i \in \mathcal{I}$ indicating whether to include or not the i-th item in the knapsack, as follows:

$$\max \left\{ \sum_{i \in \mathcal{I}} p_i x_i \ : \ \sum_{i \in \mathcal{I}} w_i x_i \leq C, x_i \in \{0, 1\}, \ \forall i \in \mathcal{I} \right\} \qquad \text{(KP-IP)}$$

For the purpose of this paper, a BDD is a layered-acyclic arc-weighted digraph with node set U and arc set A, where the paths encode subsets of $\mathcal{I}$. U is partitioned into $n + 1$ layers $L_1, \ldots, L_{n+1}$, and each node $u \in U$ belongs to the layer indexed by $\ell(u) \in \{1, 2, \ldots, n + 1\}$; i.e., $L_i = \{u \mid \ell(u) = i\}$. In particular, $L_1 = \{\mathbf{r}\}$ and $L_{n+1} = \{\mathbf{t}\}$, where $\mathbf{r}$ and $\mathbf{t}$ are referred to as the *root* and *terminal* node, respectively. The *width* of a layer L_i is the number of nodes in L_i, i.e. $|L_i|$, and the width $W(B)$ of B is $W(B) = \max_{1 \leq i \leq n+1} |L_i|$. Each arc $a = (t(a), h(a)) \in A$ has *tail* $t(a) \in U$ and a *head* $h(a) \in U$, with $\ell(h(a)) - \ell(t(a)) = 1$; i.e. arcs connect only nodes in adjacent layers. A node $u \in U \backslash \{\mathbf{t}\}$ is the tail of at most one 0-arc and at most one 1-arc. The layer $\ell(a)$ of an arc a corresponds to the layer of the tail of a: $\ell(a) = \ell(t(a))$. Moreover, a is associated with an *arc value* $v(a) \in \mathbb{Z}^*$ and an *arc domain* $d(a) \in \{0, 1\}$. An arc a is referred to as a *0-arc* and *1-arc* when $d(a) = 0$ and $d(a) = 1$, respectively.

A 0-arc a in layer $\ell(a)$ corresponds to *not* including the item $\ell(a)$ in the knapsack and has a value of $v(a) = 0$. In turn, an 1-arc a in layer $\ell(a)$ corresponds to including item $\ell(a)$ in the knapsack and has a value of $p_{\ell(a)}$. Each path $p = (a_1, \ldots, a_k)$ therefore encodes a subset $\mathcal{I}(p) \subseteq \mathcal{I}$ through the arc domains of its arcs: $\mathcal{I}(p) = \{\ell(a_j) : d(a_j) = 1, 1 \leq j \leq k\}$. As such, the sum of the arc values in p yields the profit of the subset $\mathcal{I}(p)$.

Let $\mathcal{P}$ be the set of $\mathbf{r} - \mathbf{t}$ paths in a BDD B. The *solution set* $\mathcal{S}(B)$ is the collection of subsets encoded by paths in $\mathcal{P}$: $\mathcal{I}(B) = \bigcup_{p \in \mathcal{P}} \mathcal{I}(p)$. B is called *exact* if $\mathcal{S}(B) = \mathcal{S}$ and is called *relaxed* if $\mathcal{S}(B) \supseteq \mathcal{S}$. For an exact BDD, any longest $\mathbf{r} - \mathbf{t}$ path p^* in terms of the values $v(\cdot)$ therefore encodes an optimal solution $\mathcal{I}(p^*)$ and its total value is the optimal value $w(p^*)$. For a relaxed BDD, $\mathcal{I}(p^*)$ may be infeasible, but $w(p^*) \geq w(I^*)$ and we obtain a bound on $w(I^*)$. Because B is directed and acyclic, the longest path can be found in $O(|U|)$.

Note that in our definition the i-th layer of the BDD is associated directly with the i-th item of the knapsack. However, any one-to-one mapping between layers and variables (i.e., a *variable ordering*) yields a valid exact or relaxed BDD, but each with possibly distinct size or optimization bound [6]. For the purposes of this paper we restrict our attention to the variable ordering given by the index of the items, which is assumed without loss of generality.

Motivating Example. We now provide a KP instance for which building a relaxed BDD via the standard approach from [6] yields a weaker bound than the best possible relaxed BDD of a fixed width. Consider the following instance:

$$\max \quad 94x_1 + 98x_2 + 97x_3 + 75x_4 \tag{E}$$

$$\text{s.t.} \quad 51x_1 + 56x_2 + 66x_3 + 93x_4 \leq 133, \ x \in \{0,1\}^4$$

The exact BDD for problem (E) is depicted in Fig. 1(a) and has a width of 3. The longest path is depicted by the shaded arcs in the figure and corresponds to the solution $x = (0,1,1,0)$ with a value of 195. Existing works in BDDs for optimization (e.g., [6,8]) consider a top-down construction where layers are built one at a time, "merging" nodes if a given maximum width of the layer is exceeded. The chief merging strategy in practice is to aggregate nodes with the least longest path values from the root $\mathbf{r}$ to those nodes (denoted by $minLP$ procedure in [6]). This procedure outputs the 2-width BDD in Fig. 1(b), where the longest path corresponds to the infeasible solution $x = (1,1,0,1)$ with a value of 267. An optimal 2-width BDD relaxation is depicted in Fig. 1(c) and its optimal longest path yields the optimal solution $x = (0,1,1,0)$. Notice that it is still a relaxed DD since it contains the infeasible solution $x = (0,1,0,1)$.

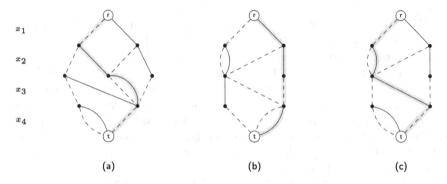

Fig. 1. (a) Exact BDD, (b) relaxed BDD using the procedure from [6], and (c) an optimal relaxed BDD for the KP example (E). 0-arcs and 1-arcs are depicted by dashed and solid arcs, respectively. Shaded arcs represent the longest path in each BDD.

3   IP Model for the Optimal Relaxed BDD

The relaxations bounds obtained by relaxed BDDs is perhaps the most crucial ingredient to the successful application of BDDs for optimization. Since there are problems for which the exact BDD will be exponential in size, relaxed BDDs are used to approximate the feasible set and provide relaxation bounds. The variable ordering chosen for the layers can have a significant impact on the bound, but we focus here on finding the relaxed BDD with width less than or equal to W that provides the strongest (i.e., lowest) relaxation bound.

The problem can be cast as the IP model (1), as follows. Let $[t] = \{1, \ldots, t\}$ for any number $t \in \mathbb{Z}^+$. For each layer ℓ, we define W variables $y_{\ell,i}, i \in [W]$, which indicates whether or not a node with index i is used on layer ℓ. Additionally, binary variables $x^d_{\ell,i,j}$ are defined for every $d \in \{0,1\}$, every pair of indices in $[W]$, and for every layer in $[n]$. This variable indicates if a d-arc between a node with index $i \in L_\ell$ and a node with index $j \in L_{\ell+1}$ exists.

A node u in the resulting BDD can be equivalently written as $u = (\ell, i)$, indicating that u is the i-th node at layer ℓ. Variable $S_{\ell,i}$ defines a *state* for each node $u = (\ell, i)$, which is used to enforce that no feasible solution in S is lost. The variables $p_{\ell,i}$ define the longest path from that node to the terminal. The size of the resulting model is approximately $O(nW^2)$ (both in terms of variables and constraints).

$$\min \quad p_{1,1} \tag{1a}$$

$$\text{s.t.} \quad \sum_{d \in \{0,1\}} \sum_{j \in [W]} x^d_{1,1,j} \geq 1 \tag{1b}$$

$$\sum_{d \in \{0,1\}} \sum_{i \in [W]} x^d_{n+1,i,1} \geq 1 \tag{1c}$$

$$x^d_{\ell,i,j} \leq y_{\ell,i}, \; x^d_{\ell,i,j} \leq y_{\ell,j}, \qquad \forall d \in \{0,1\}, \forall \ell \in [n], \forall i,j \in [W] \tag{1d}$$

$$\sum_{j \in [n]} x^0_{\ell,i,j} = y_{\ell,i}, \qquad \forall \ell \in [n] \, \forall i \in [W] \tag{1e}$$

$$\sum_{j \in [n]} x^1_{\ell,i,j} \leq y_{\ell,i}, \qquad \forall \ell \in [n] \, \forall i \in [W] \tag{1f}$$

$$S_{1,1} = 0 \tag{1g}$$

$$S_{\ell,i} \leq C \tag{1h}$$

$$S_{\ell+1,i} \leq S_{\ell,j} + d \cdot w_\ell + \left(1 - x^d_{\ell,i,j}\right) \cdot M, \quad \forall d \in \{0,1\}, \forall \ell \in [n], \forall i,j \in [W] \tag{1i}$$

$$S_{\ell,i} + w_\ell \geq \left(1 - \sum_{j \in [W]} x^1_{\ell,i,j}\right) \cdot (C+1), \quad \forall \ell \in [n], \forall i \in [W] \tag{1j}$$

$$p_{\ell,i} \geq p_{\ell+1,j} + d \cdot p_\ell - \left(1 - x^d_{\ell,i,j}\right) \cdot M, \quad \forall d \in \{0,1\}, \forall \ell \in [n], \forall i,j \in [W] \tag{1k}$$

$$y_{\ell,i} \in \{0,1\}, S_{\ell,i}, p_{\ell,i} \in \mathbb{Z}^+, \qquad \forall \ell \in [n+1] \, \forall i \in [W] \tag{1l}$$

$$x^d_{\ell,i,j} \in \{0,1\}, \qquad \forall d \in \{0,1\}, \forall \ell \in [n], \forall i,j \in [W] \tag{1m}$$

$$p_{n+1,1} = 0 \tag{1n}$$

Theorem 1. *The optimal solution to model (1) yields a relaxed BDD B with node set $U = \{(\ell, i) : y_{\ell,i}{}^* = 1\}$, partitioned into $n+1$ layers lexicographically by ℓ, and arcs $A \subseteq U \times U$, defined by arcs $a = ((\ell,i), (\ell+1,j)) \in A \leftrightarrow x^d_{\ell,i,j}{}^* = 1$, with $d(a) = d$, for $d \in \{0,1\}$.*

Proof. We first claim that the digraph is connected. Associate $\mathbf{r}$ with node $(1,1)$ and $\mathbf{t}$ with node $(n+1,1)$. Constraints (1b) and (1c) required that at least one node is directed out of $\mathbf{r}$ and into $\mathbf{t}$. Constraints (1d) require that if an arc is defined, so are both of its endpoints, and the result follows. It is layered and acyclic by definition, and so the resulting graph is a BDD.

We claim the following: for any path p consisting of arcs $a_1, \ldots, a_{\ell'-1}$ traversing nodes $u_1 = \mathbf{r}, u_2, \ldots, u_{\ell'}$, with $u_\ell = (\ell, j_\ell)$ for $\ell = 1, \ldots, \ell'$, we have

$$S_{\ell', j_{\ell'}} \leq \sum_{\ell \in [\ell'-1]:d(a_\ell)=1} w_\ell. \tag{2}$$

We proceed by induction on ℓ', showing that for any possible path of length less than or equal to $\ell' - 1$, inequality (2) is satisfied. For $\ell' = 1$, $S_{1,1} = 0$ by constraint (1h). For $\ell' = 2$, consider any index j for which $y_{2,j}{}^* = 1$. Any arc directed to this node has tail $\mathbf{r}$ with $S_{1,1} = 0$. If a 0-arc is directed at this node, constraint (1i) ensures that $S_{2,j} = 0$ so that the inequality is trivially satisfied. Otherwise, there is a single arc directed at this node which is a 1-arc, and constraint (1i) enforces that $S_{2,j} \leq 0 + w_1$, as desired.

By induction, suppose that for all $k \leq \ell' - 1$, the inequality is satisfied. Then, $S_{\ell'-1,j_{\ell'-1}} \leq \sum_{\ell \in [\ell'-2]:d(a_\ell)=1} w_\ell$. If $d(a_{\ell'-1}) = 0$, then constraint (1h) will enforce that

$$S_{\ell,j_{\ell'}} \leq S_{\ell'-1,j_{\ell'-1}} \leq \sum_{\ell \in [\ell'-2]:d(a_\ell)=1} w_\ell = \sum_{\ell \in [\ell'-1]:d(a_\ell)=1} w_\ell,$$

and if $d(a_{\ell'-1}) = 1$, then constraint (1h) will enforce that

$$S_{\ell,j_{\ell'}} \leq S_{\ell'-1,j_{\ell'-1}} + w_{\ell'+1} \leq \sum_{\ell \in [\ell'-2]:d(a_\ell)=1} w_\ell + w_{\ell'+1} = \sum_{\ell \in [\ell'-1]:d(a_\ell)=1} w_\ell.$$

We now show that the BDD is relaxed using inequality (2). This requires establishing that any set $I \in S$ is encoded by some path of the BDD. By way of contradiction, suppose I' is a set with no corresponding path. Consider the ordered set of nodes $(u_1, \ldots, u_{n+1})$, with $u_\ell = (\ell, j_\ell)$, and arcs $(a_1, \ldots, a_n)$ defined inductively as follows. Initialize the path with $u_1 = (1,1)$ and $a_1 = (u_1, (2, j_2))$, with j_2 determined by, if $1 \in I'$, the index for which $x_{1,1,j_1}^1{}^* = 1$, and if $1 \notin I'$, the index for which $x_{1,1,j_1}^0{}^* = 1$. This also establishes u_2 as $(2, j_2)$. For $\ell' = 1, \ldots, n$, having defined $u_1, \ldots, u_{\ell'}$ and $a_1, \ldots, a_{\ell'-1}$ (therefore also $j_1, \ldots, j_{\ell'}$), arc $a_{\ell'}$ and node $u_{\ell'+1}$ can be written as follows: If $\ell' \in I'$, let $a_{\ell'} = (u_{\ell'}, (\ell'+1, j_{\ell'+1}))$ where $j_{\ell'+1}$ is the index for which $x_{\ell',j_{\ell'},j_{\ell'+1}}^1{}^* = 1$, and if $\ell' \notin I'$, let $a_{\ell'} = (u_{\ell'}, (\ell'+1, j_{\ell'+1}))$ where $j_{\ell'+1}$ is the index for which $x_{\ell',j_{\ell'},j_{\ell'+1}}^0{}^* = 1$.

If this procedure successfully identifies a node and arc on each layer, then the resulting path p' satisfies that $\mathcal{I}(p') = I'$. Therefore, there must be some index ℓ' for which the procedure fails to find an arc $a_{\ell'}$. This cannot be a 0-arc, because, if $u_{\ell'}$ exists, constraints (1e) require that a 0-arc is directed out of every node. It must thus fail on a layer with $\ell' \in I'$ where there is no index $j_{\ell'+1}$ for which $x_{\ell',j_{\ell'},j_{\ell'+1}}^1{}^* = 1$. Because $I' \in S, \sum_{i \in I' \cap [\ell']} w_i = \sum_{i \in I' \cap [\ell'-1]} w_i + w_{\ell'} \leq C$. Also, by constraint (1j), $S_{\ell',j_{\ell'}} + w_{\ell'} \geq (1 - (0)) \cdot (C+1) = C+1$, resulting in $S_{\ell',j_{\ell'}} > \sum_{i \in I' \cap [\ell'-1]} w_i$. However, this contradicts inequality (2) because for p', the set $I' \cap [\ell'-1]$ is identically the set $\{i \in [\ell'-1] : d(a_i) = 1\}$.   □

The following theorem establishes that the results bounds are valid.

Theorem 2. *At the optimal solution for (1), the values* $p_{\ell,i}{}^*$ *will be an upper bound on the length of the longest path from node* (ℓ, i) *to* t *in the BDD defined by the solution, as in the statement of Theorem 1.* $\qquad\square$

Extensions. We briefly comment on some possible extensions of the model. To incorporate variable ordering, we can introduce a new binary variable $z_{\ell,j}$ that indicates whether layer ℓ is associated with item j. The constant w_ℓ should then be replaced by $\sum_j w_\ell z_{\ell,j}$ (similarly for p_j), and exactly one variable $\{z_{\ell,j}\}_{\forall j}$ must have a value of 1 for any layer ℓ. One could also impose a maximum number of *nodes* as opposed to a maximum *width* by introducing a new binary variable $u_{\ell,i}$ indicating if a node i is added to a layer ℓ, adjusting the constraints as necessary (e.g., that an arc must link nodes from adjacent layers).

Finally, to adapt for other discrete optimization problems, one must modify the state variable $S_{\ell,i}$ and associated constraints to enforce sufficient conditions for the *removal* of arcs, i.e., any condition that certifies that the paths crossing the arc are infeasible. This would yield valid relaxed BDDs for the problem, albeit not the optimal ones. Existing sufficient conditions have been proposed for a variety of global constraints in CP [8]. However, conditions that are also *necessary*, i.e., that certifies that at least one path crossing the arc is feasible, will yield the strongest possible relaxed BDD for the problem, but are typically NP-Hard to verify [19].

4 Experimental Results

The experiments ran on an Intel(R) Xeon(R) CPU E5-2640 v3 at 2.60 GHz with 128 GB RAM. The BDD method was implemented in C++ and compiled with GCC 4.8.4. We used ILOG CPLEX 12.6.3 with 2 cores to solve all IPs. Our source code and all tested instances will be made available at http://www.andrew. cmu.edu/user/vanhoeve/mdd/. We generated random KP instances following the procedure adapted from [20]. The values p_i and w_i were drawn uniformly at random from the set $\{1, \ldots, 100\}$, and $C = \lceil r \times \sum_{i=1}^{n} w_i \rceil$, where r is a ratio such that $r \in \{0.1, \ldots, 0.9\}$. We considered instances of size $|\mathcal{I}| \in \{15, 20\}$ and generated 10 samples for each pair (n, r).

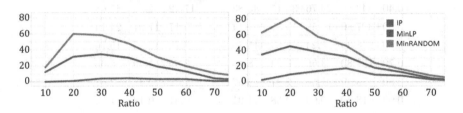

Fig. 2. Percentage gap $\times$ scaled ratio ($r \times 10$) for $|\mathcal{I}| = 15$ (left) and $|\mathcal{I}| = 20$ (right).

Table 1 shows the results considering widths of $W \in \{4, 8\}$ for the relaxed BDDs. The column $|\mathcal{B}|$ represents the average width of the exact BDD, while IP represents the optimality gap of the best solution found by our IP model (1) in 1,800 s (i.e., $gap = 100*(best\ ub - optimal)/optimal$). The column MinLP refers to the optimality gap of the standard relaxed BDD construction from [6,7], and MinRandom is the best gap obtained by 50 runs of a completely random relaxed BDD construction, as investigated in [5]. None of the IP models were solved to optimality within the time limit (the average IP optimality gap was 35% and 52% for instances with $|\mathcal{I}| = 15$ and $|\mathcal{I}| = 20$, respectively).

The bound provided by the IP model was significantly superior in almost all cases but for $r = 0.9$ and $|\mathcal{I}| = 20$. Figure 2 depicts the bound comparison, where the IP model was particularly tighter when the ratio was small (as intuitively this results in a much smaller search space).

Table 1. Results for W-relaxed BDDs

$|\mathcal{I}| = 15$							
		$W = 4$			$W = 8$		
r	$|\mathcal{B}|$	IP	MinLP	MinRandom	IP	MinLP	MinRandom
0.1	7.50	0.71	37.50	36.34	0.18	12.34	18.22
0.2	21.50	6.08	47.68	69.34	1.12	31.40	59.89
0.3	43.30	11.82	45.24	61.86	4.02	34.43	58.33
0.4	73.00	14.09	36.03	49.84	4.31	29.86	47.59
0.5	80.70	7.83	23.47	31.28	3.23	18.93	30.66
0.6	73.30	4.25	15.64	19.40	3.11	12.72	19.40
0.7	44.40	1.48	6.74	10.63	0.67	3.98	10.63
0.8	23.20	0.80	3.47	6.13	0.71	1.76	6.13
0.9	8.70	0.12	0.16	1.69	0.00	0.00	1.69
$|\mathcal{I}| = 20$							
		$W = 4$			$W = 8$		
r	$|\mathcal{B}|$	IP	MinLP	MinRand	IP	MinLP	MinRand
0.10	19.20	5.89	66.03	85.13	2.67	34.92	62.94
0.20	67.70	14.52	62.49	85.57	9.55	45.34	81.62
0.30	161.90	19.42	46.44	57.43	13.86	37.95	57.62
0.40	243.80	23.76	38.15	45.83	17.22	32.33	45.68
0.50	307.80	11.59	19.85	24.14	9.17	17.91	24.14
0.60	248.30	8.24	13.67	15.70	7.79	12.15	15.70
0.70	157.80	3.66	6.24	8.05	3.09	4.69	8.05
0.80	73.30	1.55	2.37	3.67	1.32	1.46	3.67
0.90	18.40	0.63	0.58	1.91	0.61	0.13	1.91

5   Conclusions and Future Work

This paper opens the door for a new stream of research on BDD minimization. In particular, it introduces a mathematical programming model for finding the relaxed BDD with strongest relaxation bound for knapsack problems, given a maximum width. The BDDs identified by these mathematical programming models provide much stronger bounds than the BDDs obtained by previously investigated compilation techniques, showing promise for the approach. The model developed, however, is simplistic and computationally challenging. Many enhancements are possible, for example through the use of symmetry breaking constraints, developing search heuristics, and designing stronger formulations. Also, this model presented is specific for knapsack problem, but can be extended to other problem classes and lead to similar bound improvements.

References

1. Akers, S.B.: Binary decision diagrams. IEEE Trans. Comput. **27**, 509–516 (1978)
2. Andersen, H.R., Hadzic, T., Hooker, J.N., Tiedemann, P.: A constraint store based on multivalued decision diagrams. In: Bessière, C. (ed.) CP 2007. LNCS, vol. 4741, pp. 118–132. Springer, Heidelberg (2007). doi:10.1007/978-3-540-74970-7_11. http://dx.doi.org/10.1007/978-3-540-74970-7_11
3. Behle, M.: On threshold BDDs and the optimal variable ordering problem. J. Comb. Optim. **16**(2), 107–118 (2007). http://dx.doi.org/10.1007/s10878-007-9123-z
4. Bergman, D., Cire, A.A., van Hoeve, W.J.: MDD propagation for sequence constraints. J. Artif. Intell. Res. **50**, 697–722 (2014). http://dx.doi.org/10.1613/jair.4199
5. Bergman, D., Cire, A.A., van Hoeve, W.-J., Hooker, J.N.: Variable ordering for the application of BDDs to the maximum independent set problem. In: Beldiceanu, N., Jussien, N., Pinson, É. (eds.) CPAIOR 2012. LNCS, vol. 7298, pp. 34–49. Springer, Heidelberg (2012). doi:10.1007/978-3-642-29828-8_3
6. Bergman, D., Cire, A.A., van Hoeve, W.J., Hooker, J.N.: Optimization bounds from binary decision diagrams. INFORMS J. Comput. **26**(2), 253–268 (2014). http://dx.doi.org/10.1287/ijoc.2013.0561
7. Bergman, D., Cire, A.A., van Hoeve, W.J., Hooker, J.N.: Discrete optimization with decision diagrams. INFORMS J. Comput. **28**(1), 47–66 (2016)
8. Bergman, D., Cire, A.A., van Hoeve, W.J., Hooker, J.: Decision Diagrams for Optimization. Artificial Intelligence: Foundations, Theory, and Algorithms, 1st edn. Springer, Switzerland (2016)
9. Bergman, D., Hoeve, W.-J., Hooker, J.N.: Manipulating MDD relaxations for combinatorial optimization. In: Achterberg, T., Beck, J.C. (eds.) CPAIOR 2011. LNCS, vol. 6697, pp. 20–35. Springer, Heidelberg (2011). doi:10.1007/978-3-642-21311-3_5. http://dx.doi.org/10.1007/978-3-642-21311-3_5
10. Bertsekas, D.P.: Dynamic Programming and Optimal Control, 4th edn. Athena Scientific, Belmont (2012)
11. Bryant, R.E.: Graph-based algorithms for boolean function manipulation. IEEE Trans. Comput. **35**, 677–691 (1986)

12. Bryant, R.E.: Symbolic boolean manipulation with ordered binary decision diagrams. ACM Comput. Surv. **24**, 293–318 (1992)
13. Drechsler, R., Drechsler, N., Günther, W.: Fast exact minimization of BDD's. IEEE Trans. CAD Integr. Circ. Syst. **19**(3), 384–389 (2000). http://dx.doi.org/10.1109/43.833206
14. Felt, E., York, G., Brayton, R.K., Sangiovanni-Vincentelli, A.L.: Dynamic variable reordering for BDD minimization. In: Proceedings of the European Design Automation Conference 1993, EURO-DAC 1993 with EURO-VHDL 1993, Hamburg, Germany, 20–24 September 1993. pp. 130–135. IEEE Computer Society (1993). http://dx.doi.org/10.1109/EURDAC.1993.410627
15. Gogate, V., Domingos, P.M.: Approximation by quantization. CoRR abs/1202.3723 (2012). http://arxiv.org/abs/1202.3723
16. Gogate, V., Domingos, P.M.: Structured message passing. In: Proceedings of the Twenty-Ninth Conference on Uncertainty in Artificial Intelligence, UAI 2013, Bellevue, WA, USA, 11–15 August 2013 (2013). https://dslpitt.org/uai/displayArticleDetails.jsp?mmnu=1&smnu=2&article_id=2386&proceeding_id=29
17. Günther, W., Drechsler, R.: Linear transformations and exact minimization of BDDs. In: 8th Great Lakes Symposium on VLSI (GLS-VLSI 1998), 19–21 February 1998, Lafayette, LA, USA, pp. 325–330. IEEE Computer Society (1998). http://dx.doi.org/10.1109/GLSV.1998.665287
18. Hadzic, T., Hooker, J.N., O'Sullivan, B., Tiedemann, P.: Approximate compilation of constraints into multivalued decision diagrams. In: Stuckey, P.J. (ed.) CP 2008. LNCS, vol. 5202, pp. 448–462. Springer, Heidelberg (2008). doi:10.1007/978-3-540-85958-1_30
19. Hoda, S., Hoeve, W.-J., Hooker, J.N.: A systematic approach to MDD-based constraint programming. In: Cohen, D. (ed.) CP 2010. LNCS, vol. 6308, pp. 266–280. Springer, Heidelberg (2010). doi:10.1007/978-3-642-15396-9_23. http://dl.acm.org/citation.cfm?id=1886008.1886034
20. Kirlik, G., Sayın, S.: A new algorithm for generating all nondominated solutions of multiobjective discrete optimization problems. Eur. J. Oper. Res. **232**(3), 479–488 (2014). http://www.sciencedirect.com/science/article/pii/S0377221713006474
21. Lee, C.Y.: Representation of switching circuits by binary-decision programs. Bell Syst. Tech. J. **38**, 985–999 (1959)
22. Shiple, T.R., Hojati, R., Sangiovanni-Vincentelli, A.L., Brayton, R.K.: Heuristic minimization of BDDs using don't cares. In: DAC, pp. 225–231 (1994). http://doi.acm.org/10.1145/196244.196360
23. Soeken, M., Große, D., Chandrasekharan, A., Drechsler, R.: BDD minimization for approximate computing. In: 21st Asia and South Pacific Design Automation Conference, ASP-DAC 2016, Macao, Macao, 25–28 January 2016, pp. 474–479. IEEE (2016). http://dx.doi.org/10.1109/ASPDAC.2016.7428057
24. St-Aubin, R., Hoey, J., Boutilier, C.: APRICODD: approximate policy construction using decision diagrams. In: Proceedings of Conference on Neural Information Processing Systems, pp. 1089–1095 (2000)

Design and Implementation of Bounded-Length Sequence Variables

Joseph D. Scott[1]([✉]), Pierre Flener[1], Justin Pearson[1], and Christian Schulte[2]

[1] Department of Information Technology, Uppsala University, Uppsala, Sweden
{Pierre.Flener,Justin.Pearson}@it.uu.se, JosephDScott@gmail.com
[2] KTH Royal Institute of Technology, Stockholm, Sweden
cschulte@kth.se

Abstract. We present the design and implementation of bounded - length sequence (BLS) variables for a CP solver. The domain of a BLS variable is represented as the combination of a set of candidate lengths and a sequence of sets of candidate characters. We show how this representation, together with requirements imposed by propagators, affects the implementation of BLS variables for a copying CP solver, most importantly the closely related decisions of data structure, domain restriction operations, and propagation events. The resulting implementation outperforms traditional bounded-length string representations for CP solvers, which use a fixed-length array of candidate characters and a padding symbol.

1 Introduction

String variables are useful for expressing a wide variety of real-world problems, such as test generation [7], program analysis [4], model checking [10], security [3], and data mining [16]. Despite this usefulness, string variables have never received an optimized implementation in a modern constraint programming (CP) solver.

We describe the design and implementation of a string variable type for a copying CP solver. Properly designed, a string variable can be much more efficient, in both time and space complexity, than the decompositions commonly used to model strings in CP. Additionally, string variables greatly simplify the modeling of problems including strings, with benefits to both readability and correctness. We choose to implement a *bounded*-length sequence (BLS) variable type, as it provides much more flexibility [9,18] than fixed-length approaches, but avoids the blow-ups that plague unbounded-length representations.

We make contributions in three dimensions. First, we select a data structure for the BLS variable domain, namely a dynamic list of bitsets, to represent domains in reasonable space. The data structure is designed for efficient implementation of domain restriction operations, which BLS variables expose to propagators; we develop a correct and minimal set of these operations. Variables in turn notify propagators of domain changes using propagation events; we design a monotonic set of propagation events that are useful for propagators on BLS variables. Second, the BLS variable is implemented on top of GECODE [25],

© Springer International Publishing AG 2017
D. Salvagnin and M. Lombardi (Eds.): CPAIOR 2017, LNCS 10335, pp. 51–67, 2017.
DOI: 10.1007/978-3-319-59776-8_5

a mature CP toolkit with state-of-the-art performance and many deployments in industry. Third, we show that BLS variables outperform other string solving methods in CP, as demonstrated upon both a string problem in the CSPlib repository and a representative benchmark from the field of software verification.

In this paper, we summarize [22, Chaps. 10 and 11]: this is an improvement of our [23], as discussed in the experiments of Sect. 7. We forego discussion in this paper of several interesting and important theoretical aspects of the BLS representation, as previous publications cover these theoretical aspects in detail: [22] provides an extended discussion of the logical properties of the representation, a justification for the utility of BLS variables in string constraint problems, a comparison of different modeling strategies, a survey of related work in CP such as [11], and a discussion of intensional representations; these topics are also presented in briefer form in [23]. Hence we choose to deal specifically with the implementation of an efficient BLS variable, rather than repeating material that has been presented elsewhere.

Plan of the Paper. Section 2 discusses the design of variable implementations in general. Section 3 defines a domain representation appropriate for BLS variables and motivates the design choices of subsequent sections. Section 4 considers several representations for each of the components of a BLS variable and motivates why particular choices have been made. Section 5 defines restriction operations, which allow propagators to modify the domain of such a variable. Section 6 describes BLS-specific systems of propagation events, which describe restrictions of the domain of a variable. Section 7 justifies these design choices by an experimental evaluation. Section 8 concludes the paper.

2   Variables and Propagators

We summarize key concepts relating to propagators as implementations of constraints and three fundamental choices of how to implement a variable type.

Propagators. A *propagator* implements a constraint and is executed by a *propagation loop*. The execution of a propagator results in the restriction of some variable domains pruning values for variables that are in conflict with the implemented constraint. The propagation loop executes the propagators until no propagator can prune any values, the propagator is said to be at *fixpoint*. For a more detailed discussion of propagators see for example [19].

Variables. The central decision in implementing a variable type is how to represent the domain of a variable. With this representation decided, the implementation of the variable type must take into account three primary choices: how a domain is stored, how a propagator prunes a domain, and how a propagator is informed of a restriction of a domain. We consider each of these choices in turn.

No *data structure* exists in a vacuum, and every choice of data structure represents a tradeoff between memory requirements and computational complexity.

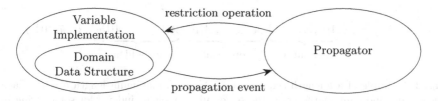

Fig. 1. Interactions of a variable implementation and a propagator: a propagator requests a change to the domain of a variable via a *restriction operation*, and a variable notifies a propagator via a *propagation event* that its domain has been restricted.

Evaluating the choices made in designing a data structure, therefore, requires considering the environment in which the data structure will function. For a variable implementation, that means considering how it interacts with the CP solver, and most importantly how it interacts with propagators. A simplistic view of this interaction is shown in Fig. 1.

In an implementation, propagators work by modifying the domains of variables. These domains are not exposed directly to propagators; rather the variable implementation encapsulates the domain representation and exposes some *restriction operations* by which propagators can request that the domain be modified. Restriction operations provide an interface to the domain that allows the removal of candidate values; no other modifications are allowed, as a propagator is required to be contracting on domains.

When the domain of a variable is restricted by a propagator, any other propagator that implements a constraint on the corresponding variable must be notified, as it is possible that the latter propagator is no longer at fixpoint. Most CP solvers make use of more fine-grained information than simply which variable has a newly restricted domain. This more detailed information is called a *propagation event* (for an extended discussion of events, see [20]); for an integer variable X, for example, a propagation event might indicate that the upper bound of dom(X) has shrunk, or that dom(X) has become a singleton.

3   BLS Variables

The domain of a string variable is a set of strings. Even in the bounded-length case, sets of strings are difficult to represent in extension, due to high space complexity; on the other hand, intensional representations, such as finite automata or regular expressions, generally have a higher time complexity for operations. However, both the time complexity and space complexity may be reduced if the representation of the set of strings is not required to be exact. For example, an integer interval is a compact and efficient representation for a set of integers. Of course, many sets of integers cannot be exactly represented by an interval; however, every set of integers may be *over-approximated* by some interval.

We now define an over-approximation that is appropriate for representing the domain of a bounded-length set of strings. We then give an example of how a propagator for a string constraint would interact with domains thus represented.

$$\Big\langle \boldsymbol{\mathcal{A}}^b, \mathcal{N} \Big\rangle = \Big\langle \big\langle \underbrace{\mathcal{A}[1], \dots, \mathcal{A}[5]}_{\text{mandatory}}, \underbrace{\mathcal{A}[6], \dots, \mathcal{A}[10]}_{\text{optional}}, \underbrace{\overset{\varnothing}{\mathcal{A}[11]}, \dots, \overset{\varnothing}{\mathcal{A}[15]}}_{\text{forbidden}} \big\rangle, \big\{ 5, 7, 10 \big\} \Big\rangle$$

Fig. 2. Example of a b-length sequence $\langle \boldsymbol{\mathcal{A}}^b, \mathcal{N} \rangle$ with maximum length $b = 15$. Each set $\mathcal{A}[i]$ of candidate characters is the set of all symbols at index i for some string in the domain; the set $\mathcal{N}$ of candidate lengths is $\{5, 7, 10\}$. The lower bound, 5, and upper bound, 10, of the set of candidate lengths partition the sets of candidate characters into *mandatory*, *optional*, and *forbidden* regions, as indicated. A set of candidate characters is empty if and only if it is found in the forbidden region.

Representation of a String Domain. For any finite set $\mathcal{W}$ of strings over an alphabet Σ there exists an upper bound on the lengths of the strings in $\mathcal{W}$, i.e., some number $b \in \mathbb{N}$ such that for every string $w \in \mathcal{W}$ the length of w is not greater than b. Such a set $\mathcal{W}$ may be over-approximated by a pair $\langle \boldsymbol{\mathcal{A}}^b, \mathcal{N} \rangle$, consisting of a sequence $\boldsymbol{\mathcal{A}}^b$ of sets $\langle \mathcal{A}[1], \dots, \mathcal{A}[b] \rangle$ over Σ, where every $\mathcal{A}[i] \subseteq \Sigma$ is the set of all *candidate characters* occurring at index $i \in 1, b$ for some string in $\mathcal{W}$, and a set $\mathcal{N} \subseteq [0, b]$ of *candidate lengths* of the strings in $\mathcal{W}$. This over-approximation, illustrated in Fig. 2, is called the *bounded-length sequence representation*. A pair $\langle \boldsymbol{\mathcal{A}}^b, \mathcal{N} \rangle$ is referred to as a *b-length sequence*, and is an abstract representation of the set of all strings that have a length $\ell \in \mathcal{N}$ and a character at each index $i \in [1, \ell]$ taken from the set $\mathcal{A}[i]$ of symbols.

The domain of a string variable X_j, written dom (X_j), is a finite set of strings; thus, the domain of any string variable X_j can be over-approximated by a b-length sequence $\langle \boldsymbol{\mathcal{A}}^b_j, \mathcal{N}_j \rangle$ for some appropriately large value of b. In this context, we will sometimes write dom$(\boldsymbol{\mathcal{A}}^b_j, \mathcal{N}_j)$ to mean the domain of a string variable X_j that is represented by the b-length sequence $\langle \boldsymbol{\mathcal{A}}^b_j, \mathcal{N}_j \rangle$; thus we have:

$$\text{dom}(\boldsymbol{\mathcal{A}}^b_j, \mathcal{N}_j) = \bigcup_{\ell \in \mathcal{N}_j} \{ w \in \Sigma^\ell | \forall i \in [1, \ell] : w[i] \in \mathcal{A}_j[i] \} \tag{1}$$

Bounded-length sequences are not unique: if the sequences $\langle \boldsymbol{\mathcal{A}}^b_1, \mathcal{N} \rangle$ and $\langle \boldsymbol{\mathcal{A}}^b_2, \mathcal{N} \rangle$ differ only in one or more pair of sets $\mathcal{A}_1[i]$ and $\mathcal{A}_2[i]$ where $i > \max(\mathcal{N})$, then $\langle \boldsymbol{\mathcal{A}}^b_1, \mathcal{N} \rangle$ and $\langle \boldsymbol{\mathcal{A}}^b_2, \mathcal{N} \rangle$ represent the same set of strings. Uniqueness may be imposed by a *representation invariant* that only allows bounded-length sequences of a canonical form, in which a set $\mathcal{A}[i]$ of candidate characters is the empty set if and only if the index i is greater than $\max(\mathcal{N})$. As illustrated in Fig. 2, the set $\mathcal{N}$ of candidate lengths divides the sequence $\boldsymbol{\mathcal{A}}^b$ into three regions: the *mandatory* candidate characters, with an index at most $\min(\mathcal{N})$; the *optional* candidate characters, with an index greater than $\min(\mathcal{N})$ but at most $\max(\mathcal{N})$; and the *forbidden* candidate characters, with an index greater than $\max(\mathcal{N})$.

Propagation. A properly designed variable implementation must take into account the operation of propagators implementing constraints for the corresponding variable type. For string variables, expected constraints include regular language membership (REGULAR$_{\mathcal{L}}$), string length (LEN), reversal (REV),

$$\Big\langle \mathcal{A}_3[1], \mathcal{A}_3[2], \mathcal{A}_3[3], \mathcal{A}_3[4], \boxed{\mathcal{A}_3[5]}, \mathcal{A}_3[6], \dots \Big\rangle$$
$$\subseteq$$
$$\mathcal{N}_1 = \{3\} : \Big\langle \mathcal{A}_1[1], \mathcal{A}_1[2], \mathcal{A}_1[3], \mathcal{A}_2[1], \boxed{\mathcal{A}_2[2]}, \mathcal{A}_2[3], \dots \Big\rangle$$
$$\cup$$
$$\mathcal{N}_1 = \{4\} : \Big\langle \mathcal{A}_1[1], \mathcal{A}_1[2], \mathcal{A}_1[3], \mathcal{A}_1[4], \boxed{\mathcal{A}_2[1]}, \mathcal{A}_2[2], \dots \Big\rangle$$
$$\cup$$
$$\mathcal{N}_1 = \{5\} : \Big\langle \mathcal{A}_1[1], \mathcal{A}_1[2], \mathcal{A}_1[3], \mathcal{A}_1[4], \boxed{\mathcal{A}_1[5]}, \mathcal{A}_2[1], \dots \Big\rangle$$
$$\cup$$
$$\mathcal{N}_1 = \{6\} : \Big\langle \mathcal{A}_1[1], \mathcal{A}_1[2], \mathcal{A}_1[3], \mathcal{A}_1[4], \boxed{\mathcal{A}_1[5]}, \mathcal{A}_1[6], \dots \Big\rangle$$

Fig. 3. The post-condition, implied by the constraint $\text{CAT}(X_1, X_2, X_3)$, on the set of candidate characters $\mathcal{A}_3[5]$, is determined by the four candidate lengths of X_1, each yielding a possible alignment between the three string variables.

and concatenation (CAT); many others are possible (see, e.g., [22]). The following example provides a sample of the inference required by typical propagators for such constraints, and gives context to the design decisions that follow.

Example 1 (Concatenation). The constraint $\text{CAT}(X_1, X_2, X_3)$ holds if the concatenation of string variables X_1 and X_2 is equal to the string variable X_3. Consider string variables X_1, X_2, and X_3 with domains represented by the b-length sequences $\langle \mathcal{A}_1^b, \mathcal{N}_1 \rangle$, $\langle \mathcal{A}_2^b, \mathcal{N}_2 \rangle$, and $\langle \mathcal{A}_3^b, \mathcal{N}_3 \rangle$, each with maximum length $b = 15$, and sets of candidate lengths $\mathcal{N}_1 = [3, 6]$, $\mathcal{N}_2 = [4, 7]$, and $\mathcal{N}_3 = [5, 14]$.

A propagator is a contracting function on tuples of variable domains. A propagator implementing the constraint CAT is a function of the following form:

$$\text{CAT}(\text{dom}(\mathcal{A}_1^b, \mathcal{N}_1), \text{dom}(\mathcal{A}_2^b, \mathcal{N}_2), \text{dom}(\mathcal{A}_3^b, \mathcal{N}_3))$$
$$= \langle \text{dom}(\mathcal{A}_1'^b, \mathcal{N}_1'), \text{dom}(\mathcal{A}_2'^b, \mathcal{N}_2'), \text{dom}(\mathcal{A}_3'^b, \mathcal{N}_3') \rangle \quad (2)$$

The simplest inference relevant to the propagation of CAT is the initial arithmetic adjustment of the candidate lengths. For example, every candidate length for X_3 should be the sum of some candidate lengths for X_1 and X_2:

$$\mathcal{N}_3' := \{ n_3 \in \mathcal{N}_3 \mid \exists n_1 \in \mathcal{N}_1, n_2 \in \mathcal{N}_2 : n_3 = n_1 + n_2 \} = [7, 13] \quad (3)$$

For sets of candidate characters, there are dependencies not only upon sets of candidate characters from the other string variables constrained by CAT, but also on the sets of candidate lengths. For example, Fig. 3 illustrates the four possible alignments of X_1, X_2, and X_3, corresponding to the four candidate lengths in the set $\mathcal{N}_1 = [3, 6]$. If, on the one hand, the length of X_1 were fixed to either of the two smallest of these candidates, then any symbol in the set $\mathcal{A}_3'[5]$ of candidate characters would also have to be an element in either $\mathcal{A}_2[2]$ or $\mathcal{A}_2[1]$. If, on the other hand, the length of X_1 were fixed to either of the two largest of these candidates, then in either case any symbol in the set $\mathcal{A}_3'[5]$ of candidate characters would have to be an element of the set $\mathcal{A}_1[5]$. Hence:

$$\mathcal{A}_3'[5] := \mathcal{A}_3[5] \cap (\mathcal{A}_2[1] \cup \mathcal{A}_2[2] \cup \mathcal{A}_1[5]) \quad (4)$$

The resulting sets of candidate lengths also depend upon the sets of candidate characters. For example, if the intersection of $\mathcal{A}_2[2]$ and $\mathcal{A}_3[5]$ were empty, then the uppermost alignment illustrated in Fig. 3, where $\mathcal{N}_1 = \{3\}$ could be ruled out: for any combination of strings X_1, X_2, and X_3 satisfying $\text{CAT}(X_1, X_2, X_3)$ such that the length of X_1 was 3, the maximum length of X_3 would be $4 \notin [7, 13]$, since the set $\mathcal{A}_3[5]$ of candidate characters would be empty.   □

Solver-Based Requirements. In a CP solver, the propagation loop described in Sect. 2 is interleaved with a backtracking search. If that loop ends with some variables as yet unassigned, then the search tree is grown by partitioning the search space to create two or more subproblems, obtained by mutually exclusive *decisions* on the domain of an unassigned variable; the propagation loop is then executed on the chosen subproblem. On the other hand, if the propagation loop results in a failed domain, then some prior decision is undone: the search returns to a previously visited node, from which the tree is grown, if possible, by choosing another of the mutually exclusive decisions that were determined at that node.

There are two main CP techniques for restoring a previously visited node during backtracking [19]: *trailing*, in which changes to the variable domains during search are stored on a stack, and backtracking consists of reconstructing a previously visited node; and *copying*, in which the domains are copied before a choice is applied, and every new node of the search tree begins with a fresh copy.

The choice of a copying solver has two main impacts on the design of a variable type. First, under copying, the search procedure and the fixpoint algorithm are orthogonal, while under trailing all the solver components are affected by backtracking; thus, under copying, design choices can be made without reference to the details of the search procedure. Second, for a copying system, memory management is critical. The amount of memory required by a data structure used to represent the domain of a variable (as well as data structures used o store the states of other solver components) should therefore be minimal.

4   Data Structure

For a string variable X with a domain represented by a bounded-length sequence, there are three largely orthogonal structural choices to be considered: the representation of the set $\mathcal{N}$ of candidate lengths, the representation of the sets $\mathcal{A}[i]$ of candidate characters, and the construction of the sequence $\langle \mathcal{A}[1], \ldots, \mathcal{A}[b] \rangle$ itself. Lengths are natural numbers, and any finite alphabet may be mapped to the natural numbers (upon imposing an ordering over the alphabet); therefore the possible implementations for both the set $\mathcal{N}$ of candidate lengths and the sets $\mathcal{A}[i]$ of candidate characters at all indices i of X are the same. However, lengths and characters have different characteristics and benefit from different choices of representation. Each of these choices is now considered in turn.

The Set of Candidate Lengths. The set $\mathcal{N}$ of candidate lengths may be exactly represented or over-approximated. An exact representation could utilize a data structure such as those used in CP solvers to exactly represent the domain of an integer variable (e.g., a range sequence or a bitset). Alternately, the set of candidate lengths can be over-approximated as an interval: a pair of integers representing the lower and upper bounds of the set.

Recall from Example 1 that a propagator implementing the constraint CAT must calculate sums of all candidate lengths for its string variables. This computation is quadratic in the size of the set representation (i.e., in the number of elements, ranges, or bounds used to represent the set). For the interval representation, the representation size is constant, but for exact set representations it is not. Example 1 also shows that the upper and lower bounds of $\mathcal{N}$ are slightly more useful during propagation than interior values of $\mathcal{N}$. The *lower* bound of the set of candidate lengths defines the smallest sequence of sets of candidate characters that an implementation should represent explicitly, as *every* candidate solution must have some symbol at those *mandatory* indices. In contrast, all, some, or none of the other sets of candidate characters might be explicitly represented; the *upper* bound of the set of candidate lengths divides these sets of candidate characters into those at *optional* indices that participate in some of the candidate solutions, and those at *forbidden* indices that participate in none.

An exact representation would allow some extra inference during the propagation of string constraints, and could be sensible given enough string constraints with efficient propagators performing non-trivial reasoning on the interior values of $\mathcal{N}$. Otherwise, an interval representation appears to be most suitable.

Sets of Candidate Characters. The exact composition of an alphabet is problem dependent; however, several properties generally apply. First, alphabet sizes are typically manageable; this is certainly true for many interesting classes of string constraint problems, such as occur in computational biology, natural-language processing, and the verification and validation of software. Second, the symbols of an alphabet often have a known total ordering, in which case the strings of any language on that alphabet have a corresponding lexicographic order; even when no meaningful order exists, such an order may be imposed as needed. Third, intervals of symbols often have little or no inherent meaning, as, in contrast to numeric types, there is generally no logical relationship between consecutive symbols in even a totally-ordered alphabet. Finally, as shown in Example 1, the propagation of constraints over string variables with bounded-length sequence domain representations depends heavily on standard set operations (i.e., union and intersection) between sets of candidate characters.

Upon considering these properties, we propose to implement each set of candidate characters as a bitset. A *bitset* for a finite set $\mathcal{S} \subset \mathbb{N}$ is a sequence of $\max(\mathcal{S})$ bits (i.e., values in $\{0, 1\}$) such that the i-th bit is equal to 1 if and only if $i \in \mathcal{S}$. With word size w, a bitset representation of $\mathcal{S}$ requires $k = \lceil \max(\mathcal{S})/w \rceil$ words, or a total of kw bits. Bitsets allow for very fast set operations (typically, a small constant number of operations per word is required): specifically, the complexity of the union and intersection operations is linear in k. Furthermore, as long as $|\Sigma|$ is not significantly larger than w, the memory requirement of a bitset representation is competitive with other common exact set representations.

Sequence. The sequence $\langle \mathcal{A}[1], \ldots, \mathcal{A}[b] \rangle$ of sets of candidate characters could be implemented in two ways: as an array-based structure, requiring minimal memory overhead and affording direct access to the elements; or as a list-based structure, allowing for better memory management of dynamic-sized lists, but no direct access. A key observation is that the maximum length of any b-length sequence is already given, namely b. We propose a hybrid array-list implementation: a fixed length array of n pointers to blocks of m bitsets each, where $n = \lceil b/m \rceil$. This design offers more efficient access to individual nodes than a traditional linked list, and easier memory (de)allocation than a static-sized array. Blocks are allocated as needed: at a minimum, a number of blocks sufficient to accommodate $\min(\mathcal{N})$ sets of candidate characters (i.e., the mandatory region) is allocated; additional blocks are allocated when $\min(\mathcal{N})$ increases, or when a restriction operation is applied to a set of candidate characters in the optional region. Deallocation is performed as needed upon a decrease of $\max(\mathcal{N})$. For an evaluation of alternate sequence implementations, see [22].

5   Domain Restriction Operations

As seen in Sect. 2, propagators interact with domains via restriction operations that are exposed by the variable implementation. A restriction operation should satisfy three important properties. First, the operation should be *useful* for the implementation of some propagator. Second, the operation should be *efficient* when executed on the data structure that implements the variable domain. Finally, the operation should be *correct*, meaning that the resulting domain is actually restricted according to the semantics of the operation.

We now describe restriction operations appropriate for a BLS variable type. The data structure chosen in Sect. 4 allows an efficient implementation of these operations. Interestingly, though, the correct semantics of BLS variable restriction operations is not obvious, so before proceeding we define the semantics of string equality and disequality that is most suited to string variables.

Restriction Operations on String Variables. Table 1 lists several restriction operations that might be provided by a string variable implementation. The operations are divided into two categories, affecting either lengths or characters.

The first category of operations works on sets of candidate lengths: the equality (`leq`) operation restricts the domain of the variable to strings of a given length, and two inequality operations tighten the lower (`lgr`) or upper (`lle`) bound of the set of candidate lengths. The second category of operations works on sets of candidate characters: the equality and disequality operations restrict the domain of the variable to strings with a given symbol at the specified index (`ceq`) or to exclude all strings with a given symbol at the specified index (`cnq`), two inequality operations tighten the lower (`cgr`) or upper (`cle`) bound of the set of candidate characters, and the set intersection and set subtraction operations restrict the set of candidate characters to their intersection with a given set of symbols `cin`) or to exclude a given set of symbols (`cmi`). Note that strict

Table 1. Restriction operations for bounded length string variables.

Restriction		Operator
Length	Equality	$\texttt{leq}(\text{dom}(X), \ell) = \{x \in \text{dom}(X) || x | = \ell\}$
	$\leq$	$\texttt{lle}(\text{dom}(X), \ell) = \{x \in \text{dom}(X) || x | \leq \ell\}$
	$>$	$\texttt{lgr}(\text{dom}(X), \ell) = \{x \in \text{dom}(X) || x | > \ell\}$
Character	Equality	$\texttt{ceq}(\text{dom}(X), i, c) = \{x \in \text{dom}(X) | x[i] = c\}$
	Disequality	$\texttt{cnq}(\text{dom}(X), i, c) = \{x \in \text{dom}(X) | x[i] \neq c\}$
	$\leq$	$\texttt{cle}(\text{dom}(X), i, c) = \{x \in \text{dom}(X) | x[i] \leq c\}$
	$>$	$\texttt{cgr}(\text{dom}(X), i, c) = \{x \in \text{dom}(X) | x[i] > c\}$
	Intersection	$\texttt{cin}(\text{dom}(X), i, \mathcal{C}) = \{x \in \text{dom}(X) | x[i] \in \mathcal{C}\}$
	Subtraction	$\texttt{cmi}(\text{dom}(X), i, \mathcal{C}) = \{x \in \text{dom}(X) | x[i] \notin \mathcal{C}\}$

greater than ($>$) and non-strict less than ($\leq$) operations have been chosen for each category based solely on the complementarity of the operations; the pairs could just as well have been $<$ and $\geq$, $<$ and $>$, or $\leq$ and $\geq$.

Table 1 omits restriction operations of length disequality, intersection, and subtraction, as these would be incorrect here. For example, a length disequality operation that attempts to remove an interior value from the set of candidate lengths will result in no change to the domain. As sets of candidate characters are assumed to be explicitly represented, the character disequality and intersection operations are included.

Some of the restriction operations in Table 1 can be rewritten using the other operations. For example, the character intersection and subtraction operations are equivalent to a series of character disequality operations. In practice, an implementation using these decompositions may be inefficient; however, for the purpose of *defining* the restriction operations for domains represented by bounded-length sequences, a set of four *required* operations is sufficient: $\texttt{lle}$, $\texttt{lgr}$, $\texttt{ceq}$, and $\texttt{cnq}$, allowing a propagator to increase the lower bound of the string length ($\texttt{lgr}$), decrease the upper bound of the string length ($\texttt{lle}$), fix the character at an index i ($\texttt{ceq}$), or forbid a character at index i ($\texttt{cnq}$).

Restriction Operations as Update Rules. Table 2 defines, for string variable domains represented by a b-length sequence, the four chosen restriction operations. They are given as a series of update rules on the components of the affected b-length sequence; for each restriction operation, only those components of the resulting b-length sequence $\langle \mathcal{A}'^b, \mathcal{N}' \rangle$ that differ from the corresponding component of the initial b-length sequence are defined; all other components are unchanged.

Representation Invariant. All of the restriction operations in Table 2 are designed to respect the representation invariant of the b-length sequence representation (see Sect. 3), which enforces the relationship between empty sets of candidate characters and the upper bound of the set of candidate lengths. Hence, the restriction operation $\texttt{lle}(\langle \mathcal{A}^b, \mathcal{N} \rangle, \ell)$, which reduces the upper bound of $\mathcal{N}$,

Table 2. Restriction operations for a string variable with a domain represented by a b-length sequence, expressed as update rules. Primed set identifiers ($\mathcal{A}'[i]$ and $\mathcal{N}'$) indicate changed component values in the b-length sequence resulting from the restriction operation; all other components in the resulting b-length sequence are identical to the corresponding component in the original b-length sequence.

Restriction operation	Update rule
$\mathtt{lle}(\langle \mathcal{A}^b, \mathcal{N} \rangle, \ell)$	$\mathtt{forall}\, k \in [\ell + 1, b] : \mathcal{A}'[k] := \varnothing; \mathcal{N}' := \mathcal{N} \cap [0, \ell]$
$\mathtt{lgr}(\langle \mathcal{A}^b, \mathcal{N} \rangle, \ell)$	$\mathcal{N}' := \mathcal{N} \cap [\ell + 1, b]$
$\mathtt{ceq}(\langle \mathcal{A}^b, \mathcal{N} \rangle, i, c)$	$\mathcal{A}'[i] := \mathcal{A}[i] \cap \{c\}; \mathcal{N}' := \mathcal{N} \cap [i, b]$
$\mathtt{cnq}(\langle \mathcal{A}^b, \mathcal{N} \rangle, i, c)$	$\mathcal{A}'[i] := \mathcal{A}[i] \setminus \{c\}; \mathtt{if}\, \mathcal{A}'[i] = \varnothing\, \mathtt{then}$
	$(\mathcal{N}' := \mathcal{N} \cap [0, i - 1]; \mathtt{forall}\, k \in [i, b] : \mathcal{A}'[k] := \varnothing)$

also updates all sets of candidate characters at indices greater than ℓ to be the empty set. The restriction operation $\mathtt{lgr}(\langle \mathcal{A}^b, \mathcal{N} \rangle, \ell)$ requires no corresponding update on sets of candidate characters with indices less than or equal to ℓ: if there existed an empty set $\mathcal{A}[i] = \varnothing$ of candidate characters such that i was less than or equal to ℓ, then the upper bound of $\mathcal{N}$ would already be at most i; hence $\mathcal{N} \cap [\ell + 1, b]$ would also be the empty set, and the domain would be failed.

Equality Semantics. $\mathtt{ceq}$ and $\mathtt{cnq}$ have very different effects on the considered set of candidate lengths: after $\mathtt{ceq}$ the set $\mathcal{N}'$ only contains lengths that are at least i; but if $\mathcal{A}'[i]$ is not empty after $\mathtt{cnq}$, then the set $\mathcal{N}'$ may contain lengths greater than or equal to i. This appears to run counter to the intuition that equality and disequality should be complementary operations. However, a restriction operation is a function on the *domain* of a variable, not a function on a *component* of that domain; the operations $\mathtt{ceq}$ and $\mathtt{cnq}$ are complementary in that together they always define a partition of a string variable domain:

$$\mathrm{dom}(\mathtt{ceq}(\langle \mathcal{A}^b, \mathcal{N} \rangle, i, c)) \cup \mathrm{dom}(\mathtt{cnq}(\langle \mathcal{A}^b, \mathcal{N} \rangle, i, c)) = \mathrm{dom}(\langle \mathcal{A}^b, \mathcal{N} \rangle) \quad (5)$$

$$\mathrm{dom}(\mathtt{ceq}(\langle \mathcal{A}^b, \mathcal{N} \rangle, i, c)) \cap \mathrm{dom}(\mathtt{cnq}(\langle \mathcal{A}^b, \mathcal{N} \rangle, i, c)) = \varnothing \quad (6)$$

For full details on the equality semantics, see [22].

6   Propagation Events

As seen in Sect. 2, information about a domain restriction is communicated to a propagator via a *propagation event* [19]. In practice, a propagation event should be useful to some propagator; that is, the propagation event should distinguish between a change to the domain that leaves the propagator at fixpoint, versus one that does not. The set of propagation events exposed by a variable implementation is called a *propagation event system*. Larger propagation event systems come with a commensurate cost to efficiency; see [20] for a detailed analysis of event-based propagation.

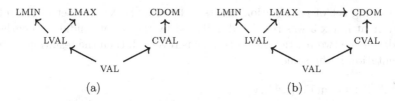

(a)   (b)

Fig. 4. Implications in a minimal string variable event system (omitting transitive implications). If the changes to the set of candidate lengths and the sets of candidate characters are viewed independently (a), then the resulting propagation event system is not monotonic. The corrected propagation event system (b) is monotonic.

A minimal event system for a string variable X with a domain represented by a b length sequence $\langle \mathcal{A}^b, \mathcal{N} \rangle$ is shown in Fig. 4. The VAL propagation event indicates that the string variable domain has been reduced to a single string. The next three propagation events indicate changes to the set $\mathcal{N}$ of candidate lengths: either the lower bound has increased (LMIN), or the upper bound has decreased (LMAX), or the set has become a singleton (LVAL). The remaining propagation events indicate changes to a set $\mathcal{A}[i]$ of candidate characters: either some symbol has been removed (CDOM) or the symbol of the character at index i has become known (CVAL). Note that CVAL indicates that the character at some index i is a symbol $c \in \Sigma$, and *not* that the set of candidate characters at i is a singleton. If i is a mandatory index, then the two conditions are equivalent: for every string x in the domain of X, the character $x[i]$ is c. However, when an optional set of candidate characters $\mathcal{A}[i]$ is a singleton, then there are additionally strings in the domain for which the character at index i is undefined.

Some propagation events are implied by others. Figure 4(a) shows an event system in which length and character propagation events are isolated, making it possible to report changes to the set of candidate lengths and changes to sets of candidate characters independently. Unfortunately, this design violates one of the properties required of propagation events [24]: the set of propagation events generated by a sequence of restriction operations must be *monotonic*, that is, it must not depend on the order of the restriction operations.

In a non-monotonic propagation event system, the set of events generated by a series of restriction operations is dependent on their order, as shown now:

Example 2. Let X be a string variable with a domain represented by the b-length sequence $\langle \mathcal{A}^4, \mathcal{N} \rangle = \langle \langle \{1,2\}, \{1,2\}, \{5,6\}, \varnothing \rangle, [1,3] \rangle$, where $b = 4$. The following sequence of restriction operations restricts first the set of candidate characters at index 3, and then the set of candidate lengths:

$$\langle \mathcal{A}'^4, \mathcal{N}' \rangle := \texttt{lle}(\texttt{cnq}(\langle \mathcal{A}^b, \mathcal{N} \rangle, 3, 6), 2)$$
$$:= \texttt{lle}(\langle \langle \{1,2\}, \{1,2\}, \{5,6\} \setminus \{6\}, \varnothing \rangle, [1,3] \rangle, 2)$$
$$:= \texttt{lle}(\langle \langle \{1,2\}, \{1,2\}, \{5\}, \varnothing \rangle, [1,3] \rangle, 2)$$
$$:= \langle \langle \{1,2\}, \{1,2\}, \{5\} \cap \varnothing, \varnothing \rangle, [1,3] \cap [0,2] \rangle = \langle \langle \{1,2\}, \{1,2\}, \varnothing, \varnothing \rangle, [1,2] \rangle$$

The resulting set of propagation events is {CDOM,LMAX}: the set of candidate characters at index 3 was restricted, followed by the set of candidate lengths. If the order of the two restriction operations is reversed, then the resulting domain representation is the same:

$$\langle \mathcal{A}''^4, \mathcal{N}'' \rangle := \mathtt{cnq}(\mathtt{lle}(\langle \mathcal{A}^b, \mathcal{N} \rangle, 2), 3, 6)$$
$$:= \mathtt{cnq}(\langle\langle\{1,2\},\{1,2\},\{5,6\} \cap \varnothing, \varnothing\rangle, [1,3] \cap [0,2]\rangle, 3, 6)$$
$$:= \mathtt{cnq}(\langle\langle\{1,2\},\{1,2\},\varnothing,\varnothing\rangle, [1,2]\rangle, 3, 6)$$
$$:= \langle\langle\{1,2\},\{1,2\},\varnothing \setminus \{6\}, \varnothing\rangle, [1,2]\rangle = \langle\langle\{1,2\},\{1,2\},\varnothing,\varnothing\rangle, [1,2]\rangle$$

but the set of propagation events generated by this sequence of restriction operations is different, namely only {LMAX}: no CDOM propagation event is generated, because the **cnq** restriction operation did not change the domain.   □

The monotonicity of event systems is important because propagation events are used in the fixpoint algorithm to schedule propagators for execution. A non-monotonic event system makes propagator scheduling non-monotonic in the sense that executing propagators in different orders can generate different propagation events possibly resulting in different amounts of pruning; see [21,24] for a complete explanation of event systems and efficient propagator scheduling.

Figure 4(b) shows a monotonic version of our propagation event system, in which LMAX implies CDOM. Intuitively, this implication arises from the representation invariant: any change to the upper bound of the set $\mathcal{N}$ of candidate lengths must empty at least one set $\mathcal{A}[i]$ of candidate characters. A proof of the monotonicity of this propagation event system is omitted for reasons of space: see [22] for further discussion.

7   Experimental Evaluation

Experimental Methodology. Experiments were carried out on a VirtualBox 4.3.10 virtual client with 1024 MB of RAM, running Xubuntu 14.04. The host machine was a 2.66 GHz Intel Core 2 Duo with 4 GB of RAM, running OpenSUSE 13.1. Code for implemented propagators was written in C++ for the GECODE 4.4.0 constraint solving library, using 64-bit bitsets, and compiled with GCC 4.8.4.

For each problem including strings of unknown length, the same initial maximum length b was used for every string variable, and the experiments were run for several possible values of b when possible. Timeout always was at 10 min.

Models and Implementations. BLS variables are compared with two other bounded length string representations for finite domain CP solvers.

The first of these methods, the *de facto* standard for solving bounded-length string constraint problems using a CP solver, is the *padded-string method* [12], which requires no proper string variable type; instead, each string unknown in a problem is modeled as a pessimistically large array of integer variables, allowing multiple occurrences of a null or padding symbol at the end of each string.

In the padded-string method there are no propagators implementing string constraints. Instead, each string constraint is modeled as a decomposition consisting of a conjunction of reified constraints over the sequence of integer variables corresponding to each string in the scope of the constraint, which express the relationship between the length of the modeled string and the occurrences of the padding character in the corresponding sequence.

The second method, the *aggregate string method* [23], is similar to the padded string method, but an integer variable is added for each string unknown, modeling its set of candidate lengths. Thus, the model of a string unknown in the aggregate-string method is isomorphic to the bounded-length sequence representation in Sect. 3, modulo the inclusion of the padding character. The aggregate-string method also differs from the padded-string method in that each string constraint is implemented by a single propagator. This method is implemented with the aid of the indexical compiler [15].

All implementations, as well as the corresponding models for the benchmarks listed below, can be found at https://bitbucket.org/jossco/gecode-string. Note that these experiments do not use the machinery of the recent string extension [2] of the MiniZinc modeling language [17].

Search. A serviceable, if not compelling, branching heuristic for string variables applies a value selection heuristic for integer variables to either a set of candidate characters or the set of candidate lengths. More interesting branching heuristics for string variables should be explored; however, a simple heuristic is sufficient for the purpose of comparing our string variable type with the two alternate methods described above. We used the following heuristic: at each choice point, the first unassigned string variable is selected; the sets of candidate characters at the mandatory indices are evaluated, and the set with the lowest cardinality is selected; if there exist no mandatory indices, then the minimum of the set of candidate lengths is increased instead. Character value selections are made by splitting the selected set at its median element, and taking the lower half first.

Benchmarks. The well-known string benchmarks of HAMPI [9], KALUZA [18], and SUSHI [8] have previously been shown to be trivial for CP solvers even without sequence variables, see [12,23] for instance, hence we do not revisit them here.

Word Design for DNA Computing on Surfaces. This problem [6], with origins in bioinformatics and coding theory, is to find the largest set of strings S, each of length 8 and with alphabet $\Sigma = \{A, T, C, G\}$, such that:

- Each string $s \in S$ contains exactly four characters from the set $\{C, G\}$.
- For all $x, y \in S$ such that $x \neq y$, x and y differ in at least four positions.
- For all $x, y \in S$ (including when $x = y$), the strings x^{rev} and $Comp(y)$ differ in at least four positions, where $Comp$ is the permutation $(AT)(CG)$.

Despite its name, this problem is actually rather a weak candidate for modeling with bounded-length string variables. Every word in S has the same fixed length,

Table 3. Time, in seconds, either to find a solution if one exists, or to prove b-bounded unsatisfiability otherwise; t indicates that the instance timed out (>10 min).

Benchmark	inst \ b	Padded 256	512	1024	Aggregate 256	512	1024	BLS 256	512	1024	SAT\UNSAT
WordDesign	80		216.0			221.3			25.8		sat
WordDesign	85		290.0			t			32.9		sat
WordDesign	112		t			t			92.5		sat
ChunkSplit	16	t	t	t	t	t	t	0.2	2.0	21.9	sat
ChunkSplit	25	t	t	t	t	t	t	1.4	15.0	215.2	sat
ChunkSplit	29	t	t	t	50.8	t	t	0.3	2.0	21.7	sat
Levenshtein	2	0.1	0.1	0.1	0.1	0.1	0.1	0.1	0.1	0.1	sat
Levenshtein	37	0.1	0.1	0.1	0.1	0.1	0.1	0.1	0.1	0.1	sat
Levenshtein	84	0.1	0.1	0.1	0.1	0.1	0.1	0.1	0.1	0.1	sat
$a^n b^n$	89	2.0	t	t	0.2	0.5	1.7	0.1	0.3	2.3	sat
$a^n b^n$	102	2.0	2.1	2.0	0.1	0.1	0.1	0.5	4.5	50.3	unsat
$a^n b^n$	154	0.1	0.1	0.4	0.1	0.2	0.5	0.4	3.1	34.0	unsat
StringReplace	20	t	t	t	t	t	t	t	t	t	—
StringReplace	136	0.9	3.5	t	0.9	3.6	t	0.1	0.2	1.6	unsat
StringReplace	142	21.0	t	t	0.1	0.4	1.5	0.1	0.3	1.8	sat
Hamming	389	t	t	t	t	t	t	0.4	3.9	59.5	unsat
Hamming	1005	8.5	t	t	7.9	t	t	0.7	5.8	61.8	unsat
Hamming	1168	t	t	t	t	t	t	0.4	3.7	55.9	unsat

and most of the constraints are binary constraints on characters; hence, the problem is easily modeled as an $|\mathcal{S}| \times 8$ matrix of integer variables. In such a fixed length string case, a proper string variable type seems to have relatively little to offer over a fixed-length array of integer variables.

Experimental results for the Word Design problem are shown in Table 3. We treat each value (shown in the instance column) of the cardinality of $\mathcal{S}$ as a satisfaction problem. No implementation times out for $|\mathcal{S}| \leq 80$, only the aggregate implementation times out for $80 < |\mathcal{S}| \leq 85$, only the BLS implementation does not time out for $85 < |\mathcal{S}| \leq 112$, and all implementations time out for $112 < |\mathcal{S}|$. As all strings in the problem are of fixed length, there is nothing to be gained by varying the maximum string length b for the problem; hence, we report only one run for $b = 8$ per cardinality for each implementation.

The model using BLS variables performs comparatively well. As all strings in the problem are of fixed length, the padded and aggregate methods collapse into a single method; the propagators provided by the aggregate model for the purpose of treating fixed-length arrays of variables as bounded-length strings are unhelpful for true fixed-length strings. In this fixed-length context, the superior performance of BLS variables must be attributed to the choice of a bitset representation for the alphabet.

Benchmark of NORN. A set of approximately 1000 string constraint problems were generated for the *unbounded*-length string solver NORN [1]. These instances do not require Unicode and have regular-language membership constraints, generated by a model checker, based on counter-example-guided abstraction refinement (CEGAR), for string-manipulating programs, as well as concatenation and length constraints on string variables, and linear constraints on integer variables.

Instances of the benchmark of NORN are written in the CVC4 dialect of the SMTLIB2 language [14]. These instances were translated into GECODE models using the three methods described above. Regular languages in the instances are specified as regular expressions; these were directly translated into GECODE'S regular-expression language and modeled by REGULAR$_{\mathcal{L}}$ constraints, with the exception of expressions of the form $X \in \epsilon$, which were instead modeled by a constraint LEN$(X, 0)$. Approximately three quarters of all instances in the benchmark include a negated regular expression. As GECODE does not implement negation for regular expressions, these instances were omitted from the experiments as a matter of convenience; adding support for taking the complement of regular languages to GECODE is straightforward (e.g., [13]).

We solve the remaining 255 instances in a *bounded*-length context; our results are therefore incomparable with those of an unbounded-length solver such as NORN. For satisfiable instances, NORN generates a language of satisfying assignments for each string variable, whereas a CP-based method returns individual satisfying strings; furthermore, NORN can determine that an instance is unsatisfiable for strings of *any* length, whereas CP solvers are limited to determining b-bounded unsatisfiability, for some practical upper bound b on string length.

Nevertheless, the benchmark of NORN remains interesting in a *bounded*-length context, as there are several challenging instances to be found. Complete results for the 255 evaluated instances are omitted for space; for further discussion, see [22]. Table 3 shows results for three instances in each of the five categories of the benchmark of NORN; these instances were selected as they appear to be the hardest, in their respective categories, for solving in a bounded-length context.

As shown in Table 3, the BLS variable implementation is either significantly faster than the aggregate and padding implementations, or tied with them, except (for reasons we failed so far to understand) on the unsatisfiable instances of the $a^n b^n$ benchmark. For large upper bounds on string sizes ($b > 256$), both of the decomposition-based implementations are prone to time outs, mostly likely as the search space is exhausting the available memory.

8 Conclusion

We have designed a new variable type, called bounded-length sequence (BLS) variables. Implemented for the copying CP solver GECODE, BLS variables ease the modeling of string constraint problems while simultaneously providing considerable performance improvements over the alternatives. The described extension is agreed to become official part of GECODE. It would be interesting to see how our ideas transpose to a trailing CP solver.

Bitsets are less appropriate for large alphabets, say when full Unicode (16 bit) coverage is required: future work includes adapting the choice of character-set representation based on alphabet size, possibly using BDDs, or a sparse-bitset representation similar to that of [5].

BLS variables, together with propagators for string constraints [22], are part of an emerging ecosystem of string solving in CP. The recent string extension [2] of the MiniZinc modeling language [17] — for which we have extended the FlatZinc interpreter of GECODE in order to support the BLS variable extension — can only serve to encourage further development.

Acknowledgements. We thank the anonymous referees for their helpful comments. The authors based at Uppsala University are supported by the Swedish Research Council (VR) under grant 2015-4910.

References

1. Abdulla, P.A., Atig, M.F., Chen, Y.-F., Holík, L., Rezine, A., Rümmer, P., Stenman, J.: Norn: an SMT solver for string constraints. In: Kroening, D., Păsăreanu, C.S. (eds.) CAV 2015. LNCS, vol. 9206, pp. 462–469. Springer, Cham (2015). doi:10.1007/978-3-319-21690-4_29. http://user.it.uu.se/~jarst116/norn

2. Amadini, R., Flener, P., Pearson, J., Scott, J.D., Stuckey, P.J., Tack, G.: MiniZinc with strings. In: Hermenegildo, M., López-García, P. (eds.) LOPSTR 2016. LNCS, vol. 10184, pp. 52–67 (2016). Post-proceedings, Pre-proceedings Version at Computing Research Repository. https://arxiv.org/abs/1608.03650

3. Bisht, P., Hinrichs, T., Skrupsky, N., Venkatakrishnan, V.N.: WAPTEC: white-box analysis of web applications for parameter tampering exploit construction. In: Chen, Y., Danezis, G., Shmatikov, V. (eds.) Computer and Communications Security (CCS 2011), pp. 575–586. ACM (2011)

4. Bjørner, N., Tillmann, N., Voronkov, A.: Path feasibility analysis for string-manipulating programs. In: Kowalewski, S., Philippou, A. (eds.) TACAS 2009. LNCS, vol. 5505, pp. 307–321. Springer, Heidelberg (2009). doi:10.1007/978-3-642-00768-2_27

5. Demeulenaere, J., Hartert, R., Lecoutre, C., Perez, G., Perron, L., Régin, J.-C., Schaus, P.: Compact-table: efficiently filtering table constraints with reversible sparse bit-sets. In: Rueher, M. (ed.) CP 2016. LNCS, vol. 9892, pp. 207–223. Springer, Cham (2016). doi:10.1007/978-3-319-44953-1_14

6. van Dongen, M.: CSPLib problem 033: word design for DNA computing on surfaces. http://www.csplib.org/Problems/prob033

7. Emmi, M., Majumdar, R., Sen, K.: Dynamic test input generation for database applications. In: Rosenblum, D.S., Elbaum, S.G. (eds.) Software Testing and Analysis (ISSTA 2007), pp. 151–162. ACM (2007)

8. Fu, X., Powell, M.C., Bantegui, M., Li, C.C.: Simple linear string constraints. Formal Aspects Comput. **25**, 847–891 (2013)

9. Ganesh, V., Kieżun, A., Artzi, S., Guo, P.J., Hooimeijer, P., Ernst, M.: HAMPI: a string solver for testing, analysis and vulnerability detection. In: Gopalakrishnan, G., Qadeer, S. (eds.) CAV 2011. LNCS, vol. 6806, pp. 1–19. Springer, Heidelberg (2011). doi:10.1007/978-3-642-22110-1_1

10. Gange, G., Navas, J.A., Stuckey, P.J., Søndergaard, H., Schachte, P.: Unbounded model-checking with interpolation for regular language constraints. In: Piterman, N., Smolka, S.A. (eds.) TACAS 2013. LNCS, vol. 7795, pp. 277–291. Springer, Heidelberg (2013). doi:10.1007/978-3-642-36742-7_20

11. Golden, K., Pang, W.: A constraint-based planner applied to data processing domains. In: Wallace, M. (ed.) CP 2004. LNCS, vol. 3258, p. 815. Springer, Heidelberg (2004). doi:10.1007/978-3-540-30201-8_93

12. He, J., Flener, P., Pearson, J., Zhang, W.M.: Solving string constraints: the case for constraint programming. In: Schulte, C. (ed.) CP 2013. LNCS, vol. 8124, pp. 381–397. Springer, Heidelberg (2013). doi:10.1007/978-3-642-40627-0_31

13. Hopcroft, J., Motwani, R., Ullman, J.D.: Introduction to Automata Theory, Languages, and Computation, 3rd edn. Addison Wesley, Redwood City (2007)

14. Liang, T., Reynolds, A., Tinelli, C., Barrett, C., Deters, M.: A DPLL(T) theory solver for a theory of strings and regular expressions. In: Biere, A., Bloem, R. (eds.) CAV 2014. LNCS, vol. 8559, pp. 646–662. Springer, Cham (2014). doi:10.1007/978-3-319-08867-9_43

15. Monette, J.-N., Flener, P., Pearson, J.: Towards solver-independent propagators. In: Milano, M. (ed.) CP 2012. LNCS, pp. 544–560. Springer, Heidelberg (2012). doi:10.1007/978-3-642-33558-7_40.   http://www.it.uu.se/research/group/astra/software\#indexicals

16. Negrevergne, B., Guns, T.: Constraint-based sequence mining using constraint programming. In: Michel, L. (ed.) CPAIOR 2015. LNCS, vol. 9075, pp. 288–305. Springer, Cham (2015). doi:10.1007/978-3-319-18008-3_20

17. Nethercote, N., Stuckey, P.J., Becket, R., Brand, S., Duck, G.J., Tack, G.: MiniZinc: towards a standard CP modelling language. In: Bessière, C. (ed.) CP 2007. LNCS, vol. 4741, pp. 529–543. Springer, Heidelberg (2007). doi:10.1007/978-3-540-74970-7_38. http://www.minizinc.org

18. Saxena, P., Akhawe, D., Hanna, S., Mao, F., McCamant, S., Song, D.: A symbolic execution framework for JavaScript. In: Security and Privacy (S&P 2010), pp. 513–528. IEEE Computer Society (2010)

19. Schulte, C., Carlsson, M.: Finite domain constraint programming systems. In: Rossi, F., van Beek, P., Walsh, T. (eds.) Handbook of Constraint Programming, pp. 495–526. Elsevier, Amsterdam (2006). Chap. 14

20. Schulte, C., Stuckey, P.J.: Efficient constraint propagation engines. Trans. Program. Lang. Syst. **31**(1), 2:1–2:43 (2008)

21. Schulte, C., Tack, G.: Implementing efficient propagation control. In: Proceedings of TRICS: Techniques for Implementing Constraint programming Systems, a Conference Workshop of CP 2010, St Andrews, UK (2010)

22. Scott, J.: Other things besides number: abstraction, constraint propagation, and string variable types. Ph.D. thesis, Uppsala University, Sweden (2016). http://urn.kb.se/resolve?urn=urn:nbn:se:uu:diva-273311

23. Scott, J.D., Flener, P., Pearson, J.: Constraint solving on bounded string variables. In: Michel, L. (ed.) CPAIOR 2015. LNCS, vol. 9075, pp. 375–392. Springer, Cham (2015). doi:10.1007/978-3-319-18008-3_26

24. Tack, G.: Constraint propagation: models, techniques, implementation. Ph.D. thesis, Saarland University, Germany (2009)

25. The Gecode Team: Gecode: a generic constraint development environment (2006). http://www.gecode.org

In Search of Balance: The Challenge of Generating Balanced Latin Rectangles

Mateo Díaz[1]([⊠]), Ronan Le Bras[2], and Carla Gomes[2]

[1] Center for Applied Mathematics, Cornell University,
Ithaca, NY 14853, USA
md825@cornell.edu
[2] Computer Science Department, Cornell University,
Ithaca, NY 14853, USA
{lebras,gomes}@cs.cornell.edu

Abstract. Spatially Balanced Latin Squares are combinatorial structures of great importance for experimental design. From a computational perspective they present a challenging problem and there is a need for efficient methods to generate them. Motivated by a real-world application, we consider a natural extension to this problem, balanced Latin Rectangles. Balanced Latin Rectangles appear to be even more defiant than balanced Latin Squares, to such an extent that perfect balance may not be feasible for Latin rectangles. Nonetheless, for real applications, it is still valuable to have well balanced Latin rectangles. In this work, we study some of the properties of balanced Latin rectangles, prove the nonexistence of perfect balance for an infinite family of sizes, and present several methods to generate the most balanced solutions.

Keywords: Latin Rectangles · Experimental design · Local search · Constraint satisfaction problem · Mixed-integer programming

1 Introduction

In recent years we have seen tremendous progress in search and optimization procedures, driven in part by hard challenging structured problems, such as those from combinatorial design. As an example, Spatially Balanced Latin squares (SBLSs) have led to the notion of streamlining constraints [7], in which additional, non-redundant constraints are added to the original problem to increase constraint propagation and to focus the search on a small part of the subspace, and other notions such as XOR constraints, a powerful general purpose streamlining technique, with applications in search, probabilistic reasoning, and stochastic optimization [1,5,6,8], and local search procedures [9].

SBLSs have been studied in detail, e.g., [4,10,11,14]. In the recent work, by combining streamlining reasoning and human-computation, Le Bras et al. [10,11] discovered the first polynomial time construction for the generation of SBLSs, for any size n such that $2n + 1$ is prime, and formally proved correctness of the

© Springer International Publishing AG 2017
D. Salvagnin and M. Lombardi (Eds.): CPAIOR 2017, LNCS 10335, pp. 68–76, 2017.
DOI: 10.1007/978-3-319-59776-8_6

procedure. This constructive solution is the first example of the use of automated reasoning and search to find a general constructive algorithm for a combinatorial design problem for an unlimited range of sizes. Previous automated reasoning results in combinatorial mathematics were restricted to finding constructions for a particular (fixed) problem size.

The demand for balanced Latin squares arises from agronomic field experiments, where one wants to compare n treatments consisting of n fertilizers applied in different orderings. Such experiments can be modeled using a Latin Square – a standard procedure in the field of experimental design. However, for agronomic experiments there is a caveat, geometric imbalance can potentially bias the results, see for example [14]. In this paper, we propose as a new benchmark problem Partially Balanced Latin Rectangles (PBLRs), a natural extension SBLSs. PBLRs are also used in experimental design; in fact, in practice rectangular sizes are more common than squares, one of the key reasons why we study them. Despite significant progress concerning SBLSs, little is known about PBLRs. Here, we present some of the properties of balanced Latin rectangles, prove the nonexistence of perfect balance for an infinite family of sizes, and present several methods to generate the most balanced solutions. We also identify an interesting easy-hard-easy pattern in the run times of a MIP algorithm, for proving optimality, and a CSP approach, for finding feasible solutions. We believe that the PBLR problem is ideal as a benchmark for the study of different search and optimization approaches.

2   Preliminaries

Definition 1. *Let $k, n \in \mathbb{N}$ be positive numbers such that $k \leq n$ and A a $k \times n$ matrix. We say that A is a Latin rectangle if every cell of A contains a number in $\{1, \ldots, n\}$ and there are no repetitions in a row or a column.*

Let $u, v \in \{1, \ldots, n\}$ be different symbols and $i \leq k$. Define $d_i(u, v)$ as the distance between the symbols u and v in row i. We define the distance $d(u, v) := \sum_{i \leq k} d_i(u, v)$.

Definition 2. *We say that a $k \times n$ rectangle A is spatially balanced if all the distances $d(u, v)$ are the same. In order to abbreviate we refer to them as $SBLR(k, n)$.*

The following proposition provides novel necessary and sufficient conditions for the existence of SBLRs.

Proposition 1. *If there exists a solution for $SBLR(k, n)$ then the distance between any pair of symbols is equal to $\dfrac{k(n + 1)}{3}$.*

Proof. If there is a solution for $SBLR(k, n)$ then the sum of all the possible distances divided by all the possible pairs of symbols must be integer. For a fixed row, we have $n - d$ pairs with distance d. Since we have k rows, the sum of all the distances is equal to $k \sum_{d=1}^{n} d(n - d) = k \left(n \sum_{d=1}^{n} d - \sum_{d=1}^{n} d^2 \right) = \frac{kn(n-1)(n+1)}{6}$. Dividing this by the number of pairs, $\binom{n}{2}$, gives the result.

As a corollary, $k \equiv_3 0$ or $n \equiv_3 2$ are necessary conditions to have balance, since the distances are integers, $k(n+1)$ must be multiple of 3 and one of the two conditions must hold.

We note that two Latin rectangles are in the same *isotopy class* if it is possible to obtain one from the other by permuting its rows, columns or symbol labels. However, spatial balance is invariant only under row permutations or relabels of the symbols. In order to avoid redundancy, we say that a Latin rectangle is in its reduced form if the first row is the sequence $1, 2, 3, \ldots, n$ and the first column is in Lexicographic order, i.e. the digits are in ascending order.

2.1   Non-existence Results

Our experiments suggest that perfectly balanced rectangles do not exist, see Sect. 4. The next results support this claim for small ks.

Proposition 2. *For $k = 2$ the only n for which $SBLR(2, n)$ is nonempty is 2.*

Proof. When $(k, n) = (2, 2)$, the only solution is perfectly balanced. Let $n > 2$ and assume that there is a solution for $SBLR(k, n)$. By Proposition 1 we know that $n = 3l + 2$, for some integer $l > 0$. Without loss of generality we may assume that in the first row the extremal symbols are 1 and n. Then, for the pair $(1, n)$ the smallest distance that a solution could achieve is n, as shown in the rectangle.

1	...	i	$i+1$	...	n
*	...	1	n	...	*

Hence, $d(1, n)$ cannot satisfy the constraint in Proposition 1, since $\dfrac{2(n+1)}{3} = 2l + 2 < 3l + 2 = n \leq d(1, n)$. This contradiction completes the proof.

Proposition 3. *For $k = 3$ the only n for which $SBLR(3, n)$ is nonempty is 3.*

Proof. For the case $(k, n) = (3, 3)$, any solution is balanced. For the cases $(k, n) = (3, 4)$ and $(k, n) = (3, 5)$, we used a complete approach to check all the possible solutions and proved that none exists.

Suppose now that $n > 5$, by Proposition 1 we know that if $A \in SBLR(k, n)$ then for every pair of symbols we have $d(a, b) = n + 1$. Again, without loss of generality we may assume that the first row of A is the sequence $1, 2, 3, \ldots, n$. Then $d_1(1, n) = n - 1$, this forces the pair 1 and n to be next to each other in the next two rows, otherwise $d(1, n) \neq n + 1$. Analogously, since $d_1(1, n-1) = n - 2$, the pair $(1, n - 1)$ must have a distance of 1 in one of the remaining rows and 2 in the other, as it is shown in the rectangle.

1	...	*	*	*	...	*	*	*	*	...	$n-1$	n
*	...	$n-1$	1	n	...	*	*	*	*	...	*	*
*	...	*	*	*	...	$n-1$	*	1	n	...	*	*

Thus, we can bound $d(n - 1, n) = d_1(n - 1, n) + d_2(n - 1, n) + d_3(n - 1, n) \leq 1 + 3 + 2 = 6 < n + 1$, which yields a contradiction, this completes the proof for all the remaining cases.

Proposition 4. *If there exists a solution for SBLR(k, n) then* $6 - \dfrac{18}{n+1} \leq k$.

Proof. Suppose that $A \in \mathrm{SBLR}(k, n)$, without loss of generality we may assume that A is in its reduced form. With this setting we know that

$$d_1(1, n-1) = n - 2, \quad d_1(1, n) = n - 1 \quad \text{and} \quad d_1(n-1, n) = 1. \quad (1)$$

By the triangular inequality we have that for every $i = 1, \ldots, k$, $d_i(n-1, n) \leq d_i(1, n-1) + d_i(1, n)$, summing over all the $i \geq 2$,

$$\sum_{i=2}^{k} d_i(n-1, n) \leq \sum_{i=2}^{k} d_i(1, n-1) + \sum_{i=2}^{k} d_i(1, n).$$

Using the distances that we know (1), we obtain $d(n-1, n) - 1 \leq d(1, n-1) - (n-1) + d_i(1, n) - (n-2)$, since A is perfectly balanced, $\frac{k(n+1)}{3} \leq \frac{2k(n+1)}{3} - 2n + 4$. By simplifying this expression we get the stated result.

With our last result we proved that there are no solutions for $k = 4$ and $n > 8$, and there are also no solutions for $k = 5$ and $n > 17$. Thus, at least for small ks and for all the sizes that do not satisfy $k \equiv_3 0$ or $n \equiv_3 2$, perfect balanced cannot be achieved.

2.2 Partially Balanced Latin Rectangles

These results suggest that SBLR(k, n) may be an infeasible problem when $k < n$. For real applications if balance cannot be achieved, it is useful to have the most balanced solution. To measure the balance of a rectangle R we define the *imbalance function*, $\mathbb{I}(R) = \sum_{i<j} \left| d(i,j) - \frac{k(n+1)}{3} \right|$. Note that this function is non-negative and if the rectangle is perfectly balanced $\mathbb{I}(R) = 0$.

Algorithm 1. Random-restart hill climbing balanced Latin Rectangles

input : k, n

output: A well Balanced Latin Rectangle R

Generate a cyclic $n \times n$ Latin Square L and define R as a $k \times n$ rectangle of zeros;

for *a fixed number of iterations* **do**

    Take k random rows of L and assign them to R;

    for *a fixed number of iterations* **do**

        Draw a pair of columns (i, j) at random and assign to $\bar{R}$ a copy of R with the columns (i, j) switched;

        if $\mathbb{I}(\bar{R}) < \mathbb{I}(R)$ **then**

            $R = \bar{R}$;

            if $\mathbb{I}(R) = 0$ **then**

                Break all the loops, we found perfect balance;

Definition 3. *Let R be a k×n Latin rectangle, we say that it is partially spatially balanced if* $\mathbb{I}(R)$ *is minimum. For a fixed size, we denote by PBLR(k, n) the set of these rectangles.*

Thus, if perfect balance is achieved, PBLR and SBLR are equivalent. This presents a new challenge, since the minimum value of the imbalance is unknown a priori. Now, we need to determine the minimum imbalance, that we call I_{kn}, and at the same time generate the associated partially balanced rectangle.

3   Proposed Solutions

We developed three different approaches for generating rectangles with minimum imbalance: a local search algorithm, a constraint satisfaction encoding and a mixed-integer programming encoding. In this section we describe these methods and discuss some advantages and disadvantages of each encoding.

3.1   Local Search

The objective of the algorithm is to minimize the imbalance function, by performing random-restart hill climbing local search on the space of Latin rectangles. Our algorithm starts with a cyclic solution, i.e., every row is obtained by shifting by one position the previous row, and it was adapted from one of the best known algorithms for balanced Latin squares proposed in [13]. One issue with this algorithm is that it starts with an initial solution and never leaves its isotopic class. So if there is no solution for that particular class, the algorithm will only explore that class. Nevertheless, this technique was used before with successful results for Latin squares. In that case, the stopping criterion is clear, $\mathbb{I}(R) = 0$; but for partially balance Latin rectangles the value of $\min_R \mathbb{I}(R)$ is unknown. Thus, while this incomplete approach can only be used to find upper bounds on the minimum imbalance, it constitutes a relevant baseline and in practice, it is able to find good solutions fast.

3.2   Constraint Satisfaction Problem and Mixed-Integer Programming Encodings

For any $m \in \mathbb{N}$ define $[m] = \{1, \dots, m\}$. Let Δ_n be the set of possible pairs (i, j) with $i, j \in [n]$ such that $i < j$. We consider a simple constraint satisfaction problem that encodes the problem, where the variables R_{ij} indicate the position (column number) of the symbol j in row i. The first set of *alldifferent* constraints guarantees that the symbols cover all the columns, while the second one guarantees the non-repetition of symbols in columns. The last constraint assures that the imbalance of solutions is upper bounded by the constant M.

$$
\begin{aligned}
&R_{ij} \in [n] &&\forall i \in [k], j \in [n] \\
&alldiff(R_{i1}, \dots, R_{in}) &&\forall i \in [k] \\
&alldiff(R_{1j}, \dots, R_{kj}) &&\forall j \in [n] \\
&\sum_{(i,j) \in \Delta_n} \left| \frac{k(n+1)}{3} - \sum_{l=1}^{k} |R_{li} - R_{lj}| \right| \leq M.
\end{aligned} \tag{2}
$$

A classical approach to boost the performance of this algorithm is to use symmetry breaking constraints [2,3,12], which is done with the use of the following constraints:

$$R_{1j} = j \qquad \forall j \in \{1, \ldots, n\}$$
$$R_{i1} < R_{(i+1)1} \quad \forall i \in \{1, \ldots, k-1\} \tag{3}$$

These constraints enforce that the Latin rectangle is in its reduced form, i.e., the first row is the sequence $1, 2, 3, \ldots, n$, and the first column is in Lexicographic order, i.e. the digits are in ascending order. By imposing these constraints we are shrinking the size of the space of solutions by a factor of $n!k!$ and only eliminating equivalent elements from isotopic classes. For balanced squares, previous formulations use other streamlining constraints such as symmetry, that have proven to be very effective and, in fact, the solutions found with those constraints served as basis to uncover the underlying pattern that gave rise to the polynomial time construction in [10]. Unfortunately, the rectangular shape prevents us from applying such constraints.

The main drawback with the CSP formulation is that its performance depends on the scheme chosen for the values of M and it does not incrementally learn from one iteration to the next. To tackle this issue we transform this formulation into a mixed-integer program, using $\mathbb{I}(\cdot)$ as an objective function. The MIP formulation can be written as:

$$\min \sum_{(i,j) \in \Delta_n} D_{ij}$$
$$\text{s. t. } R_{ij} \in [n] \qquad\qquad \forall i \in [k], j \in [n],$$
$$D_{ij} = \left| \frac{k(n+1)}{3} - \sum_{l=1}^{k} |R_{li} - R_{lj}| \right| \qquad \forall (i,j) \in \Delta_n$$
$$R_{ih} \neq R_{il} \qquad\qquad \forall i \in [k], (h,l) \in \Delta_n,$$
$$R_{il} \neq R_{jl} \qquad\qquad \forall (i,j) \in \Delta_k, l \in [n]$$

The formulation in this case is almost the same as in (2), with the only difference that the left-hand side of the last constraint becomes the function that we are minimizing. Also, note that the alldifferent constraints translate to a conjunction of binary disequalities.

Both the CSP and MIP encoding can be used to certify optimality, namely they can prove that they find the minimal imbalance I_{kn}. Nevertheless, the MIP solver does not require an initial guess for the objective function bound. In addition, through branch and cut methods, it generates lower bounds on the optimal value of imbalance using a hierarchy of linear relaxations. In particular, even before proving optimality, the MIP solver can certify that there is no perfect balance if it finds a nontrivial lower bound, which could be potentially practical even for the square case.

4   Experimental Results

We identified an interesting **easy-hard-easy** pattern in the runtimes of the MIP solver, for proving optimality, and the CSP, for finding a PBLR given I_{kn}, when

k/n varies. For a fixed n, the problem becomes harder as k increases; it is easier when $k = n$, see Fig. 1.

All the experiments were run using 12 cores on a single compute server with 2.93 GHz Intel(R) Xeon(R) x5670 processors and 48 GB RAM. For the optimization and satisfiability problems we use IBM ILOG CPLEX-CP Optimizer.

Proving Optimality. To compare the CSP and MIP formulations, we ran experiments for proving optimality, each method with $6 \leq n \leq 8$, $2 \leq k \leq n$ and a time limit of 35 h. The MIP solver was able to solve the problem for all the instances up to $n = 8$ and $k = 4$, see Table 1. To prove optimality using CSP

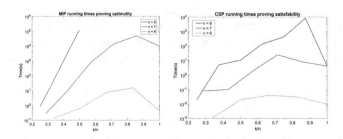

Fig. 1. Easy-hard-easy patterns in running times of the MIP solving for proving optimality and the CSP proving satisfiability of I_{kn} with $n = 6, 7, 8$.

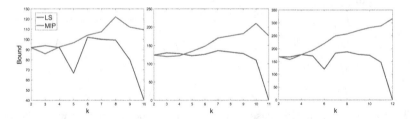

Fig. 2. Bounds found by the local search algorithm and the MIP solver, from left to right $n = 10, 11, 12$.

Table 1. Minimum imbalance found with our algorithms. The column corresponds to k and the row to n. Grey cells indicate a proven optimal value.

	2	3	4	5	6	7	8	9	10	11	12
4	2.$\bar{6}$	4	5.$\bar{3}$								
5	8	6	8	0							
6	16	12	13.$\bar{3}$	16	0						
7	28	22	22.$\bar{6}$	22.$\bar{6}$	20	18.$\bar{6}$					
8	40	36	32	30	24	28	0				
9	65.$\bar{3}$	56	56.$\bar{6}$	56	52	66	60	0			
10	92	86	92	66.$\bar{6}$	102	100	99.$\bar{3}$	80	40		
11	124	120	122	122	126	136	132	128	110	0	
12	168	158	162.$\bar{6}$	170.$\bar{6}$	120	183	184.$\bar{6}$	178	174.$\bar{6}$	147.$\bar{3}$	0

we start with an initial feasible guess for the bound M and decrease it until the problem is unsatisfiable. A natural way to measure the time is to take the sum of the last satisfiable case and the first infeasible one. The CSP solver was good at proving optimality for the square cases where balance is feasible. In almost any other case, if $n \leq 6$ it takes more than the time limit to run the unsatisfiable case. Thus, the MIP seems to be a better method for proving optimality in rectangles. However, the CSP formulation is fast at proving satisfiability once we have the minimum imbalance a priori. **Finding well balanced solutions for larger sizes.** For larger sizes, it is very difficult to prove optimality. As a consequence, we use our local search algorithm to find upper bounds. To compare this approach with the MIP formulation, we ran the MIP solver with a time limit of 1500 s and execute Algorithm 1 with 1000 iterations on both loops (it takes less than 1200 s). For large sizes the pattern is similar, if $k > 3$ the local search approach outperforms the MIP one, see Fig. 2.

5 Conclusions

We study the problem of generating balanced Latin rectangles. We conjecture (and prove for several cases) that unlike squares, it is impossible to have perfect balance for Latin rectangles. We examined the problem of finding the most balanced solutions, namely partially balanced Latin rectangles. We developed different methods to prove optimality and to find well balanced solutions when it is too expensive to guarantee optimality. We identified an interesting easy-hard-easy pattern in the run times of the MIP, for proving optimality, and CSP, for finding a feasible solution. For larger problems, our simple local search approach outperforms the MIP approach.

Finding Latin rectangles with minimum imbalance seems to be a very challenging computational problem. We believe that this problem is ideal as a benchmark for different search and optimization approaches. We leave as an open challenge for future researchers to improve the results in Table 1.

Acknowledgments. This work was supported by the National Science Foundation (NSF Expeditions in Computing awards for Computational Sustainability, grants CCF-1522054 and CNS-0832782, NSF Computing research infrastructure for Computational Sustainability, grant CNS-1059284).

References

1. Ermon, S., Gomes, C.P., Sabharwal, A., Selman, B.: Low-density parity constraints for hashing-based discrete integration. In: Proceedings of the 31th International Conference on Machine Learning, ICML 2014, Beijing, China, 21–26 June 2014, pp. 271–279 (2014)
2. Fahle, T., Schamberger, S., Sellmann, M.: Symmetry breaking. In: Walsh, T. (ed.) CP 2001. LNCS, vol. 2239, pp. 93–107. Springer, Heidelberg (2001). doi:10.1007/3-540-45578-7_7

3. Gent, I.P., Smith, B.M.: Symmetry breaking in constraint programming. In: Proceedings of ECAI-2000, pp. 599–603. IOS Press (2000)
4. Gomes, C., Sellmann, M., Van ES, C., Van Es, H.: The challenge of generating spatially balanced scientific experiment designs. In: Régin, J.-C., Rueher, M. (eds.) CPAIOR 2004. LNCS, vol. 3011, pp. 387–394. Springer, Heidelberg (2004). doi:10.1007/978-3-540-24664-0_28
5. Gomes, C.P., Hoffmann, J., Sabharwal, A., Selman, B.: Short XORs for model counting: from theory to practice. In: Marques-Silva, J., Sakallah, K.A. (eds.) SAT 2007. LNCS, vol. 4501, pp. 100–106. Springer, Heidelberg (2007). doi:10.1007/978-3-540-72788-0_13
6. Gomes, C.P., Sabharwal, A., Selman, B.: Model counting: a new strategy for obtaining good bounds. In: Proceedings, The Twenty-First National Conference on Artificial Intelligence and the Eighteenth Innovative Applications of Artificial Intelligence Conference, Boston, Massachusetts, USA, 16–20 July 2006, pp. 54–61 (2006)
7. Gomes, C.P., Sellmann, M.: Streamlined constraint reasoning. In: Principles and Practice of Constraint Programming - CP 2004, 10th International Conference, CP 2004, Toronto, Canada, 27 September–1 October 2004, Proceedings, pp. 274–289 (2004)
8. Gomes, C.P., van Hoeve, W.J., Sabharwal, A., Selman, B.: Counting CSP solutions using generalized XOR constraints. In: AAAI, pp. 204–209 (2007)
9. Van Hentenryck, P., Michel, L.: Differentiable invariants. In: Principles and Practice of Constraint Programming - CP 2006, 12th International Conference, CP 2006, Nantes, France, 25–29 September 2006, Proceedings, pp. 604–619 (2006)
10. Le Bras, R., Gomes, C.P., Selman, B.: From streamlined combinatorial search to efficient constructive procedures. In: AAAI (2012)
11. Le Bras, R., Perrault, A., Gomes, C.: Polynomial time construction for spatially balanced Latin squares (2012)
12. Rossi, F., Van Beek, P., Walsh, T.: Handbook of Constraint Programming. Elsevier, Amsterdam (2006)
13. Smith, C., Gomes, C., Fernandez, C.: Streamlining local search for spatially balanced Latin squares. In: IJCAI, vol. 5, pp. 1539–1541. Citeseer (2005)
14. Van Es, H., Van Es, C.: Spatial nature of randomization and its effect on the outcome of field experiments. Agron. J. 85(2), 420–428 (1993)

Debugging Unsatisfiable Constraint Models

Kevin Leo[(✉)] and Guido Tack

Data 61/CSIRO, Faculty of IT, Monash University, Melbourne, Australia
{kevin.leo,guido.tack}@monash.edu

Abstract. The first constraint model that you write for a new problem is often unsatisfiable, and constraint modelling tools offer little support for debugging. Existing algorithms for computing Minimal Unsatisfiable Subsets (MUSes) can help explain to a user which sets of constraints are causing unsatisfiability. However, these algorithms are usually not aimed at high-level, structured constraint models, and tend to not scale well for them. Furthermore, when used naively, they enumerate sets of solver-level variables and constraints, which may have been introduced by modelling language compilers and are therefore often far removed from the user model.

This paper presents an approach for using high-level model structure to, at the same time, speed up computation of MUSes for constraint models, present meaningful diagnoses to users, and enable users to identify different sources of unsatisfiability in different instances of a model. We discuss the implementation of the approach for the MiniZinc modelling language, and evaluate its effectiveness.

1 Introduction

Modelling languages for constraint programming such as ESSENCE [6] and MiniZinc [18] allow decision problems to be modelled in terms of high-level constraints. Models are combined with instance data and are then compiled into input programs for solving tools (solvers). The goal of these languages is to allow users to solve constraint problems without expert knowledge of targeted solvers.

Unfortunately, real-world problems typically exhibit a level of complexity that makes it difficult to create a correct model. The first attempt at modelling a problem often results in an incorrect model. There are multiple ways in which a model may be incorrect. In this paper we focus on the case of over-constrained models where the conjunction of all constraints are *unsatisfiable* for any instance.

When faced with unsatisfiability, the user has few tools to help with debugging. The main strategy usually consists in activating and deactivating constraints in an attempt to locate the cause of the problem, but this approach is tedious and often impractical due to the fact that the fault may involve a non-trivial combination of groups of constraints and instance data.

Several techniques for debugging unsatisfiable constraint programs exist, some are designed for debugging specific kinds of constraint programs [7,9,17],

---

Partly sponsored by the Australian Research Council grant DP140100058.

D. Salvagnin and M. Lombardi (Eds.): CPAIOR 2017, LNCS 10335, pp. 77–93, 2017.
DOI: 10.1007/978-3-319-59776-8_7

while others focus on diagnosis of unsatisfiability during search [11,19,21]. We are concerned with approaches that are constraint-system agnostic [4,10,14]. Many of these approaches focus on the search for *Minimal Unsatisfiable Subsets* (MUSes): sets of constraints that together are unsatisfiable, but become satisfiable with the removal of any one of the constraints. MUSes can therefore help explain the *sources* of unsatisfiability in a constraint program.

Existing tools for finding MUSes focus on program level constraints, i.e., the constraints at the level of the solver. When combining MUS detection with high-level modelling, this has two main drawbacks. Firstly, the solver-level program typically contains hundreds or thousands of constraints, even for relatively simple high-level models. MUS detection algorithms do not scale well to these problem sizes[1]. Secondly, the user may find it difficult to interpret the resulting sets of constraints, because they have lost all connection to the original model and may involve variables that were introduced by the compilation.

The main contribution of this paper is an approach that **uses high-level model structure to guide the MUS detection algorithm**. We show that using the structure available in a high-level MiniZinc model can speed up the search for MUSes. We also demonstrate how these can be presented to the user in a meaningful and useful way, in terms of the high-level model instead of the solver-level program. Finally, we show how MUSes found across a *set* of instances can be expressed in terms of the high-level (parametric) model, allowing us to *generalise* the detected conflicts and distinguish between genuine modelling bugs and unsatisfiability that arises from faulty instance data.

Structure. The next section presents some of the background techniques. Section 3 introduces a MUS detection algorithm that can take advantage of high-level model structure. Section 4 discusses implementation aspects and presents some experiments that show promising speed-ups. Section 5 shows how the additional information about model structure can be used to present more meaningful diagnoses to the user, and Sect. 6 discusses how the presented approach can help generalise the results found across several instances to the model-level. Finally, Sect. 7 discusses related approaches and Sect. 8 concludes the paper.

2   Background

A *constraint program* is a formal representation of an instance of a decision or optimisation problem. Constraint programs consist of a set of variables representing the decisions to be made, a set of domains representing the possible assignments to these variables, and a set of constraints, in the form of predicates which describe the relationship between variables. Optionally, an objective function which is to be minimised or maximised can be defined. A solution is a set of assignments to the variables such that all constraints are satisfied and the value of the objective function is optimal. If no assignment exists that satisfies all constraints, the program is said to be unsatisfiable.

---

[1] [4] reports runtimes higher than 10 s for models of only 1000 constraints.

2.1   Constraint Models

MiniZinc is a high-level language for describing parametric *models* of problems. These models can be combined with instance data and compiled into concrete constraint programs that target specific solvers by the MiniZinc compiler.

In this paper we will use the Latin Square problem as a running example. A Latin Square is an n by n matrix of integers where each row and column contain permutations of the numbers 1 to n. Listing 1.1 presents a MiniZinc model for this problem. Line 1 declares that the model has an integer parameter 'n', the dimension of the matrix. Line 2 creates the $n \times n$ matrix named X which is a matrix of integer variables which must take values in the interval $[1, n]$. Line 4 introduces the first set of constraints. It states that the variables in each row of the matrix must take distinct values. Following this on line 5 the second constraint states that the variables in each column must also take distinct values.

```
1  int: n;
2  array[1..n, 1..n] of var 1..n: X;
3
4  forall (r in 1..n) (alldifferent(row(X, r)));
5  forall (c in 1..n) (alldifferent(col(X, c)));
```

Listing 1.1. MiniZinc Model for the Latin Squares Problem

During compilation, MiniZinc performs a bottom-up translation, replacing each argument to a predicate or function call with an auxiliary variable bound to the result of the call. MiniZinc provides a set of standard decompositions that encode constraints in terms of simpler predicates. Solver-specific MiniZinc libraries can define custom decompositions, or declare a predicate as a built-in, in which case it is simply added to the FlatZinc. For illustration purposes, a compilation that decomposes the alldifferent constraint is shown, although most MiniZinc solvers provide it as a built-in.

The following is an example trace through the compiler that introduces an int_ne predicate to the FlatZinc. Starting with line 4 from Lisiting 1.1 the compiler starts to evaluate the forall predicate. The argument to the forall constraint is an array comprehension. To evaluate the comprehension the compiler must loop through values for r in the set $[1, n]$ and evaluate the expression alldifferent(row(X, r)). With r set to 1 it evaluates row(X, 1) which returns an array containing the variables corresponding to the first row of the matrix X. Next the compiler evaluates alldifferent(A) where A is the returned row array. The standard library contains a definition for alldifferent with an array of integer variables, rewriting it to the more specific all_different_int(A) predicate. The compiler must now evaluate this predicate call, resulting in a decomposition into a set of not-equal constraints (int_ne). The compiler then assigns 2 to r and compilation proceeds. Compiling this model for n = 3 results in 18 int_ne constraints being added to the program.

2.2   Program Level Diagnosis

The Latin Squares model presented earlier has a set of symmetries which we can break to find solutions faster. One such symmetry is the ordering of values in consecutive rows or columns. A naive user may try to break several of these symmetries by adding the following constraints:

```
7  forall (r in 1..n-1) (lex_less(row(X, r), row(X, r+1)));
8  forall (c in 1..n-1) (lex_greater(col(X, c), col(X, c+1)));
```

Unfortunately, these constraints, combined with the `alldifferent` constraints are in conflict since they enforce different orderings on the rows and columns. With these constraints added to the model, the decomposed, flattened program contains 58 constraints.

Current approaches for fault diagnosis in constraint programs work at the level of individual constraints in a compiled program. These approaches typically aim to enumerate all or some subset of Minimal Unsatisfiable Subsets (MUSes). Having a selection of MUSes gives the user a better chance of discovering the root cause of unsatisfiability, although in some cases a single MUS may be sufficient.

Enumerating the MUSes of a program can be achieved by exploring the power-set, or all combinations, of constraints, performing a satisfiability check for each combination, and collecting all unsatisfiable subsets, discarding all strict supersets. The satisfiability check is typically delegated to an external solver, thus making the MUS detection algorithm itself agnostic to the concrete type of problem that is being diagnosed.

Most MUS algorithms avoid enumerating the entire exponential-size power-set in an attempt to minimise the number of satisfiability checks required. For example, they will avoid the exploration of any superset of an already discovered MUS. A detailed survey of MUS enumeration approaches can be found in [15]. Good techniques for pruning the search space can reduce the number of satisfiability checks considerably. However, depending on the time taken by each satisfiability check, this pruning may still not be enough to make these approaches practical for large constraint programs.

When used with constraint programs generated by a compiler like MiniZinc, the generated MUSes are likely to include constraints and variables introduced during compilation. These can be difficult to map back to their source in the model. Presenting these MUSes to a user who is unfamiliar with the workings of the compiler may not be conducive to fixing modelling mistakes. A MUS for the faulty Latin Squares model might include constraints such as `int_lin_le_reif`(...) and `array_bool_or`(...). Finding where these constraints came from in the original model is possible but not simple.

3   Exploiting Model Structure for MUS Detection

This section introduces our approach to augment any existing MUS detection algorithm so that it can take advantage of high-level model structure. Our approach combines the idea to *group* constraints into a hierarchy, as explored in [11], with an extension of the *variable path* concept first introduced in [13], and the idea to find MUSes based on select groups of constraints from [1,16] in order to speed up MUS detection and provide more meaningful diagnoses.

3.1 Constraint Paths

We introduced variable paths in [13] in order to match variables across different compilations of an instance, enabling the compiler to perform some whole-program optimisation to produce simpler, more efficient programs.

A variable path describes the path the compiler took from the model to the point where a new variable is introduced during compilation. Paths contain information on the location of each syntactic construct, as well as the bindings of all loop variables that lead to the introduction of a variable.

For the purposes of this paper, we will extend the concept of variable paths to also apply to constraints, in order to be able to identify the origins and high-level structure of program-level constraints.

Listing 1.2 shows simplified paths for the first 4 constraints in the compiled Latin Squares program (Listing 1.1), with a graphical representation of the entire compilation in Fig. 1. The path for the first `int_ne` constraint encodes that it came from a `forall` call on line 4 (4 : forall). Next we see the index variable r was set to 1 (r=1). After this, the `alldifferent` call is replaced with a call to `all_different_int`. At this point the compiler has decomposed `all_different_int` into simpler constraints. It begins with a `forall`, for which the index variables i and j are set to 1 and 2. Finally we see that the `int_ne` constraint was added for the binary disequality != .

The amount of detail in the paths presented here has been reduced for illustrative purposes, actual MiniZinc paths retain information about what file a call is in, the span of text the call covers, and the full name of each call.

```
int_ne(X[1],X[2]);     | 4:forall:r=1 | all_different_int | forall:i=1 j=2 | != |
int_ne(X[1],X[3]);     | 4:forall:r=1 | all_different_int | forall:i=1 j=3 | != |
int_ne(X[1],X[4]);     | 5:forall:c=1 | all_different_int | forall:i=1 j=2 | != |
int_ne(X[1],X[7]);     | 5:forall:c=1 | all_different_int | forall:i=1 j=3 | != |
```

Listing 1.2. Simplified MiniZinc paths with depth 2 prefixes marked in bold

Constraint paths make the model-level hierarchy explicit at the program level. The root of this hierarchy (or depth 1) for our Latin Squares model would have 2 abstract constraints with prefixes 4 : forall and 5 : forall, representing the `alldifferent` constraints on rows or columns of X. The child constraints of these (at depth 2) would be the n possible assignments for both r and c, giving us the $2n$ individual `alldifferent` constraints. Jumping forward to a depth of 5 the prefixes finally distinguish between specific `int_ne` constraints.

3.2 Grouping Constraints by Paths

The idea to perform diagnosis based on groups of constraints can be seen in [1,16] where MUSes are found in a given set of grouped constraints. The hierarchical

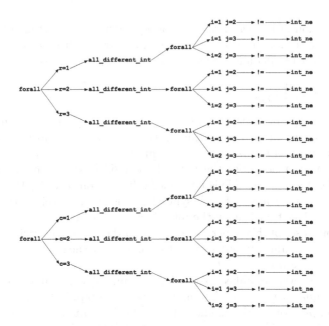

Fig. 1. Constraint paths for the Latin Squares model

grouping of model constraints goes back to [11] where users define names for groups of lower-level constraints so that more useful feedback can be presented to users. We refer to these groupings as *abstract constraints*. Constraint paths explicitly encode the hierarchy of constraints and give us access to model-level structure in the solver-level programs. This enables us to group constraints together *automatically* and *intelligently* when searching for MUSes. Instead of relying on the user to define the abstract constraints, or the MUS detection working on the full set, we can automatically generate abstract constraints based on the model structure, starting from the top-level constraints and iteratively refining them down to individual solver-level constraints.

Take for example Listing 1.2. Grouping all constraints whose paths are equal up to a depth of 1, a MUS enumeration algorithm will have to explore the power-set of only two abstract constraints, a much smaller search space than that of the full set of 18 program level constraints. Of course MUSes found at a depth of 1 can only give a user a hint of what might be wrong with their model by drawing their attention to the correct lines in their model. We may have to use longer prefixes to find more useful MUSes.

In Algorithm 1 the grouping is managed by a *checker* which is aware of the constraints (C) and their depth in the hierarchy. The *checker* is initialised by a call to create_checker which takes arguments for the initial depth in the hierarchy where the approach should start looking for MUSes (d_{min}) and also the target depth of the procedure (d_{max}). The *checker* provides a procedure called increase_depth which takes a set of abstract constraints and splits these constraints at the next depth (splitting them into a larger number of more specific constraints). This procedure returns *false* if the depth cannot be increased any further (either due to d_{max} or simply reaching the program level constraints).

Algorithm 1 can be used with any solver-independent MUS detection algorithm. We just assume functions detectMUSes.finished() that returns true when the algorithm has fully enumerated the MUSes at the current depth and detectMUSes.next() that returns the next MUS.

Algorithm 1. Procedure for reporting MUSes found at different depths

1: **procedure** diagnose(C, d_{min}, d_{max}, *maxMUSes*, *complete*)
2:     *checker* ← create_checker(C, d_{min}, d_{max}, *complete*)
3:     **do** *MUSes* ← ∅
4:         **while** $|MUSes|$ < *maxMUSes* **and not** detectMUSes(*checker*).finished( ) **do**
5:             *MUSes* ← *MUSes* ∪ detectMUSes(*checker*).next( )
6:         *checker*.report(*MUSes*)
7:     **while** *checker*.increase_depth($\bigcup MUSes$)

Selective Deepening. After detecting a set of MUSes for a certain grouping depth, we may want to expand the depth to get more fine-grained information. Of course we could simply increase the grouping depth for all abstract constraints, which will split each of them again according to the next part of the paths. However, we do not need to increase the depth for abstract constraints that did not take part in any MUS. This allows us to restrict the MUS detection at the next depth to the abstract constraints that actually appear in MUSes at the current level. This can speed up enumeration considerably since the abstract constraints that are not involved in any MUS can be made up of very large sets of constraints. In Algorithm 1 this is represented by the call to the *checker*.increase_depth procedure on line 7 with the union of constraints occurring in MUSes.

Selective deepening allows for the discovery of MUSes at deep levels much quicker than trying to find MUSes directly at those depths. For example, if running a MUS enumeration algorithm on a set of abstract constraints $\{a, b\}$ results in a MUS $\{a\}$, we need only increase the depth for a resulting in the new set: $\{a_1, a_2, a_3, b\}$. If b contains five lower level constraints this approach, without affecting completeness (since b is never removed), can lead to a significant speedup when compared to searching in the set $\{a_1, a_2, a_3, b_1, b_2, b_3, b_4, b_5\}$.

Figure 2 shows several iterations of the approach when applied to a program for the faulty Latin Squares model. The program has 58 constraints (the leaves of the tree). In this figure, ovals represent abstract constraints that are to be evaluated, with filled ovals representing a discovered MUS. Note that the frontier only deepens under the abstract constraints that occurred partake in a MUS. We see that the first iteration can at most search through combinations of 4 abstract constraints. Upon finding a MUS, the depth is increased only for the three responsible abstract constraints. At the next depth we see that the algorithm must only search through combinations of 7 constraints. This continues until, at the program level, the algorithm searches through just 36 constraints, instead of the full set of 58 which traditional approaches would have to explore.

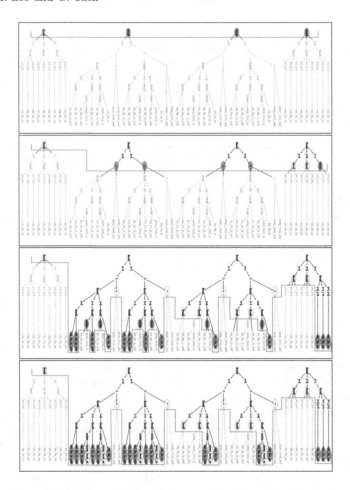

Fig. 2. Automatic deepening, from top: iterations 1, 2, 5 and 6

Incomplete enumeration. A further optimisation that can improve scalability is to omit the abstract constraints that did not occur in any MUS at the current depth. However, this

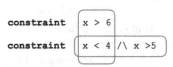

removal may occasionally cut off MUSes that can only be found at lower depths. In the example shown here, it is clear that there are two MUSes: {x<4, x>5} and {x>6, x<4}. A MUS detection algorithm searching at a depth of 1, i.e., only looking at each constraint item as a whole, will find the conflict {x>6, x<4 /\ x>5 }. However, this is not minimal as the set is still unsatisfiable when constraint x>6 is removed. Incomplete enumeration will then seek to find MUSes in the set {x<4, x>5} discarding constraint x>6 and thus cutting off the MUS {x>6, x<4}. This approach sacrifices completeness for scalability allowing users to quickly discover a subset of MUSes. In Algorithm 1, this behaviour is enabled by setting *complete* to *false*.

4 Implementation and Evaluation

The approach was implemented by extending an existing implementation of an enumeration algorithm called MARCO, developed and implemented by Liffiton and Malik [14,15], and evaluated on a set of MiniZinc benchmark models that were augmented to make them unsatisfiable.

4.1 Extending MARCO

Our proof of concept implementation consists of a FlatZinc satisfaction checker that is aware of MiniZinc paths, and a new frontend to MARCO that controls the grouping of FlatZinc constraints during the enumeration.

To enumerate MUSes the MARCO algorithm maintains a CNF formula (called a map) that encodes the problem of selecting from the remaining unchecked subsets of abstract constraints. The algorithm proceeds by finding a solution to the map, which represents a subset of abstract constraints. If this set is satisfiable it is expanded by adding constraints that do not cause unsatisfiability. The map is then updated to reflect that all subsets of this "grown" subset are satisfiable and should not be explored except in the presence of an abstract constraint from outside this subset. If the subset was unsatisfiable, a "shrink" method is called which removes constraints that do not affect the subset's unsatisfiability. In this case the map is updated to mark all supersets as explored. The algorithm continues as long as the map is satisfiable, indicating that unexplored subsets remain.

The frontend supports running MARCO on a set of abstract constraints of a target depth, as well as starting at some depth and deepening after enumeration of MUSes at that depth until a target depth is reached. Removing abstract constraints that do not occur in a MUS at deeper enumerations (which makes enumeration incomplete as discussed in the previous section) is also supported. A further configuration option for improving the performance of the tool is to limit the number of MUSes to be discovered at each depth. This setting combined with the setting for removing abstract constraints from the map allows the tool to focus on discovering a specific fault as quickly as possible.

4.2 Experiments

To demonstrate the usefulness of the new approach two experiments were performed. The first explores the viability of the approach for finding sets of MUSes at different depths. The second experiment explores the amount of time it takes to return a single MUS. Typically, MUS detection algorithms are evaluated on sets of unsatisfiable program-level *instances*[2], not on parametric *models*. Currently there are no collections of models that contain bugs that make them unsatisfiable. Collections such as CSPLib [8] or the MiniZinc benchmarks[3] contain only finished, correct models for problems. We therefore introduced artificial faults into models from the MiniZinc Challenge 2015 [22], similarly to how fault injection was applied in [12]. The faults were selected to make the models unsatisfiable in a non-trivial way (i.e., so that the compiler does not detect it during compilation). Faults added include swapping arguments to global constraints, changing operators (`<=` to `<`), changing array index offsets (`X[j] <= X[i+1]` to `X[j] < X[i]`), using the wrong variables in constraints (`X[n] = X[succ[n]]` to `X[n] = X[pred[n]]`), removing negations (`x = -v` to `x = v`), and changing constants (`x != 0` to `x != 1`)). The instances used were selected at random.

Six different configurations of the system were evaluated, paired by their target d_{max}. Each execution was given a timeout of 5 min (300 s). The columns show the duration in seconds (T), the number of abstract constraint groups (G), and the number of conflicts found (C). The configurations are presented in Fig. 3. Configurations C_2, C_3, and C_{Max} correspond to a traditional approach to MUS enumeration at depths 2, 3, and at the depth of the individual program-level constraints. Configura-

Config	d_{min}	d_{max}	complete
C_2	2	2	true
$C_{1\rightarrow2}$	1	2	false
C_3	3	3	true
$C_{1\rightarrow3}$	1	3	false
C_{Max}	Max	Max	true
$C_{1\rightarrow Max}$	1	Max	false

Fig. 3. Experiment setup

tions $C_{1\rightarrow2}$, $C_{1\rightarrow3}$ and $C_{1\rightarrow Max}$ perform incomplete enumeration (setting *complete* to *false*), enumerating MUSes first at a depth of 1 and then increasing the depth, discarding abstract constraints that are not involved in any MUS as it proceeds.

MUS enumeration. Table 1 presents a comparison of the performance of the different configurations when the goal is to enumerate multiple MUSes.

Comparing configurations C_2 and $C_{1\rightarrow2}$ we see that $C_{1\rightarrow2}$ is sometimes faster, but in three of the eleven cases it cannot find as many MUSes as the fixed depth C_2 configuration. For example, in the case of the "Free Pizza" model, configuration C_2 returns 13 MUSes before timing out, while configuration $C_{1\rightarrow2}$ can only yield a single MUS after discarding many of the other constraints. For configurations C_3 and $C_{1\rightarrow3}$, there are a few cases where configuration $C_{1\rightarrow3}$ is faster. However, we also find cases where just starting at a depth of 3 finds more MUSes faster. Finally, in configurations C_{Max} and $C_{1\rightarrow Max}$ we see that the deepening approach discovers MUSes at the level of individual program level constraints much faster than the full enumeration approach on three occasions. Again, we see an instance (MKnapsack) where increasing the depth does not change the number of abstract constraints. Since the enumeration algorithm has to essentially repeat enumeration at every depth it ends up being much slower.

---

[2] For example, from the MaxSAT competition http://www.maxsat.udl.cat/.

[3] https://github.com/MiniZinc/minizinc-benchmarks.

Table 1. Comparison of enumeration behaviour

Configuration	C_2			$C_{1\to2}$			C_3			$C_{1\to3}$			C_{Max}			$C_{1\to Max}$		
Model	T	G	C	T	G	C	T	G	C	T	G	C	T	G	C	T	G	C
Costas array	300.0	56	88	300.0	54	11	300.0	83	0	300.0	0	0	300.0	83	0	300.0	0	0
CVRP	300.0	34	1	**0.5**	7	1	300.0	66	2	**0.7**	7	1	300.0	101	3	**1.4**	14	4
Free pizza	300.0	81	13	**0.7**	9	1	300.0	82	12	**0.7**	1	1	300.0	553	8	**1.2**	1	1
Mapping	0.4	24	1	**0.3**	2	1	28.9	69	70	**10.9**	28	66	300.0	254	70	300.0	0	0
MKnapsack	7.6	31	49	**7.2**	30	49	**7.8**	31	49	14.3	30	49	**9.5**	32	65	58.2	31	65
NMSeq	300.0	40	0	300.0	40	0	300.0	40	0	300.0	0	0	300.0	3240	0	300.0	0	0
Open stacks	300.0	802	5	300.0	800	5	300.0	841	4	300.0	0	0	300.0	4421	0	300.0	0	0
p1f	113.6	89	77	**89.3**	77	77	**113.9**	89	77	170.6	77	77	300.0	947	10	300.0	0	0
Radiation	0.5	5	1	0.5	4	1	12.0	68	12	**4.1**	25	12	73.1	388	12	**20.4**	12	12
Spot5	300.0	4998	0	300.0	4406	1	300.0	5227	0	300.0	0	0	300.0	5457	0	300.0	0	0
TDTSP	300.0	46	1	300.0	0	0	300.0	62	1	300.0	0	0	300.0	170	1	300.0	0	0

This indicates that the deepening approach may be less useful when most of the constraints are in conflict.

Time to First MUS. In practice a user will often start trying to fix the first few MUSes that are presented rather than waiting for a full enumeration. This is similar to debugging in traditional programming languages, where one would rarely try to fix multiple reported errors at once but instead fix one and then check whether it was also the cause of the other errors. With this in mind the second experiment, presented in Table 2, explores how quickly the configurations can report the first MUS found at a target depth.

Comparing configurations C_2 and $C_{1\to2}$ in this case shows that the incomplete enumeration configuration ($C_{1\to2}$) is almost always faster even with a d_{max} of 2. For configurations C_3 and $C_{1\to3}$ the effect is even more pronounced with $C_{1\to3}$ outperforming C_3 in almost all cases. The impact of incomplete enumeration can be seen in the results for the Free Pizza problem, where C_2 must find MUSes given 82 abstract constraints at depth 2 whereas configuration $C_{1\to2}$ has narrowed the set of constraints to a single abstract constraint. Finally, in configurations C_{Max} and $C_{1\to Max}$, we compare the traditional program-level approach for finding a MUS against the incomplete approach. Here again we see the approach outperforming C_{Max} in most cases by honing in on a single MUS.

5 Displaying Diagnoses

The approach presented in Sect. 4 produces diagnoses as sets of MiniZinc paths. While paths contain the information that is required for debugging, they are not very easy to read. What a user may need to extract from a set of paths to help interpret a diagnosis is just the set of syntactic positions that it relates to, and the specific assignments to loop index variables during compilation that make

Table 2. First MUS found for different depths

Configuration	C_2			$C_{1\to 2}$			C_3			$C_{1\to 3}$			C_{Max}			$C_{1\to Max}$		
Model	T	G	C	T	G	C	T	G	C	T	G	C	T	G	C	T	G	C
Costas array	**2.5**	56	1	3.9	54	1	300.0	83	0	**4.7**	33	1	300.0	83	0	**6.0**	33	1
CVRP	0.5	34	1	**0.3**	7	1	0.9	66	1	**0.4**	7	1	1.3	101	1	**0.8**	14	1
Free pizza	1.6	81	1	**0.6**	9	1	1.7	82	1	**0.7**	1	1	9.4	553	1	**0.9**	1	1
Mapping	0.4	24	1	**0.3**	2	1	0.6	69	1	**0.4**	28	1	1.6	254	1	**0.6**	4	1
MKnapsack	0.5	31	1	0.5	30	1	0.5	31	1	0.5	1	1	**0.6**	32	1	0.7	1	1
NMSeq	300.0	40	0	300.0	40	0	300.0	40	0	300.0	0	0	300.0	3240	0	300.0	0	0
Open stacks	53.9	802	1	52.1	800	1	73.7	841	1	**55.4**	7	1	300.0	4421	0	**73.4**	14	1
plf	3.9	89	1	**1.9**	11	1	4.2	89	1	**2.1**	1	1	29.6	947	1	**2.4**	1	1
Radiation	**0.5**	5	1	0.5	4	1	1.4	68	1	**0.9**	25	1	6.5	388	1	**1.2**	1	1
Spot5	300.0	4998	0	**258.9**	4406	1	300.0	5227	0	**264.2**	1	1	300.0	5457	0	**264.2**	1	1
TDTSP	1.2	46	1	**0.5**	1	1	1.4	62	1	**0.4**	1	1	3.0	170	1	**0.7**	1	1

up the diagnosis. To make things easier for the user, we can present diagnoses in a more useful form. We have extended the MiniZinc IDE to be able to interpret and display MiniZinc paths directly in the source code editor.

Case Study: Over-constrained Latin Squares

In Fig. 4 we see the Latin Squares model from Sect. 2 in the MiniZinc IDE with some conflicting symmetry breaking constraints added. Enumerating MUSes for this model instantiated with $n = 3$ produces 12 diagnoses. In Fig. 4 we show how a selection of these are presented to the user. Each MUS is displayed in the output section, with the number of abstract constraints involved and a list of parameter values that lead to this diagnosis.

Looking at the highlighting of the model presented in Fig. 4a we see that the selected diagnosis involves some combination of the row `alldifferent` constraints, the row `lex_less` constraints, and the column `lex_greater` constraints. All of the MUSes for this problem involve some combination of the `alldifferent` constraints and the `lex` constraints. Looking at the *intersection of abstract constraints* involved in MUSes it is easy to find that at least 3 `lex` constraints are involved in every MUS and must be the source of the bug.

If the user cannot immediately deduce what may be causing the issue they can look at what specific rows and columns are involved by examining the list of assignments to parameters. From the list in Fig. 4a we can see that the bug involves the first two iterations of the loop which introduce `lex_greater` constraints for the first three columns (`c=1; c=2`), the `alldifferent` constraints on the first two rows (`i=1; i=2`) and the `lex_less` over the first two rows `r=1`.

If the enumeration algorithm is configured to find MUSes at greater depths, the diagnoses will involve constraints introduced by decompositions of the model-level constraints. This can be seen in Fig. 4b where a user has clicked on one such

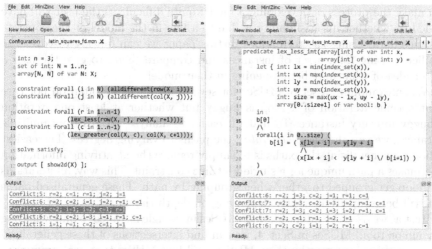

(a) MUS with first **alldifferent**   (b) MUS at library level

Fig. 4. Prototype MiniZinc IDE UI showing different diagnoses.

diagnosis. This opens tabs for displaying the MiniZinc standard decompositions of the lex_less_int and all_different_int constraints. The highlighting shows a specific less-than-or-equal constraint that is unsatisfiable. Tracking down constraints across multiple MiniZinc files is made much easier by this feature.

6 Generalising to the Model Level

MiniZinc paths were introduced to identify common structure across multiple compilations of the same instance. With a slight modification they can provide insights into common structures found across instances, the intersection of which can be considered to be instance-independent.

In some cases multiple instances of a problem will have MiniZinc paths in common. For example, the paths for the first few iterations of a forall loop will often be the same. For example in the Latin Square model presented in Sect. 2 all valid values for the parameter n result in the first forall always being evaluated for r = 1. Different instances will have similar paths for these constraints. Since the instances are different the constraints are not the same and so we cannot perform any automatic reasoning on these. However, we will now show how grouping the constraints can still provide useful insights.

6.1 Cross-Instance MUSes

The user may have a set of instances of their problem, some of which should be satisfiable, while others may be unsatisfiable (even assuming a correct model). If the user does not know which instances are unsatisfiable to begin with, this can make the process of developing a model quite difficult as the user will have to deduce whether unsatisfiability is due to a bug in the model or the concrete instance data. Using the techniques in this paper we can quickly discover the conflicts arising from a set of instances and compare them, to give the user a better idea of what may be happening in their model.

When examining a set of MUSes for several instances we often do not care what specific iteration of a loop is buggy but whether at least one iteration is buggy in every instance. To make it easier to analyse MUSes from multiple instances, we can therefore *generalise* the paths to varying degrees. The easiest option for generalising the paths is to simply remove the identifying information that makes a path unique for a single FlatZinc constraint. This way, all iterations of a loop get grouped together when grouping by path. These generalised paths can still be grouped differently depending on depth though.

Just as we used the intersection of MUSes in a single instance to find what bug is common to all MUSes in Sect. 4, we can find intersections of these generalised paths to group instances by their MUSes, and find MUSes that are common to all instances, indicating a modelling bug.

Given a model and a set of instances (some of which are unsatisfiable), we can enumerate MUSes for each instance. The MUSes can then be presented to the user, ranked by the number of instances that they occur in. Looking at this ranked list, a user can discern whether a MUS occurs in all or most instances, which indicates that it may be a bug in the model. Similarly the user can also see the types of unsatisfiability triggered by different instances, allowing them to classify their instances into different groups.

6.2   Case Study: Cyclic-RCPSP

To demonstrate how this approach would work in practice we will look at a relatively complex model along with a set of satisfiable and unsatisfiable instances. The model we selected was the Cyclic Resource-Constrained Project Scheduling Problem [2]. This problem is a project scheduling problem where tasks are repeated infinitely. The objective is to find a cyclic schedule that first minimises the period of the schedule and then minimises the makespan.

We introduced a bug to the model by swapping arguments of the `cumulative` constraints that represent start times and durations of tasks. Since both are arrays of integer variables, the compiler does not detect this mistake.

The unsatisfiable instances fall into two groups: two instances that have unsatisfiable resource capacities and two in which the precedences of some tasks are cyclical (task a depends on task b which in turn depends on task a). In addition to these unsatisfiable instances there are three satisfiable instances.

The FlatZinc programs for these instances can be quite large, making full enumeration of MUSes a time consuming task even for a single instance. We therefore use our selective deepening approach and instruct it to finish at the relatively shallow depth of 8. Since we are interested in comparing diagnoses across instances, we generalise the paths by removing all specific assignments from them. This allows us to group larger sets of program constraints together into even fewer abstract constraints, to avoid searching at an unnecessary level of detail. In this model, this grouping means that the MUS enumeration algorithm only needs to look at combinations of twelve abstract constraints regardless of instance data. The instance data will only change which abstract constraints fail. MUSes for each instance can be discovered very quickly with these settings and can then be reported in terms of the model.

For these instances the algorithm discovers five distinct MUSes. These MUSes occur in instances $\{0, 1, 2, 5, 6\}, \{3, 4\}, \{1, 5\}, \{2, 6\}$ and $\{2\}$. The first MUS occurs in 5 of the 7 instances and as such is a strong candidate for being a model-level bug. Indeed, this MUS involves the incorrect arguments to cumulative. The instance that does not include this exact MUS also relates to cumulative but it fails in a slightly different way leading to a different MUS. Once the bug has been fixed the user can run the analysis again which will show that there are only two MUSes remaining. These remaining MUSes occur in instances $\{3, 4\}$ and $\{2, 6\}$ which correspond to the instances with the two classes of faults.

7   Related Work

MUS enumeration at the program level has been researched extensively. Some approaches focus on specific constraint systems, taking advantage of e.g. properties of linear systems in MIPs [9,17], or structure inherent to numerical CSPs [7].

Several algorithms have been proposed for constraint agnostic MUS enumeration [15]. QuickXplain [10] attempts to discover MUSes using a divide and conquer approach. Later approaches such as DAA [4] have more powerful techniques for pruning the search space. DAA's main drawback was that it has to enumerate very large hitting sets and as a result had a high memory and time cost. The MARCO algorithm [14] and the more recent MCS-MUS-BT [3] provide much more efficient approaches for finding MUSes (see Sect. 4).

At the modelling system level there has been some effort to provide more meaningful explanations of unsatisfiability. In [11] users can explicitly group their constraints, giving a user friendly name to each sub-group. When a constraint system is found to be unsatisfiable these names are used to provide feedback. CPTEST [12] is a modelling system level framework that can aide a user in correcting several types of faults in iterations of an initially correct model. Many of these approaches could be combined with, and benefit from, the hierarchical structure that can be extracted from high-level models.

8   Conclusion

This paper presented a novel approach that can find Minimal Unsatisfiable Subsets (MUSes) faster and present them to the user in a way that allows them to quickly identify the source of the unsatisfiability. Our approach is based on automatically grouping related constraints together during MUS enumeration, which reduces the search space and can speed up the discovery of diagnoses.

MiniZinc paths enable both this automatic grouping as well as the presentation of the extracted MUSes in terms of the high-level model the user wrote.

We also presented a methodology for deducing, given a set of satisfiable and unsatisfiable instances, whether the model has a bug. Further we can help a user group unsatisfiable instances by the types of unsatisfiability that they introduce.

Future work. The approaches explored here will be better integrated into the MiniZinc compiler and IDE, providing a more consistent interface for users. Additionally, user studies focussing on the debugging of constraint models could provide valuable insights into how these tools can be further improved. Finally, integrating solvers based on Lazy Clause Generation [5,20] more tightly with the MARCO algorithm seems a promising direction for further speed ups.

References

1. Andraus, Z.S., Liffiton, M.H., Sakallah, K.A.: Refinement strategies for verification methods based on datapath abstraction. In: Asia and South Pacific Conference on Design Automation, January 2006
2. Ayala, M., Benabid, A., Artigues, C., Hanen, C.: The resource-constrained modulo scheduling problem: an experimental study. Comput. Optim. Appl. **54**(3), 645–673 (2013)
3. Bacchus, F., Katsirelos, G.: Finding a collection of MUSes incrementally. In: Quimper, C.-G. (ed.) CPAIOR 2016. LNCS, vol. 9676, pp. 35–44. Springer, Cham (2016). doi:10.1007/978-3-319-33954-2_3
4. Bailey, J., Stuckey, P.J.: Discovery of minimal unsatisfiable subsets of constraints using hitting set dualization. In: Hermenegildo, M.V., Cabeza, D. (eds.) PADL 2005. LNCS, vol. 3350, pp. 174–186. Springer, Heidelberg (2005). doi:10.1007/978-3-540-30557-6_14
5. Feydy, T., Stuckey, P.J.: Lazy clause generation reengineered. In: Gent, I.P. (ed.) CP 2009. LNCS, vol. 5732, pp. 352–366. Springer, Heidelberg (2009). doi:10.1007/978-3-642-04244-7_29
6. Frisch, A.M., Grum, M., Jefferson, C., Martnez, B., Miguel, H.I.: The design of ESSENCE: a constraint language for specifying combinatorial problems. In: IJCAI 2007, pp. 80–87 (2007)
7. Gasca, R.M., Valle, C., Gómez-López, M.T., Ceballos, R.: NMUS: structural analysis for improving the derivation of all MUSes in overconstrained numeric CSPs. In: Borrajo, D., Castillo, L., Corchado, J.M. (eds.) CAEPIA 2007. LNCS (LNAI), vol. 4788, pp. 160–169. Springer, Heidelberg (2007). doi:10.1007/978-3-540-75271-4_17
8. Gent, I.P., Walsh, T.: CSPlib: a benchmark library for constraints. In: Jaffar, J. (ed.) CP 1999. LNCS, vol. 1713, pp. 480–481. Springer, Heidelberg (1999). doi:10.1007/978-3-540-48085-3_36
9. Gleeson, J., Ryan, J.: Identifying minimally infeasible subsystems of inequalities. INFORMS J. Comput. **2**(1), 61–63 (1990)
10. Junker, U.: QuickXplain: Conflict detection for arbitrary constraint propagation algorithms. In: IJCAI01 Workshop on Modelling and Solving problems with constraints (2001)
11. Jussien, N., Ouis, S.: User-friendly explanations for constraint programming. In: Kusalik, A.J. (ed.) Proceedings of the Eleventh Workshop on Logic Programming Environments (WLPE 2001), Paphos, Cyprus, 1 December 2001 (2001)

12. Lazaar, N., Gotlieb, A., Lebbah, Y.: A CP framework for testing CP. Constraints **17**(2), 123–147 (2012)
13. Leo, K., Tack, G.: Multi-pass high-level presolving. In: Yang, Q., Wooldridge, M. (eds.) Proceedings of the Twenty-Fourth International Joint Conference on Artificial Intelligence, IJCAI 2015, Buenos Aires, Argentina, 25–31 July 2015, pp. 346–352. AAAI Press (2015)
14. Liffiton, M.H., Malik, A.: Enumerating infeasibility: finding multiple MUSes quickly. In: Gomes, C., Sellmann, M. (eds.) CPAIOR 2013. LNCS, vol. 7874, pp. 160–175. Springer, Heidelberg (2013). doi:10.1007/978-3-642-38171-3_11
15. Liffiton, M.H., Previti, A., Malik, A., Marques-Silva, J.: Fast, flexible MUS enumeration. Constraints **21**(2), 223–250 (2015)
16. Liffiton, M.H., Sakallah, K.A.: Algorithms for computing minimal unsatisfiable subsets of constraints. J. Autom. Reasoning **40**(1), 1–33 (2008)
17. van Loon, J.: Irreducibly inconsistent systems of linear inequalities. Eur. J. Oper. Res. **8**(3), 283–288 (1981)
18. Nethercote, N., Stuckey, P.J., Becket, R., Brand, S., Duck, G.J., Tack, G.: MiniZinc: towards a standard CP modelling language. In: Bessière, C. (ed.) CP 2007. LNCS, vol. 4741, pp. 529–543. Springer, Heidelberg (2007). doi:10.1007/978-3-540-74970-7_38
19. O'Callaghan, B., O'Sullivan, B., Freuder, E.C.: Generating corrective explanations for interactive constraint satisfaction. In: Beek, P. (ed.) CP 2005. LNCS, vol. 3709, pp. 445–459. Springer, Heidelberg (2005). doi:10.1007/11564751_34
20. Ohrimenko, O., Stuckey, P.J., Codish, M.: Propagation = lazy clause generation. In: Bessière, C. (ed.) CP 2007. LNCS, vol. 4741, pp. 544–558. Springer, Heidelberg (2007). doi:10.1007/978-3-540-74970-7_39
21. Ouis, S., Jussien, N., Boizumault, P.: k-relevant explanations for constraint programming. In: Russell, I., Haller, S.M. (eds.) Proceedings of the Sixteenth International Florida Artificial Intelligence Research Society Conference, 12–14 May 2003, St. Augustine, Florida, USA, pp. 192–196. AAAI Press (2003)
22. Stuckey, P., Becket, R., Fischer, J.: Philosophy of the MiniZinc challenge. Constraints **15**(3), 307–316 (2010)

Learning Decision Trees with Flexible Constraints and Objectives Using Integer Optimization

Sicco Verwer[1] and Yingqian Zhang[2(✉)]

[1] Delft University of Technology, Delft, The Netherlands
s.e.verwer@tudelft.nl
[2] Eindhoven University of Technology, Eindhoven, The Netherlands
yqzhang@tue.nl

Abstract. We encode the problem of learning the optimal decision tree of a given depth as an integer optimization problem. We show experimentally that our method (DTIP) can be used to learn good trees up to depth 5 from data sets of size up to 1000. In addition to being efficient, our new formulation allows for a lot of flexibility. Experiments show that we can use the trees learned from any existing decision tree algorithms as starting solutions and improve the trees using DTIP. Moreover, the proposed formulation allows us to easily create decision trees with different optimization objectives instead of accuracy and error, and constraints can be added explicitly during the tree construction phase. We show how this flexibility can be used to learn discrimination-aware classification trees, to improve learning from imbalanced data, and to learn trees that minimise false positive/negative errors.

1 Introduction

Decision trees [3] have gained increasing popularity these years due to their effectiveness in solving classification and regression problems. As the problem of learning optimal decision trees is a NP-complete problem [11], greedy based heuristics such as CART [3] and ID3 [15] are widely used to construct suboptimal trees. Greedy decision tree algorithm builds a tree recursively starting from a single node. At each decision node, an optimization problem is solved to determine the locally best split decision based on a subset of the training data such that the training data is further split into two subsets. Decisions are determined in turn for each of these subsets on children nodes of the starting node. The advantage of such a greedy approach is its computational efficiency. The limitation is that the constructed trees may be far from optimal.

In this paper, we aim to build optimal decision trees by optimizing all decisions concurrently. We formulate the problem of constructing the optimal decision tree of a given depth as an integer linear program. We call our method DTIP. One benefit of this formulation is that we can take advantage of the powerful mixed-integer linear programming (i.e., MIP) solver to find good trees.

© Springer International Publishing AG 2017
D. Salvagnin and M. Lombardi (Eds.): CPAIOR 2017, LNCS 10335, pp. 94–103, 2017.
DOI: 10.1007/978-3-319-59776-8_8

Researcher have previously investigated using solvers for learning different kinds of models and rules, see e.g., [4, 6, 8, 10]. We are not the first who try such an approach for decision tree learning. Bennett and Blue [1] proposed a formulation to solve the problem of constructing binary classification trees with fixed structure and labels, where paths of the tree are encoded as disjunctive linear inequalities, and non-linear objective functions are introduced to minimize errors. Norouzi et al. [14] linked the decision tree optimization problem with the problem of structured prediction with latent variables. To the best of our knowledge, our method is the first that encodes decision tree learning entirely in an integer program. A similar approach for more general models is given in [2]. This method, however, is quadratic instead of linear in the data set size and therefore requires a lot of preprocessing in order to reduce the number of generated constraints. [4] discusses modeling the problem of finding the smallest size decision tree that perfectly separates the training data as a constraint program.

Section 2 shows the proposed encoding of learning optimal decision trees as an integer optimization problem only requires $O(2dn)$ constraints for regression and $O(nu + nv)$ constraints for classification, where n is the size of the dataset, v is the number of leafs, and u is the number of tree nodes. In addition, it requires $O(mu + nk + vy)$ variables, where k is the tree depth and y is the unique target values in the dataset. This makes the encoding linear in the dataset size for fixed size trees. Moreover, the number of binary variables depends on the dataset size up to a small constant factor (the tree depth). The formulated problem can be directly solved by any MIP solver such as CPLEX. In Sect. 3, we show experimentally that our method can be used to learn good trees up to depth 5 from datasets of size up to 1000. In addition to being efficient, our new formulation allows for a lot of flexibility. Our formulation enables that the trees obtained from existing greedy algorithms from Scikit-learn can be used as starting solutions for the CPLEX optimizer. Experiments with several real datasets show that our method improves the starting solutions. Moreover, the proposed formulation allows us to create decision trees with different optimization objectives other than standard objectives such as accuracy and error, and constraints can be added explicitly when learning trees. We show how this flexibility can be used to learn discrimination-aware trees and to improve learning from imbalanced data. Starting from a solution given by Scikit-learn, our method can find trees of good performance that are discrimination-free, and trees that return zero false-positives on training data.

2 Learning Decision Trees as Integer Programs (DTIP)

We assume the reader to be familiar with decision trees. We refer to [9] for more information. The optimization problem that we aim to solve is to find an optimal classification/regression tree of depth exactly k for a given dataset of n rows (samples) and m features. The Boolean decisions and predictions are variables and need to be set such that accuracy, absolute error, or any other linear measures, is optimized. We solve this problem by translating/encoding it entirely into linear constraints and providing this to an off-the-shelf MIP solver.

Table 1. Feature values from first few rows from the Iris data before (left) and after (right) our data transform. The feature values are first sorted, and then identical values are mapped to integers in increasing order. Several feature values (such as SepLen {4.9, 4.7, 4.6}, PetLen, PetWid) are combined into a single integer because these only occur for Iris-setosa flowers. A SepWid value of 3.0 occurs most frequently in the data and is mapped to 0. The target values are not transformed.

SepLen	SepWid	PetLen	PetWid	SepLen	SepWid	PetLen	PetWid
5.1	3.5	1.4	0.2	−9.0	5.0	−3.0	−3.0
4.9	3.0	1.4	0.2	−11.0	0.0	−3.0	−3.0
4.7	3.2	1.3	0.2	−11.0	2.0	−3.0	−3.0
4.6	3.1	1.5	0.2	−11.0	1.0	−3.0	−3.0
5.0	3.6	1.4	0.2	−10.0	6.0	−3.0	−3.0

It has been shown in our previous work (e.g., [16]) that it is beneficial to keep encoding from machine learning models to linear constraints as small as possible, thereby increasing the data size it can handle. In contrast to earlier decision tree encodings (e.g. [1,4,16]), we therefore encode the leaf every data row ends up in using a binary instead of a unary encoding, i.e., using k variables instead of 2^k. We start our encoding with a transformation of the input data.

2.1   Data Transformation

Earlier encodings of decision trees [1] or similar classification/regression models [2] linearly scale the input data to the interval $[0.0, 1.0]$. There are good reasons for doing so. For instance, this avoids large values in so-called big-M formulations (a way to encode binary decisions in integer programming), which can lead to numerical issues and long run-times. In spite of the benefits of a linear scaling, we advocate the use of a non-linear transform that assigns every unique value of every feature to a unique integer, only maintaining the ordering of these values. Using this transform:

- The thresholds in decision nodes can be represented by integer values instead of continuous ones. This allows MIP solvers to branch on these values instead of whether a certain row takes a left or right branch, reducing and balancing the search tree used by these algorithms.
- When all rows having successive integers as feature values also have the same class label, these values can be merged into a single integer.
- The most frequently occurring feature value can be mapped to the value 0, reducing the number of non-zero coefficients in the linear constraints.
- The ranges of feature values can all be centered around 0, see Table 1 for an example. This reduces the size of M values used in the big-M formulations.

These benefits all affect the MIP solvers capacity to solve problems efficiently, and therefore they are important considerations when encoding decision tree

Table 2. Summary of notation, constants, and variables used in the encoding.

Symbol	Type	Definition
n	Constant	Number of rows in data file
u	Constant	Number of nodes in tree, excluding leaf nodes
m	Constant	Number of features in data
v	Constant	Number of leaves in tree
k	Constant	Number of depths in tree
y	Constant	Number of unique target values in data
$d(j)$	Constant	Depth of node j of tree; the root node has depth 0
$v(r,i)$	Constant	Feature value for data row r and feature i
$t(r)$	Constant	Target value of data row r
LF, UF	Constant	Minimum, maximum feature value over all features
$f_{i,j}$	Binary	Decision variable, feature i is used in decision rule of node j
c_j	Integer	Decision variable, threshold of decision rule of node j
$d_{h,r}$	Binary	Decision variable, path of data row r goes right/left at depth h
$p_{l,t}$	Binary	Classifier prediction of leaf l and target t
p_l	Continuous	Regressor prediction of leaf l
e_r	Continuous	Prediction error for data row r

learning problems. Although a simple linear scaling can provide better results for some problem instances, we have experienced significant improvements in the obtained solutions using the non-linear transform. In our experiments, we demonstrate that our encoding is capable of producing good results when there are 1000 rows in the input data, which is significantly greater than previous works on integer programming encodings for decision trees and would not have been possible without this data transform.

2.2 Encoding Classification Trees

Our encoding for classification and regression trees is partly based on earlier work where we translated already learned models into linear constraints in order to deal with optimization under uncertainty [16]. Two key differences between this work and our new encoding are: *(1)* the coefficients denoting which constant threshold and feature to use in a node's binary decision are free variables, and *(2)* the leaf that a data row ends up in is represented in binary instead of unary. Table 2 summarizes the notation and variables that we use to encode trees. The objective function minimizes the total prediction error for all data rows:

$$\min \sum_{1 \le r \le n} e_r, \quad \text{for all } r \in [1, n] \tag{1}$$

where n is the number of input rows and $e_r \in \mathbb{R}$ is the error for data row r, defined in Eq. 5. Every node j, excluding leaf nodes, in the tree needs a binary decision variable $f_{i,j} \in \{0,1\}$ to specify whether feature $i \in [1,m]$ is used in the decision rule on node j:

$$\sum_{1 \leq i \leq m} f_{i,j} = 1 \quad \text{for all } j \in [1,u] \tag{2}$$

Every node j requires an integer decision variable $c_j \in [LF, UF]$ that represents the threshold, where $LF = \min\{v(r,i) | i \in [1,m], r \in [1,n]\}$, $UF = \max\{v(r,i) | i \in [1,m], r \in [1,n]\}$, and $v(r,i)$ denoting feature value for row r and feature i. For each row r, we encode whether it takes the left or right branch of a node using a variable $d_{h,r} \in \{0,1\}$ for every depth h, for all $j \in [1,u], r \in [1,n]$:

$$\sum_{1 \leq h \leq d(j)} M_r \mathrm{dlr}(h,j,r) + M_r d_{d(j),r} + \sum_{1 \leq i \leq m} v(r,i) f_{i,j} \leq M_r d(j) + c_j$$

$$\sum_{1 \leq h \leq d(j)} M'_r \mathrm{dlr}(h,j,r) - M'_r d_{d(j),r} - \sum_{1 \leq i \leq m} v(r,i) f_{i,j} \leq M'_r (d(j)-1) - c_j \tag{3}$$

where $M_r = \max\{(v(r,i) - LF) | i \in [1,m]\}$ and $M'_r = \max\{(UF - v(r,i)) | i \in [1,m]\}$ are tight big-M values, $d(j)$ is the depth of node j, and $\mathrm{dlr}(h,j,r)$ returns the path directions from the root node required to reach node j:

$$\mathrm{dlr}(h,j,r) = \begin{cases} d_{h,r} & \text{if the path to node } j \text{ goes \textbf{left} at depth } h \\ 1 - d_{h,r} & \text{if the path to node } j \text{ goes \textbf{right} at depth } h \end{cases} \tag{4}$$

The formulation is essentially a big-M formulation for the constraint that if row r takes the left (right) branch at depth $d(j)$ of node j, denoted by $d_{d(j),r}$, and row r takes the path to node j (i.e., $\sum_{1 \leq h < d(j)}(\mathrm{dlr}(h,j,r)) = d(j) - 1$), then the feature value $v(r,i)$ for which $f_{i,j}$ is true has to be smaller (greater) than threshold c_j. This encodes all possible paths through the tree for all rows using only $O(nu)$ constraints and $O(nk)$ binary variables, where k is the depth of the tree. What remains is the computation of the classification error:

$$\sum_{1 \leq t \leq y} p_{l,t} = 1 \quad \text{for all } l \in [1,v]$$

$$\sum_{1 \leq h \leq k} \mathrm{dlr}'(h,l,r) + \sum_{t \neq t(r)} p_{l,t} \leq e_r + k \quad \text{for all } l \in [1,v], r \in [1,n] \tag{5}$$

where $p_{l,t} \in \{0,1\}$ is the prediction on leaf l, y is the number of unique target values (2 for binary classification), v is the number of leafs in the tree, $\mathrm{dlr}'(h,l,r)$ is the same as $\mathrm{dlr}(h,j,r)$ but for leafs l instead of internal nodes j, and $t(r)$ is the target value for row r. These constraints force that if a row ends in a leaf l (setting $\mathrm{dlr}'(h,l,r)$ to 1 along the path to l), then the error e_r for row r is 1 if the leaf prediction type (for which $p_{l,t}$ is 1) is different from the rows target $t(r)$. This adds $O(vn)$ constraints and $O(vy)$ to the encoding. This fully encodes classification tree learning in $O(n(u+v))$ constraints, $O(mu + nk + vy)$ binary variables, and u integers. A nice property is that the number of variables grows with a very small constant factor (the depth of the tree k) in the dataset size n.

$$5f_{1,1} + 8f_{2,1} + 10f_{3,1} + d_{1,r}M_r \leq c_1 + M_r$$
$$-5f_{1,1} - 8f_{2,1} - 10f_{3,1} - d_{1,r}M_r' \leq -c_1$$

$\text{dlr}(1,2,r) = d_{1,r} = 0 \Rightarrow \top$ | $\text{dlr}(1,3,r) = 1 - d_{1,r} = 1$

$$5f_{1,3} + 8f_{2,3} + 10f_{3,3} + d_{2,r}M_r + (1 - d_{1,r})M_r \leq c_3 + 2M_r$$
$$-5f_{1,3} - 8f_{2,3} - 10f_{3,3} - d_{2,r}M_r' + (1 - d_{1,r})M_r' \leq -c_3 + M_r'$$

$\text{dlr}(1,6,r) + \text{dlr}(2,6,r) = 1 - d_{1,r} + d_{2,r} = 2$ | $\text{dlr}(1,7,r) + \text{dlr}(2,7,r) = 1 \Rightarrow \top$

$$5f_{1,6} + 8f_{2,6} + 10f_{3,6} + d_{3,r}M_r + (1 - d_{1,r} + d_{2,r})M_r \leq c_6 + 3M_r$$
$$-5f_{1,6} - 8f_{2,6} - 10f_{3,6} - d_{3,r}M_r' + (1 - d_{1,r} + d_{2,r})M_r' \leq -c_6 + 2M_r'$$

$\sum_{1 \leq h \leq 3} \text{dlr}'(h,5,r) = 1 - d_{1,r} + d_{2,r} + d_{3,r} = 3$ | $\sum_{1 \leq h \leq 3} \text{dlr}'(h,6,r) = 2 \Rightarrow \top$

$$1 - d_{1,r} + d_{2,r} + d_{3,r} + p_{1,5} + p_{3,5} \leq e_r + 3$$

Fig. 1. The encoding of an example row r, with features values $(5, 8, 10)$ and target value 2 out of three possible targets $\{1, 2, 3\}$. We show the path taken by an assignment of $(d_{1,r}, d_{2,r}, d_{3,r}) = (0, 1, 1)$, i.e., the row first takes the right branch to node 3, and then two left branches to node 6 and leaf 5. Observe that the big-M formulation forces many of the path constraints to be satisfied, e.g., $5f_{1,1} + 8f_{2,1} + 10f_{3,1} + d_{1,r}M_r \leq c_1 + M_r$ reduces to $5f_{1,1} + 8f_{2,1} + 10f_{3,1} \leq c_1 + M_r$, which is always true. The same holds for all constraints in subtrees rooted under a $\top$ symbol. The constraint $-5f_{1,1} - 8f_{2,1} - 10f_{3,1} - d_{1,r}M_r' \leq -c_1$ reduces to $-5f_{1,1} - 8f_{2,1} - 10f_{3,1} \leq -c_1$, which is true only if the feature value of the feature type of node 1 ($f_{i,1}$) is greater than the constraint value for node 1 (c_1). The same reasoning holds for the constraints on other depths. This thus encodes the node constraints of a decision tree. The depth values $(d_{1,r}, d_{2,r}, d_{3,r})$ have a similar effect on the leaf constraints. Only leaf 5 reached by r forces the error e_r of row r to be 1 when the prediction type ($p_{t,5}$) of leaf 5 is unequal to 2 (the target type of r). All other leaf constraints are true for any error ≥ 0.

We strengthen the above encoding by bounding the node thresholds between the minimum and maximum values of the features used in the binary decisions:

$$\sum_{1 \leq i \leq m} LF \cdot f_{i,j} \leq c_j \leq \sum_{1 \leq i \leq m} UF \cdot f_{i,j} \quad \text{for all } j \in [1, u]$$

In addition, it does not make sense for the two leaf nodes l and l' of the same parent node to have the same values. The following breaks this symmetry:

$$p_{l,t} + p_{l',t} = 1 \quad \text{for all } t \in [1, y] \text{ and all such pairs } (l, l').$$

Figure 1 shows the encoding for an example row.

2.3 Encoding Regression Trees

Our regression tree formulation is identical to the classification tree formulation. We only replace the error computation in Eq. 5 with the following constraints, for all $l \in [1, v], r \in [1, n]$:

$$\sum_{1 \leq h \leq k} M_t \mathtt{dlr'}(h, l, r) + p_l - t(r) \leq e_r + M_t k$$

$$\sum_{1 \leq h \leq k} M_t' \mathtt{dlr'}(h, l, r) + t(r) - p_l \leq e_r + M_t' k \qquad (6)$$

where $p_l \in [LT, UT]$ is the prediction value of leaf l, $LT = \min\{t(r)|r \in [1, n]\}$, $UT = \max\{t(r)|r \in [1, n]\}$, $M_t = \max\{UT - t(r)|r \in [1, n]\}$ and $M_t' = \max\{t(r) - LT|r \in [1, n]\}$. This computes the absolute error for each row r from the prediction value p_l of the leaf it ends up in, depending on the path variables from $\mathtt{dlr'}$, using $O(2vn)$ constraints.

3   Experiments

We conducted experiments on several benchmark datasets for both classification and regression tasks from the UCI machine learning repository [12]. We compared the performance of the following three methods: (1) the classification and regression method from sciki-learn (i.e., optimized version of CART), (2) the proposed decision tree as linear programs (DTIP) method that is solved by CPLEX, and (3) DTIP solved by CPLEX with starting trees learned from CART (DTIPs). The time limit for solving each problem is set to 30 min. We learn decision trees of various depths, ranging from 1 to 5.

Classification. We tested our method on three real datasets. The "Iris" data have 4 attributes and 150 data points with 3 classes. The "Diabetes" data are from the Pima Indian Diabetes database, which have 8 attributes and 768 data points to two classes. The "Bank" data are from direct marketing campaigns of a Portuguese banking institution [13]. The "Bank" dataset is considerable larger than Iris and Diabetes, with 51 attributes and 4521 instances. As the purpose of this paper is to demonstrate the performance of DTIP, we use all data points for constructing the trees and use the classification accuracy of the method on all data points as the performance measurement. Table 3 reports the results.

With Iris data, CART is able to find the optimal trees for depths 1, 2 and 5. Our proposed methods (DTIP and DTIPs) can always find the optimal trees with depths 1 to 5, no matter whether it starts with initial trees returned from CART or not. For Diabetes, DTIP and DTIPs can construct the optimal trees with depth 1. When the trees are larger, the encoded MIP models become more and more difficult to solve. This can be seen for depths 4 and 5, the performances of DTIP are worse than CART when CPLEX tries to solve from scratch within the limited running time (i.e., 30 min). This difficulty of CPLEX in solving large instances becomes very obvious when DTIP builds the classification trees of depth 4 for the Bank data. The accuracy drops below 0.2, it is essentially still preprocessing the data. However, when CPLEX starts with initial solutions, DTIPs always improves the initial trees that are found by CART, resulting higher or equal accuracies on all datasets and all different sized trees.

Table 3. Classification accuracy (top) and absolute error (bottom) of three regression methods with trees of depths 1–5. The values with * indicate the optimal solutions.

		d = 1	d = 2	d = 3	d = 4	d = 5
Iris	CART	0.6667*	0.96*	0.9733	0.9933	1*
	DTIP	0.6667*	0.96*	0.9933*	1*	1*
	DTIPs	0.6667*	0.96*	0.9933*	1*	1*
Diabetes	CART	0.7357	0.7721	0.776	0.7930	0.8372
	DTIP	0.75*	0.7773	0.7969	0.7852	0.7852
	DTIPs	0.75*	0.7773	0.7943	0.8255	0.8503
Bank	CART	0.8848	0.9009*	0.9044	0.9124	0.9206
	DTIP	0.8929*	0.8956	0.8213	0.1152	0.1152
	DTIPs	0.8929*	0.9009	0.9056	0.9129	0.9208
RedWine	CART	780*	745	718	687	661
	DTIP	780*	747	749	745	996
	DTIPs	780*	745	715	686	661
Boston	CART	2518.1*	1755.6	1419.6	1230.7	1012
	DTIP	2518.1*	1783.4	1410.2	1250.6	1200.6
	DTIPs	2518.1*	1755.6	1413.6	1205	954

Regression. We used two real datasets. The first one "RedWine" is from the wine quality dataset [7]. It contains 1599 data points, each with 11 input variables, and 1 output variable indicating the wine quality with scores between 0 and 10. The "Boston" data have 13 input attributes and 1 output attribute containing median value of owned houses in suburbs of Boston. There are 506 instances. The bottom of Table 3 shows the performances, measured with absolute error. The conclusions of this set of experiments are similar to those from the classification trees. The best performed one is DTIPs, where the regression trees learned from CART are used as initial solutions to DTIP.

Discrimination-aware DTIP. In order to model the discrimination level of a learned tree, we include a simple constraint that computes the difference in positive class probability for different sets of rows (by summing and comparing errors e_r). This difference is added to the objective function with a large multiplier (the data size), in this way the solver will try find the most accurate tree with zero discrimination. For this experiment, we assume that `married` is a sensitive attribute in the bank data set. Since bank is too large to solve efficiently using DTIP, we only use the top 1000 rows. After running DTIP for 15 min from a Scikit-learn starting solution with accuracy 0.86 and 0.05 discrimination, we obtain 0 discrimination for a depth 3 tree, with 0.85 accuracy. For comparison, we also ran DTIP without discrimination constraints, which gives an accuracy of 0.81 after 15 min. This result demonstrates the flexibility of DTIP: adding a single constraint gives solutions satisfying a different objective.

Imbalanced DTIP. For imbalanced data problems, such as the first 1000 rows from the bank set, depending on the problem context it can be important to find solutions with very few false positives or very few false negatives. In order to demonstrate the flexibility of DTIP, we again add a single constraint for counting false positives or negatives (by summing e_r values) and add it to the objective function with a large multiplier. We ran DTIP for 15 min, starting from the Scikit-learn solution, and obtain a depth 3 tree with 0 false positives and 101 false negatives, or one with 672 false positives and 0 false negatives.

4   Conclusion

We give an efficient encoding of decision tree learning in integer programming. Experimental results demonstrate the strengths and limitations of our approach. Decision trees of depth up to 5 can be learned from data sets of size up to 1000. Larger data sets create to many constraints to be solved effectively using a MIP solver. We show how to use our approach to improve existing solutions provided by a standard greedy approach. Moreover, we demonstrate the flexibility of our approach by modelling different objective functions. In the future, we will investigate other objectives, integration with existing MIP models, and speeding up the search by fixing variables and using lazy constraints.

Acknowledgments. This work is partially funded by Technologiestichting STW VENI project 13136 (MANTA).

References

1. Bennett, K.P., Blue, J.A.: Optimal decision trees. Technical report., R.P.I. Math Report No. 214, Rensselaer Polytechnic Institute (1996)
2. Bertsimas, D., Shioda, R.: Classification and regression via integer optimization. Oper. Res. **55**(2), 252–271 (2007)
3. Breiman, L., Friedman, J., Olshen, R., Stone, C.: Classification and Regression Trees. Wadsworth International Group, Belmont (1984)
4. Bessiere, C., Hebrard, E., O'Sullivan, B.: Minimising decision tree size as combinatorial optimisation. In: Gent, I.P. (ed.) CP 2009. LNCS, vol. 5732, pp. 173–187. Springer, Heidelberg (2009). doi:10.1007/978-3-642-04244-7_16
5. Bruynooghe, M., Blockeel, H., Bogaerts, B., De Cat, B., De Pooter, S., Jansen, J., Labarre, A., Ramon, J., Denecker, M., Verwer, S.: Predicate logic as a modeling language: modeling and solving some machine learning and data mining problems with idp3. Theor. Pract. Logic Program. **15**(06), 783–817 (2015)
6. Carrizosa, E., Morales, D.R.: Supervised classification and mathematical optimization. Comput. Oper. Res. **40**(1), 150–165 (2013)
7. Cortez, P., Cerdeira, A., Almeida, F., Matos, T., Reis, J.: Modeling wine preferences by data mining from physicochemical properties. Decis. Support Syst. **47**(4), 547–553 (2009)
8. De Raedt, L., Guns, T., Nijssen, S.: Constraint programming for data mining and machine learning. In: AAAI, pp. 1671–1675 (2010)

9. Flach, P.: Machine Learning: The Art and Science of Algorithms that Make Sense of Data. Cambridge University Press, Cambridge (2012)
10. Heule, M.J., Verwer, S.: Software model synthesis using satisfiability solvers. Empirical Softw. Eng. **18**(4), 825–856 (2013)
11. Hyafil, L., Rivest, R.L.: Constructing optimal binary decision trees is np-complete. Inf. Proc. Lett. **5**(1), 15–17 (1976)
12. Lichman, M.: UCI machine learning repository. http://archive.ics.uci.edu/ml
13. Moro, S., Cortez, P., Rita, P.: A data-driven approach to predict the success of bank telemarketing. Decis. Support Syst. **62**, 22–31 (2014)
14. Norouzi, M., Collins, M.D., Johnson, M., Fleet, D.J., Kohli, P.: Efficient non-greedy optimization of decision trees. In: NIPS, pp. 1729–1737. MIT Press (2015)
15. Quinlan, J.R.: Induction of decision trees. Mach. Learn. **1**(1), 81–106 (1986)
16. Verwer, S., Zhang, Y., Ye, Q.C.: Auction optimization using regression trees and linear models as integer programs. Artif. Intell. **244**, 368–395 (2017)

Relaxation Methods for Constrained Matrix Factorization Problems: Solving the Phase Mapping Problem in Materials Discovery

Junwen Bai[1], Johan Bjorck[1(✉)], Yexiang Xue[1], Santosh K. Suram[2],
John Gregoire[2], and Carla Gomes[1]

[1] Department of Computer Science, Cornell University, Ithaca, NY 14850, USA
ujb225@cornell.edu
[2] Joint Center for Artificial Photosynthesis, California Institute of Technology,
Pasadena, CA 91125, USA

Abstract. Matrix factorization is a robust and widely adopted technique in data science, in which a given matrix is decomposed as the product of low rank matrices. We study a challenging constrained matrix factorization problem in materials discovery, the so-called phase mapping problem. We introduce a novel "lazy" Iterative Agile Factor Decomposition (IAFD) approach that relaxes and postpones non-convex constraint sets (the lazy constraints), iteratively enforcing them when violations are detected. IAFD interleaves multiplicative gradient-based updates with efficient modular algorithms that detect and repair constraint violations, while still ensuring fast run times. Experimental results show that IAFD is several orders of magnitude faster and its solutions are also in general considerably better than previous approaches. IAFD solves a key problem in materials discovery while also paving the way towards tackling constrained matrix factorization problems in general, with broader implications for data science.

Keywords: Constrained matrix factorization · Relaxation methods · Multiplicative updates · Phase-mapping

1 Introduction

Matrix factorization has become a ubiquitous technique in data analysis, with applications in a variety of domains such as computer vision [10], topic modeling [6], audio signal processing [11], and crystallography [12]. Often the phenomena considered is naturally non-negative. In non-negative matrix-factorization, the goal is to explain a non-negative signal as the product of (typically) two non-negative low rank matrices. Nonnegative matrix factorization is known to be NP-Hard [13], so a general algorithm for matrix factorization most likely scales exponentially in the worst case.

We consider a challenging and central problem in materials discovery, so-called phase-mapping, an inverse problem whose goal is to infer the materials'

© Springer International Publishing AG 2017
D. Salvagnin and M. Lombardi (Eds.): CPAIOR 2017, LNCS 10335, pp. 104–112, 2017.
DOI: 10.1007/978-3-319-59776-8_9

crystal structure based on X-ray sample data, see Fig. 1(Left). Phase-mapping was shown to be NP-Hard [5]. Existing approaches to phase mapping, discussed in the next section, do not satisfy all the problem constraints. Furthermore, approaches that explicitly try to incorporate the main problem constraints have prohibitive run times on typical real-world data, hours or days, while still not producing solutions that are completely physically meaningful.

We propose a novel **Interleaved Agile Factor Decomposition (IAFD)** approach that "lazily" relaxes and postpones non-convex constraint sets (the lazy constraints), iteratively enforcing them when violations are detected, see Fig. 1(Right). IAFD uncovers the main underlying problem structure revealed by the sample data by rapidly performing a large number of lightweight gradient-based moves. In order to incorporate more intricate combinatorial constraints, the algorithm interleaves the multiplicative gradient-based updates with efficient modular algorithms that detect and repair constraint violations, while still ensuring fast run times, scaling up to large scale real-world problems. Our experimental results show that IAFD is several orders of magnitude faster and its solutions are also in general considerably better than previous approaches. Our work provides an efficient approach to solving a central problem in materials discovery, while paving the way towards tackling constrained matrix factorization problems in general, with broader implications for data science.

2   The Phase Mapping Problem

In search of new materials a common experimental method is to deposit several elements onto a sample wafer at different angles. The sample locations on the wafer receive different concentrations of the elements. As a result, distinct and potentially undiscovered materials are formed at different locations. All materials can be characterized by a one-dimensional X-ray diffraction pattern $F(q)$, which can be measured at high energy accelerators. However, several phases might be present at one sample location and the X-ray diffraction pattern at that location then becomes a linear combination of a set of basis patterns, each corresponding to the pattern of one pure phase. Figure 1(Left) illustrates this phenomenon.

In the mathematical model of the problem, a matrix A representing a set of X-ray measurements on a sample wafer is obtained. Each column of A is a vector representing the pattern $F(q)$ obtained at one sample location, sampled for Q fixed values of q. The phase mapping problem entails factorizing A into the product of W and H such that $A \approx WH$.

The matrix W encodes the characteristic patterns of pure phases while H represents how much of the different phases are present at individual sample location. A complicating factor of the phase-mapping problem is that the laws of thermodynamics induce a set of physical constraints on the possible underlying low rank representation. The solutions must satisfy these constraints, defined below, and must additionally be nonnegative as the physical quantities described by the matrices cannot be negative. Efficient methods of solving this problem accelerates materials science and enables automatic experimentation in search of tomorrow's semiconductor and photovoltaic materials.

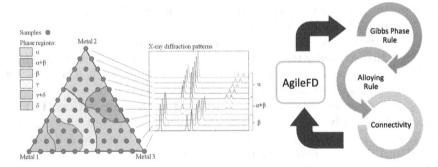

Fig. 1. (Left) The goal of the phase mapping problem is to explain observed X-ray diffraction patterns at multiple sample locations in terms of the underlying phases or crystal structures of the materials. Here the X-ray diffraction patterns of sample locations on the right edge of the triangle are shown in the middle plot. The top four sample locations only have phase α, the bottom three only have phase β, while the middle four sample locations have both α and β. In addition, the X-ray diffraction patterns of both phase α and β are shifting to the right. **(Right)** At a high level, our Interleaved Agile Factor Decomposition (IAFD) algorithm starts with solving a relaxed problem using the multiplicative update rules of AgileFD [14], without enforcing combinatorial constraints. Violations of the Gibbs' phase rule, the alloying rule, and the connectivity constraint in the relaxed solutions are then addressed by efficient modular algorithms, in an interleaving manner. This procedure is iterated, creating a closed loop involving AgileFD and the three modules.

Shifting. A phenomenon that complicates the matrix factorization is "shifting", where the X-ray patterns are changed in the sense $F(q) \rightarrow F(\lambda_k q)$, for some real number λ_k that is fixed for each phase k and column in A. For example, the X-ray patterns in Fig. 1 are shifting to the right. The problem can be circumvented by resampling the signal uniformly on a logarithmic scale, where multiplicative shifts becomes additive. For fixed m and k, the vector $(0, \ldots, 0, W_{1,k}, \ldots, W_{Q-m,k})^T$ formed by shifting the k-th column of W down by m entries (and filling 0 for remaining entries) describes basis pattern of phase k shifted by an amount controlled by m. We can then allow λ_k to attain M different discrete values by letting $m \in 0, 1 \ldots M - 1$. By characterizing the H matrix with three indices, one per phase k, sample point n, and allowed discrete value of λ_k m, we can now express a linear combination of shifted basis patterns as $A_{qn} \approx \sum_{km} W_{q-m,k} H_{kmn}$. Since this specific formulation will be used, the constraints of the phase mapping problem will be given in terms of W_{qk} and H_{kmn}, however other formulations of the rules are possible [3].

Gibbs' Phase Rule. In a setting with three elements deposited, such as in Fig. 1, Gibbs' phase rule [1] states that the number of phases present at each sample location is at most three. Mathematically, it is equivalent to constraining the number of non-zero elements in vector $(\sum_m H_{1mn}, \sum_m H_{2mn}, \ldots)$ for any phase k to be no more than three. Thus, for fixed n we have $\| \sum_m H_{kmn} \|_0 \leq 3$.

Connectivity. The connectivity rule requires that the sample points where a specific phase is present form a continuous domain on the sample wafer. For example, in Fig. 1, each pattern occupies a continuous region. Mathematically, since we have a discrete set of measurements we describe the constraint via a graph G where sample points are nodes and nearby sample points are connected with an edge. This graph is obtained through Delauney triangulation [7] of the sample points. A continuous domain then corresponds to a connected component on this graph, and we require that all sample points n with phase k present, i.e. $\sum_m H_{kmn} > 0$, form a connected component on G.

Alloying Rule. The shifting parameter λ_k for phase k may shift continuously across the sample points as a result of so called alloying. The alloying rule states that for points where λ_k is changing, Gibbs' phase rule becomes even stricter and requires $\| \sum_m H_{kmn} \|_0 \le 2$. In this discrete setting we interpret λ_k of a point n as $\sum_m H_{nkm} m / \sum_m H_{nkm}$, which can be thought of as the expectation of m when we normalize H_{kmn} to a probability distribution. Two neighboring sample points n and n' with phase k present, which means $\sum_m H_{kmn} > 0$ and $\sum_m H_{kmn'} > 0$, are considered shifting if

$$\left\| \frac{\sum_m H_{kmn} m}{\sum_m H_{kmn}} - \frac{\sum_m H_{kmn'} m}{\sum_m H_{kmn'}} \right\| > \epsilon, \tag{1}$$

The alloying rule states that if Eq. 1 is satisfied for any phase k and neighbouring sample points n' and n, then we must have $\| \sum_m H_{kmn} \|_0 \le 2$.

2.1   Previous Approaches

Many algorithms have been proposed for solving the phase mapping problem, for example [5,8,9]. Recently an efficient algorithm called AgileFD [14], based on coordinate descent using multiplicative updates, has been proposed. If we let the matrix R represent the product of H and W, i.e. $R_{qn} = \sum_m W_{q-m,k} H_{kmn}$ these updates are

$$H_{kmn} \leftarrow H_{kmn} \frac{\sum_q W_{q-m,k}(A_{qn}/R_{qn})}{\sum_q W_{q-m,k} + \gamma}, \tag{2}$$

$$W_{qk} \leftarrow W_{qk} \frac{\sum_{mn} \frac{A_{q+m,n}}{R_{q+m,n}} H_{kmn} + W_{qk} \sum_{q'nm} H_{kmn} W_{q'k}}{\sum_{nk} H_{kmn} + W_{qk} \sum_{q'nm} \frac{A_{q'+m,n}}{R_{q'+m,n}} H_{kmn} W_{q'k}}. \tag{3}$$

The algorithm relies on manual refinement by domain experts to enforce combinatorial constraints, which makes it problematic to use in a scalable fashion.

Another approach called combiFD, able to express all constraints, has been proposed [2]. It relies on a combinatorial factor decomposition formulation, where iteratively H or W are frozen while the other is updated by solving a MIP. This formulation allows all constraints to be expressed upfront, however solving the complete MIP programs is infeasible in practice.

3   Interleaved Agile Factor Decomposition

Given a non-negative Q-by-N measurement matrix A and the dimensions K and M of the factorization, the phase mapping problem entails explaining A as a generalized product of two low rank non-negative matrices W, H. The entire mathematical formulation becomes:

$$\min \sum_{qn} |A_{qn} - \sum_{mk} W_{q-m,k} H_{kmn}|, \quad s.t. \quad H \in \mathbb{R}_+^{K \times M \times N}, \quad W \in \mathbb{R}_+^{Q \times K},$$

$$H, W \text{ satisfies Gibbs' phase rule, Connectivity, Alloying rule.} \quad (4)$$

Representing the combinatorial rules as integer constraints has previously been tried [2], however the resulting large MIP formulations are not feasible to solve in practice. Instead, we propose a novel iterative framework that interleaves efficient multiplicative updates with compact subroutines able to address specific constraints, called Interleaved Agile Factor Decomposition (IAFD). The algorithm is illustrated, at a high level, in Fig. 1 (Right). The central insight is that our constraints are too expensive to explicitly encode and maintain, however finding and rectifying individual violations can be done efficiently. This motivates a lazy approach that relaxes and postpones non-convex constraint sets (the lazy constraints), iteratively enforcing them only as violations are detected. For each constraint we provide an efficient method to detect violations and repair them through much smaller optimization problems.

The IAFD algorithm starts with solving the relaxed problem, with only the convex non-negativity constraint, using the multiplicative updating rules (2) and (3) of AgileFD [14]. This relaxed solution is then slightly refined by three subroutines which sample and rectify violations of Gibbs' phase rule, the alloying rule, and the connectivity constraint respectively, by solving small scale optimization problems. The refined solution is then relaxed again and improved through the multiplicative updates. This process is repeated in an interleaving manner which creates a closed loop involving AgileFD and the three refining modules. A reason why this interleaving can be expected to not produce much duplicate effort is due to the following observation:

Proposition 1. *The number of non-zero entries in H : $\|\{(n,k)| \sum_m H_{nkm} > 0\}\|$ is nonincreasing under updates (2) of AgileFD.*

This comes from the fact that every component is updated through multiplication with itself in (2), which ensures that zero-components stay zero. Thus, if Gibbs' phase rule is satisfied before the multiplicative updates, it will still be satisfied after. We now describe the subroutines handling the constraints.

Gibbs' Phase Rule Refinement. After obtaining the matrix W and H, we find violations of Gibbs' phase rule by scanning sample points and noting which ones have more than three phases present. One key insight is that the problem of enforcing Gibbs' phase rule decouples between sample points once the matrix W is fixed. In order to represent the constraint that no more than three phases are present, we introduce a binary variable δ_{kn} denoting whether phase k is present at sample location n (i.e., $\sum_m H_{kmn}$ is nonzero). The constraint is now enforced by solving the following mixed integer program with W fixed for each violated sample point, which results in a very light-weight refinement:

$$\min_{\delta, H_{kmn} \forall k, m} \sum_q |A_{qn} - \sum_{mk} W_{q-m,k} H_{kmn}|,$$

$$\text{s.t. } \forall k, m \ H_{kmn} \leq M\delta_{nk}, \quad \sum_k \delta_{nk} \leq 3. \quad (5)$$

Here, $H_{kmn} \leq M\delta_{nk}$ is a big-M constraint, which enforces that phase k is zero if δ_{nk} is zero. We use $\sum_k \delta_{nk} \leq 3$ to enforce that only three phases are allowed. These compact programs typically contains two orders of magnitude fewer variables then the complete program, and can be quickly solved in parallel.

Alloying Rule Refinement. Violations of the alloying rule can be found by comparing the shift parameter λ_k of some sample point n, here interpreted as $\sum_m H_{kmn} m / \sum_m H_{kmn}$, to that of its neighbors in graph G. This simply amounts to a linear scan through all sample points. It is again possible to decouple the constraint by taking W and n as fixed, which allows for a compact mixed integer program formulation. We fix the violating sample point n, denote the set of its neighbors as $N(n)$, and then calculate $\lambda_{kn'} = \sum_m m H_{kmn'} / \sum_m H_{kmn'}$ for all neighbors $n' \in N(n)$ where phase k is present. In the MIP the binary variable δ_{kn} is used to denote whether phase k is present at sample point n, another binary variable τ_n is then introduced to denote whether the sample point undergoes shift. By using a large M-constraint as in Gibbs' phase rule module we can encode that unless the sample point is shifting or doesn't contain the phase k, the λ_k has to be close to that of it's neighbors as follows:

$$\min_{\tau, \delta, H_{kmn} \forall k, m} \sum_q |A_{qn} - \sum_{mk} W_{q-m,k} H_{kmn}|,$$

$$\text{s.t. } |\sum_m H_{kmn} m - \lambda_{kn'} \sum_m H_{kmn}| \leq \epsilon \sum_m H_{kmn} + M\tau_n + M(1 - \delta_{kn}),$$

$$\forall k, m, n' \in N(n), \quad H_{kmn} \leq M\delta_{nk}, \quad \sum_k \delta_{nk} + \tau_n \leq 3. \quad (6)$$

Connectivity Refinement. While explicitly encoding the constraint is computationally expensive, finding violations can be done in a lightweight manner. For each phase k we find all continuous regions containing phase k by simply finding the connected components of our graph G where phase k is present. To rectify the constraint, every connected component C is then weighted by the total amount of present phase, which amounts to calculating the quantity $\sum_{n \in C, m} H_{kmn}$. This weight corresponds to the amount of present signal. We then zero out components in H corresponding to phase k and sample points in the least weighted connected components. This procedure ensures that all the phases correspond to a single contiguous regions, without deteriorating much (if at all) the objective function in general.

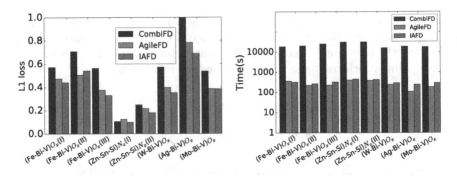

Fig. 2. (Left) Normalized L1 Loss of the difference between ground-truth and recon-structed X-ray patterns for the algorithms on 8 real world systems. IAFD performs best, with combiFD lagging behind the two other methods. (Right) Runtime for CombiFD, AgileFD, IAFD to solve 8 real systems, note the logarithmic time scale. We can clearly see that the heavy duty MIP formulation of combiFD results running times of hours, while the two lightweight methods runs in a matter of minutes.

4   Experimental Results

IAFD is evaluated on several real world instances of the phase mapping problem, available at [4]. We randomly initialize the matrices, and as the interleaving with the connectivity-subroutine and the alloying-subroutine assumes structured data, the whole algorithm starts with several rounds of AgileFD interleaving with Gibbs' rule followed by the other two subroutines. The diffraction patterns are probed at around 200 locations of the respective wafers with approximately 1700 values of q sampled, we set $K = 6$ and $M = 8$ which gives us around two million variables per problem. More rounds of interleaving lead to better results but of course it takes more time. We chose to do three rounds of AgileFD interleaving with Gibbs' rule followed by enforcing the other two constraints to balance these tradeoffs. Our method is compared against CombiFD [2], with a mipgap of 0.1 and 15 iterations. Due to its poor scaling properties only the Gibbs' phase rule is enforced for CombiFD. We also compare IAFD against AgileFD [14], with termination constant set to 10^{-5}.

The most important metric when comparing different methods is the solution quality, measured by L1 loss. Results shown in Fig. 2 (Left). It is evident that CombiFD in [2] has subpar performance, while IAFD wins by a slight margin over AgileFD. This suggests that enforcing the constraints actually improves the reconstruction error. The area where we expect IAFD to perform the best is in terms of enforcing the physical constraints, which is illustrated in Table 1. Here IAFD consistently performs the best with zero violations, which results in physically meaningful solutions to the phase mapping problem.

The smaller subroutines are evidently able to handle all constraints and additionally provide a low loss, which might lead one to suspect that IAFD has long run times. That is not the case. The run times can be viewed in Fig. 2 (Right). While AgileFD is slightly faster than IAFD, the difference is very small.

CombiFD, which explicitly enforces the constraints [2], has prohibitive long run times in practice, which suggests that a complete MIP encoding is both inefficient and unnecessary. These results show that IAFD can enforce all physical rules, without sacrificing much in either reconstruction error or running time.

Table 1. To the left we see the fraction of sample points violating the alloying rule for different algorithms, where IAFD consistently has no violations. The right side gives the average number of connected components per phase, and here only IAFD always contain a single continuous region as required by the connectivity constraint.

System	Alloying constraint			Connectivity constraint		
	CombiFD	AgileFD	IAFD	CombiFD	AgileFD	IAFD
$(Fe-Bi-V)O_x(I)$	0.57	0.15	**0.00**	**1.00**	1.65	**1.00**
$(Fe-Bi-V)O_x(II)$	0.55	0.30	**0.00**	2.40	1.65	**1.00**
$(Fe-Bi-V)O_x(III)$	0.18	0.03	**0.00**	2.50	2.18	**1.00**
$(Zn-Sn-Si)N_x(I)$	0.06	0.01	**0.00**	**1.00**	2.38	**1.00**
$(Zn-Sn-Si)N_x(II)$	0.05	0.02	**0.00**	2.00	1.38	**1.00**
$(W-Bi-V)O_x$	0.54	0.08	**0.00**	1.67	2.31	**1.00**
$(Ag-Bi-V)O_x$	0.84	0.16	**0.00**	3.60	1.96	**1.00**
$(Mo-Bi-V)O_x$	0.46	0.08	**0.00**	1.60	1.72	**1.00**

5   Conclusions

We propose a novel Interleaved Agile Factor Decomposition (IAFD) framework for solving the phase mapping problem, a challenging constrained matrix factorization problem in materials discovery. IAFD is a lightweight iterative approach that lazily enforces non-convex constraints. The algorithm is evaluated on several real world instances and outperforms previous solvers both in terms of run time and solution quality. IAFD's approach, based on efficient multiplicative updates from unconstrained nonnegative matrix factorization and lazily enforced constraints, performs much better compared to approaches that enforce all constraints upfront, using a large mathematical program. This approach opens up a new angle for efficiently solving more general constrained factorization problems. We anticipate deploying IAFD at the Stanford Synchrotron Radiation Lightsource in the near future to the benefit of the materials science community.

Acknowledgements. We thank Ronan Le Bras and Rich Bernstein for fruitful discussion. This material is supported by NSF awards CCF-1522054, CNS-0832782, CNS-1059284, IIS-1344201 and W911-NF-14-1-0498. Experiments were supported through the Office of Science of the U.S. Department of Energy under Award No. DE-SC0004993. Use of the Stanford Synchrotron Radiation Lightsource, SLAC National Accelerator Laboratory, is supported by the U.S. Department of Energy, Office of Science, Office of Basic Energy Sciences under Contract No. DE-AC02-76SF00515.

References

1. Atkins, P., De Paula, J.: Atkins' Physical Chemistry, p. 77. Oxford University Press, New York (2006)
2. Ermon, S., Bras, R.L., Suram, S.K., Gregoire, J.M., Gomes, C., Selman, B., Van Dover, R.B.: Pattern decomposition with complex combinatorial constraints: application to materials discovery. arXiv preprint arXiv:1411.7441 (2014)
3. Ermon, S., Bras, R., Gomes, C.P., Selman, B., Dover, R.B.: SMT-aided combinatorial materials discovery. In: Cimatti, A., Sebastiani, R. (eds.) SAT 2012. LNCS, vol. 7317, pp. 172–185. Springer, Heidelberg (2012). doi:10.1007/978-3-642-31612-8_14
4. Le Bras, R., Bernstein, R., Suram, S.K., Gregoire, J.M., Selman, B., Gomes, C.P., van Dover, R.B.: A computational challenge problem in materials discovery: synthetic problem generator and real-world datasets (2014)
5. LeBras, R., Damoulas, T., Gregoire, J.M., Sabharwal, A., Gomes, C.P., Dover, R.B.: Constraint reasoning and kernel clustering for pattern decomposition with scaling. In: Lee, J. (ed.) CP 2011. LNCS, vol. 6876, pp. 508–522. Springer, Heidelberg (2011). doi:10.1007/978-3-642-23786-7_39
6. Lee, D.D., Seung, H.S.: Learning the parts of objects by non-negative matrix factorization. Nature 401(6755), 788–791 (1999)
7. Lee, D.T., Schachter, B.J.: Two algorithms for constructing a delaunay triangulation. Int. J. Comput. Inf. Sci. 9(3), 219–242 (1980)
8. Long, C., Bunker, D., Li, X., Karen, V., Takeuchi, I.: Rapid identification of structural phases in combinatorial thin-film libraries using X-ray diffraction and non-negative matrix factorization. Rev. Sci. Instrum. 80(10), 103902 (2009)
9. Long, C., Hattrick-Simpers, J., Murakami, M., Srivastava, R., Takeuchi, I., Karen, V.L., Li, X.: Rapid structural mapping of ternary metallic alloy systems using the combinatorial approach and cluster analysis. Rev. Sci. Instrum. 78(7), 072217 (2007)
10. Shashua, A., Hazan, T.: Non-negative tensor factorization with applications to statistics and computer vision. In: Proceedings of the 22nd International Conference on Machine Learning, pp. 792–799. ACM (2005)
11. Smaragdis, P.: Non-negative matrix factor deconvolution; extraction of multiple sound sources from monophonic inputs. In: Puntonet, C.G., Prieto, A. (eds.) ICA 2004. LNCS, vol. 3195, pp. 494–499. Springer, Heidelberg (2004). doi:10.1007/978-3-540-30110-3_63
12. Suram, S.K., Xue, Y., Bai, J., Bras, R.L., Rappazzo, B., Bernstein, R., Bjorck, J., Zhou, L., van Dover, R.B., Gomes, C.P., et al.: Automated phase mapping with agilefd and its application to light absorber discovery in the V-Mn-Nb oxide system. arXiv preprint arXiv:1610.02005 (2016)
13. Vavasis, S.A.: On the complexity of nonnegative matrix factorization. SIAM J. Optim. 20(3), 1364–1377 (2009)
14. Xue, Y., Bai, J., Le Bras, R., Rappazzo, B., Bernstein, R., Bjorck, J., Longpre, L., Suram, S., van Dover, B., Gregoire, J., Gomes, C.: Phase mapper: an AI platform to accelerate high throughput materials discovery. In: Twenty-Ninth International Conference on Innovative Applications of Artificial Intelligence (2016)

Minimizing Landscape Resistance for Habitat Conservation

Diego de Uña[1](✉), Graeme Gange[1], Peter Schachte[1], and Peter J. Stuckey[1,2]iD

[1] Department of Computing and Information Systems, The University of Melbourne,
Melbourne, Australia
{gkgange,schachte,pstuckey}@unimelb.edu.au,
d.deunagomez@student.unimelb.edu.au
[2] Data 61, CSIRO, Melbourne, Australia

Abstract. Modeling ecological connectivity is an area of increasing interest amongst biologists and conservation agencies. In the past few years, different modeling approaches have been used by experts in the field to understand the state of wildlife distribution. One of these approaches is based on modeling land as a resistive network. The analysis of electric current in such networks allows biologists to understand how random walkers (animals) move across the landscape. In this paper we present a MIP model and a Local Search approach to tackle the problem of minimizing the effective resistance in an electrical network. This is then mapped onto landscapes in order to decide which areas need restoration to facilitate the movement of wildlife.

1 Introduction

In the past decades, the natural habitat of different species across the globe have become disrupted and fragmented. This is considered to be a major threat to the conservation of biodiversity by biologists and conservation experts [24]. As pointed out by Rosenberg *et al.* [24] and Pimm *et al.* [22] among other experts, the isolation of groups of animals can easily lead to extinction. For this reason, restoration needs to be undertaken in order to maintain a suitable environment where wildlife can move, feed and breed. Landscape restoration is part of Computational Sustainability, which has been increasingly attractive for researchers in our community as it presents interesting computational problems that happen to be NP-hard [10,14,17,26,27].

In ecology, patches of landscape are characterized by their *resistance*. This corresponds to how hard it is for wildlife to traverse a land patch. For instance, a low resistance value might be caused by soil of good quality that allows plants to spread more easily, or an open field might have high resistance for small rodents (that may be targets for birds of prey). Of course, this measure depends on the species being studied: some animals may be able to cross rivers more easily than others. Lower resistance means the land patch is more suitable for the species studied. Therefore, a suitable environment for wildlife would be one where their

D. Salvagnin and M. Lombardi (Eds.): CPAIOR 2017, LNCS 10335, pp. 113–130, 2017.
DOI: 10.1007/978-3-319-59776-8_10

core habitats are connected by low resistance patches, so that animals can freely travel between core habitats to breed and feed.

To improve landscape connectivity, ecologists have used the idea of *corridors* [7,15]. A corridor connects core habitats of animals using uninterrupted paths through which the animals can move. A common measure of the quality of a corridor is know as the Least-Cost Corridor (LCC) [1,6]. LCCs are chosen to minimize the total sum of the resistance in the corridor. There has been extensive work in the Constraint Programming, AI and OR communities in helping identify the best corridors to be built (e.g. [10,17,27]).

Other work has addressed the problem of habitat conservation without enforcing connectivity. For instance, Crossman *et al.* [9] wrote a MIP model for habitat conservation. In their case, the intention was to minimize the number of sites to be restored while keeping a desired area for the animals and maintaining safe distances to roads and other dangers.

Although LCC is a valid model that is broadly used, it has been criticized for over simplifying the actual movement of species [20]. McRae [19,21] proposed the use of electric circuit theory to measure the total connectivity of a landscape. This model is called Isolation By Resistance (IBR). In particular, it was shown [20] how the IBR model better matches the empirical evidence of genetic distance amongst a distributed species measured with two standard statistical models (fixation index and a step-wise mutation model). The IBR model has since been used to study the effects of habitat loss in birds [3,4], for instance.

In the IBR model, the land patches are modeled as nodes in an electric circuit. The transition between contiguous patches is modeled by a branch of the electric circuit carrying a resistor. The resistance of the resistor in Ohms gives the resistance of moving between adjacent patches. The circuit is then connected to a 1 A current source in a core habitat, while another core habitat is connected to the ground (0 V). The *effective resistance* between two habitats in the electric circuit is the measure used for connectivity in the IBR model. It physically corresponds to the real resistance between two points of the circuit. An example of such circuit with 16 land patches (i.e. 16 nodes) can be seen on the left of Fig. 2. Experts use the tool Circuitscape [25] for this model. Its task is to compute the currents, voltages and effective resistance that are then viewable by experts in geographic visualization software. Nevertheless, Circuitscape does not make conservation decisions, it only builds a linear system and solves it for the experts.

The model is justified by the fact that the commute time between two nodes s and t is given by $2mR_{st}$ where m is the number of edges in the graph, and R_{st} is the effective resistance between s and t in the underlying electrical network (Theorem 4.1 in [18]). The goal is therefore to lower the effective resistance between core habitats, as this directly translates into decreasing the commute time between habitats. Previous work by Doyle and Snell [12] also proved this property, and gave an interpretation of current i_{xy} through a branch (x, y) of the underlying electrical network. The current gives the net number of times a random walker would walk from x to y using that branch. This is exactly what is used by biologists to detect areas where wildlife concentrates most in the IBR model [21]. As an example, consider Fig. 1. On the right side the landscape is plotted showing the conductance values in Ω^{-1} (inverse of resistance). On the left, a heatmap of the current at each patch of land. We can observe that, when moving out of their habitats, animals tend to walk using high conductance areas: indeed the $[0.2, 1]$ areas of the heatmap tend to coincide with high conductance areas on the conductance map.

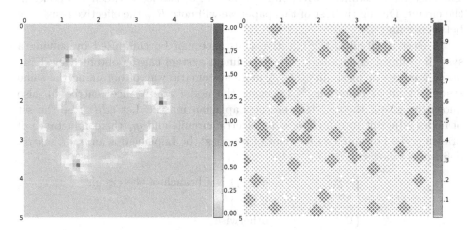

Fig. 1. Example of landscape. Left: The heatmap shows the current (in Amperes) in the electric circuit. The three darkest/red patches correspond to core habitats. This model predicts that animals will walk on the high current areas ($\gtrsim 0.2$ A, yellow/orange) often and less often in low current ($\lesssim 0.2$ A, green) areas. Right: Map of conductance (in Ω^{-1}). Darker/green areas are land where the resistance is low and easy for animals to move across. (Color figure online)

For habitat and ecologic planning using the Isolation By Resistance model, we are interested in minimizing the effective resistance between habitats. To do so, biologists need to decide where improving the habitat is more beneficial. An improvement could mean planning for reforestation, building highway bridges for animals or improving the quality of the soil, among other actions. Our goal here will be to choose the right spots for these investments, subject to a budget. Note how the technical term *effective resistance* is often substituted by *resistance distance* in the biology literature. Also, in a non-reactive circuit with no alternating current, effective resistance and equivalent resistance are equal. For consistency we will always refer to effective resistance.

A similar problem was addressed by Ghosh *et al.* [13], but their problem was continuous. The continuous variant does not apply in our habitat conservation planning, as it is not possible to invest an infinitesimal fraction of budget in one location. In the real world, reforestation areas are typically well defined discrete interventions.

Section 2 is a brief summary on some basic electric circuit theory that we use later on. In Sect. 3 we discuss our Mixed-Integer Programming (MIP) model in detail. Section 4 discusses the problems of our model and presents our Local Search (LS) approach. Lastly, Sect. 5 presents our experimental results.

2   Preliminaries

The effective resistance corresponds to the real resistance between two nodes in the circuit. During the rest of this paper we will note R_{xy} the effective resistance between x and y.

In electric circuits, the effective resistance can be computed by solving a system of linear equations [2,11]. Such linear system can be obtained by doing nodal analysis or mesh analysis. A more systematic way of obtaining the same system of linear equations is by using the nodal admittance matrix [5], also known by graph theoreticians as the Laplacian matrix. Let $adj(n)$ be the set of nodes adjacent to a node n in an electric circuit, and g_{ij} the conductance of branch (i,j). For an electric circuit of n nodes, the Laplacian is an $n \times n$ matrix defined as follows:

$$L_{i,j} = \begin{cases} -g_{ij} & \text{if } (i,j) \text{ is a branch of the circuit} \\ \sum_{k \in adj(i)} g_{ik} & \text{if } i = j \\ 0 & \text{otherwise} \end{cases}$$

For any two nodes s and t in an electric circuit of Laplacian L, we can calculate the effective resistance R_{st} between those nodes by simply solving the linear system given by $L_t v = f$ where L_t is L without the tth row and column, and f is a vector with all its elements equal to 0 except for the sth, which is equal to 1. The effective resistance we are after will be found in v_s.

This is strictly equivalent to nodal analysis. In classic nodal analysis, we would connect a source of 1 A to s and connect t to the ground. Then we obtain an equation for each node of the form $\sum_{y \in adj(x)} i_{xy} = 0$ (flow conservation of the current), except for the node s where the flow of current is 1 since it's connected to the source, and the node t where the flow is -1 for being connected to the ground. Ohm's law gives us that, for any branch, $i_{xy} = g_{xy}(v_x - v_y)$ where v_x and v_y are the voltages at nodes x and y and g_{xy} is the conductance of the resistor between x and y. Then any node can be selected as *reference* or *datum* node, setting its voltage to 0. We select t as the datum node. Then by substituting with Ohm's law and the value of the datum node in the nodal analysis equations we obtain the same system as with the Laplacian. Solving the system yields the voltage of node s, among others. Since the source of current was 1 A, Ohm's law gives us that $R_{st} = (v_s - v_t)/1 = v_s$. It is therefore only necessary to look at the sth component of the solution vector to obtain the effective resistance. This is exactly the method implemented in Circuitscape [25], software used nowadays for landscape planning by experts.

Note how the definition of effective resistance only allows the computation of one effective resistance at a time: the current source needs to be connected to exactly one node, s, and only one node, t, must be connected to the ground. Therefore, the effective resistance between three or more nodes is undefined. Instead, we will consider the total effective resistance for three nodes a, b and c to be $R_{ab} + R_{ac} + R_{bc}$. These have to be computed by solving three different equation systems. In general, for f nodes, we will need $f(f-1)/2$ linear systems.

3 Problem Formulation

We define the problem of finding the Minimum Effective Resistance in a Circuit with Binary Investments (MERCBI) as follows.

We are given an electric network $G^{\emptyset} = (N, E, g^{\emptyset})$ where N are the nodes, E are the edges (i.e. *branches* in circuit theory) and $g^{\emptyset} : E \mapsto \mathbb{R}^{+}$ is a function giving the value of the original conductances that are placed on each edge. We are also given a function $g^{E} : E \mapsto \mathbb{R}^{+}$ of new improved conductances for each edge. These functions are such that $\forall e \in E, g^{\emptyset}(e) \leq g^{E}(e)$. Lastly, we are given a set of pairs of *focal* nodes (i.e. core habitats) $P = \{\langle s_1, t_1 \rangle, ..., \langle s_{|F|}, t_{|F|} \rangle\}$, a budget B, and a cost function $c : E \mapsto \mathbb{R}^{+}$. The pairs in P are the pairs of core habitats between which we want to improve the resistance. It typically contains all the possible pairs of habitats in the studied landscape, but not necessarily.

We note G^{A} for $A \subseteq E$ the network given by $G^{A} = (N, E, g^{A})$ where g^{A} is a function defined by $g^{\emptyset}$ for all $e \notin A$ and g^{E} for all $e \in A$. Similarly, R_{xy}^{A} will refer to the effective resistance between x and y in G^{A}.

Our goal is to find a set $S \subseteq E$ such that we minimize $R^{S} = \sum_{i=1}^{|F|} R_{s_i t_i}^{S}$ while keeping the sum of the investment costs below B. We say that the edges in S are *investments*. The edges in $E \backslash S$ are *wild* edges. Note that it may be the case that g^{S} forms an open circuit (i.e. there is no way for current to go from some s_i to some t_i due to 0-conductance branches). In such case, $R^{S} = \infty$ by definition, which can happen if there is not enough budget to ensure connectivity.

Formally, the model is translated into a Mixed Integer Program with the following additional variables:

- b_e is a binary decision on whether the edge e of the circuit is part of the selected solution S.
- $v_x^{\langle s,t \rangle}$ is the voltage at node x when s is connected to a 1 A source and t to the ground.
- $p_{(a,b),c}^{\langle s,t \rangle}$ is an intermediate variable for the product $g^{S}((a,b)) * v_c^{\langle s,t \rangle}$.

$$\text{Minimize} \quad \sum_{i=1}^{|F|} v_{s_i}^{\langle s_i, t_i \rangle} \tag{1}$$

$$\text{s.t.} \quad \sum_{e=0}^{e=|E|} b_e c(e) \leq B \tag{2}$$

$$\forall \langle s_i, t_i \rangle \in P, \quad \forall x \in N \backslash \{s_i, t_i\}, \quad \sum_{y \in adj(x)} p_{(x,y),x}^{\langle s_i, t_i \rangle} - \sum_{y \in adj(x) \backslash \{t_i\}} p_{(x,y),y}^{\langle s_i, t_i \rangle} = 0 \tag{3}$$

$$\forall \langle s_i, t_i \rangle \in P, \quad \sum_{y \in adj(s_i)} p_{(s_i,y),s_i}^{\langle s_i,t_i \rangle} - \sum_{y \in adj(s_i) \backslash \{t_i\}} p_{(s_i,y),y}^{\langle s_i,t_i \rangle} = 1 \tag{4}$$

$$\forall \langle s_i, t_i \rangle \in P, \quad \forall e = (x,y) \in E, \forall z \in \{x,y\},$$
$$p_{e,z}^{\langle s_i,t_i \rangle} = b_e g^E(e) v_z^{\langle s_i,t_i \rangle} + (1 - b_e) g^\emptyset(e) v_z^{\langle s_i,t_i \rangle} \tag{5}$$

Equation 1 is our objective: minimize the sum of effective resistances between focal nodes. Equation 2 is our budget constraint. Equation 3 constrains the flow of current to be 0 in all nodes, except the nodes directly connected to the source. Equation 4 indicates that the flow at the source nodes has to be 1. Note how the equations at the sinks with flow -1 have been removed as they are linearly dependant from the others. Equation 5 chooses the value of the p variables based on the values of the Booleans.

In Eqs. 3 and 4, the first sum correspond to the diagonal terms in the Laplacian matrix, whereas the second sum is the non-diagonal terms of the Laplacian matrix.

Furthermore, Eqs. 3, 4 and 5 are repeated for each pair of focal nodes in P. Therefore, the equations obtained by nodal analysis are repeated $C = |P|$ times, one per pair in P. Nonetheless, these C systems are not equivalent, as the source of 1 A is connected at different nodes, and the voltages are therefore different in each system. That is, it is not necessarily the case that $v_x^{\langle s_i,t_i \rangle} = v_x^{\langle s_j,t_j \rangle}$, $\forall i,j,x, i \neq j$. We say that our model contains C circuits.

3.1 Complexity

We now prove that the MERCBI problem is NP-hard by reduction from the Steiner Tree Problem on graphs (STP). The STP, as formulated by Karp [16] in his paper where he proved its NP-completeness, is: Given a graph $G = (N, E)$, a set $R \subseteq N$, weighting function w on the edges and positive integer K, is there a subtree of G weight $\leq K$ containing the set of nodes in R?

We apply the following reduction:

- The electric circuit is the graph G.
- The cost function is w.
- The original resistance of edges is infinite (i.e. $g^\emptyset(e) = 0, \forall e \in E$).
- The resistance upon investment of all edges is 1 (i.e. $g^E(e) = 1, \forall e \in E$).
- The budget is K.
- The set of pairs of focal nodes P is the set of all pairs of distinct nodes in R, which can be built in $\mathcal{O}(|R|^2)$.

Assume we have an algorithm to solve the MERCBI problem that gives a solution S. Clearly, by investing in the selected edges S we will obtain a resistance $R^S \in \mathbb{R}$ iff there is enough budget K, and $R^S = \infty$ otherwise:

1. If the resistance is ∞, then there is no Steiner Tree of cost less than K, since we could not connect focal nodes with $1\,\Omega$ resistors.

2. If the resistance is 0 then we can obtain a graph G^* by restricting G^S to the edges that have been invested. Clearly, G^* is of cost $\leq K$ and a subgraph of G that connects all pairs in P. Because P is the set of all pairs of nodes from R, this means that all nodes in R are connected pairwise in G^*. Although G^* may not be a tree, we can extract a tree T from it by breaking any cycle G^* contains while maintaining its connectivity. The tree T is a Steiner tree of cost at most K.

We have therefore shown that if we have a solver for the MERCBI problem, we have one for the STP. Thus, the MERCBI problem is NP-hard.

4   Solving Approach

4.1   Greedy Algorithm

We first devise a greedy algorithm for the problem based on the following observation: increasing the conductance of an edge with low current has little impact in the overall effective resistance. This intuition is easily justifiable: if the current of an edge corresponds to the net number of times a random walker uses that edge, the lower that number, the less impact improving that area would have on the total commute time. Clearly edges near focal nodes or near low resistance areas tend to concentrate more current and be used by more random walkers. Thus increasing the conductance of those edges will likely have a stronger impact in lowering the effective resistance. Our greedy algorithm is presented in Algorithm 1.

Algorithm 1. Greedy algorithm

1: **procedure** GREEDY($\mathcal{G} = (N, E), F, g^0, g^E, c, B$)
2:    $lp \leftarrow$ BUILDMODEL(G, F, g^0, g^E, c, B)   ▷ Build the model from Sect. 3
3:    $lp.setBinariesFalse(E)$ ▷ Sets all the binaries to false
4:    $lp.solve()$ ▷ Solve the LP, with all binary variables fixed
5:    $i[e] \leftarrow 0, \ \forall e \in E$ ▷ Array of currents
6:    **for all** $(s_i, t_i) \in P$ **do**
7:        **for all** $e = (x, y) \in E$ **do**
8:            $i[e] \leftarrow i[e] + |g^0(e) * (v_x^{\langle s_i, t_i \rangle} - v_y^{\langle s_i, t_i \rangle})|$ ▷ Ohm's law
9:    $se \leftarrow$ REVERSE(SORTBY(E, i))   ▷ Array of edges sorted by decreasing current
10:    $S \leftarrow \emptyset; b \leftarrow 0; j \leftarrow 0$
11:    **while** $b < B \wedge j < |E|$ **do**
12:        **if** $g^0(se[j]) < g^E(se[j]) \wedge b + c(se[j]) \leq B$ **then**
13:            $S \leftarrow S \cup se[j]; \ b \leftarrow b + c(se[j])$ ▷ Select edges with high current.
14:        $j \leftarrow j + 1$
    return S

The algorithm solves the conductance LP once to find the voltages at each node, then calculates the currents in each edge (with no investments being made), and greedily selects the edges with most current to invest in, that fit within the budget.

As we will see in the experiments, this algorithm performs surprisingly well despite not providing optimal solutions. In next sections we will try to obtain better solutions than the ones provided by this algorithm.

4.2   Performance of the Pure MIP Model

The MIP model expressed in Sect. 3 is a direct mapping of the nodal analysis performed in the electric circuit into a MIP model. In our implementation Eq. 5 was split into two *indicator constraints* which are supported by the IBM ILOG CPLEX 12.4. solver as follows:

$$\forall \langle s_i, t_i \rangle \in P, \quad \forall e = (x, y) \in E, \forall z \in \{x, y\},$$

$$b_e \implies p_{e,z}^{\langle s_i, t_i \rangle} = g^E(e) v_z^{\langle s_i, t_i \rangle} \tag{6}$$

$$\neg b_e \implies p_{e,z}^{\langle s_i, t_i \rangle} = g^0(e) v_z^{\langle s_i, t_i \rangle} \tag{7}$$

Finding Bounds for the Variables. To help CPLEX tackle the problem, we need to compute bounds on the variables. We apply basic circuit analysis to find these bounds. To do so, we need an initial assignment for the binary variables: this could be assigning all to false, or to the value of some initial solution that respects the budget constraint (either a random solution, or obtained with Algorithm 1). Without loss of generality, let us assume that all the binaries are set to false, thus g^0 gives the conductance for all edges.

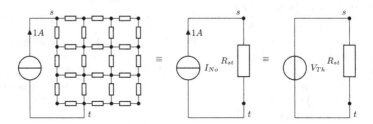

Fig. 2. Conversion from the circuit built for calculating effective resistance to an equivalent circuit with one resistor, and application of Thévenin's theorem.

When we are computing the effective resistance between nodes s and t of a circuit, we connect a current source between s and t as in Fig. 2. This can be converted into an equivalent circuit as seen in the center of the figure. The value of R_{st} is actually the value of the effective resistance when all our binary variables are fixed. To obtain bounds for the voltages at the nodes marked •, we observe that the circuit in the center is a Norton circuit, thus we can apply Thévenin's theorem [8] to it to obtain an equivalent circuit with a voltage source instead of a current source (circuit on the right). Because the current in that Norton's circuit is 1A, the voltage of Thevenin's equivalent source will be $V_{Th} = I_{No}R_{st} = R_{st}$. Thus, the upper bound of v_s is R_{st}. Since t is connected to the ground, $v_t = 0V$. All the voltage at points between s and t are necessarily bounded by the voltages at these points, because voltage can only drop. Therefore, the bounds for all voltage variables are: $\forall \langle s_i, t_i \rangle \in P, \forall x \in N, \quad 0 \leq v_x^{\langle s_i, t_i \rangle} \leq R_{s_i t_i}^{\emptyset}$. From these bounds it is easy to derive bounds for the p variables: $\forall \langle s_i, t_i \rangle \in P, \quad \forall e = (x, y) \in E, \forall z \in \{x, y\}, \quad g^{\emptyset}(e)v_z^{\langle s_i, t_i \rangle} \leq p_{e,z}^{\langle s_i, t_i \rangle} \leq g^E(e)v_z^{\langle s_i, t_i \rangle}$.

Motivation for Local Search Approaches. We attempted to solve the MIP model using IBM ILOG CPLEX 12.4. The major challenge we discovered while running our experiments was that the linear relaxation of the model performed poorly. The gap between the LP relaxation and the best integer solution found in less than 5 hours was usually bigger than 40% even for small 10×10 grid-like networks, as reported by CPLEX. As a small example, consider a 3×3 grid with $\forall e, \ g^{\emptyset}(e) = 0.01$ and $g^E(e) = 1$, a budget of 2 and 2 core habitats. The proven optimal solution gives a resistance of 49.15, whereas the linear relaxation achieves a resistance as low as 0.50. With such a big gap, the linear relaxation cannot help pruning branches. We observed these gaps both when using indicator constraints and without them (using an alternative encoding of the product of continuous and binary variables as linear inequalities).

Since the linear relaxation is so weak, we decided to try a Local Search approach.

4.3 Local Search

Although solving the MIP problem is time consuming, once all binary variables are fixed, the problem becomes a Linear Program, which can be solved efficiently. The idea of local search is to repeatedly destroy part of the solution and build a new one from the remaining solution.

Our Local Search (LS) moves are based on two stages: first choose areas where we invested but we are no longer interested in investing, then choose new areas where we would like to invest. In this section we will discuss different techniques we use to make moves in the LS.

Destroying Investments. We start from a solution s where we have selected a set S of edges to be investments. We call d the *destruction rate* of investments.

To destroy investments, we select a subset S_d of S such that $|S_d| = d$. We will convert these d investments into wild edges, thus freeing part of the budget to be used elsewhere. We implemented 3 ways of selecting those d edges:

1. INVRAND: Choose d invested edges randomly.
2. INVLC: Choose the d invested edges with lowest current in s.
3. INVLCP: Choose d invested edges based on a probability distribution that favors low current edges in s being selected.

The way to compute the current of an edge is by simply using Ohm's law, as in line 8 of Algorithm 1. Because we are solving C linear systems for C circuits at once, the current across an edge will be the sum of the currents of that edge in the C different circuits.

As for the probability distribution, the total current across an edge is proportional to the inverse of its probability. Therefore edges with low current will have a higher tendency to be chosen than edges with high current:

$$\forall e = (x, y) \in E, \quad Pr(e) \propto \left(\sum_{\langle s_i, t_i \rangle \in P} |g^S(e) * (v_x^{\langle s_i, t_i \rangle} - v_y^{\langle s_i, t_i \rangle})| \right)^{-1}.$$

Making New Investments. Let s' be the solution s with investments $S \backslash S_d$, that is all investments of s except those selected for destruction. After destroying investments S_d, we have gotten part $b = \Sigma_{e \in S_d} c(e)$ of our budget back, so we are ready to choose new places to invest. To do so, we have 4 different strategies to choose new edges to invest in:

1. WILRAND: Choose a set W of wild edges randomly.
2. WILBFS: Choose a set W of wild edges by doing a Breath-First Search (BFS) in the graph. The origin of the BFS is chosen according to a probability distribution that favors high current nodes. This ensures that the chosen edges are close to each other.
3. WILHC: Choose a set W of wild edges that have the highest current in s.
4. WILHCP: Choose a set W of wild edges based on a probability distribution that favors high-current wild edges.

For all these strategies, we ensure that $\sum_{e \in W} c(e) = b - \epsilon$ and we select those edges greedily to get the smallest slack ϵ.

The second strategy, WILBFS, requires computing the current at a node. We define it as $c_x = \sum_{(s_i, t_i) \in P} \frac{1}{2} \sum_{y \in adj(x)} |g^S((x, y)) * (v_x^{\langle s_i, t_i \rangle} - v_y^{\langle s_i, t_i \rangle})|$. That is, the sum of the accumulated current of the surrounding edges of the node.

As for the probability distributions used by both WILBFS and WILHCP, we could not use a distribution as simple as for INVLCP. The reason is that, even though the probability of choosing a particular node (or edge) of high current is much higher than the probability of choosing particular node (or edge) of low current, because there are many more nodes and edges with low current, the probability of choosing a low current node or edge would be much higher.

To overcome this, we will only allow the choice of nodes or edges that have a current higher than 10% of the current of the highest node or edge, plus a certain number γ of low current elements (i.e. below that 10%). We choose γ to be equal to the number of nodes or edges above 10% of the maximum current. The probability of choosing an element with a current below 10% of the maximum is now the same as the probability of choosing an element with a current above that threshold.

Initial Solution. Local Search needs to start with an initial solution. Our approach was to take the output of the greedy algorithm as the initial solution for the LS.

Using Simulated Annealing. As it is well known, Local Search can get stuck into a local minimum if we are only accepting improving solutions. Therefore, we implement the simulated annealing approach introduced by Ropke *et al.* for the Pickup and Delivery Problem with Time Windows [23].

Here the idea is that we will always have 3 solutions at hand: s_{best} (for a set of invested edges S_{best}), s, and s_{new} (for S_{new}). The first is the solution found so far with lowest resistance. The second is the current solution maintained in the search (called *accepted* in [23]). The last, is the new solution which we obtained by modifying the solution s using the destruction and investment phases described earlier. If we accept a solution s_{new} only when it is better than the current solution we might get trapped in a local minimum. With simulated annealing, we accept a solution s_{new} with probability $e^{-(s_{new}-s)/T}$, where T is the temperature. The temperature decreases at each iteration by a factor called the *cooling rate*: $T_{i+1} = c_r * T_i$. This allows us to accept solutions that are worse than our currently accepted solution, thus allowing us to get out of local minima.

Initially we will have a higher tendency to accept worsening solutions, and as the temperature drops slowly over the iterations, we will only accept improving solutions. We discuss the value of the initial temperature in Sect. 5.

Even using simulated annealing, it is easy to see how the combined usage of INVLC and WILHC can perform the exact same destruction over and over once it hits a local minimum. To avoid that, when we are using these two destruction techniques together and we don't accept a solution, we switch to WILHCP for one iteration, which will put us in another neighborhood to explore (like a restart).

5   Experiments

All the following experiments use IBM ILOG CPLEX 12.4 on an Intel® Core™ i7-4770 CPU machine @ 3.40 GHz with 15.6 GB of RAM running Linux 3.16.

5.1   Instances

We could not obtain real-world instances. Nonetheless we generated artificial landscapes while trying to keep them realistic. In papers like [3,4] where the IBR model is used on empirical data, high resistance areas have an approximate resistance between 10 and 100 times bigger than areas with low resistance.

Our instances were built in the form of grids with 4 neighbor nodes (except at the border of the landscape where nodes have 2 or 3 neighbors). The values of conductances (i.e. inverse of resistances) are chosen following a probabilistic beta-distribution with parameters $\alpha = 5$ and $\beta = 80$ (Fig. 3).

Using this distribution, most conductances are in the desired range, still allowing the unlikely possibility of having some high conductances scattered on the map. Then, for each instance of size $n \times n$ we select n nodes, and their neighbors, to be *oases*. All the edges coming out of oasis nodes have a conductance of 1, that is the same as if we had invested. This is to account for areas that are not

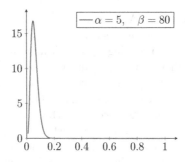

Fig. 3. Plot of beta distribution for $\alpha = 5, \beta = 80$.

considered core habitats but are friendly to the animals: wild trees, fresh water lakes, etc. Figure 1 shows, on the right, an example of such map in a 50×50 grid (the scale being in Ω^{-1}).

Regarding the budget cost, we created instances with homogeneous costs (labelled HOMO) where each edge costs 1. The reason for homogeneous costs is that the main application for our problem is in wild land and specifically on National Parks, where rehabilitation costs are typically fairly uniform across a large area. Typically these kind of projects are used to recover land that has been damaged by bush fires, flooding or landslides or that has been fragmented by urbanization that cannot be torn down or purchased. We also constructed instances where the cost of an investments is a uniform random number between 0.1 and 10 (labeled RAND).

Furthermore, the locations of core habitats are selected from a uniform distribution. For each size of grid and number of habitats, we generated 20 different instances, totalling 500 instances. Values reported are arithmetic means.

5.2   Improvement over the Initial Landscape

To measure the quality of a solution, we look at the ratio between the effective resistance in the solution and the original resistance when no investment is made. Table 1 shows the average of these ratios. All the results reported on that table had a budget equal to twice the side of a grid (e.g., for a grid of 50×50, we have a budget of 100). The ratio of the budget to the number of edges where we could invest is shown in the third column. The destruction rate here is fixed to 10 for instances of size 20×20 and 25×25, and 25 for bigger instances. The number of iterations the LS does for these tests is 200. We do not show the results obtained by random selections for destruction as they perform worse than the ones shown here. The initial temperature is chosen such that a solution 50% worse than the initial solution is accepted with a probability of 10%. The cooling rate is 0.98 for all experiments.

Our first observation is that even with a very low budget, proportionally to the mass of land, we manage to have significant drops of resistance. We infer that our approach is placing the investments in the right places. The greedy algorithm

Table 1. Averages of the ratios of updated resistance over original resistance. Budgets are always twice the side of a grid.

Cost type	Size	Budget ratio	Habitats	GREEDY	INVLC			INVLCP		
					WILBFS	WILHC	WILHCP	WILBFS	WILHC	WILHCP
HOMO	20 × 20	0.11	2	0.30	**0.24**	0.25	0.25	0.29	0.26	0.27
HOMO	25 × 25	0.09	2	0.30	**0.22**	0.23	0.24	0.28	0.24	0.26
HOMO	25 × 25	0.09	3	0.38	**0.29**	0.30	**0.29**	0.37	0.31	0.35
HOMO	30 × 30	0.07	2	0.37	**0.26**	0.27	0.28	0.35	0.29	0.31
HOMO	30 × 30	0.07	3	0.42	**0.33**	**0.33**	**0.33**	0.42	0.35	0.41
HOMO	50 × 50	0.04	2	0.44	**0.34**	0.35	0.36	0.44	0.37	0.40
HOMO	50 × 50	0.04	3	0.48	0.42	**0.40**	0.41	0.48	0.43	0.48
HOMO	50 × 50	0.04	4	0.52	0.46	**0.44**	0.45	0.52	r0.47	0.52
HOMO	100 × 100	0.02	2	0.49	0.39	**0.38**	0.41	0.42	0.39	0.37
HOMO	100 × 100	0.02	3	0.53	**0.42**	0.43	**0.42**	0.53	**0.42**	0.50
HOMO	100 × 100	0.02	4	0.55	0.49	0.47	**0.46**	0.55	0.49	0.55
HOMO	100 × 100	0.02	5	0.58	0.49	**0.46**	0.50	0.58	0.48	0.58
RAND	20 × 20	0.11	2	0.44	**0.34**	0.34	0.38	0.39	0.35	0.40
RAND	25 × 25	0.09	2	0.45	0.33	**0.32**	0.36	0.39	0.33	0.38
RAND	25 × 25	0.09	3	0.53	0.41	0.39	**0.38**	0.47	0.40	0.44
RAND	30 × 30	0.07	2	0.52	0.45	**0.42**	0.44	0.51	0.45	0.48
RAND	30 × 30	0.07	3	0.57	0.53	**0.51**	0.52	0.57	0.52	0.56
RAND	50 × 50	0.04	2	0.56	0.54	**0.51**	0.53	0.56	0.53	0.56
RAND	50 × 50	0.04	3	0.60	0.59	**0.55**	0.58	0.60	0.58	0.60
RAND	50 × 50	0.04	4	0.63	0.62	**0.60**	0.63	0.63	0.62	0.63
RAND	100 × 100	0.02	2	0.61	**0.51**	0.53	0.52	0.66	0.61	0.65
RAND	100 × 100	0.02	3	0.64	0.59	**0.57**	0.60	0.63	0.60	0.64
RAND	100 × 100	0.02	4	0.65	**0.58**	0.60	0.62	0.62	0.61	0.64
RAND	100 × 100	0.02	5	0.68	0.60	**0.59**	0.62	0.63	0.62	0.65

performs very well too. The Local Search is able to reduce the resistance of the greedy algorithm by up to 11%.

We can observe how in general the INVLC approach to destroy bad investments outperforms INVLCP. Regarding the selection of new investment locations it seems there is no clear winner, although WILBFS and WILHC tend to choose investments in places nearby or directly connected, and that seems to give them the advantage.

As an example, Fig. 4 shows the solution found to the instance of Fig. 1. On the left, the heatmap shows the places of investment (marked lines) which redirect most of the animals, as these become more suitable for them. On the right, we see how the best investments tend to be aligned to connect low resistance areas that already existed. Notice how they do not force a complete corridor: some investments may not be used to maintain connectivity and are instead moved to areas where they might be more useful.

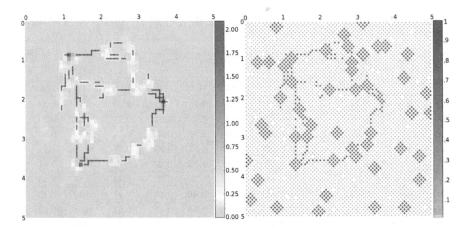

Fig. 4. Example of a solution obtained with INVLC+WILHC for the same instance as shown in Fig. 1

5.3 Optimality Gap

In this section we look at the difference between the solution obtained and the proven optimum for some instances. We could not obtain the optimal value for instances of the size given in the previous subsection. For this reason, we ran this experiment on 10×10 grids. These took, in some cases, up to 8 h to prove optimality, even given the small size. It is clearly impractical to use the pure MIP formulation.

Table 2. Average (across 20 instances) of ratio between LS and proven optimum for 10×10 instances (200 iterations)

	INVLC			INVLCP		
	WILBFS	WILHC	WILHCP	WILBFS	WILHC	WILHCP
Average	1.06	1.07	1.07	1.11	**1.05**	1.07

As it can be seen in Table 2, the LS approach does not deviate much from the optimum. Once again we see no clear domination between the different techniques applied, although INVLCP+WILBFS seems slightly worse than others. The LS even managed to find optimal solution of two instances.

5.4 Number of Iterations

We now study the effect of the number of iterations. We are interested to know how quickly we get improvements in our solutions. To evaluate this, we looked at the 50×50 instances with homogeneous investment cost and a budget of 100 (same instances as in Table 1). We compute the ratio between the best solution found so far at the xth iteration to the non-investment resistance. We average this ratio across the 60 HOMO 50 × 50 instances (2, 3 and 4 habitats) for each iteration. The results are plotted in Fig. 5. Based on the results seen in Table 1 and this plot, we can see that INVLCP+WILBFS and INVLCP+WILHCP are not only the ones that perform the worse, but we also only find better solutions very rarely. The INVLC+WILHC shows the quickest drop, but then slows down. Since the drop happens in the first 20 iterations, the temperature is still high, thus suggesting that it does not get trapped in a local minimum, instead it is most likely moving towards the global optimum (as we also saw on the optimality gap results). All other techniques have a slower slope.

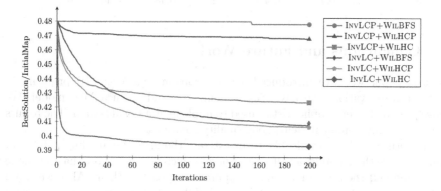

Fig. 5. Ratio between the best solution found and the value of the initial resistance over the number of iterations for 6 destruction strategies (50 × 50 HOMO grids, 2 to 4 habitats).

Regarding the time spent to perform these 200 iterations: no instance needed more than 5 min in our machine. Therefore this number of iterations is also suitable for use by habitat planners during their work.

5.5   Destruction Rate

In this subsection we will look at the effects of the choice of destruction rate. Recall, the destruction rate corresponds to how many edges are converted from investments into wild edges for the next iteration of the LS. We ran these experiments on the 50 × 50 instances for 2, 3 and 4 habitats with a budget of 100 and homogeneous costs. Table 3 reports the average ratio between our solution and the original resistance across the 60 instances depending on the destruction rate. The number of iterations is once again 200.

As we can see in Table 3, the choice of destruction rate only marginally affects the quality of the solution. This shows that our local search approach is robust and no matter what parameter is used (within a reasonable range), the algorithm is likely to give similarly good solutions.

Table 3. Ratio of solution found over initial landscape resistance for 50×50 grids with different destruction rates, over 200 iterations.

Destruction rate	InvLC			InvLCP		
	WilBFS	WilHC	WilHCP	WilBFS	WilHC	WilHCP
5	0.40	**0.38**	0.40	0.47	0.39	0.45
10	0.40	**0.38**	0.39	0.48	0.39	0.46
15	0.40	**0.39**	0.39	0.48	0.40	0.46
20	0.40	**0.39**	0.40	0.48	0.41	0.47
25	0.41	**0.39**	0.41	0.48	0.42	0.47
30	0.42	**0.40**	0.41	0.48	0.43	0.47
35	0.43	**0.40**	0.42	0.48	0.44	0.47

6   Conclusions and Future Work

In this paper we have introduced a new problem, the MERCBI problem, in habitat conservation. We have clearly defined it and given a MIP model. We have provided a series of heuristics used to implement a Local Search. The results show that this approach yields good quality solutions.

In future work we would like to investigate the idea of extending patches of land to more than a node: a golf course, or a farm could cover more than one node at a time and then we might consider purchasing part of them. Also, we could take into account distance for patches with different shapes where purchasing a thin part of it may be more interesting than purchasing the entire land. Also, we could look at having more than one species and take into account their predatory behavior or possibly favor endangered species.

Acknowledgements. We would like to thank Julian Di Stefano and Holly Sitters from the School of Ecosystems and Forest Sciences at the University of Melbourne as well as Nevil Amos from the Department of Environment, Land, Water and Planning of Victoria for meeting with us and introducing us to this problem.

References

1. Adriaensen, F., Chardon, J., De Blust, G., Swinnen, E., Villalba, S., Gulinck, H., Matthysen, E.: The application of 'least-cost' modelling as a functional landscape model. Landscape Urban Plann. **64**(4), 233–247 (2003)
2. Alexander, C.K., Sadiku, M.N.: Electric circuits (2000)
3. Amos, J.N., Bennett, A.F., Mac Nally, R., Newell, G., Pavlova, A., Radford, J.Q., Thomson, J.R., White, M., Sunnucks, P.: Predicting landscape-genetic consequences of habitat loss, fragmentation and mobility for multiple species of woodland birds. PLoS One **7**, 1–12 (2012)
4. Amos, J.N., Harrisson, K.A., Radford, J.Q., White, M., Newell, G., Nally, R.M., Sunnucks, P., Pavlova, A.: Species- and sex-specific connectivity effects of habitat fragmentation in a suite of woodland birds. Ecology **95**(6), 1556–1568 (2014)

5. Balabanian, N., Bickart, T.A., Seshu, S.: Electrical Network Theory. Wiley, New York (1969)

6. Beier, P., Majka, D.R., Spencer, W.D.: Forks in the road: choices in procedures for designing wildland linkages. Conserv. Biol. **22**(4), 836–851 (2008). http://dx.doi.org/10.1111/j.1523-1739.2008.00942.x

7. Beier, P., Noss, R.F.: Do habitat corridors provide connectivity? Conserv. Biol. **12**(6), 1241–1252 (1998). http://dx.doi.org/10.1111/j.1523-1739.1998.98036.x

8. Brittain, J.E.: Thevenin's theorem. IEEE Spectr. **27**(3), 42 (1990)

9. Crossman, N.D., Bryan, B.A.: Systematic landscape restoration using integer programming. Biol. Conserv. **128**(3), 369–383 (2006)

10. Dilkina, B., Gomes, C.P.: Solving connected subgraph problems in wildlife conservation. In: Lodi, A., Milano, M., Toth, P. (eds.) CPAIOR 2010. LNCS, vol. 6140, pp. 102–116. Springer, Heidelberg (2010). doi:10.1007/978-3-642-13520-0_14

11. Dorf, R.C., Svoboda, J.A.: Introduction to Electric Circuits. Wiley, New York (2010)

12. Doyle, P.G., Snell, J.L.: Random Walks and Electric Networks. Mathematical Association of America, Washington, D.C. (1984)

13. Ghosh, A., Boyd, S., Saberi, A.: Minimizing effective resistance of a graph. SIAM Rev. **50**(1), 37–66 (2008)

14. Gomes, C.P.: Computational sustainability: computational methods for a sustainable environment, economy, and society. Bridge **39**(4), 5–13 (2009)

15. Aars, J.: R.A.I.: The effect of habitat corridors on rates of transfer and interbreeding between vole demes. Ecology **80**(5), 1648–1655 (1999). http://www.jstor.org/stable/176553

16. Karp, R.M.: Reducibility among combinatorial problems. In: Miller, R.E., Thatcher, J.W., Bohlinger, J.D. (eds.) Complexity of Computer Computations. The IBM Research Symposia Series, pp. 85–103. Springer, US (1972)

17. LeBras, R., Dilkina, B.N., Xue, Y., Gomes, C.P., McKelvey, K.S., Schwartz, M.K., Montgomery, C.A., et al.: Robust network design for multispecies conservation. In: AAAI (2013)

18. Lovász, L.: Random walks on graphs. Combinatorics, Paul erdos is eighty **2**, 1–46 (1993)

19. McRae, B.H.: Isolation by resistance. Evolution **60**(8), 1551–1561 (2006). http://dx.doi.org/10.1111/j.0014-3820.2006.tb00500.x

20. McRae, B.H., Beier, P.: Circuit theory predicts gene flow in plant and animal populations. Proc. Nat. Acad. Sci. **104**(50), 19885–19890 (2007). http://www.pnas.org/content/104/50/19885.abstract

21. McRae, B.H., Dickson, B.G., Keitt, T.H., Shah, V.B.: Using circuit theory to model connectivity in ecology, evolution, and conservation. Ecology **89**(10), 2712–2724 (2008). http://dx.doi.org/10.1890/07-1861.1

22. Pimm, S.L., Jones, H.L., Diamond, J.: On the risk of extinction. Am. Nat. **132**(6), 757–785 (1988). http://dx.doi.org/10.1086/284889

23. Ropke, S., Pisinger, D.: An adaptive large neighborhood search heuristic for the pickup and delivery problem with time windows. Transp. Sci. **40**(4), 455–472 (2006). http://dx.doi.org/10.1287/trsc.1050.0135

24. Rosenberg, D.K., Noon, B.R., Meslow, E.C.: Biological corridors: form, function, and efficacy. BioScience **47**(10), 677–687 (1997). http://bioscience.oxfordjournals.org/content/47/10/677.short

25. Shah, V., McRae, B.: Circuitscape: a tool for landscape ecology. In: Proceedings of the 7th Python in Science Conference, vol. 7, pp. 62–66 (2008)

26. Urli, T., Brotánková, J., Kilby, P., Van Hentenryck, P.: Intelligent habitat restoration under uncertainty. In: Thirtieth AAAI Conference on Artificial Intelligence (2016)
27. Williams, J.C.: Delineating protected wildlife corridors with multi-objective programming. Environ. Model. Assess. **3**(1–2), 77–86 (1998)

A Hybrid Approach for Stator Winding Design Optimization

Alessandro Zanarini[✉] and Jan Poland

ABB Corporate Research Center, Baden-Dättwil, Switzerland
{alessandro.zanarini,jan.poland}@ch.abb.com

Abstract. Large electrical machines (e.g. hydro generators or motors for mining applications) are typically designed and built specifically for their site of installation. Engineering these one-of-a-kind designs requires a considerable amount of time and effort, with no guarantees that the final blueprint is cost-optimal. One part of the design involves the stator windings, i.e. the static part of an electric motor generating an electro-magnetical field. In order to achieve low harmonic losses during operation, fractional windings are preferred, where the number of slots per phase in the stator is not a multiple of the number of poles. This renders the layout of the stator winding a challenging combinatorial optimization problem. The paper proposes a decomposition approach in which the problem can be reformulated as a rich variant of the Travelling Salesman Problem in which both MIP and CP are employed to tackle the problem.

1 Introduction

Three-phase electrical rotating machines used in different industries are composed by a non-moving part called stator, and a moving part known as rotor. The stator has a cylindrical shape where the curved surface presents a set of parallel bars running along the cylindrical surface in axial direction that conduct alternating electrical current in order to generate a magnetic field capable of moving the rotor. The rotor is also cylindrical in shape and concentric to the stator and it rotates thanks to a number of magnetic poles that are subject to the magnetic field generated by the stator. An example of such machines are mills employed in the mining industry for crushing rocks from which, after some intermediate steps, material is extracted. The conducting bars and their connections (windings) made of copper contribute to the final material cost for constructing a stator. Depending on the stator design and machine dimensions, the number of bars ranges from few dozens to few hundreds. As the number of bars grows, the combinatorial complexity of the stator design (how to connect the different bars) becomes challenging for the engineer, leading to several hours spent in the design with no guarantee of cost-optimality of the final blueprint.

This paper formalizes the problem of designing the stator windings subject to the electro-mechanical constraints with the objective of minimizing the material cost. We show how this problem can be formulated as a rich variant of the well-known Travelling Salesman Problem (TSP). It can be efficiently solved with

© Springer International Publishing AG 2017
D. Salvagnin and M. Lombardi (Eds.): CPAIOR 2017, LNCS 10335, pp. 131–146, 2017.
DOI: 10.1007/978-3-319-59776-8_11

Mixed Integer Programming and Constraint Programming. The solution developed has been successfully employed internally for the design of stator windings leading both to savings in term of engineering labor and material costs.

The paper is organized as follows: Sect. 2 presents the literature background; Sect. 3 formalizes the problem; Sect. 4 proposes a solution; experiments are reported in Sect. 5. Conclusions are drawn in the last section.

2   Background

The three phases will be referred to as U, V, and W. Each stator bar is located in a slot and each bar is assigned to a specific phase. The number of bars per pole and phase is:

$$q = \frac{n_s}{3n_p}, \tag{1}$$

where n_s is the number of slots, n_p is the number of poles, and we restrict to three phase machines to simplify the presentation (note that our methods easily generalize to arbitrarily many phases).

If q is integral, the design is referred to as integral-slot winding. If q is fractional, the design is instead known as fractional-slot winding. Fractional-slot windings are known to reduce the unwanted high-frequency harmonics in the electromagnetical forces and give greater flexibility to the designer in choosing the number of slots to reach the target magnetic flux density [1]. Each slot may host a single bar (single-layer winding), two bars (two-layers winding) or more. Two-layer windings have the advantages of easing the production and assembly and also to reduce unwanted magnetomotive force harmonics [2]. In this paper, we will focus on two-layer fractional-slot windings. In two-layer windings, the two bars located in a slot are often referred to as top and bottom bars.

In Fig. 1, a typical stator winding blueprint is shown. In the example, the stator presents 96 slots (numbered from 0 to 95), each one having two bars. For simplicity, the drawing only shows phase U (bars 0, 1, 2 then 7, 8, 9 then 14, 15 and so on). The solid and dashed radial lines grouped by pair represent respectively the top and bottom bars of each slot; Fig. 1b highlights the top and bottom bars placed in slot 0. The number of poles is 14, therefore $q = 2 + \frac{2}{7}$.

The connections drawn in the inner circle are located on one side of the stator, and the connection drawn in the outer circle are located on the opposite side. The former presents a regular structure and in the following it will be referred to as the *no jumper side*, whereas the latter shows some regular patterns interleaved by some long jumpers (referred to as *jumper side*). Figure 1a shows the jumper and no jumper sides; examples of jumpers can be seen in the top left side or in the bottom right side. Figure 1b highlights a *coil* (in the top right part of the stator): a coil is a pair of top and bottom bars linked via a regular connection in the no jumper side. The highlighted coil is formed by a top bar in slot 77 and a bottom bar in slot 83 (referred in the following to as *[XX]T([YY]B)*, in the specific example 77T(83B)). In a conventional two layer winding, each top bar

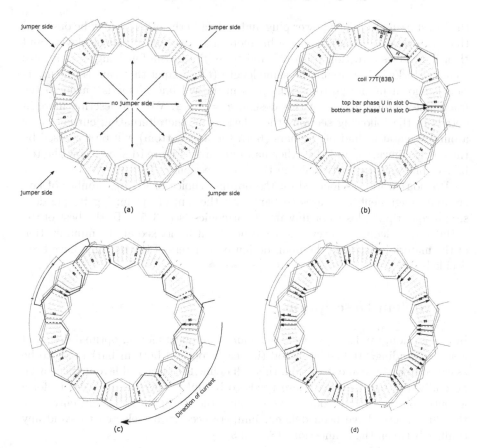

Fig. 1. Stator winding drawing of phase U: (a) no jumper side and jumper side are shown; (b) top and bottom bars, coil and coil pitch are highlighted; (c) direction of current for a sub-part of phase U; (d) current direction for all the bars of phase U

is connected with a bottom bar y_1 slots away via the no jumper side, forming as many coils as there are slots. The number y_1 is referred to as *coil pitch*.

On the jumper side, there is more flexibility. Basically, there are two different types of winding. In *lap winding*, adjacent coils are always connected. This is easy to construct, but needs a large number of jumpers, in particular for fractional windings. In this paper, we focus on *wave winding*, where each coil by default connects to a coil of the subsequent group of coils. This gives rise to the opportunity of solving a complex routing problem and thus significantly reducing the winding length. Figure 1b shows a wave winding with coil pitch $y_1 = 6$. The complementary coil pitch on the jumper side is chosen as $y_2 = 8$.

On the right side of Fig. 1c, two dangling connections are shown: these represent the plug to the external electrical power. The figure shows how the current flows starting from the external connector and continues clock-wise around the stator. Only one full round is highlighted, however by following the overall thread, starting from one power plug and ending in the other, it may be observed that the overall connection goes a bit more than two full rounds clockwise, and then it reverses direction (also referred to as *flip* in the following) and it goes counter-clockwise for about the same length (the jumper that causes this inversion is shown in the bottom right part in dashed line). This is linked to the fractional-winding nature of this design, $q = 2 + \frac{2}{7}$. We will describe more in details in the following sections the relation. As the alternating current flows around the stator, half of the bars (both top and bottom) will be traversed by the current inwards, and the other half outwards. Figure 1d shows how the top bars of phase U are partitioned in the two directions.

Previous work has focused on the optimal choice of n_s, n_p, number of layers and assignment of phases to bars with the aim of optimizing torque density, torque ripple, losses or unwanted harmonics (see [3–5]). To the best of the authors' knowledge, no previous scientific literature looked at the minimization of the material cost once the main design parameters of the machine have been decided.

3   Problem Description

In this section, we describe the three main components for an optimal design of a stator winding. At first each coil (i.e. each top and bottom bar) needs to be assigned to a phase and direction: this will define the electrical layout (Sect. 3.1). Secondly, the bars of each phase needs to be linked together in order to form a continuous thread (or route) going around the stator (Sect. 3.2). Finally, once the three routes have been defined, jumpers need to be selected to avoid any conflict between the connections (Sect. 3.3).

3.1   Bar Assignment

Assigning the top and bottom bars to phases defines the main electrical properties of the machine, hence this step in the design is referred to as *electrical layout design*. We restrict our focus to symmetric layouts between the phases, i.e. where phase V and phase W follow rotated layouts of phase U. Moreover, for a configuration with $q = a + \frac{b}{c}$ (with $0 < b < c$), we restrict to layouts where phase U is distributed to $(c - b)$ many blocks of a contiguous slots and b many blocks of $(a + 1)$ contiguous slots. E.g., in Fig. 1, focusing on the top bars, we will note the contiguous bars forming blocks of 3 and 2, respectively; starting from slot 0, the pattern is "3 3 2 2 2 2 2", that is, the bars belonging to phase U are grouped as follows: 3 bars (slots 0,1,2) then another group of 3 (slots 7,8,9) followed by 5 groups of 2 bars (14,15 then 21,22, and so on).

Symmetric layouts between the phases are constructed by rotating the layout of phase U by c blocks and $2c$ blocks, respectively. This is illustrated in the Table 1 for the example above.

The table is filled by first filling phase U with the given pattern and then traversing the table column by column, counting the number of elements

Table 1. Illustration of the construction algorithm for an electrical layout which is symmetric between the phases.

Phase U	3	3	2	2	2	2	2
Phase V	2	2	3	3	2	2	2
Phase W	2	2	2	2	3	3	2

(i.e. blocks) traversed and filling the pattern to phase V and W after c and $2c$ blocks have been traversed, respectively. This is illustrated by the bold face entries which indicate starts of the pattern. Also, from this it is clear that c must not be a multiple of 3, since otherwise after traversing c blocks, we would end up again in phase U and not V or W.

The following combinatorial arguments answer the question of how many possible electrical layouts exist modulo flipping and rotation. The layout where all $(a + 1)$-blocks are next to each other is referred to as the default electrical layout. It often minimizes the number of jumpers needed for a wave winding and therefore is a good candidate. If $b = 1$ or $b = c - 1$, there is (modulo rotation) only the default electrical layout. In all other fractional windings, there are more electrical layouts.

In order to count these layouts, we distinguish between *flip-symmetric* and *flip-asymmetric* ones. The flip-symmetric layouts are those for which *some* flip exists that preserves the pattern. For example, in the layout above, "3 3 2 2 2 2 2", is flip symmetric, which becomes clear when we write it as "3 2 2 2 2 2 3".

We can count the number of flip-symmetric layouts: For odd c (as in our example), it is

$$n_{sym} = \binom{\frac{1}{2}(c-1)}{\lfloor \frac{1}{2}b \rfloor},$$

since the center block is here of length $a + 1$ if b is odd and length a if b is even, the blocks right of the center are assigned freely, and the blocks left of the center are flipped copies. For even c, similar arguments show that

$$n_{sym} = \begin{cases} \binom{c/2}{b/2} & \text{for even } b, \\ \binom{c/2-1}{(b-1)/2} & \text{for odd } b. \end{cases}$$

Finally, $\frac{1}{c}\binom{c}{b} = 2n_{asym} + n_{sym}$ counts the asymmetric layouts twice and the symmetric ones once (the factor $\frac{1}{c}$ accounts for rotational symmetry). Hence, the total number of electrical layouts is

$$\frac{1}{2}\left(\frac{1}{c}\binom{c}{b} + n_{sym}\right).$$

Once the electrical layout is fixed, the coils belonging to phases U, V and W are defined and referred to as C^U, C^V, and C^W, with $C^U = \{c_k^U, k = 1, \ldots, \frac{1}{3}n_s\}$ etc., and $c_k^U \in \{0 \ldots n_s - 1\}$ being the slot number of the corresponding top bar. We will also use the sign of the electrical current in each coil $\sigma_k^U \in \{1, -1\}$.

Harmonics. The electrical layout has to be chosen in order to result in a large fundamental harmonic and at the same time small higher order harmonics of the magnetomotive force (MMF). For phase U (and by rotational symmetry, also the other phases) and harmonic order $h = 1, 2, \ldots$, the normalized harmonic content is given by the absolute of a sum of complex vectors:

$$\frac{1}{\frac{2}{3}n_s} \left| \sum_{n=1}^{n_s/3} \sigma_k^U \exp\left(i\pi h \cdot \frac{n_p}{n_s} c_k^U\right) - \sigma_k^U \exp\left(i\pi h \cdot \frac{n_p}{n_s}(c_k^U + y_1)\right) \right|,$$

where $i = \sqrt{-1}$. (The first exp corresponds to the top bars, the second exp to the bottom bars.)

Multiple Branches. Not every three phase electrical machine has three windings on the stator, some machines have windings with *parallel branches*. Distributing the bars of a phase to parallel branches adds one more criterion to the optimization, namely the *unbalanced magnetic pull* in case of rotor eccentricities. Optimizing multiple branches can be tackled with our methods, but we restrict the presentation to single branch layouts in the following.

3.2   Coil Routing

Once each bar has been assigned to a phase and sign of the current, the problem is then to connect[1] all the coils of the same phase together while respecting the direction their bars have been assigned to. The minimization of the total length of one phase connection leads to reduced electrical losses and reduced material consumption hence translating to a direct cost benefit.

In the following, we will focus the description on phase U, analogous observations can be drawn about the other phases. By construction each coil c_k^U is formed by a top bar and a bottom bar; connecting the top bar (resp. bottom bar) of c_k^U to the bottom bar (resp. top bar) of another coil c_j^U will keep the sign of the current. Conversely, whenever the top bar (resp. bottom bar) of a coil c_k^U will be connected to the top bar (resp. bottom bar) of a coil c_j^U, the sign of the current will be reversed. A change in sign is referred to as a *flip*. Depending on the design and manufacturing constraints, the engineer may want to force only one flip per phase or allow multiple flips.

Given a coil c_k^U, in order to respect the phase and direction assignment, it can be connected to a coil c_j^U in an electrically valid way: top to bottom if their sign coincides and top to top or bottom to bottom in case of different signs.

For illustration, we consider the example shown in Fig. 2. The stator has $n_s = 66$ slots, the number of poles is $n_p = 10$, and the coil pitches are $y_1 = 6$ and $y_2 = 7$. Two sets of coils are emphasized with black dotted lines, the

---

[1] For the sake of clarity, in this paper we focus at the case in which connections can be placed only at the *jumper side*; the problem can be generalized to account also connections at the opposite side.

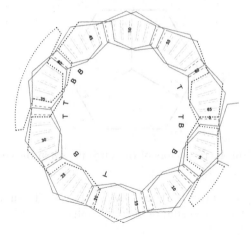

Fig. 2. Coils, jumpers and flip connection for phase U.

corresponding bars are marked to be top (T) or bottom (B). The left group of
coils starts at coil $20T(26B)$, i.e. the coil which connects the top bar in slot 20
with the bottom bar in slot 26. It connects, in clockwise direction and without
a jumper, to coil $33T(39B)$. The following coil is $35T(41B)$, connected with a
jumper. In the right group of coils, we observe a flip between coil $61T(1B)$ and
coil $0T(6B)$. That is, the current flow in coil $0T$ is reversed wrt. coil $61T$.

The problem to traverse each coil exactly once with the total minimum cost
clearly resembles to a Travelling Salesman Problem instance where coils represent cities and jumpers represent the travelling arcs between the cities. We will
formalize the model in Sect. 4.

Sum of Currents. The jumpers connecting coils have undesired losses and
decrease of efficiency. It is favorable to combine jumpers in a way that their
magnetic fields cancel as much as possible, thereby reducing the stray losses.

To formalize the problem, let

$$J = \{J^U \cup J^V \cup J^W\} = \{j_k^p, k = 1, \ldots, n_J^p, \ p = U, V, W\}$$

be the set of jumpers of the three phases. Each jumper j_k^p is associated with
its sign of the electrical current $\tilde{\sigma}(j_k^p) \in \{1, -1\}$ relative to clockwise rotation,
which follows from the signs of the coils it connects and the relative position of
these coils. For each jumper j_k^p, we introduce the set of slots it spans over to be
$\mathcal{S}(j_k^p)$. Then, conversely for each slot s, we define the set of jumpers going over
this slot to be $\mathcal{J}(s) = \{j \in J : s \in \mathcal{S}(j)\}$.

For each slot s, the relative amount of stray losses is approximated by the
absolute of the sum of complex current vectors corresponding to the jumpers in
$\mathcal{J}(s)$. Each jumper $j = j_k^p$ corresponds to a phase p and thus to a current vector

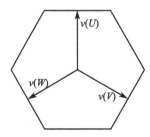

Fig. 3. Electric/magnetic vectors of the three phases in the complex plane.

$v(j) = v(j_k^p) = v(p) = \exp(2\pi i p/3)$, as shown in Fig. 3. The sum of the stray loss terms over all slots yields the sum of currents value:

$$\text{SumCur} = \sum_{s=0}^{n_s-1} \left| \sum_{j \in \mathcal{J}(s)} \tilde{\sigma}(j)v(j) \right| \tag{2}$$

In the optimization model described later, the sum of currents criterion may either enter the cost function or be constrained.

3.3   Jumper Placement

Lastly, jumpers need to be placed in a conflict-free manner, where potential conflicts arise from two or more jumpers of the same phase or different phases, which cross the same slot. An illustration is given in Fig. 4, where we revisit the example from Fig. 1, now with all phases shown.

In the top left corner of the stator, three jumpers j_1, j_2, and j_3 are highlighted with thick black lines: j_1 (dotted line) connects coil $50T(56B)$ and coil $48T(54B)$, j_2 (dashed line) connects coil $56T(62B)$ and coil $55T(61B)$, and j_3 (solid line) connects coil $66T(72B)$ and coil $64T(70B)$. The jumpers j_1 and j_2 belong to phase U, j_3 belongs to phase V, their respective jumper slot ranges are $\mathcal{S}(j_1)$, $\mathcal{S}(j_2)$ and $\mathcal{S}(j_3)$. Jumpers j_1 and j_3 alone never cause conflicts, but because of the presence of j_2, the potential conflicts in placement need to be resolved.

Conflicts are resolved by using different radial and axial placement of the jumpers in the 3-dimensional space. This is illustrated in Fig. 5 for two jumper geometries. For the optimization problem, a fixed set of *axial and radial levels* is pre-defined, depending on the electrical constraints (minimum distance between jumpers because of voltage) and the manufacturing constraints.

In the example in Fig. 4, conflicts are resolved by using jumpers at different radial levels. This is indicated by dotted concentric circles outside the stator, corresponding to the available radial levels: a jumper outside the innermost dotted circle is placed on radial level 1, a jumper outside the second dotted circle is placed on radial level 2.

The set of conflicting jumpers is defined as:

$$\Omega = \{\{j_j, j_k\} \mid j \neq k \ \wedge \ \mathcal{S}(j_j) \cap \mathcal{S}(j_k) \neq \emptyset\} \tag{3}$$

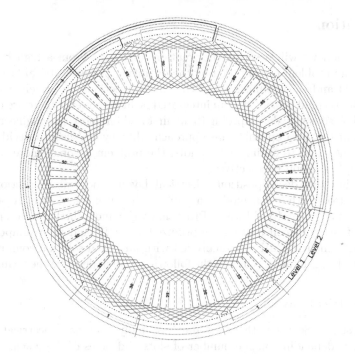

Fig. 4. Jumpers on different radial levels

In other words two jumpers are conflicting if their respective slot ranges intersect.

Ultimately, the jumper selection sub-problem consist in choosing the appropriate jumper geometries for all the pair of jumpers that have conflicts.

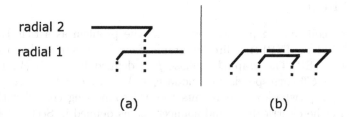

Fig. 5. Jumper geometries avoiding conflict: (a) using different radial levels, where the jumpers cross at different axial levels, (b) using different axial levels, where the longer jumper bridges over the shorter one. The line styles indicate the axial level: axial level 0 = dotted, axial level 1 = solid, axial level 2 = dashed. For simplicity, the typical curvature of the tangential part (left to right) of the jumpers is not shown.

4   Solution

In this section we will describe the solution approach. From a high level perspective, the problem is composed by the definition of the electrical layout (bar assignment) and the optimization of the coil connections. In principle, the two sub-problems could be solved in an integrated fashion, however there are two fundamental reasons why decomposing them: first and foremost, the engineer prefers a human in the loop optimization approach whereby he/she can decide which electrical design(s) to optimize. Secondly, the problem becomes more tractable from an optimization perspective.

Following this decomposition (electrical layout and coil connections), we developed two methods: a model in which coil routing and jumper selection is tackled in a single Mixed Integer Program (MIP) and a further decomposition in which firstly the coil routing is optimized via MIP and then jumper selection is performed via MIP optionally fed with an initial solution computed via Constraint Programming (CP). In the following we will describe the formulation.

4.1   Electrical Layout

As discussed in Sect. 3.1, there are possibly many candidate electrical layouts to consider, depending on the number of slots and poles of the machine. Some of these may already disqualify for their harmonics properties. Those which are harmonically feasible may be completely enumerated, or Monte-Carlo sampled, for further individual optimization. Alternatively, a multi-objective optimization can be performed, offering different trade-offs of harmonics properties and material consumption to the engineer. These and other alternatives are out of the technical scope of the paper. In the following, we will focus on solving the routing and jumper placement efficiently for a *given electrical layout*.

4.2   Coil Routing

To solve the coil routing problem, we model the problem as a rich Travelling Salesman Problem (TSP) on a directed graph. Each phase is treated as a separate instance of a TSP. The graph for phase p is defined by $G^p = (V, A)$ where each coil $c_i^p \in C^p$ corresponds to a node $v_i \in V$ with $i = 1, \dots, n_V$. The set of arcs $A = a_{ij}$ (where a_{ij} represents the arc connecting coil c_i^p with c_j^p) is composed by the electrically sound connections as defined in Sect. 3.2. The set of arcs is partitioned in two subset $A = A_d \cup A_f$ where A_d contains all the arcs that keep the direction of the stator (i.e. bottom bar-top bar or top-bottom bar connections), and A_f contains all the flip connectors (i.e. top-top bar or bottom-bottom bar connections). Each arc a_{ij} has an associated weight $w_{a_{ij}}$ that is proportional to its length. The first and last coils of the stator have dangling endpoints as they will be connected to the power source. In the model, we force these two coils to be connected together in order to form a closed loop.

The TSP is formulated via a typical model where each arc is associated with a binary decision variable x_{ij} that will be equal to one iff it will be part of the final solution.

The objective function is (with a slight abuse of notation):

$$\min \sum_{a_{ij} \in A} w_{ij} x_{ij} \tag{4}$$

Here, weights w_{ij} are set to the exact material costs of the corresponding jumper in the integrated MIP formulation, i.e. where TSP and jumper placement are solved together. In the decomposed approach, the exact cost for the jumper is not known at the TSP solving stage, so w_{ij} is assigned the cost of the cheapest corresponding jumper. This may introduce small but perceivable sub-optimality of the final solution.

The set of constraints is listed below:

- the incoming degree and outgoing degree must be equal to one:

$$\sum_{j} x_{ij} = \sum_{j} x_{ji} = 1 \qquad \forall i = 1, \ldots, n_V \tag{5}$$

- the solution should not contain sub-tours. This constraint is handled by a typical cut generation procedure that add a linear constraint during the branch and bound procedure to avoid the creation of sub-tours (see [6–8])
- optionally, the number of flip should be equal to 1:

$$\sum_{a_{ij} \in A_f} x_{ij} = 1 \tag{6}$$

- The sum of currents objective/constraint has been linearized: Instead of considering the complex absolute = Euclidean norm in (2), we use a hexagonal norm where $|v|_{\text{hex}}$ is the radius of the scaled hexagon as depicted in Fig. 3 on which v lies. Each $|v|_{\text{hex}}$ is implemented in the optimization by introducing a corresponding slack variable and lower bounding it with the projection of v onto all six vectors $\{\exp(i\pi j/3) : 0 \le j < 6\}$.

The model has been developed in SCIP 3.0 [8], using a customized variant of the 3-opt heuristic for the solution process.

4.3 Jumper Placement

The jumper placement sub-problem is about choosing the correct conflict-free jumper geometry for each connection that has been chosen by the coil routing. According to the jumper geometry definition, we introduce for each jumper $j \in J$ its possible geometrical variants as $\mathcal{G}_j = \{g_q, q = 1, \ldots, n_{\mathcal{G}_j}\}$; each variant has an associated material cost w_{jq} (with $j \in J$ and $q \in \{1, \ldots, n_{\mathcal{G}_j}\}$). Finally, for each pair of potentially conflicting jumpers $\{j', j''\} \in \Omega$, we introduce the set of unordered pair of conflicting geometry variants as $\Lambda_{j',j''} = \{\{q', q''\} \mid$ iff $g_{q'}$ conflicts with $g_{q''}\}$.

In the following, we present two different models: a MIP based formulation and a CP based formulation.

MIP Model. We use binary variables y_{jq} to represent the variant q of jumper j that will be assigned value 1 iff that variant is chosen.

The objective function is:

$$\min \sum_{j \in J, q \in \{1, \dots, n_{\mathcal{G}_j}\}} w_{jq} y_{jq} \tag{7}$$

The constraints are:

– for each jumper only one variant can be chosen:

$$\sum_{q \in \{1, \dots, n_{\mathcal{G}_j}\}} y_{jq} = 1 \qquad \forall j \in J \tag{8}$$

– for each conflicting jumper pair $\{j', j''\} \in \Omega$, the respective geometrical variant should not collide:

$$y_{j'q'} \leq (1 - y_{j''q''}) \qquad \forall \{q', q''\} \in \Lambda_{j', j''}, \forall \{j', j''\} \in \Omega \tag{9}$$

CP Model. The CP model is employed optionally on those instances where the feasibility aspect of the jumper selection is hard for the MIP model.

The CP model resembles closely the MIP model with the main difference being the use of integer variables instead of binary ones. Specifically, each jumper j is associated with a variable $y_j \in \{1, \dots, n_{\mathcal{G}_j}\}$. For brevity, we do not report the full model as it follows trivially from the MIP model.

The search strategy chooses the variable to branch on based on minimum domain size, and it breaks ties based on the cost regret. The value heuristic selects the jumper geometry variant based on the minimum cost. The first search phase is followed by a Large Neighborhood Search. The neighborhood is chosen randomly, and the search strategy is the same as the one employed in the first phase, except that ties are broken randomly.

5   Experimental Evaluation

We develop the solution proposed using SCIP 3.0 [8] and Google Optimization Tools ('OR-Tools') [9]. All the experiments have been conducted on a PC with an Intel Core i5-5300U (2.3 GHz) and 16 GB of RAM.

5.1   Electrical Layout

We firstly tested the performance of the electrical layout generation by fixing the machine parameters for some stator setups and enumerating all the possible layouts. Results are reported in Table 2; the first four columns report the machine parameters (the number of slots, number of poles, the coil pitch and the number of slots per pole and phase); the last two columns show the total number of layouts generated and the time (in milliseconds) to enumerate them all. Layout enumeration has been implemented in Python 2.7 with SciPy. The method takes less than a second to enumerate hundreds of thousands of possible electrical layouts. In practice, some of these layouts may be discarded a priori based on the harmonics of the MMF and in a second stage a selection is presented to the user. Finally, the user will decide which design(s) to optimize.

Table 2. Performance test for generating all the electrical layouts

n_s	n_p	y_1	$q = a + \frac{b}{c}$	# layouts	Time (ms)
102	14	8	$2 + \frac{3}{7}$	10	4
264	34	8	$2 + \frac{10}{17}$	5005	282
384	46	8	$2 + \frac{18}{23}$	5985	251
480	58	8	$2 + \frac{22}{29}$	296010	601
576	76	6	$2 + \frac{10}{19}$	24310	770

5.2   Coil Routing

We tested the coil routing on a subset of the electrical layouts found in the previous section. Specifically, 20 instances were selected randomly for each machine design defined above (10 instances for the 102 slots machine). We then compared the decomposition with and without the usage of CP, and the integrated approach. We setup the jumper geometry in order to have 10 radial levels and 5 axial levels. The time limits were set to be of 90 s for the TSP problem and 30 s for the jumper placement problem. For the integrated approach we set a time limit of 300 s. For the MIP formulations, we additionally set the optimality gap tolerance to 1%. As for the CP model (optionally used to feed a first solution to the jumper placement MIP model), the search is limited to 2 s (1 s to find the first solution, plus an additional one for the Large Neighborhood Search).

In a first test the sum of currents was left unconstrained. Results are reported in Table 3. Each row presents aggregated results for a given subset of instances. The first column indicates the number of slots of the machine; next there are three groups of columns summarizing the results for the decomposed MIP+CP approach, the decomposed MIP approach and the integrated MIP approach. Each group presents the average time and standard deviation, the value of the objective function and the percentage of the solved instances. For the decomposed MIP and the integrated approach the objective function has been normalized to the value obtained in the MIP+CP approach.

The MIP+CP shows clearly the best robustness in term of number of instances solved, with a drawback of adding few seconds to the overall solution process; the objective function is overall very similar.

The integrated MIP model manifests its limits: only small-size instances are solved and the overall gain in the objective function is minimal. Please note that the optimality gap reached was below the threshold of 1% except for 3 instances

Table 3. Results for coil routing and jumper placement with no sum of current constraint.

n_s	Decomposed MIP+CP				Decomposed MIP				MIP			
	t (μ)	t (σ)	Obj_{CP}	%Sol	t (μ)	t (σ)	$\frac{Obj}{Obj_{CP}}$	%Sol	t (μ)	t (σ)	$\frac{Obj}{Obj_{CP}}$	%Sol
102	4.4	1.0	12.18	100%	2.4	1.2	100.0%	100%	177.6	112.2	98.2%	90%
264	28.6	28.7	23.57	100%	26.0	28.9	100.0%	95%	340.7	2.0	101.7%	5%
384	23.2	19.5	25.39	100%	19.4	19.4	99.9%	95%	342.1	3.2	–	0%
480	42.0	35.6	32.34	100%	38.8	34.8	100.1%	100%	339.8	2.2	–	0%
576	65.0	33.4	43.56	70%	60.4	32.7	99.8%	30%	341.2	2.4	–	0%

that timed out with a gap slightly higher than 2%. Out of the 10 instances solved by the integrated approach only one showed a relevant objective function improvement of 11% w.r.t. the decomposition. This suggests that the decomposition is in most cases able to capture well the overall problem structure. One of the main reason why the integrated MIP model struggles is that all the jumper geometries need to be considered even for arcs that will not be part of the final solution, leading therefore to an explosion of number of variables.

In a second test, we run the same instance set from the previous experimentation and constrained the sum of currents to be about half of what it was initially found. In Table 4, we report for each instance set: the average sum of current found by the decomposition in the first benchmark, the constraint used for the second benchmark, and the resulting average sum of current found in the second benchmark when constrained.

Table 4. Average sum of currents found by the decomposition approach.

n_s	SumCur (unconstrained)	Upper bound	SumCur (constrained)
102	2757.8	1370	130.6
264	8100.7	4050	304.5
384	13561.2	6780	343.6
480	16721.0	8360	458.9
576	11883.9	8490	492.2

The results of the second test are presented in Table 5. Observations are similar to the instances with no sum of current constraint: the CP-supported model is more consistent throughout the instance set, while adding little overhead to the solution process. Note that for some instance sets (e.g. $n_s = 384$) the average objective function is slightly lower: this is due partly to the optimality gap tolerance of 1% and partly to sub-optimality caused by the decomposition. The integrated model is again bringing limited improvements in term of solution quality while not being able to solve medium and large-size instances.

Table 5. Results for coil routing and jumper placement with sum of current constraint.

n_s	Decomposed MIP+CP				Decomposed MIP				MIP			
	t (μ)	t (σ)	Obj_{CP}	%Sol	t (μ)	t (σ)	$\frac{Obj}{Obj_{CP}}$	%Sol	t (μ)	t (σ)	$\frac{Obj}{Obj_{CP}}$	%Sol
102	4.9	1.0	12.18	100%	2.6	1.3	100.0%	100%	180.4	113.2	99.7%	100%
264	28.5	29.1	23.69	100%	25.8	29.7	99.9%	95%	345.9	21.2	100.9%	5%
384	23.4	19.4	25.29	100%	20.3	19.6	100.0%	95%	346.3	20.1	–	0%
480	43.7	33.5	33.41	95%	40.0	35.0	100.1%	95%	341.8	16.2	–	0%
576	68.1	34.7	42.43	75%	62.7	34.2	98.8%	45%	343.7	19.4	–	0%

6 Conclusion

In this paper, we formalized a design optimization problem for stator windings of rotating machines. We proposed a hybrid approach that solves real-life instances in a satisfactory manner. The most efficient and robust solution decomposes the problem in three parts: firstly, it generates the electrical layout, then it routes optimally the connections of each phase, and lastly it selects jumpers in order to avoid physical and electrical conflicts. The routing and jumper placement sub-problems are solved via a MIP model supported by a CP model to bootstrap the search for conflict-free jumpers. The MIP model for the TSP performs remarkably well on this problem class and allows to enrich the TSP easily with additional requirements, e.g. the sum of currents. The decomposition provides a solution quality that is close to the one of a fully integrated model, with the advantage of scaling to large-size instances, and being more efficient. The tool has been in use for the last two years in the engineering departments of ABB for gearless mill drives and hydro-generators.

References

1. Stromberg, T.: Alternator voltage waveshape with particular reference to the higher harmonics. ASEA J. **20**, 139–148 (1947)
2. Jones, G.R.: Electrical Engineer's Reference Book. Section 20, 3 edn. Elsevier (2013)
3. Fornasiero, E., Alberti, L., Bianchi, N., Bolognani, S.: Considerations on selecting fractional-slot nonoverlapped coil windings. IEEE Trans. Ind. Appl. **49**(3), 1316–1324 (2013)
4. Alberti, L., Bianchi, N.: Theory and design of fractional-slot multilayer windings. IEEE Trans. Ind. Appl. **49**(2), 841–849 (2013)
5. Prieto, B.: Design and analysis of fractional-slot concentrated-winding multiphase fault-tolerant permanent magnet synchronous machines. Ph.D. thesis, University of Navarra, Spain (2015)
6. Gusfield, D.: Very simple algorithms and programs for all pairs network flow analysis. SIAM J. Comput. **19**(1), 143–155 (1990)
7. Gomory, R.E., Hu, T.C.: Multi-terminal network flows. SIAM J. Appl. Math. **9**, 551–570 (1961)

8. Gamrath, G., Fischer, T., Gally, T., Gleixner, A.M., Hendel, G., Koch, T., Maher, S.J., Miltenberger, M., Müller, B., Pfetsch, M.E., Puchert, C., Rehfeldt, D., Schenker, S., Schwarz, R., Serrano, F., Shinano, Y., Vigerske, S., Weninger, D., Winkler, M., Witt, J.T., Witzig, J.: The SCIP Optimization Suite 3.2. ZIB Institute (2016)
9. Google Optimization Tools. https://developers.google.com/optimization/. Accessed 21 Nov 2016

A Distributed Optimization Method for the Geographically Distributed Data Centres Problem

Mohamed Wahbi$^{(\boxtimes)}$, Diarmuid Grimes, Deepak Mehta, Kenneth N. Brown, and Barry O'Sullivan

Insight Centre for Data Analytics, School of Computer Science and IT, University College Cork, Cork, Ireland
{mohamed.wahbi,diarmuid.grimes,deepak.mehta, ken.brown,barry.osullivan}@insight-centre.org

Abstract. The *geographically distributed data centres* problem (GDDC) is a naturally distributed resource allocation problem. The problem involves allocating a set of virtual machines (VM) amongst the data centres (DC) in each time period of an operating horizon. The goal is to optimize the allocation of workload across a set of DCs such that the energy cost is minimized, while respecting limitations on data centre capacities, migrations of VMs, etc. In this paper, we propose a distributed optimization method for GDDC using the distributed constraint optimization (DCOP) framework. First, we develop a new model of the GDDC as a DCOP where each DC operator is represented by an agent. Secondly, since traditional DCOP approaches are unsuited to these types of large-scale problem with multiple variables per agent and global constraints, we introduce a novel semi-asynchronous distributed algorithm for solving such DCOPs. Preliminary results illustrate the benefits of the new method.

1 Introduction

Distributed constraint reasoning (DCR) gained an increasing interest in recent years due to its ability to handle cooperative multi-agent problems that are naturally distributed. DCR has been applied to solve a wide range of applications in multi-agent coordination such as distributed scheduling [20], distributed planning [7], distributed resource allocation [24], target tracking in sensor networks [23], distributed vehicle routing [17], optimal dispatch in smart grid [22], etc. These applications can be solved by a centralized approach once the knowledge about the problem is delivered to a centralized authority. However, in such applications, it may be undesirable or even impossible to gather the whole problem knowledge into a single authority. In general, this restriction is mainly due to

This work is funded by the European Commission under FP7 Grant 608826 (GENiC - Globally Optimised Energy Efficient Data Centres).
This work is funded by Science Foundation Ireland (SFI) under Grant Number SFI/12/RC/2289.

D. Salvagnin and M. Lombardi (Eds.): CPAIOR 2017, LNCS 10335, pp. 147–166, 2017.
DOI: 10.1007/978-3-319-59776-8_12

privacy and/or security requirements: constraints may represent strategic information that should not be revealed to other agents that can be seen as competitors, or even to a central authority. The cost or the inability of translating all information to a single format may be another reason: in many cases, constraints arise from complex decision processes that are internal to an agent and cannot be articulated to a central authority. More reasons why distributed methods may be desirable for such applications and often make a centralized process inadequate have been listed in [11].

In DCR, a problem is expressed as a distributed constraint network. A distributed constraint network is composed of a group of autonomous agents where each agent has control of some elements of information about the problem, that is, variables and constraints. Each agent owns its local constraint network, and variables in different agents are connected by constraints. Traditionally, there are two large classes of distributed constraint networks. The first class considers problems where all constraints are described by boolean relations (*hard* constraint) on possible assignments of variables they involve. They are called *distributed constraint satisfaction problems* (DisCSP). The second class of problems are *distributed constraint optimization problems* (DCOP) where constraints are described by a set of *cost* functions for combinations of values assigned to the variables they connect. In DisCSP, the goal is to find assignments of values to variables such that all (hard) constraints are satisfied while in DCOP the goal is to find assignments that minimize the objective function defined by the sum of all constraint costs.

Researchers in the DCR field have developed a range of different constraint satisfaction and optimisation algorithms. The main algorithms and protocols include synchronous [10,15,37], asynchronous [6,23,36,38] and semi-synchronous [13,21,34] search, dynamic programming methods [25], and algorithms which respect privacy and autonomy [8,35] versus those which perform local search [16,40]. In order to simplify the algorithm specification, most of these algorithms assume that all constraints are binary, and that each agent controls exactly one variable. Such assumptions simplify the algorithm specification but represent a limitation in the adoption of distributed constraint reasoning techniques in real-world applications.

The first assumption was justified by techniques which translated non-binary problems into equivalent binary ones; however, recent research has demonstrated the benefits of handling non-binary constraints directly in distributed algorithms [5,32]. The second assumption is justified by the fact that any DCR problem with several variables per agent can be solved by those algorithms once transformed using one of the following reformulations [9,37]: (i) compilation, where for each agent we define a new variable whose domain is the set of solutions to the original local problem; (ii) decomposition, where for each agent we create a virtual agent for each of its variables. However, neither of these methods scales up well as the size of the local problems increase, either (i) because of the space and time requirements of the reformulation, or (ii) because of the extra communication overhead and the loss of a complete view on the local problem structure. Only a few algorithms for handling multiple local variables in DisCSP have been proposed [3,19,39]. These algorithms are specific to DisCSP, since they reason about hard constraints, and cannot be applied directly to DCOP, which are concerned with costs. Another limitation is that most DCOP algorithms do not actively exploit hard constraints as they require all constraints to be expressed as cost functions.

In this paper, we present a general distributed model for solving real-life DCOPs, including hard constraints, in order to model the geographically distributed data centres problem [4,27–29]. After finding that traditional DCOP approaches are unsuited to these types of large-scale problem with multiple variables per agent and global (hard) constraints, we introduce, AGAC-ng, a novel optimization distributed search algorithm for solving such DCOPs. In the AGAC-ng algorithm agents assign their variables and generate a partial solution sequentially and synchronously following an ordering on agents. However, generated partial solutions are propagated asynchronously by agents with unassigned variables. The concurrent propagation of partial solutions enables early detection of a need to backtrack and saves a lot of search effort. AGAC-ng can perform bounded-error approximation while maintaining a theoretical guarantee on solution quality. In AGAC-ng, a global upper bound which represents the cost of the best known solution to the problem is maintained by all agents. When propagating partial solutions agents ensure that the current lower bound on the global objective augmented by the bounded-error distance does not exceed the global upper bound.

Data centres consume considerable amounts of energy: an estimated 416.2 terawatt-hours was consumed in 2014 [2] with the US accounting for approximately 91 terawatts [1]. While many industries and economies are dependent on such infrastructure, the increase in high-computing requirements for cloud-based services around internet-of-things, big data, etc. have lead to some experts predicting that this consumption could treble in the next decade [2] unless significant breakthroughs are made in reducing consumption.

Geographically distributed data centres present many possible benefits in terms of reducing energy costs through global, rather than local, optimisation. In particular each location may have different unit energy costs, external weather conditions, local renewable sources, etc. Therefore reasoning at a global level can exploit these differences through optimally reallocating workload in each time period through migration of *virtual machines* (VMs), subject to constraints on number of migrations, virtual machine sovereignty, data centre capacities, etc.

Gathering the whole knowledge about the problem into a centralized location may not be feasible in the geographically distributed data centres because of the large amount of information (problem specifications) each data centre would need to communicate to the centralized solver to solve the problem. In addition, data centres may wish to keep some information about their local constraints, costs, and topology confidential and not share it with other data centres.

The maximum energy that can be consumed by a data centre and its running capacity might also be confidential information that data centres want to keep private from operators of other data centres. Thus, the geographically distributed data centres is a naturally distributed resource allocation problem where data centres may not be willing to reveal their private information to other data centres. In addition, sending the whole knowledge about the problem to a centralized location will create a bottleneck on the communication towards that location. Thus, a distributed solving process is preferred for the geographically distributed data centres problem.

This paper is structured as follows. Section 2 gives the necessary background for distributed constraint reasoning and the general definition of the geographically distributed data centres. We present our model of the GDDC problem as DCOP in Sect. 3. Section 4 introduces our new algorithm, AGAC-ng, for solving DCOP with global hard constraints and multiple variables per agent. We report experimental results in Sect. 5. Section 6 gives a brief overview of related works on distributed constraint reasoning. Finally, we conclude the paper in Sect. 7.

2   Preliminaries/Background

2.1   Distributed Constraint Optimization Problem

A *constraint satisfaction problem* (CSP) has been defined for a centralized architecture by a triple $(\mathcal{X}, \mathcal{D}, \mathcal{C})$, where $\mathcal{X} = \{x_1, \ldots, x_p\}$ is a set of p variables, $\mathcal{D} = \{d_1, \ldots, d_p\}$ is the set of their respective finite domains, and $\mathcal{C}$ is a set of (hard) constraints. A constraint $c(X) \in \mathcal{C}$ is a relation over the ordered subset of variables $X = (x_{j_1}, \ldots, x_{j_k})$, which defines those value combinations which may be assigned simultaneously to the variables in X. $|X|$ is the *arity* of $c(X)$, and X is its *scope*. A tuple $\tau = (v_{j_1}, \ldots, v_{j_k}) \in c(X)$ is a *support* for $c(X)$, and $\tau[x_i]$ is the value of x_i in τ. A solution to a CSP is an assignment to each variable of a value from its domain, such that all constraints are satisfied.

A *global constraint* captures a relation over an arbitrary number of variables. For example, the ALLDIFF constraint states that the values assigned to the variables in its scope must all be different [30]. Filtering algorithms which exploit the specific structure of global constraints are one of the main strengths of constraint programming. During search, any value $v \in d_i$ that is not *generalized arc-consistent* (GAC) can be removed from d_i. A value $v_i \in d_i$, $x_i \in X$ is **generalized arc-consistent** with respect to a constraint $c(X)$ iff there exists a support τ for $c(X)$ such that $v_i = \tau[x_i]$, and for every $x_j \in X$, $x_i \neq x_j$, $\tau[x_j] \in d_j$. Variable x_i is GAC if all its values are GAC with respect to every constraint in $\mathcal{C}$. A CSP is GAC if all its variables are GAC.

A *distributed constraint satisfaction problem* (DisCSP) is a CSP where the variables, domains and constraints of the underlying network are distributed over a set of autonomous agents. Formally, a distributed constraint network (DisCSP) is defined by a 5-tuple $(\mathcal{A}, \mathcal{X}, \mathcal{D}, \mathcal{C}, \varphi)$, where $\mathcal{X}, \mathcal{D}$ and $\mathcal{C}$ are as above. $\mathcal{A}$ is a set of n agents $\{a_1, \ldots, a_n\}$, and $\varphi : \mathcal{X} \to \mathcal{A}$ is a function specifying an agent to control each variable. Each variable belongs to one agent and only the agent who owns a variable has control of its value and knowledge of its domain. Let X_i denotes the set of variables belonging to agent a_i, i.e. $X_i = \{\forall x_j \in \mathcal{X} : \varphi(x_j) = a_i\}$. Agent a_i knows all constraints, $\mathcal{C}_i$, involving its variables X_i. X_i can be partitioned in two disjoint subsets $P_i = \{x_j \mid \forall c(X) \in \mathcal{C}_i, x_j \in X \to X \subseteq X_i\}$ and $E_i = X_i \backslash P_i$. P_i

is a set of private variables, which only share constraints with variables inside a_i. Conversely, E_i is a set of variables linked to the outside world and sometimes referred to as external (or negotiation) variables.

This distribution of variables divides C in two disjoint subsets, C^{intra} which are between variables of same agent (i.e., $c(X) \in C^{intra}, X \subseteq X_i$), and C^{inter} which are between variables of different agents called intra-agent (private) and inter-agent constraint sets, respectively. An intra-agent constraint $c(X)$ is known by the agent owner of X, and it is unknown by other agents. Usually, it is considered that an inter-agent constraint $c(X)$ is known by every agent owning a variable in X. Two agents are *neighbours* if they control variables that share a constraint; we denote by $\mathcal{N}_i$ the set of neighbours of agent a_i.

As in the centralized case, a solution to a DisCSP is an assignment to each variable of a value from its domain, satisfying all constraints. A *distributed constraint optimisation problem* (DCOP) is defined by a DisCSP $(\mathcal{A}, \mathcal{X}, \mathcal{D}, \mathcal{C}, \varphi)$, together with an objective function. A solution to a DCOP is a solution for the DisCSP which is optimal with respect to the objective function.[1]

For the rest of the paper we consider a generic agent $a_i \in \mathcal{A}$. Agent a_i stores a unique **order** on agents, i.e. an ordered tuple of all the agent IDs, denoted by $\prec_o$. $\prec_o$ is called the current order of a_i. Agents appearing before agent a_i in $\prec_o$ are the higher priority agents (predecessors) and conversely the lower priority agents (successors) are agents appearing after a_i in $\prec_o$. For sake of clarity, we assume that the order is the lexicographic ordering $[1, 2, \ldots, n]$. Each agent maintains a counter, and increments it whenever it changes its assignment. The current value of the counter *tags* each generated assignment.

Definition 1. *An **assignment** for an agent $a_i \in \mathcal{A}$ is an assignment for each external variable of a value from its domain. That is, a tuple $(\langle E_i, V_i \rangle, t_i)$ where $V_i \in \underset{x_j \in E_i}{\times} d_j$, $V_i[x_j] \in d_j$ and t_i is the tag value. When comparing two assignments, the most up to date is the one with the greatest tag t_i.*

Definition 2. *A **current partial assignment** CPA is an ordered set of assignments of external variables $[(\langle E_1, V_1 \rangle, t_1), \ldots, (\langle E_k, V_k \rangle, t_k)]$. Two CPAs are **compatible** if the values of each common variable amongst the two CPAs are equal. ag(CPA) is the set of all agents with assigned variables in the CPA.*

Definition 3. *A **timestamp** associated with a CPA is an ordered list of counters $[t_1, \ldots, t_k]$ where $\forall j \in 1..k$, t_j is the tag of the agent a_j. When comparing two CPAs, the **strongest** one is that associated with the lexicographically greater timestamp. That is, the CPA with greatest value on the first counter on which they differ, if any, otherwise the longest one.*

During search agents can infer inconsistent sets of assignments called no-goods.

---

[1] Cost functions can be implemented using element constraints [31].

Definition 4. *A **no-good** or conflict set is an assignment set of variables that is not contained in any solution. A no-good is a clause of the form* $\neg[(x_i = v_i) \wedge (x_j = v_j) \wedge \ldots \wedge (x_k = v_k)]$, *meaning that these assignments cannot be extended to a solution. We say that a no-good is **compatible** with a* CPA *if every common variable is assigned the same value in both.*

Agents use these no-goods to prune the search space. To stay polynomial we only keep no-goods that are compatible with the current state of the search. These no-goods can be a direct consequence of propagating constraints (e.g., any assignment set that violates a constraint is a no-good) or can be derived from a set of no-goods. Literally, when all values of the variable x_k are ruled out by some no-good, they are resolved computing a new no-good as follows. The new generated no-good is the conjunction of all these no-goods for values of x_k removing variable x_k. If the generated no-good contains assignments of local variables, agent a_i uses this no-good locally to prune the search space. Otherwise, agent a_i reports the no-good to the agent having the lowest priority among those having variables in the new generated no-good.

2.2   Geographically Distributed Data Centres Problem

The geographically distributed data centres (GDDC) problem considered here can be defined as follows. We are given a set of data centre locations L, and a set of virtual machines that must be assigned to physical servers in each time period of an operating horizon T. Each location has its own unit energy price ep_t^l for each time period (sample real time prices are shown in Fig. 1). The cost of executing a virtual machine (VM) in a location can differ for each time period for the same location, and for each location in the same time period. This cost constitutes not only the electricity price from the local utility, but is also affected by external factors such as the outdoor temperature, equipment quality, etc. The latter can be estimated using the *Power Usage Effectiveness* (PUE) which is the ratio of the total power consumption of a DC to the power consumption of the IT equipment alone.

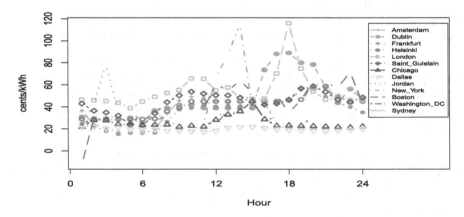

Fig. 1. Sample real time prices for 13 different locations over the same 24 h period.

Furthermore each location has an associated region k that a subset of *sovereign* VMs can only be performed in (e.g., for security reasons). Therefore for some VMs the set of locations where they can be performed is restricted. A data centre has a maximum capacity on the IT power consumption $pmax^l$ (aggregated across all servers), and a maximum number of migrations in/out, $mmax^l$, that can occur in each time period across all VM types.

There is a set of VM types V, where each type v has an associated power consumption p_v, and quantity n_v that are running in each time period. The type of a VM type can be further subdivided into those that can be run everywhere and those that can only be run in a specific region, i.e. that can only be run at its current location or in one of a limited set of other locations (e.g., a subset of VMs can only be run in European data centres). Let R^l be the set of VM types that can be run in location l.

We must then decide how many VMs of each type to allocate to each data centre in each time period such that the total energy cost of performing the VMs over the horizon, plus the costs of all migrations (where emi_v/emo_v is the energy cost of migrating a vm in/out), is minimized. This allocation is subject to capacity restrictions in the data centres, limitations on the number of migrations per data centre per time period, and limitations on migration of sovereign VMs.

3   GDDC as DCOP

The geographically distributed DCs can naturally be modeled as a DCOP as follows. (For simplicity we consider the time periods to be of duration one hour and thus energy and power values can be used interchangeably.)

Agents.

- Each DC is represented by an agent $\mathcal{A} = L$.

(External) Variables.

- In each DC/agent there is an integer variable x_{vt}^l that represents the number of VMs of type v allocated to DC/agent a_l in period t.
- In each DC/agent there is an integer variable c^l that indicates the total energy cost for running and migrating VMs in R^l over all time-periods.

Note here that agents only have variables x_{vt}^l if $v \in R^l$. This is sufficient for enforcing the sovereignty constraint.

(Private) Variables.

- In each DC/agent there is an integer variable mi_{vt}^l that indicates how many VMs of type v are migrated *in* to the DC a_l in time-period t.
- In each DC/agent there is an integer variable mo_{vt}^l that indicates how many VMs of type v are migrated *out* of the DC a_l in time-period t.

- In each DC/agent there is a real variable u_t^l that indicates the energy consumption of DC l for time-period t.
- In each DC/agent there is an integer variable r^l that indicates the total energy cost of running VMs in R^l over the entire horizon.
- In each DC/agent there is an integer variable m^l that indicates the total energy cost for migrating VMs in R^l over all time-periods.

(Intra-agent) Constraints.
We present here the intra-agent constraints held by agent $a_l \in \mathcal{A}$. Only agent/DC a_l is aware of the existence of its intra-agent constraints.

Allocation. The number of VMs of a given type running in time period are equal to the number running in the previous time period plus the incomings, minus the outgoings.

$$\forall_{v \in R^l}, \forall_{t \in T} : \quad x_{v(t+1)}^l = x_{vt}^l + mi_{vt}^l - mo_{vt}^l \tag{3.1}$$

Capacity. The total energy consumed by a DC a_l for running VMs is bounded:

$$\forall_{t \in T} : \quad u_t^l = \sum_{v \in R^l} x_{vt}^l \cdot p_v \leq pmax^l \tag{3.2}$$

Migration. The amount of incoming/outgoing VMs per time period for each DC is limited by a given threshold:

$$\forall_{t \in T} : \quad \sum_{v \in R^l} mi_{vt}^l + mo_{vt}^l \leq mmax^l \tag{3.3}$$

We further add the following redundant constraint on migration in/out variable pairs, enforcing that at least one must be 0 in every location in every time period for every type:

$$\forall_{v \in R^l}, \forall_{t \in T} : \quad mi_{vt}^l = 0 \lor mo_{vt}^l = 0 \tag{3.4}$$

Running cost. The total energy cost of running VMs in a_l over the entire horizon is:

$$r^l = \sum_{t \in T} ep_t^l \cdot u_t^l \tag{3.5}$$

Migration cost. The total energy cost for migrating VMs in R^l over all time-periods is:

$$m^l = \sum_{v \in R^l} \sum_{t \in T} (mi_{vt}^l \cdot emi_v + mo_{vt}^l \cdot emo_v) \cdot ep_t^l \tag{3.6}$$

Internal cost. The total energy cost for running and migrating VMs in R^l over all time-periods is:

$$c^l = r^l + m^l \tag{3.7}$$

(Inter-agent) Constraints.
An inter-agent constraint is totally known by all agents owning variables it involves.

Assignment. VMs of type v must all be assigned to DCs in each time-period where v is in R^l:

$$\forall_{v \in V}, \forall_{t \in T} : \sum_{l \in L | v \in R^l} x_{vt}^l = n_v \tag{3.8}$$

Objective. The global objective is to minimize the sum of the total energy cost of running VMs in all DCs together with total energy cost for migrating VMs over the entire horizon:

$$obj = \sum_{l \in L} c^l$$

4 Nogood-Based Asynchronous Generalized Arc-Consistency AGAC-ng

To solve a challenging distributed constraint optimization problem such as GDDC, we propose a new DCOP algorithm, called AGAC-ng (*nogood-based asynchronous generalized arc-consistency*). To the best of our knowledge, AGAC-ng is the first algorithm for solving DCOP with multiple variables per agent and non-binary and hard constraints that can find the optimal solution, or a solution within a user-specified distance from the optimal using polynomial space at each agent.

When solving distributed constraint networks, the solution process is restricted: each agent a_i is only responsible for making decisions (assignments) of the variables it controls (X_i). Thus, agents must communicate with each other exchanging messages about their variable assignments and conflicts of constraints in order to find a global (optimal) solution. Several distributed algorithms for solving the DCOP have been designed by the distributed constraint reasoning community. Regarding the manner on which assignments are processed and search performed on these algorithms, they can be categorized into synchronous, asynchronous, or semi-synchronous.

The first category consists of those algorithms in which agents assign values to their variables in a synchronous and sequential way. Although synchronous algorithms do not exploit the parallelism inherent from the distributed system, their agents receive consistent information from each other. The second category consists of algorithms in which the process of proposing values to the variables and exchanging these proposals is performed concurrently and asynchronously between agents. Agents take advantage from the distributed formalism to enhance the degree of concurrency. However, in asynchronous algorithms, the global assignment state at any particular agent is in general inconsistent. The third category is that of algorithms combining both sequential value assignments by agents together with concurrent computation. Agents take advantage from both the above-mentioned categories: they perform concurrent computation while exchanging consistent information between agents.

In this section we propose a novel semi-synchronous search algorithm for optimally solve DCOPs called AGAC-ng (*nogood-based asynchronous generalized arc-consistency*). In AGAC-ng algorithm, agents assign their variables and generate a partial solution sequentially and synchronously following an ordering on agents. However, generated partial solutions are propagated asynchronously by neighbours with unassigned variables. The concurrent propagation of partial solutions enables an early detection of a need to backtrack and saves search effort.

AGAC-ng incorporates an asynchronously generalized arc-consistency phase in a synchronous search procedure. AGAC-ng follows an ordering on agents to perform the sequential assignments by agents. Agents assign their variables only when they hold the CPA. The CPA is a unique message (token) that is passed from one agent to the next in the ordering. The CPA message carries the current partial assignment that agents attempt to extend into a complete solution by assigning their variables in it.[2] When an agent succeeds in assigning its variables on the CPA, it sends this CPA (token) to the next agent on the ordering. Furthermore, copies of the CPA are sent to all neighbors whose assignments are not yet on the CPA. These agents maintain the generalized arc-consistency asynchronously in order to detect as early as possible inconsistent partial assignments. The generalized arc-consistency process is performed as follows. When an agent receives a CPA, it updates the domain of its variables and copies of neighbors variables, removing all values that are not GAC using the no-goods as justification of value deletions.

When an agent generates an empty domain as a result of maintaining GAC, it resolves the no-goods ruling out values from that domain, producing a new no-good. Then, the agent backtracks to the agent with the lowest priority in the conflict by sending the resolved no-good. Hence, multiple backtracks may be performed at the same time coming from different agents having an empty domain. These backtracks are sent concurrently by these different agents to different destinations. The reassignments of the destination agents then happen simultaneously and generate several CPAs. However, the strongest CPA coming from the highest level in the agent ordering will eventually dominate all others. Agents use timestamps attached to CPAs to detect the strongest one (see Definition 3). Interestingly, the search process of higher levels with stronger CPA can use no-goods reported by the (killed) lower level processes, so that it benefits from their computational effort.

4.1   The Algorithm Description

Each agent $a_i \in \mathcal{A}$ executes the pseudo-code shown in Fig. 2. The agent a_i has a local solver where it stores and propagates the most up-to-date assignments received from higher priority agents (*solver*.CPA) w.r.t. the agent ordering ($\prec_o$). In AGAC-ng, agents exchange the following types of messages (where a_i is the sender):

---

[2]   A complete solution here is a complete assignments of all agents' external variables.

```
procedure AGAC-ng()
01. initialize();
02. while ( ¬end ) do
03.    msg ← getMsg();
04.    switch ( msg.type ) do
05.       ok?: processOK(msg.cpa, msg.next);
06.       ngd: processNgd(msg.nogood);
07.       sol : storeSolution(msg.sol, msg.b);
08.       stp : end ← true ;

procedure initialize()
09. end ← false; solution ← ∅; UB ← +∞;
10. solver.initialize();
11. solver.setObjective(obj);
12. foreach ( a_j ∈ A ) do t_j ← 0;
13. if ( isFirstAgent(≺_o) ) then
14.    token ← a_i;
15.    assign();

procedure assign()
16. if ( solver.findSolution() ) then
17.    t_i ← t_i + 1;
18.    CPA ← {solver.CPA ∪ E_i};
19.    if ( n = |ag(CPA)| ) then
20.       reportSolution(CPA);
21.    else
22.       sendForward(CPA, nextAgent());
23. else backtrack() ;

procedure reportSolution(cpa)
24. B ← obj.getValue();
25. sendMsg:sol⟨cpa, B⟩ to A \ a_i;
26. storeSolution(cpa, B);
27. assign();

procedure sendForward(cpa, a_j)
28. token ← a_j;
29. foreach ( a_k ∈ {N_i \ ag(CPA)} ) do
30.    sendMsg:ok?⟨cpa, a_j⟩ to a_k;

procedure processOK(cpa, next)
31. s ← compareTimeStamp(cpa);
32. if ( s > 0 ) then
33.    token ← next;
34.    foreach ( t_j ∈ cpa ) do
35.       t_j ← cpa.t_j;
36.    if ( solver.update(cpa[s..]) ) then
37.       t_i ← 0;
38.       if ( token = a_i ) then assign();
39.    else backtrack();
```

```
procedure backtrack()
40. ng ← solver.explainFailure() ;
41. if ( ng = ∅ ) then
42.    sendMsg:stp⟨sol, UB⟩ to {A \ a_i};
43. else
44.    token ← ng.lastAgent();
45.    if ( token ≠ a_i ) then
46.       sendMsg:ngd⟨ng⟩ to token;
47.    else
48.       solver.post(ng);
49.       assign();

procedure processNgd(ng)
50. if ( isCompatible(ng) ) then
51.    token ← a_i;
52.    solver.post(ng);
53.    assign();

procedure storeSolution(cpa, bound)
54. solution ← cpa;
55. UB ← bound;
56. solver.post(obj < UB − errorBound);

function isCompatible(assignments)
57. foreach ( x_j ∈ assignments ) do
58.    if ( x_j ≠ solver.x_j ) then
59.       return(false);
60. return(true);

function compareTimeStamp(cpa)
61. from ( j ← 1 to size(cpa) ) do
62.    Let a_k ← cpa[j];
63.    if ( cpa.t_k > t_k ) then
64.       return(j);
65.    if ( cpa.t_k < t_k ) then
66.       return(-j);
67. return(0);
```

Fig. 2. AGAC-ng algorithm running by agent a_i.

ok?: agent a_i passes on the current partial assignment (CPA) to a lower priority agent. According to the ID of the agent that has the token attached to the message by a_i, the receiver will try to extend the CPA (when it is the next agent on the ordering) or maintain generalized arc-consistency phase.

ngd: agent a_i reports the inconsistency to a higher priority agent. The inconsistency is reported by a no-good (i.e., a subset of the CPA).

sol: agent a_i informs all other agents of the new best solution (CPA) and new better bound.

stp: agent a_i informs agents if either an optimal solution is found or the problem is found to be unsolvable.

Agent a_i running the AGAC-ng algorithm starts the search by calling procedure `initialize` in which a_i sets the upper bound to $+\infty$ and initializes its local *solver* and sets the objective variable to minimize (lines 9 to 11) before setting counter tags of other agents to zero (line 12). If agent a_i is the initialising agent (the first agent in the agent ordering $\prec_o$), it initiates the search by calling procedure `assign` (line 15) after setting itself as the agent that has the token (i.e., the privilege to make decisions) (line 14). Then, a loop considers the receiving and processing of the possible message types (lines 2 to 8).

When calling procedure `assign`, agent a_i tries to find a local solution, which is consistent with the assignments of higher agents (solver.CPA). If agent a_i fails to find a consistent assignment, it calls procedure `backtrack` (line 23). If a_i succeeds, it increments its counter t_i and generates a CPA from higher agents assignments (solver.CPA) augmented by the assignments of its external variables (E_i), lines 17 to 18.[3] If the CPA includes assignments of all agents (a_i is the last agent in the order), agent a_i calls procedure `reportSolution()` to report a new solution (line 20). Otherwise, agent a_i sends forward the CPA to every neighboring agent $(\mathcal{N}_i \backslash ag(\text{CPA}))$ whose assignments are not yet on the CPA including the next agent that will extend the CPA (i.e., the agent that will have the token) (lines 28 to 30) by calling procedure `sendForward()`, line 22.

When agent a_i (the last agent) calls procedure `reportSolution`, a complete assignment has been reached, with a new global cost of B (line 24). Agent a_i sends the full current partial assignment (CPA), i.e. solution, with the new global cost to all other agents (line 25). Agent a_i calls then procedure `storeSolution()` (line 26) to set the upper bound to new global cost value and the best solution to the newly found one, lines 54 to 55, and to post a new constraint requiring that the global cost should not exceed the cost of the best solution found so far (taking into account the error-bound). Finally, agent a_i calls procedure `assign()` to continue the search for new solutions with better cost (line 27). Whenever agent a_i receives a **sol** message it calls procedure `storeSolution()`.

Whenever a_i receives an **ok?** message, procedure `processOK` is called (line 5). Agent a_i checks if the received CPA is stronger than its current CPA (solver.CPA) by comparing the timestamp of the received CPA to that stored locally, function `compareTimeStamp` call (line 31). If it is not the case, the received CPA is discarded. Otherwise, a_i sets the token to the newly received one and updates the local counters of all agents in the CPA by those freshly received (lines 33 to 35). Next, agent a_i updates its solver (*solver*.`update`) to include all assignments of agents in the received CPA and propagates their effects locally (line 36).

---

[3] Only external variables linked to unassigned neighbors are needed.

If agent a_i generates an empty domain as a result of calling *solver*.update, a_i calls procedure **backtrack** (line 39), otherwise, a_i checks if it has to assign its variables (if it has the token) and then calls procedure **assign** if it is the case (line 38).

When agent a_i generates an empty domain after propagating its constraints, the procedure **backtrack** is called to resolve the conflict. Agent a_i requires its local solver to explain the failure by generating a new no-good (ng), that is, the subset of assignments that produced the inconsistency, (line 40). If the new no-good (ng) is empty, agent a_i terminates execution after sending a **stp** message to all other agents in the system meaning that the last solution found is the optimal one (line 42). Otherwise, agent a_i sets the token to be the agent having the lowest priority among those having variables in the newly generated no-good ng. If the token is different than a_i, the no-good ng is reported to *token* through a **ngd** message, line 46. Otherwise, a_i has to seek a new local solution for its variables by calling procedure **assign** after storing the generated no-good in its local solver (lines 48 to 49).

When a **ngd** message is received by an agent a_i, it checks the validity of the received no-good (line 50). A no-good is valid if the assignments on it are consistent with those stored locally in *solver*. If the received no-good (ng) is valid, a_i sets itself as the agent having the token (line 51). Next, agent a_i stores ng and propagates it in its local solver. The procedure **assign** is then called to find a new local solution for the variables of a_i (line 53). Finally, when a **stp** message is received, a_i marks the *end* flag as true to stop the main loop (line 8).

5 Evaluation of Distributed Approach

In this section we experimentally evaluate AGAC-ng, the distributed COP algorithm we proposed in Sect. 4 for solving the DCOP model of the GDDC problem presented in Sect. 3. All experiments were performed based on the DisChoco 2.0 platform[4] [33], in which agents are simulated by Java threads that communicate only through message passing. We use the Choco-4.0.0 solver as local solver of each agent in the system [26].

5.1 Empirical Setup

Instances were generated for a scenario with 13 data centres across 3 continents, each with a capacity defined to be 40 MW. Real time price data was gathered for each location for each hour for a set of five days. Dynamic PUE values of each DC were generated as a function of the temperature across a sample day. There were 5 VM *types* chosen with associated power consumption values of 20 W, 40 W, 60 W, 80 W and 100 W respectively. For each DC, VM creation petitions are randomly generated until their consumption reaches a load percentage of 40% of the DC capacity. Each VM creation petition is further randomly assigned a "sovereignty", which is either the continent the DC belongs to or the entire DC set.

---

[4] http://dischoco.sourceforge.net.

Finally, instances had a migration limit for each data centre stating what percentage of average VMs per data centre can be migrated per time period. We generated instances with a limit of 5% and with a limit of 10%. There were 3 instances generated with different seeds for each migration limit and for each of the five days of real time price data, producing a set of 30 instances. Therefore instances typically had 10000 variables, with domain sizes ranging from 15 to more than 500. For c^i variables the domains are larger and domain sizes ranges from 10^7 to more than 10^8. The number of constraints was approximately 7500 with maximum arity of 49 variables. The baseline subsequently used for comparison involves the case where there are 0 migrations, i.e. each DC just performs the load that it was initially assigned in time zero across all time periods.

5.2   Search Strategy

The AGAC-ng algorithm requires a total ordering on agents. This ordering is used to pass on the token (the privilege of assigning variables) between agents. The agent ordering can then affect the behaviour of the algorithm and our empirical results (Sect. 5.3) confirms the effect of agent ordering.

In the following we propose two agent orderings called **o1** and **o2**. In both orderings we use the same measure α_i for an agent a_i. For each agent a_i, α_i is the difference between the most expensive and the cheapest price over all time slots for the DC represented by agent a_i. For **o2**, agents are sorted using an increasing order over α_i. For **o1**, the first agent is selected to be the one with median measure α_i followed in increasing order by agents, say a_j, having the smallest distance to the measure α_i, i.e. $\mid \alpha_j - \alpha_i \mid$. For example, let $\alpha_1 = 22$, $\alpha_2 = 10$, $\alpha_3 = 44$, $\alpha_4 = 55$, $\alpha_5 = 30$, thus, **o1** $= [a_5, a_1, a_3, a_2, a_4]$ and **o2** $= [a_2, a_1, a_5, a_3, a_4]$.

We investigated search strategies for the model of each agent. We are using the *migration_in* and *migration_out* variables (mi^i_{vt} and mo^i_{vt}) as decision variables for the local solver of each agent/DC a_i. Every agent/DC a_i only communicates decisions about x^i_{vt} and c^i as they are the external variables of agent a_i while migration variables are private variables. Preliminary results suggested the best approach for solving the problems was to choose decision variables (i.e., *migration in* and *migration out* variables) according to the *domwdeg* variant of the weighted degree heuristic. Values of the *migration in* variables of the cheapest time slot and the *migration out* variables of the most expensive time slot are selected in a decreasing order (the upper bound) and other migration variables on increasing order (the lower bound).

5.3   Empirical Results

We evaluate the performance of the algorithm by communication load and solution quality. Communication load is measured by the total number of exchanged messages among agents during algorithm execution ($\#msg$) [18]. Results are presented on Table 1. Solution quality is assessed by two measures: (i) in terms

Table 1. Distributed: number of message exchanged by agents/DCs in solving each instance.

Price	Base euro	AGAC-ng (o1)		AGAC-ng (o2)	
		Mig 5%	Mig 10%	Mig 5%	Mig 10%
pr1	165.31	15,838	13,619	12,515	24,700
pr2	197.53	1,709	1,966	5,390	6,426
pr3	192.09	3,208	37,528	4,947	2,456
pr4	171.53	18,494	23,909	25,015	23,706
pr5	215.97	5,737	10,798	19,293	14,425

Table 2. Distributed: results in terms of average monetary cost (in Euros) for migration limits at 5% and 10%.

Price	Base euro	AGAC-ng (o1)		AGAC-ng (o2)	
		Mig 5%	Mig 10%	Mig 5%	Mig 10%
pr1	165.31	164.11	164.03	162.45	159.71
pr2	197.53	187.92	190.21	186.26	187.92
pr3	192.09	186.85	186.20	186.73	185.58
pr4	171.53	167.55	166.24	166.18	164.11
pr5	215.97	215.69	215.76	215.69	215.40

of average monetary cost in € (Table 2) and (ii) in terms of average percentage savings over baseline (Table 3). For all instances we use a timeout limit of one hour per instance and present the average cost (respectively $\#msg$) per price/migration-limit instance type (across the three instances).

The results are given for the both static agent ordering presented in Sect. 5.2 for the three metrics Tables 1, 2 and 3.

Regarding the communication loads shown in Table 1, the agents exchange few messages in solving the problems. In the worst case, AGAC-ng agents exchanges 37, 528 messages to solve the problems. This represents a significant result regarding the complexity and the size of the instances solved by a complete DCOP algorithm. This is mainly due to the expensive local filtering and search per each local solver and the topology of the instances solved: all instances solved have a complete constraint network. Thus, AGAC-ng produces more chronological backtracks than backjumps. This is explained by the behaviour of the algorithm where only small improvements over the first solution/UB were found. After the first solution, the AGAC-ng algorithm mainly backtracks from the last agent on the ordering to the second last that returns the token to the last agent for seeking new solutions and so on.

Comparing the solution quality (Tables 2 and 3), for instances **pr5**, AGAC-ng improvement over the baseline is insignificant. The improvement over the baseline ranges between 0.1–0.26%, which represents less than a euro per day.

Table 3. Distributed: results in terms of average percentage savings over baseline for migration limits at 5% and 10%.

Price	Base euro	AGAC-ng (o1)		AGAC-ng (o2)	
		Mig 5%	Mig 10%	Mig 5%	Mig 10%
pr1	165.31	0.73%	0.77%	1.73%	3.39%
pr2	197.53	4.87%	3.71%	5.71%	4.87%
pr3	192.09	2.73%	3.07%	2.79%	3.39%
pr4	171.53	2.32%	3.08%	3.12%	4.33%
pr5	215.97	0.13%	0.10%	0.13%	0.26%

For other instances, this improvement is more significant mainly for **pr2** where the improvement over the baseline ranges between 3.71% and 5.71%, which represents between 7.32 and 11.27 euros per day. This is an interesting result mainly because agents are solving problems only by passing messages between them, without sharing their constraints or their prices, and keeping their information private.

Comparing two different agents ordering, running AGAC-ng using agent ordering **o2** always improves AGAC-ng using **o1** for pricing. However, this is not always the case for the number of message exchanges to solve the problem. The average cost over all instances when running AGAC-ng using **o2** is 366 and 369 when using **o1**. For communication load, AGAC-ng (**o2**) requires an average of 27, 775 messages over all instances while AGAC-ng (**o2**) requires 26, 561 messages.

Regarding the migration limit, the results suggests that having a smaller migration limit is better for the distributed solving process regarding the number messages exchanges while the solution quality is slightly affected by the change on the migration limits. In the distributed problems having larger domains leads to more messages when the filtering power is limited.

6   Related Work

Many distributed algorithms for solving DisCSP/DCOP have been designed in the last two decades. Synchronous Backtrack (SBT) is the simplest DisCSP search algorithm. SBT performs assignments sequentially and synchronously. In SBT, only the agent holding a current partial assignment (CPA) performs an assignment or backtrack [41]. Meisels and Zivan [21] extended SBT to Asynchronous Forward Checking (AFC), an algorithm in which the forward checking propagation [14] is performed asynchronously [21]. In AFC, whenever an agent succeeds to extend the CPA, it sends the CPA to its successor and it sends copies of this CPA to the other unassigned agents in order to perform the forward checking asynchronously. The Nogood-Based Asynchronous Forward Checking algorithm (AFC-ng), which is an improvement of AFC, has been proposed in [34]. Unlike AFC, AFC-ng uses no-goods as justification of value removals and allows several simultaneous backtracks coming from different agents and going to different destinations. AFC-ng was shown to outperform AFC. The pioneering asynchronous algorithm for DisCSP was *asynchronous backtracking* (ABT) [6,38]. ABT is executed autonomously by each agent, and is guaranteed to converge to a global consistent solution (or detect inconsistency) in finite time.

The synchronous branch and bound (SyncBB) [15] is the basic systematic search algorithm for solving DCOP. In SyncBB, only the agent holding the token is allowed to perform an assignment while the other agents remain idle. Once it assigns its variables, it passes on the token and then remains idle. Thus, SyncBB does not make any use of concurrent computation. No-Commitment Branch and Bound (NCBB) is another synchronous polynomial-space search algorithm for solving DCOPs [10]. To capture independent sub-problems, NCBB arranges agents in constraint tree ordering. NCBB incorporates, in a synchronous search, a concurrent computation of lower bounds in non-intersecting areas of the search space based on the constraint tree structure. Asynchronous Forward Bounding (AFB) has been proposed in [12] for DCOP to incorporate a concurrent computation in a synchronous search. AFB can be seen as an improvement of SyncBB where agents extend a partial assignment as long as the lower bound on its cost does not exceed the global upper bound. In AFB, the lower bounds are computed concurrently by unassigned agents. Thus, each synchronous extension of the CPA is followed by an asynchronous forward bounding phase. Forward bounding propagates the bounds on the cost of the partial assignment by sending to all unassigned agents copies of the extended partial assignment. When the lower bound of all assignments of an agent exceeds the upper bound, it performs a simple backtrack to the previous assigned agent. Later, the AFB has been enhanced by the addition of a backjumping mechanism, resulting in the AFB_BJ algorithm [13]. The authors report that AFB_BJ, especially combined with the minimal local cost value ordering heuristic performs significantly better than other DCOP algorithms. The pioneer complete asynchronous algorithm for DCOP is Adopt [23]. Later on, the closely related BnB-Adopt [36] was presented. BnB-Adopt changes the nature of the search from Adopt best-first search to a depth-first branch-and-bound strategy, obtaining better performance.

7   Conclusions

In this paper we studied the geographically distributed data centres problem where the objective is to optimize the allocation of workload across a set of DCs such that the energy cost is minimized. We introduced a model of this problem using the distributed constraint optimization framework. We presented AGAC-ng, nogood-based asynchronous generalized arc-consistency, a new semi-asynchronous algorithm for DCOPs with multiple variables per agent and with non-binary and hard constraints. AGAC-ng can find the optimal solution, or a solution within a user-specified distance form the optimal using polynomial space at each agent. We showed empirically the benefits of the new method for solving large-scale DCOPs.

References

1. America's Data Centres Consuming and Wasting Growing Amounts of Energy (2015). https://www.nrdc.org/resources/americas-data-centers-consuming-and-wasting-growing-amounts-energy
2. Data centres to consume three times as much energy in next decade, experts warn (2016). http://www.independent.co.uk/environment/global-warming-data-centres-to-consume-three-times-as-much-energy-in-next-decade-experts-warn-a6830086.html
3. Armstrong, A.A., Durfee, E.H.: Dynamic prioritization of complex agents in distributed constraint satisfaction problems. In: Proceedings of AAAI 1997/IAAI 1997, pp. 822–822 (1997)
4. Beloglazov, A., Buyya, R.: Energy efficient resource management in virtualized cloud data centers. In: Proceedings of the 2010 10th IEEE/ACM International Conference on Cluster, Cloud and Grid Computing, pp. 826–831. IEEE Computer Society (2010)
5. Bessiere, C., Brito, I., Gutierrez, P., Meseguer, P.: Global constraints in distributed constraint satisfaction and optimization. Comput. J. **57**(6), 906–923 (2013)
6. Bessiere, C., Maestre, A., Brito, I., Meseguer, P.: Asynchronous backtracking without adding links: a new member in the ABT family. Artif. Intell. **161**, 7–24 (2005)
7. Bonnet-Torrés, O., Tessier, C.: Multiply-constrained DCOP for distributed planning and scheduling. In: AAAI Spring Symposium: Distributed Plan and Schedule Management, pp. 17–24 (2006)
8. Brito, I., Meisels, A., Meseguer, P., Zivan, R.: Distributed constraint satisfaction with partially known constraints. Constraints **14**, 199–234 (2009)
9. Burke, D.A., Brown, K.N.: Efficient handling of complex local problems in distributed constraint optimization. In: Proceedings of ECAI 2006, Riva del Garda, Italy, pp. 701–702 (2006)
10. Chechetka, A., Sycara, K.: No-commitment branch and bound search for distributed constraint optimization. In: Proceedings of AAMAS 2006, pp. 1427–1429 (2006)
11. Faltings, B., Yokoo, M.: Editorial: introduction: special issue on distributed constraint satisfaction. Artif. Intell. **161**(1–2), 1–5 (2005)
12. Gershman, A., Meisels, A., Zivan, R.: Asynchronous forward-bounding for distributed constraints optimization. In: Proceedings of ECAI 2006, pp. 103–107 (2006)
13. Gershman, A., Meisels, A., Zivan, R.: Asynchronous forward-bounding for distributed COPs. JAIR **34**, 61–88 (2009)
14. Haralick, R.M., Elliott, G.L.: Increasing tree search efficiency for constraint satisfaction problems. Artif. Intell. **14**(3), 263–313 (1980)
15. Hirayama, K., Yokoo, M.: Distributed partial constraint satisfaction problem. In: Smolka, G. (ed.) CP 1997. LNCS, vol. 1330, pp. 222–236. Springer, Heidelberg (1997). doi:10.1007/BFb0017442
16. Hirayama, K., Yokoo, M.: The distributed breakout algorithms. Artif. Intell. **161**, 89–116 (2005)
17. Léauté, T., Faltings, B.: Coordinating logistics operations with privacy guarantees. In: Proceedings of the IJCAI 2011, pp. 2482–2487 (2011)
18. Lynch, N.A.: Distributed Algorithms. Morgan Kaufmann Series (1997)
19. Maestre, A., Bessiere, C.: Improving asynchronous backtracking for dealing with complex local problems. In: Proceedings of ECAI 2004, pp. 206–210 (2004)

20. Maheswaran, R.T., Tambe, M., Bowring, E., Pearce, J.P., Varakantham, P.: Taking DCOP to the real world: efficient complete solutions for distributed multi-event scheduling. In: Proceedings of AAMAS 2004, Washington, DC, USA, pp. 310–317. IEEE Computer Society (2004)
21. Meisels, A., Zivan, R.: Asynchronous forward-checking for DisCSPs. Constraints **12**(1), 131–150 (2007)
22. Miller, S., Ramchurn, S.D., Rogers, A.: Optimal decentralised dispatch of embedded generation in the smart grid. In: Proceedings of AAMAS 2012, pp. 281–288. International Foundation for Autonomous Agents and Multiagent Systems (2012)
23. Modi, P.J., Shen, W.-M., Tambe, M., Yokoo, M.: ADOPT: asynchronous distributed constraint optimization with quality guarantees. Artif. Intell. **161**, 149–180 (2005)
24. Petcu, A., Faltings, B.: A value ordering heuristic for local search in distributed resource allocation. In: Faltings, B.V., Petcu, A., Fages, F., Rossi, F. (eds.) CSCLP 2004. LNCS, vol. 3419, pp. 86–97. Springer, Heidelberg (2005). doi:10. 1007/11402763_7
25. Petcu, A., Boi Faltings, D.: A scalable method for multiagent constraint optimization. In: Proceedings of IJCAI 2005, pp. 266–271 (2005)
26. Prud'homme, C., Fages, J.-G., Lorca, X.: Choco Documentation. TASC, INRIA Rennes, LINA CNRS UMR 6241, COSLING S.A.S. (2016)
27. Qureshi, A., Weber, R., Balakrishnan, H., Guttag, J., Maggs, B.: Cutting the electric bill for internet-scale systems. In: ACM SIGCOMM Computer Communication Review, vol. 39, pp. 123–134. ACM (2009)
28. Rahman, A., Liu, X., Kong, F.: A survey on geographic load balancing based data center power management in the smart grid environment. IEEE Commun. Surv. Tutorials **16**(1), 214–233 (2014)
29. Rao, L., Liu, X., Xie, L., Liu, W.: Minimizing electricity cost: optimization of distributed internet data centers in a multi-electricity-market environment. In: INFOCOM, 2010 Proceedings IEEE, pp. 1–9. IEEE (2010)
30. Régin, J.-C.: A filtering algorithm for constraints of difference in CSPs. In: Proceedings of AAAI 1994, pp. 362–367 (1994)
31. Van Hentenryck, P., Deville, Y., Teng, C.-M.: A generic arc-consistency algorithm and its specializations. Artif. Intell. **57**(2–3), 291–321 (1992)
32. Wahbi, M., Brown, K.N.: Global constraints in distributed CSP: concurrent GAC and explanations in ABT. In: O'Sullivan, B. (ed.) CP 2014. LNCS, vol. 8656, pp. 721–737. Springer, Cham (2014). doi:10.1007/978-3-319-10428-7_52
33. Wahbi, M., Ezzahir, R., Bessiere, C., Bouyakhf, E.H.: DisChoco 2: a platform for distributed constraint reasoning. In: Proceedings of Workshop on DCR 2011, pp. 112–121 (2011)
34. Wahbi, M., Ezzahir, R., Bessiere, C., Bouyakhf, E.H.: Nogood-based asynchronous forward-checking algorithms. Constraints **18**(3), 404–433 (2013)
35. Wallace, R.J., Freuder, E.C.: Constraint-based reasoning and privacy/efficiency tradeoffs in multi-agent problem solving. Artif. Intell. **161**, 209–228 (2005)
36. Yeoh, W., Felner, A., Koenig, S.: BnB-ADOPT: an asynchronous branch-and-bound DCOP algorithm. J. Artif. Intell. Res. (JAIR) **38**, 85–133 (2010)
37. Yokoo, M.: Distributed Constraint Satisfaction: Foundation of Cooperation in Multi-agent Systems. Springer, Berlin (2001)
38. Yokoo, M., Durfee, E.H., Ishida, T., Kuwabara, K.: Distributed constraint satisfaction for formalizing distributed problem solving. In: Proceedings of 12th IEEE International Conference on Distributed Computing Systems, pp. 614–621 (1992)

39. Yokoo, M., Hirayama, K.: Distributed constraint satisfaction algorithm for complex local problems. In: International Conference on Multi Agent Systems, pp. 372–379 (1998)
40. Zhang, W., Wang, G., Xing, Z., Wittenburg, L.: Distributed stochastic search and distributed breakout: properties, comparison and applications to constraint optimization problems in sensor networks. Artif. Intell. **161**, 55–87 (2005)
41. Zivan, R., Meisels, A.: Synchronous vs Asynchronous search on DisCSPs. In: Proceedings of EUMAS 2003 (2003)

Explanation-Based Weighted Degree

Emmanuel Hebrard[1]([✉]) and Mohamed Siala[2]

[1] LAAS-CNRS, Université de Toulouse, CNRS, Toulouse, France
`hebrard@laas.fr`
[2] Insight Centre for Data Analytics, Department of Computer Science,
University College Cork, Cork, Ireland
`mohamed.siala@insight-centre.org`

Abstract. The weighted degree heuristic is among the state of the art generic variable ordering strategies in constraint programming. However, it was often observed that when using large arity constraints, its efficiency deteriorates significantly since it loses its ability to discriminate variables. A possible answer to this drawback is to weight a conflict set rather than the entire scope of a failed constraint.

We implemented this method for three common global constraints (AllDifferent, Linear Inequality and Element) and evaluate it on instances from the MiniZinc Challenge. We observe that even with simple explanations, this method outperforms the standard Weighted Degree heuristic.

1 Introduction

When solving a constraint satisfaction problem with a backtracking algorithm, the choice of the next variable to branch on is fundamental. Several heuristics have been designed to make this choice, often relying on the concept of fail firstness: *"To succeed, try first where you are most likely to fail"* [4].

The Weighted Degree heuristic (*wdeg*) [1] is still among the state of the art for general variable ordering search heuristics in constraint programming. Its principle is to keep track of how many times each constraint has failed in the past, with the assumption that constraints that were often responsible for a fail will most likely continue this trend. It is very simple and remarkably robust, which often makes it the heuristic of choice when no dedicated heuristic exists. Depending on the application domain, other generic heuristics may be more efficient, and it would be difficult to make a strong claim that one dominates the other outside of a restricted data set. Empirical evidence from the MiniZinc Challenge and from the CSP Solver Competition,[1] however, show that *wdeg* is among the best generic heuristics for classical CSP solvers (see [16] for an analysis of the reasons of its efficiency). It is also interesting to observe that SAT and clause-learning solvers often use Variable State Independent Decaying Sum

---

M. Siala—Supported by SFI under Grant Number SFI/12/RC/2289.

[1] http://www.minizinc.org/challenge.html, http://www.cril.univ-artois.fr/CSC09.

© Springer International Publishing AG 2017
D. Salvagnin and M. Lombardi (Eds.): CPAIOR 2017, LNCS 10335, pp. 167–175, 2017.
DOI: 10.1007/978-3-319-59776-8_13

(*VSIDS*) [10], which has many similarities with *wdeg*, whilst taking advantage of the conflicts computed by these algorithms.

One drawback of Weighted Degree, however, is that the weight attributed to constraints of large arity has a weak informative value. In the extreme case, a failure on a constraint involving all variables of the problem does not give any useful information to guide search.

We empirically evaluated a relatively straightforward solution to this issue. We propose to weight the variables involved in a minimal conflict set rather than distributing it equally among all constrained variables. Such conflict sets can be computed in the same way as *explanations* within a clause learning CSP solver.

2   Background

A constraint network is a triplet $\langle \mathcal{X}, \mathcal{D}, \mathcal{C} \rangle$ where $\mathcal{X}$ is a n-tuple of variables $\langle x_1, \ldots, x_n \rangle$, $\mathcal{D}$ is a mapping from variables to finite sets of integers, and $\mathcal{C}$ is a set of constraints. A constraint C is a pair $(\mathcal{X}(C), \mathcal{R}(C))$, where $\mathcal{X}(C)$ is tuple of variables and $\mathcal{R}(C) \in \mathbb{Z}^k$. The assignment of the k-tuple of values σ to the k-tuple of variables X satisfies the constraint C iff $\exists \tau \in \mathcal{R}(C)$ such that $\forall i, j \ X[i] = \mathcal{X}(C)[j] \implies \sigma[i] = \tau[j]$. A solution of a constraint network $\langle \mathcal{X}, \mathcal{D}, \mathcal{C} \rangle$ is an assignment of a n-tuple σ to $\mathcal{X}$ satisfying all constraints in $\mathcal{C}$ with $\sigma[i] \in \mathcal{D}(x_i)$ for all $1 \leq i \leq n$. The constraint satisfaction problem (CSP) consists in deciding whether a constraint network has a solution.

Algorithm 1. SOLVE($P = \langle \mathcal{X}, \mathcal{D}, \mathcal{C} \rangle$)

---
1 **if** *not consistent(P)* **then** return **False**;
2 **if** $\forall x \in \mathcal{X}, |\mathcal{D}(x)| = 1$ **then** return **True**;
3 select $x \in \{x_i \in \mathcal{X}, |\mathcal{D}(x_i)| > 1\}$ and $v \in \mathcal{D}(x)$;
4 return Solve($P|_{x=v}$) or Solve($P|_{x \neq v}$);

We consider the class of *backtracking* algorithms, which can be defined recursively as shown in Algorithm 1. In Line 1, a certain property of consistency of the network is checked, through constraint propagation. This process yields either some domain reductions or, crucially, a *failure*. In the latter case, the last constraint that was processed before a failure is called the *culprit*. Line 2 detects that a solution was found. In Line 3, a heuristic choice of variable and value is made to branch on in Line 4, where $P|_{x=v}$ (resp. $P|_{x \neq v}$) denotes the constraint network equal to the current P except for $\mathcal{D}(x) = \{v\}$ (resp. $\mathcal{D}(x) = \mathcal{D}(x) \setminus \{v\}$).

3   General Purpose Heuristics

We first briefly survey the existing general-purpose variable ordering heuristics against which we compare the proposed improvement of Weighted Degree.

Impact-Based Search (*IBS*) [11] selects the variable with highest expected *impact*. The impact $I(x = a)$ of a decision $x = a$ corresponds to the reduction in potential search space resulting from this decision. It is defined as $1 - S_P/S_{P|_{x=a}}$ where S_P is the size of the Cartesian product of the domains of P after constraint propagation, and $P|_{x=a}$ denotes the problem P with $\mathcal{D}(x)$ restricted to $\{a\}$. This value is updated on subsequent visits of the same decision. Moreover, in order to favor recent probes, the update is biased by a parameter α: $I^a(x = a) \leftarrow ((\alpha - 1)I^b(x = a) + I^p(x = a))/\alpha$, where I^p, I^b and I^a refer to, respectively, the last recorded impact, the impact stored before, and after the decision. The preferred variable minimizes $\sum_{a \in \mathcal{D}(x)}(1 - I(x = a))$, the sum of the complement-to-one of each value's impact, i.e., the variable with fewest options, and of highest impact. This idea also defines a branching strategy since the value with lowest impact can be seen as more likely to lead to a solution. Therefore, once a variable x has been chosen, the value a with the lowest $I(x = a)$ is tried first.

Activity-Based Search (ABS) [9] selects the variable whose domain was most often reduced during propagation. It maintains a mapping A from variables to reals. After Line 1 in Algorithm 1, for every variable x whose domain has been somehow reduced during the process, the value of $A(x)$ is incremented. In order to favor recent activity, for every variable x, $A(x)$ is multiplied by factor $0 \le \gamma \le 1$ before[2]. The variable x with lowest ratio $|\mathcal{D}(x)|/A(x)$ is selected first. Similarly to IBS, one can use the activity to select the value to branch on. The activity of a decision is defined as the number of variables whose domain was reduced as its consequence. It is updated as in IBS, using the same bias α.

The Last Conflict heuristic (LC) [7] relies on a completely different concept. Once both branches ($P|_{x=a}$ and $P|_{x \ne a}$) have failed, the variable x of this choice point is always preferred, until it is successfully assigned, that is, a branch $P|_{x=b}$ (with possibly $b = a$) is tried and the following propagation step succeeds. Indeed, a non-trivial subset of decisions forming a nogood must necessarily contain x. Moreover, if there is a nogood that does not contain the previous decision (or the next, etc.) then this procedure will automatically "backjump" over it. Often, no such variable exists, for instance when the search procedure dives through left branches and therefore a default selection heuristic is required.

Conflict Ordering Search (COS) [2] also tries to focus on variables that failed recently. Here, for every failure (Line 1, Algorithm 1 returning `False`), the variable selected in the parent recursion is stamped by the total number of failures encountered so far. The variable with the highest stamp is selected first. If there are several variables with equal stamp (this can only be 0, i.e., variables which never caused a failure), then a default heuristic is used instead.

Last, Weighted Degree ($wdeg$) [1] maintains a mapping w from constraints to reals. For every failure with culprit constraint C, a decay factor γ is applied exactly as in ABS and $w(C)$ is incremented. The weight $wdeg(x_i)$ of a variable x_i is the sum of the weight of its active neighboring constraints, i.e., those constraining at least another distinct unassigned variable: $wdeg(x_i) = \sum_{C \in \mathcal{C}_i} w(C)$ where $\mathcal{C}_i = \{C \mid C \in \mathcal{C} \wedge x_i \in \mathcal{X}(C) \wedge \exists x_j \ne x_i \in \mathcal{X}(C)$ s.t. $|\mathcal{D}(x_j)| > 1\}$. The variable x with lowest ratio $|\mathcal{D}(x)|/wdeg(x)$ is selected first.

---

[2] In practice, the increment value is divided by γ, and A is scaled down when needed.

4   Explanation-Based Weight

We propose to adapt the weighting function of $wdeg$ to take into account the fact that not every variables in the scope a constraint triggering a failure may be involved in the conflict.

Consider for instance the constraint $\sum_{i=1}^{n} x_i \leq k$ where the initial domains are all in $\{0, 1\}$. When a failure is triggered, the weight of every variable is incremented. However, variables whose domain is equal to $\{1\}$ are sole responsible for this failure. It would be more accurate to increment only their weight.

When a failure occurs for a constraint C with scope $\langle x_1, \ldots, x_k \rangle$, it means that under the current domain $\mathcal{D}$, there is no tuple satisfying C in the cartesian product $\mathcal{D}(x_1) \times, \ldots, \times \mathcal{D}(x_k)$. It follows that there is no solution for the CSP under the current domain. Any subset $E = \{i_1, \ldots, i_m\}$ of $\{1, \ldots, k\}$ such that no tuple in $\mathcal{D}(x_{i_1}) \times, \ldots, \times \mathcal{D}(x_{i_m})$ satisfies C is an *explanation* of the failure of C under the domain $\mathcal{D}$. The set $\{1, \ldots, k\}$ is the trivial explanation, however explanations that are strict subsets represent a valuable information and have been used to develop highly successful search algorithms. The goal in these algorithms (e.g., Nogood Recording [12] and CDCL [6,13,14]) is to use such explanations to derive a nogood, which also entails the failure under the current domain when added to the CSP, however without falsifying individually any constraint. Another goal is to make this nogood as weak as possible since we can add the negation of the nogood as an implied constraint. For this reason, as well as other practical reasons, a more restrictive language of literals is used to represent explanations and nogoods (typically $x = a, x \neq a, x \leq a$ and $x > a$).

Our purpose is simpler and more modest: we simply aim at computing an explanation E as small as possible, and do not care about the unary domain constraints. Indeed we use this explanation only to weight the variables whose indices are in E. We keep a weight function for variables instead of constraints. When a constraint C triggers a failure, we apply a decay factor γ and by default we increment the weight $w(x)$ of every variable $x \in \mathcal{X}(C)$. However, for a small set of constraints (ALLDIFFERENT, ELEMENT, and LINEAR INEQUALITY) for which we have implemented an algorithm for computing an explanation, we update the weight only of the variables involved in the explanation.

There is one difference with $wdeg$ concerning inactive constraints, i.e., with at most one variable currently unassigned. The weight of an inactive constraint does not contribute to the selection of its last variable. We decided to ignore this, and we count the weight of active and inactive constraint alike as it did not appear to be critical. However, one can implement this easily by keeping, for each constraint C and each variable $x \in \mathcal{X}(C)$, the quantity of weight $w(C, x)$ due to C. When a constraint becomes inactive and its last unassigned variable is x, then $w(C, x)$ is subtracted from $w(x)$, and added back upon backtrack.

Notice that since we use these conflicts for informing the heuristic only, they do not actually need to be minimal if, for instance, extracting a minimal conflict is computationally hard for the constraint. Similarly, the explanation does not even need to be correct. The three very straightforward procedures that we implemented and described below produce explanations that are correct but not necessarily minimal. We chose to use fast and easy to implement explanations and found that it was sufficient to improve the behavior of *wdeg*:

ALLDIFFERENT($\langle x_1, \ldots, x_k \rangle$) $\Leftrightarrow \forall 1 \le i < j \le k$, $x_i \ne x_j$, where $\langle x_1, \ldots, x_k \rangle$ is a tuple of integer variables.

We used in our experiment two propagators for that constraint. First, with highest priority, arc consistency is achieved on binary inequalities. If a failure is obtained when processing the inequality $x_i \ne x_j$, it means that both variables are ground and equal, hence we use the conflict set $\{x_i, x_j\}$.

Then, with lower priority, we use the bounds consistency propagation algorithm described in [8]. This algorithm fails if it finds a Hall interval, i.e., a set of at least $b - a + 2$ variables whose domains are included in $\{a, \ldots, b\}$. When this happens we simply use this set of variables as conflict.

LINEAR INEQUALITY($\langle x_1, \ldots, x_k \rangle, \langle a_1, \ldots, a_k \rangle, b$) $\Leftrightarrow \sum_{i=1}^{k} a_i x_i \le b$, where b is an integer, $\langle x_1, \ldots, x_k \rangle$ a tuple of integer variables and $\langle a_1, \ldots, a_k \rangle$ of integers.

The constraint fails if and only if the lower bound of the sum is strictly larger than b. When this is true, a possible conflict is the set containing every variable x_i such that either a_i is positive and $\min(\mathcal{D}(x_i))$ is strictly larger than its initial value, or a_i is negative and $\max(\mathcal{D}(x_i))$ is strictly lower than its initial value.

ELEMENT($\langle x_1, \ldots, x_k \rangle, n, v$) $\Leftrightarrow x_n = v$, where $\langle x_1, \ldots, x_k \rangle$ is a tuple of integer variables, n, v two integer variables.

We use the conflict set $\{n, v\} \cup \{x_i \mid i \in \mathcal{D}(n)\}$, that is, we put weight only on the "index" and "value" variables n and v, as well as every variable of the array pointed to by the index variable n.

5   Experimental Evaluation

Fig. 1a illustrates the difference that explanation can make. The data comes from instance 3-10-20.dzn of the ghoulomb.mzn model. We plot, for each variable x, the number of failures of a constraints involving x (blue crosses) and the number of explanations of failures involving x (red dot). We observe a much wider distribution of the weight when using explanations. As a result, the optimal solution could be proven with *e-wdeg* in 15 s whereas the best upper bound found with *wdeg* after 1200 s was 1.83 times larger than the optimal.

Next, we experimentally evaluated the proposed variant of *wdeg*, denoted *e-wdeg*, against the state-of-the-art general-purpose branching heuristics *wdeg*, *ABS*, *IBS*, *COS* and *LC* with *wdeg* and *e-wdeg* as default heuristic for the two latter. We used lexicographic value ordering for every heuristic, except *IBS* and *ABS* since their default branching strategies were slightly better.[3] The decay factor γ was set to 0.95 for *ABS*, *wdeg* and *e-wdeg*, and the bias α was set to 8 for *ABS* and *IBS*. No probes were used to initialize weights. In all cases the initial values were set up to emulate the "minimum domain over degree" strategy. All the methods were implemented in Mistral-2.0 [5] and the same geometric restarts policy was used in all cases.

---

[3] The extra space requirement was an issue for 5 optimization instances. However the impact on the overall results is quasi null.

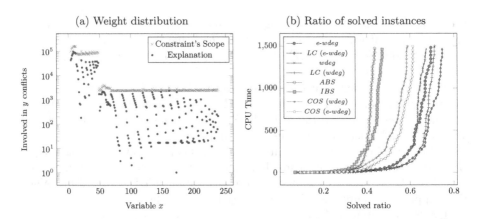

Fig. 1. Weight distribution & search efficiency on satisfaction instances (Color figure online)

We used all the instances of the Minizinc challenge from 2012 to 2015 [15]. This data set contains 399 instances, 323 are optimization problems and 76 are satisfaction problems. In this set, 72 instances have at least one ALLDIFFERENT, 147 have at least one ELEMENT and all have at least one LINEAR INEQUALITY. However, ELEMENT is often posted on arrays of constants and LINEARINEQUAL-ITY may be used to model clauses, which is not favorable to e-wdeg since the explanations are trivial. All the tests ran on Intel Xeon E5430 processors with Linux. The time cutoff was 1500 s for each instance excluding the parsing time. Heuristics were randomized by choosing uniformly between the two best choices, except COS and LC for which only the default heuristics were randomized in the same way. Each configuration was given 5 randomized runs.

5.1 Satisfaction Problems

We first report the results on satisfaction instances, where we plot, for every heuristic, the ratio of runs (among 76×5) in which the instance was solved (x-axis) over time (y-axis) in Fig. 1b. The results are clear. Among previous heuristics, wdeg is very efficient, only outperformed by LC (using wdeg as default heuristic when the testing set is empty). Conflict Ordering Search comes next, followed by IBS and ABS. Notice that IBS and ABS are often initialised using probing. However, this method could be used for other heuristics (see [3]), and is unlikely to be sufficient to close this gap. Whether used as default for COS or LC, or as stand alone, e-wdeg always solves more instances than wdeg. The difference is not significant below a 500 s time limit, but patent above this mark. Overall, e-wdeg solves 4.5%, 3.7% and 2.9% more instances than wdeg when used as default for LC, COS or as stand alone, respectively.

5.2 Optimization Problems

Second, we report the results for the 323 optimization instances (1615 runs). We first plot, for every heuristic, the ratio of instances proven optimal (x-axis) over time (y-axis) in Fig. 2a. Here the gain of explanation-based weights is less clear. *e-wdeg* can prove 2.3% more instances, however *LC* (*e-wdeg*) finds only 0.6% more proofs than *LC* and *COS* (*e-wdeg*) 1.6% more than *COS*.

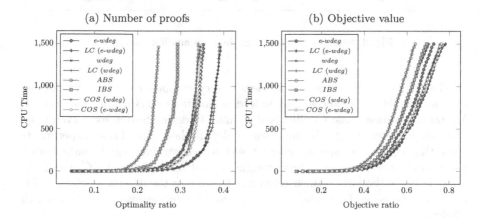

(a) Number of proofs   (b) Objective value

Fig. 2. Search efficiency, optimization instances

In Fig. 2b we plot the normalized objective value of the best solution found by heuristic h (x-axis) after a given time (y-axis). Let $h(I)$ be the objective value of the best solution found using heuristic h on instance I and $lb(I)$ (resp. $ub(I)$) the lowest (resp. highest) objective value found by any heuristic on I. The formula below gives a normalized score in the interval $[0, 1]$:

$$score(h, I) = \begin{cases} \frac{h(I) - lb(I) + 1}{ub(I) - lb(I) + 1}, & \text{if } I \text{ is a maximization instance} \\ \frac{ub(I) - h(I) + 1}{ub(I) - lb(I) + 1}, & \text{otherwise} \end{cases}$$

Notice that we add 1 to the actual and maximum gap. Moreover, if an heuristic h does not find any feasible solution for instance I, we arbitrarily set $h(I)$ to $lb(I) - 1$ for a maximization problem, and $ub(I) + 1$ for a minimization problem. It follows that $score(h, I)$ is equal to 1 if h has found the best solution for this instance among all heuristics, decreases as $h(I)$ gets further from the optimal objective value, and is equal to 0 if and only if h did not find any solution for I.

We observe that using *e-wdeg*, the solver finds significantly better solutions faster than using *wdeg*. The same observation can be made for Last Conflict and Conflict Ordering Search using *e-wdeg* outperforming their counterparts relying on *wdeg*, although the gap is slightly less important in these cases. The overall improvement (respectively 3.2%, 1.6% and 0.8%) is modest, however, the number of data points makes it statistically significant.

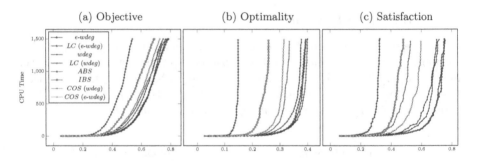

Fig. 3. Search efficiency, branching on auxiliary variables

Overall, the *wdeg* heuristic is clearly very competitive when considering a large sample of instances, and indeed it dominates *IBS* and *ABS* on the MiniZinc instances. Last Conflict and Conflict Ordering Search seem even more efficient, however, they rely on *wdeg* as default heuristic. These experiments show that computing specific conflict sets for constraints significantly boosts *wdeg*. Although the gain is less straightforward in the case of *LC* and *COS* (especially for the latter which relies less heavily on the default heuristic), this approach can be useful in those cases too, and in any case never hinders search efficiency.

Notice that we restricted the heuristic selection to the decision variables specified in the MiniZinc model. We also tried to let the heuristic branch on auxiliary variables, created internally to model expressions. The results, shown in Fig. 3 are not exactly as cleanly cut in this case, and actually difficult to understand. The only two heuristics to actually benefit from this are *wdeg* and *e-wdeg*, by 6.7% and 8.3%, respectively for the objective value criteria. Two heuristics perform much worse, again for the objective value: *ABS* loses 15.2% and *COS* (*e-wdeg*) 4.7%. Other heuristics are all marginally worse in this setting. On other criteria, such as the number of optimality proofs, there is a similar, but not identical trend, with *LC* gaining the most from the extra freedom.

6   Conclusion

We showed that the *wdeg* heuristic can be made more robust to instances with large arity constraints through a relatively simple method. Whereas *wdeg* distributes weights equally among all variables of the constraint that triggered a failure, we propose to weight only the variables of that constraint participating in an explanation of the failure. Our empirical analysis shows that this technique improves the performance of *wdeg*. In particular, the Last Conflict heuristic [7] using the improved version of Weighted Degree, *e-wdeg*, is the most efficient overall on the benchmarks from the previous MiniZinc challenges.

References

1. Boussemart, F., Hemery, F., Lecoutre, C., Sais, L.: Boosting systematic search by weighting constraints. In: Proceedings of the 16th European Conference on Artificial Intelligence - ECAI, pp. 146–150 (2004)

2. Gay, S., Hartert, R., Lecoutre, C., Schaus, P.: Conflict ordering search for scheduling problems. In: Pesant, G. (ed.) CP 2015. LNCS, vol. 9255, pp. 140–148. Springer, Cham (2015). doi:10.1007/978-3-319-23219-5_10

3. Grimes, D., Wallace, R.J.: Sampling strategies and variable selection in weighted degree heuristics. In: Bessière, C. (ed.) CP 2007. LNCS, vol. 4741, pp. 831–838. Springer, Heidelberg (2007). doi:10.1007/978-3-540-74970-7_61

4. Haralick, R.M., Elliott, G.L.: Increasing tree search efficiency for constraint satisfaction problems. In: Proceedings of the 6th International Joint Conference on Artificial Intelligence - IJCAI, pp. 356–364 (1979)

5. Hebrard, E.: Mistral, a constraint satisfaction library. In: Proceedings of the CP-08 Third International CSP Solvers Competition, pp. 31–40 (2008)

6. Bayardo Jr., R., Schrag, R.C.: Using CSP look-back techniques to solve real-world SAT instances. In: Proceedings of the 14th National Conference on Artificial Intelligence - AAAI, pp. 203–208 (1997)

7. Christophe, L., Sas, L., Tabary, S., Vidal, V.: Reasoning from last conflict(s) in constraint programming. Artif. Intell. **173**(18), 1592–1614 (2009)

8. Lopez-Ortiz, A., Quimper, C.G., Tromp, J., Beek, P.V.: A fast and simple algorithm for bounds consistency of the all different constraint. In: Proceedings of the 18th International Joint Conference on AI - IJCAI, pp. 245–250 (2003)

9. Michel, L., Hentenryck, P.: Activity-based search for black-box constraint programming solvers. In: Beldiceanu, N., Jussien, N., Pinson, É. (eds.) CPAIOR 2012. LNCS, vol. 7298, pp. 228–243. Springer, Heidelberg (2012). doi:10.1007/978-3-642-29828-8_15

10. Moskewicz, M.W., Madigan, C.F., Zhao, Y., Zhang, L., Malik, S.: Chaff: engineering an efficient SAT solver. In: Proceedings of the 38th Annual Design Automation Conference - DAC, pp. 530–535 (2001)

11. Refalo, P.: Impact-based search strategies for constraint programming. In: Wallace, M. (ed.) CP 2004. LNCS, vol. 3258, pp. 557–571. Springer, Heidelberg (2004). doi:10.1007/978-3-540-30201-8_41

12. Schiex, T., Verfaillie, G.: Nogood recording for static and dynamic constraint satisfaction problems. In: ICTAI, pp. 48–55 (1993)

13. Silva, J.P.M., Sakallah, K.A.: GRASP - a new Search algorithm for satisfiability. In: Proceedings of the IEEE/ACM International Conference on Computer-aided Design - ICCAD, pp. 220–227 (1996)

14. Silva, M.J.P., Sakallah, K.A.: GRASP: a search algorithm for propositional satisfiability. IEEE Trans. Comput. **48**(5), 506–521 (1999)

15. Stuckey, P.J., Feydy, T., Schutt, A., Tack, G., Fischer, J.: The minizinc challenge 2008–2013. AI Mag. **35**(2), 55–60 (2014)

16. Wallace, R.J., Grimes, D.: Experimental studies of variable selection strategies based on constraint weights. J. Algorithms **63**(1), 114–129 (2008)

Counting Weighted Spanning Trees to Solve Constrained Minimum Spanning Tree Problems

Antoine Delaite[1] and Gilles Pesant[1,2(✉)]

[1] École Polytechnique de Montréal, Montreal, Canada
{antoine.delaite,gilles.pesant}@polymtl.ca
[2] CIRRELT, Université de Montréal, Montreal, Canada

Abstract. Building on previous work about counting the number of spanning trees of an unweighted graph, we consider the case of edge-weighted graphs. We present a generalization of the former result to compute in pseudo-polynomial time the exact number of spanning trees of any given weight, and in particular the number of minimum spanning trees. We derive two ways to compute solution densities, one of them exhibiting a polynomial time complexity. These solution densities of individual edges of the graph can be used to sample weighted spanning trees uniformly at random and, in the context of constraint programming, to achieve domain consistency on the binary edge variables and, more importantly, to guide search through counting-based branching heuristics. We exemplify our contribution using constrained minimum spanning tree problems.

1 Introduction

Counting-based search [16] in CP relies on computing the solution density of each variable-value assignment for a constraint in order to build an integrated variable-selection and value-selection heuristic to solve constraint satisfaction problems. Given a constraint $c(x_1, \ldots, x_\ell)$, its number of solutions $\#c(x_1, \ldots, x_\ell)$, respective finite domains D_i $_{1 \leq i \leq \ell}$, a variable x_i in the scope of c, and a value $d \in D_i$, we call

$$\sigma(x_i, d, c) = \frac{\#c(x_1, \ldots, x_{i-1}, d, x_{i+1}, \ldots, x_\ell)}{\#c(x_1, \ldots, x_\ell)}$$

the *solution density* of pair (x_i, d) in c. It measures how often a certain assignment is part of a solution to c. We can exploit the combinatorial structure of the constraint to design efficient algorithms computing solution densities. Such algorithms have already been designed for several families of constraints such as alldifferent, gcc, regular, and knapsack [12], dispersion [10], and spanningTree for unweighted graphs [3].

When faced with optimization problems however, solutions should not be considered on an equal footing since they will have different costs. Let $f : \mathcal{D} \to \mathbb{R}$

© Springer International Publishing AG 2017
D. Salvagnin and M. Lombardi (Eds.): CPAIOR 2017, LNCS 10335, pp. 176–184, 2017.
DOI: 10.1007/978-3-319-59776-8_14

(with $\mathcal{D} = D_1 \times \cdots \times D_\ell$) associate a cost to each combination of values for variables $x_1, \ldots, x_\ell$ and $z \in [\min_{t \in \mathcal{D}} f(t), \max_{t \in \mathcal{D}} f(t)]$ be a bounded-domain continuous variable. An *optimization constraint* $c^\star(x_1, \ldots, x_\ell, z, f)$ holds if $c(x_1, \ldots, x_\ell)$ is satisfied and $z = f(x_1, \ldots, x_\ell)$. This is a useful concept if the objective function of the problem can be expressed using the z variables of some optimization constraints (or, even better, a single one). Without loss of generality consider minimization problems. Pesant [11] recently generalized the concept of solution density to that of cost-based solution density. Let $\epsilon \geq 0$ be a small real number and $\#c^\star_\epsilon(x_1, \ldots, x_\ell, z, f)$ denote the number of solutions to $c^\star(x_1, \ldots, x_\ell, z, f)$ with $z \leq (1 + \epsilon) \cdot \min_{t \in c(x_1, \ldots, x_\ell)} f(t)$. We call

$$\sigma^\star(x_i, d, c^\star, \epsilon) = \frac{\#c^\star_\epsilon(x_1, \ldots, x_{i-1}, d, x_{i+1}, \ldots, x_\ell, z, f)}{\#c^\star_\epsilon(x_1, \ldots, x_\ell, z, f)}$$

the *cost-based solution density* of pair (x_i, d) in $c^\star$. A value of $\epsilon = 0$ corresponds to the solution density over the optimal solutions to the constraint with respect to f and if there is a single optimal solution then this identifies the corresponding assignment to x_i with a solution density of 1. A positive ϵ gives a margin to include close-to-optimal solutions since there are likely other constraints to satisfy that will rule out the optimal ones with respect to that single optimization constraint. Algorithms to compute cost-based solution densities have already been given for the optimization versions of the `alldifferent`, `regular`, and `dispersion` constraints [11]. In this spirit we present an algorithm that, given a weighted graph, computes the exact number of spanning trees of any given weight and we contribute cost-aware solution densities for a `spanningTree` constraint.

Section 2 reviews related work on tree constraints in CP. Section 3 presents our counting algorithm for weighted spanning trees. Section 4 derives ways to compute solution densities efficiently. Section 5 gives applications of solution density information. Section 6 reports on the usefulness of this work to solve constrained minimum spanning tree problems.

2   Related Work

Previous research in CP on imposed tree structures has focused mostly on filtering algorithms but also includes branching heuristics and explanations. Beldiceanu et al. [2] introduce the tree constraint, which addresses the digraph partitioning problem from a constraint programming perspective. In their work a constraint that enforces a set of vertex-disjoint anti-arborescences is proposed. They achieve domain consistency in $\mathcal{O}(nm)$ time, where n is the number of vertices and m is the number of edges in the graph. Dooms and Katriel [6] introduce the MST constraint, requiring the tree variable to represent a minimum spanning tree of the graph on which the constraint is defined. The authors propose polytime bound consistent filtering algorithms for several restrictions of this constraint. Later Dooms and Katriel [7] propose a weighted spanning tree constraint, in which both the tree and the weight of the edges are variables, and consider several filtering algorithms. The latter is improved by Régin [13], who proposes an incremental domain consistency algorithm running in $\mathcal{O}(m + n \log n)$ time. Subsequently, Régin et al. [14] improve the time complexity further and also consider mandatory edges. De Uña et al. [5] generate explanations for cost-based failure and filtering that occurs in the weighted spanning tree constraint and feed them to a SAT solver. They show empirically that such an approach greatly improves our ability to solve diameter-constrained minimum spanning tree problems in CP. Brockbank et al. [3] build on Kirchhoff's Matrix-Tree Theorem to derive solution densities for the edges of an unweighted graph and use them in a counting-based branching heuristic. In the next sections we generalize this approach for edge-weighted graphs.

3   Counting Weighted Spanning Trees

In the case of an unweighted graph G on n vertices and m edges Kirchhoff's Matrix-Tree Theorem [15] equates the number of spanning trees of G to the absolute value of the determinant of a sub-matrix (the (i, j)-minor) of the *Laplacian matrix* of G, obtained by subtracting the adjacency matrix of G from the diagonal matrix whose i^{th} entry is equal to the degree of vertex v_i in G. Hence the number of spanning trees can be computed as the determinant of a $(n - 1) \times (n - 1)$ matrix, in $\mathcal{O}(n^3)$ time.

Let G now be an edge-weighted multigraph with vertex set V, edge set E, and weight function $w : E \to \mathbb{N}$. Following in Kirchhoff's footsteps Broder and Mayr [4] introduce a derived matrix $M = (m_{ij})_{1 \le i,j \le n}$ whose elements are univariate polynomials built from the edges of G:[1]

$$m_{ij} = \begin{cases} \sum_{e=(v_i,v_j) \in E} -x^{w(e)}, & i \ne j \\ \sum_{e=(v_i,v_k) \in E} x^{w(e)}, & i = j \end{cases} \tag{1}$$

Off-diagonal entries m_{ij} add one monomial per edge between distinct vertices v_i and v_j (there may be several since we consider multigraphs) whereas entries m_{ii} on the diagonal add one monomial per edge incident with vertex v_i.

Figure 1 shows a small graph and its derived matrix M. Observe that if all weights are 0 (or alternatively if we set x to 1) matrix M simplifies to the previously defined Laplacian matrix: it is thus a true generalization of the unweighted case. If we remove the ith row and column of M for any i and compute the determinant of the remaining matrix, we obtain a unique polynomial with very useful characteristics: each one of its monomials $a_k x^k$ indicates the number a_k of spanning trees of weight k. For example say we remove the first row and column of M in Fig. 1 and compute the determinant:

$$\begin{vmatrix} x^2 + 2x^1 & -x^1 & -x^1 & 0 \\ -x^1 & 2x^1 & -x^1 & 0 \\ -x^1 & -x^1 & 2x^3 + 2x^1 & -x^3 \\ 0 & 0 & -x^3 & x^1 + x^3 \end{vmatrix} = 2x^9 + 3x^8 + 7x^7 + 6x^6 + 3x^5$$

---

[1] We generalize slightly their definition in order to include multigraphs, which may occur when we contract edges in the context of CP search.

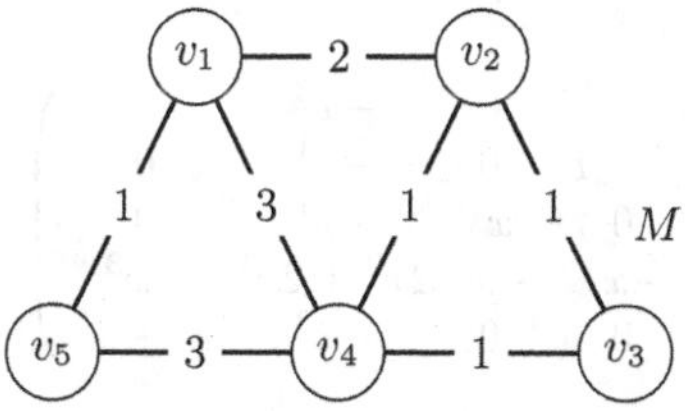

$$M = \begin{pmatrix} x^2 + x^3 + x^1 & -x^2 & 0 & -x^3 & -x^1 \\ -x^2 & x^2 + 2x^1 & -x^1 & -x^1 & 0 \\ 0 & -x^1 & 2x^1 & -x^1 & 0 \\ -x^3 & -x^1 & -x^1 & 2x^3 + 2x^1 & -x^3 \\ -x^1 & 0 & 0 & -x^3 & x^1 + x^3 \end{pmatrix}$$

Fig. 1. A graph and its derived matrix.

Thus we discover that there are two spanning trees of weight 9, three of weight 8, seven of weight 7, six of weight 6, and three of (minimum) weight 5. We will call this polynomial the *wst-polynomial* of G and denote it $P_G(x)$. It directly gives us the number of spanning trees of each weight though computing it requires $\Theta(n^3 W \max(n, W))$ time in the worst case, where $W = \max\{w(e) : e \in E\}$ [8]. Thus it is dependent on the magnitude of the edge weights.

Broder and Mayr [4] were interested in improving the efficiency of counting the number of *minimum* spanning trees, say of weight k^*, and for that purpose they provide an algorithm that, guided by an arbitrary spanning tree, takes linear combinations of some of the columns in order to factor out the corresponding power x^{k^*}, and then evaluates the remaining determinant at $x = 0$ thus isolating the coefficient a_{k^*}. The time complexity of their algorithm is the same as that of matrix multiplication (for $n \times n$ matrices).

4 Computing Cost-Aware Solution Densities

We are interested in counting spanning trees because they are instrumental in establishing the solution density of an edge $e \in E$ given a `spanningTree` constraint on G. A natural way to go about this is to divide the number of spanning trees using that edge by the total number of spanning trees. We first discuss how to compute cost-based solution densities exactly and then present an adaptation of that concept which allows us to lower the time complexity significantly.

4.1 Exact Cost-Based Solution Densities

What is the solution density of a given edge e among all spanning trees of weight k? We can compute the difference between the wst-polynomial of G and that of the same graph without that edge, $P_G(x) - P_{G-e}(x)$. The result is another polynomial giving the number of spanning trees of each weight that use e. From it we can extract the number of spanning trees of weight k which contain e and the corresponding solution densities.

Consider Fig. 2 in which we removed edge (v_2, v_3) from G. We have

$$P_{G-(v_2,v_3)}(x) = x^9 + x^8 + 3x^7 + 2x^6 + x^5$$

and therefore

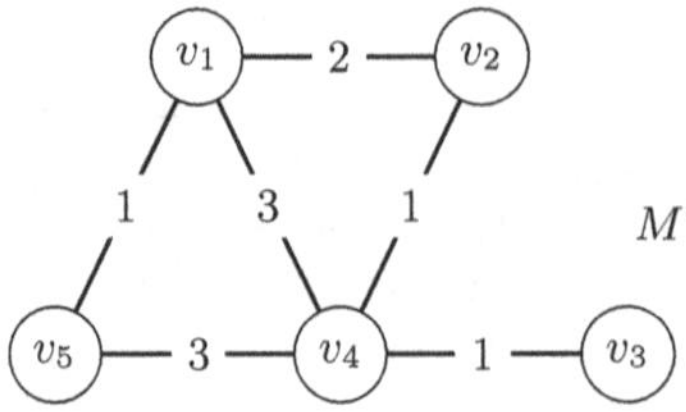

$$M' = \begin{pmatrix} x^2 + x^3 + x^1 & -x^2 & 0 & -x^3 & -x^1 \\ -x^2 & x^2 + x^1 & 0 & -x^1 & 0 \\ 0 & 0 & x^1 & -x^1 & 0 \\ -x^3 & -x^1 & -x^1 2x^3 + 2x^1 & -x^3 \\ -x^1 & 0 & 0 & -x^3 & x^1 + x^3 \end{pmatrix}$$

Fig. 2. The graph at Fig. 1 with edge (v_2, v_3) removed and its derived matrix.

$$P_G(x) - P_{G-(v_2,v_3)}(x) = x^9 + 2x^8 + 4x^7 + 4x^6 + 2x^5.$$

So for example edge (v_2, v_3) is used in four spanning trees of weight 7 and the corresponding solution density is $\frac{4}{7}$ since according to $P_G(x)$ there are seven spanning trees of that weight. For cost-based solution densities, we need to take into account the trees of weight at most $(1 + \epsilon)k^\star$. In our example $k^\star = 5$ and say we take $\epsilon = 0.2$: there are $4 + 2 = 6$ spanning trees of weight at most $(1 + 0.2) \cdot 5 = 6$ using edge (v_2, v_3) and $6 + 3 = 9$ such trees according to $P_G(x)$ (using that edge or not), yielding a cost-based solution density equal to $\frac{6}{9}$, or $\frac{2}{3}$.

4.2 Cost-Damped Solution Densities

Computing solution densities via wst-polynomials may be too time-consuming for some uses such as in a branching heuristic that is called at every node of the search tree. The success of Broder and Mayr [4] in efficiently computing the number of minimum spanning trees came in large part from the evaluation of the determinant for a fixed x, thus working with scalar values instead of polynomials.

Consider applying an exponential decay of sorts to the number a_k of spanning trees of weight k according to the difference between that weight and that of a minimum spanning tree, $k^\star$, thus giving more importance to close-to-optimal trees. We do this by choosing an appropriate value for x. Consider the wst-polynomial $P_G(x) = \sum_{k=k^\star}^{k^{max}} a_k x^k$: evaluating it at $x = 1$ yields $P_G(1) = \sum_{k=k^\star}^{k^{max}} a_k$, the total number of spanning trees regardless of their weight. Using instead some $0 < x < 1$ will have the desired damping effect on the coefficients by applying an additional $x^{k-k^\star}$ factor to a_k relative to $a_{k^\star}$ — the further weight k is from the minimum weight, the smaller the factor. As an illustration if we apply this approach to the example of the previous section with any $x \in [0.3, 0.9]$ we get a cost-damped solution density in the range $[0.62, 0.66]$ (compared to $\frac{2}{3}$ with the previous approach).

To build our matrix M we need to elevate x up to the Wth power, which can be done in $\Theta(\log W)$, and add the appropriate power in M for each edge, taking $\Theta(m \log W)$ time overall. Then we have a scalar matrix from which we proceed as in [3], requiring $\Theta(n^3)$ time to compute the solution density of all the edges incident to a given vertex. So the overall time is in $\mathcal{O}(n^2 \max(n, \log W))$ globally for these incident edges, which is truly polynomial, compared to $\Theta(n^3 W \max(n, W))$

for a single edge if we work directly with the wst-polynomials as described in Sect. 4.1, and which is pseudo-polynomial. And cost-damped solution densities can be updated incrementally as edges are selected or removed from G, in $\Theta(n^2)$ time instead of $\Theta(n^3)$ (see [3]).

5   Applications of Solution Densities

Filtering. As reviewed in Sect. 2 several filtering algorithms were previously proposed for spanning tree constraints. But note that if we already spend time computing solution densities we can achieve domain consistency at no further expense: a solution density of 0 means that the edge can be filtered; a solution density of 1 means that the edge can be fixed. This holds as well for cost-damped solution densities since every solution is still taken into account. With cost-based solution densities we can apply cost-based filtering.

Sampling. The ability to compute solution densities exactly allows us to sample uniformly at random the combinatorial structures on which they are computed. In particular if we compute cost-based solution densities on spanning trees of weighted graphs, we can sample spanning trees of a given weight. We simply consider one edge at a time, deciding whether to include it or not according to its solution density (viewed as a probability), and then updating the solution densities accordingly.

Branching. The initial motivation for this work was to provide solution-counting information from constraints in order to branch using a heuristic relying on such information (i.e. counting-based search [12]). Among them, maxSD and its optimization counterpart, maxSD*, select the variable-value pair with the highest solution density among all constraints. In the next section we report on some experiments using the latter.

6   Experiments

To demonstrate the effectiveness of using solution density information from a spanning tree optimization constraint to guide a branching heuristic, we use it to solve some degree-constrained and diameter-constrained minimum spanning tree problems. Our graph instances are the same as [5], with a number of vertices ranging from 15 to 40 and an edge density ranging from 0.13 to 1 (i.e. a complete graph). We used the IBM ILOG CP v1.6 solver for our implementation and performed our experiments on a AMD Opteron 2.2 GHz with 1 GB of memory. We applied a 20-minute timeout throughout our experiments.

We introduce one binary variable $x_e \in \{0, 1\}$ per edge of the graph and a cost variable k corresponding to the weight of the spanning tree, which we minimize. On these we post optimization constraint $\texttt{spanningTree}(G, \{x_e : e \in E\}, k, w)$ equipped to answer queries about solution densities and enforcing

$$\sum_{e \in E} w(e) \cdot x_e = k$$

$$\sum_{e \in E} x_e = n - 1$$

$$\sum_{e=(v,v') \in E} x_e \geq 1 \quad \forall v \in V$$

$$\sigma^*(x_e, 1, \texttt{spanningTree}) = 1 \quad \Rightarrow \quad x_e = 1 \quad \forall e \in E$$

$$\sigma^*(x_e, 1, \texttt{spanningTree}) = 0 \quad \Rightarrow \quad x_e = 0 \quad \forall e \in E$$

The last two constraints are only present if we use counting-based search and hence need solution densities. We omit parameter ϵ in σ^* because it is not relevant for cost-damped solution densities. We also compute a lower bound at each node of the search tree by running Kruskal's minimum spanning tree algorithm on G updated to reflect the edges that have been decided.

We compare counting-based branching heuristic maxSD* fed with cost-damped solution densities (using $x = 0.9$; the algorithm is not particularly sensitive to that choice) to a reasonable dedicated heuristic that selects edges by increasing weight.

6.1   Degree-Constrained Minimum Spanning Tree Problem

The Degree-Constrained MST Problem requires that we find a minimum spanning tree of G whose vertices have degree at most d [9]. We add to our model

$$\sum_{e=(v,v') \in E} x_e \leq d \quad \forall v \in V$$

We attempt to solve to optimality the previously-mentioned instances with $d = 2, 3,$ and 4. Our results are summarized at Fig. 3: the dedicated heuristic performs almost as well as maxSD* on the $d = 2$ instances (which corresponds to finding a minimum Hamiltonian path) but it is clearly outperformed on the larger degree bounds.

6.2   Diameter-Constrained Minimum Spanning Tree Problem

The Diameter-Constrained MST Problem requires that we find a minimum spanning tree of G such that for any two vertices the path joining them has at most p edges [1]. So in addition to the spanningTree constraint we maintain incrementally the length of a shortest path between each pair of vertices and backtrack if any exceed p. We attempt to solve to optimality the previously-mentioned instances with a diameter p ranging from 4 to 10. Our results are summarized at Fig. 4. The heuristic selecting edges by increasing weight did not solve any instance so it does not appear in the figure. maxSD* manages to solve about half of the instances but is not competitive in terms of computation time and effectiveness with EXPL, the explanation-generating approach of [5], though on

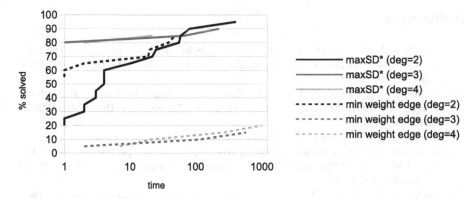

Fig. 3. Percentage of Degree-Constrained MST instances solved to optimality after a given time per instance.

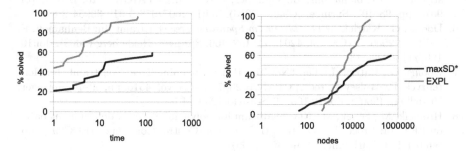

Fig. 4. Percentage of Diameter-Constrained MST instances solved to optimality after a given time (left) or size of search tree (right) per instance.

some instances it yields smaller search trees. Here the CP models are also quite different: the other one is tailored to the Diameter-Constrained MST with parent and height variables, and a dedicated branching heuristic.

7 Conclusion

We presented new algorithms to compute cost-aware solution densities for spanning tree constraints on weighted graphs and showed how they can be used in counting-based branching heuristics to improve our ability to solve constrained minimum spanning tree problems in CP. Such solution densities are also useful for domain filtering and uniform sampling.

    Financial support for this research was provided by NSERC Grant 218028/2012.

References

1. Achuthan, N.R., Caccetta, L., Caccetta, P., Geelen, J.F.: Algorithms for the minimum weight spanning tree with bounded diameter problem. Optim. Tech. Appl. **1**(2), 297–304 (1992)
2. Beldiceanu, N., Flener, P., Lorca, X.: The *tree* constraint. In: Barták, R., Milano, M. (eds.) CPAIOR 2005. LNCS, vol. 3524, pp. 64–78. Springer, Heidelberg (2005). doi:10.1007/11493853_7
3. Brockbank, S., Pesant, G., Rousseau, L.-M.: Counting spanning trees to guide search in constrained spanning tree problems. In: Schulte, C. (ed.) CP 2013. LNCS, vol. 8124, pp. 175–183. Springer, Heidelberg (2013). doi:10.1007/978-3-642-40627-0_16
4. Broder, A.Z., Mayr, E.W.: Counting minimum weight spanning trees. J. Algorithms **24**(1), 171–176 (1997)
5. de Uña, D., Gange, G., Schachte, P., Stuckey, P.J.: Weighted spanning tree constraint with explanations. In: Quimper, C.-G. (ed.) CPAIOR 2016. LNCS, vol. 9676, pp. 98–107. Springer, Cham (2016). doi:10.1007/978-3-319-33954-2_8
6. Dooms, G., Katriel, I.: The *Minimum Spanning Tree* constraint. In: Benhamou, F. (ed.) CP 2006. LNCS, vol. 4204, pp. 152–166. Springer, Heidelberg (2006). doi:10.1007/11889205_13
7. Dooms, G., Katriel, I.: The "Not-Too-Heavy Spanning Tree" constraint. In: Hentenryck, P., Wolsey, L. (eds.) CPAIOR 2007. LNCS, vol. 4510, pp. 59–70. Springer, Heidelberg (2007). doi:10.1007/978-3-540-72397-4_5
8. Hromčík, M., Šebek, M.: New algorithm for polynomial matrix determinant based on FFT. In: Proceedings of the 5th European Control Conference (ECC99), September 1–3, Karlsruhe, Germany, (1999)
9. Subhash, S.C., Ho, C.A.: Degree-constrained minimum spanning tree. Comput. Oper. Res. **7**(4), 239–249 (1980)
10. Pesant, G.: Achieving domain consistency and counting solutions for dispersion constraints. INFORMS J. Comput. **27**(4), 690–703 (2015)
11. Pesant, G.: Counting-based search for constraint optimization problems. In: Schuurmans, D., Wellman, M.P. (eds.) AAAI, pp. 3441–3448. AAAI Press, Palo Alto (2016)
12. Pesant, G., Quimper, C.-G., Zanarini, A.: Counting-based search: branching heuristics for constraint satisfaction problems. J. Artif. Intell. Res. **43**, 173–210 (2012)
13. Régin, J.-C.: Simpler and incremental consistency checking and arc consistency filtering algorithms for the weighted spanning tree constraint. In: Perron, L., Trick, M.A. (eds.) CPAIOR 2008. LNCS, vol. 5015, pp. 233–247. Springer, Heidelberg (2008). doi:10.1007/978-3-540-68155-7_19
14. Régin, J.-C., Rousseau, L.-M., Rueher, M., Hoeve, W.-J.: The weighted spanning tree constraint revisited. In: Lodi, A., Milano, M., Toth, P. (eds.) CPAIOR 2010. LNCS, vol. 6140, pp. 287–291. Springer, Heidelberg (2010). doi:10.1007/978-3-642-13520-0_31
15. Tutte, W.T.: Graph Theory. Encyclopedia of Mathematics and Its Applications. Cambridge University Press, New york (2001)
16. Zanarini, A., Pesant, G.: Solution counting algorithms for constraint-centered search heuristics. In: Bessière, C. (ed.) CP 2007. LNCS, vol. 4741, pp. 743–757. Springer, Heidelberg (2007). doi:10.1007/978-3-540-74970-7_52

The Weighted Arborescence Constraint

Vinasetan Ratheil Houndji[1,2(✉)], Pierre Schaus[1],
Mahouton Norbert Hounkonnou[2], and Laurence Wolsey[1]

[1] Université catholique de Louvain, Louvain-la-Neuve, Belgium
{vinasetan.houndji,pierre.schaus,laurence.wolsey}@uclouvain.be,
ratheil.houndji@ifri.uac.bj
[2] Université d'Abomey-Calavi, Abomey-Calavi, Benin
norbert.hounkounnou@cipma.uac.bj

Abstract. For a directed graph, a Minimum Weight Arborescence (MWA) rooted at a vertex r is a directed spanning tree rooted at r with the minimum total weight. We define the `MinArborescence` constraint to solve constrained arborescence problems (CAP) in Constraint Programming (CP). A filtering based on the LP reduced costs requires $O(|V|^2)$ where $|V|$ is the number of vertices. We propose a procedure to strengthen the quality of the LP reduced costs in some cases, also running in $O(|V|^2)$. Computational results on a variant of CAP show that the additional filtering provided by the constraint reduces the size of the search tree substantially.

1 Introduction

In graph theory, the problem of finding a Minimum Spanning Tree (MST) [13] is one of the most known problem for undirected graphs. The corresponding version for directed graphs is called Minimum Directed Spanning Tree (MDST) or Minimum Weight Arborescence (MWA). It is well known that the graphs are good structures to model some real life problems. The MWA problem has many practical applications in telecommunication networks, computer networks, transportation problems, scheduling problems, etc. It can also be considered as a subproblem in many routing and scheduling problems [9]. For example, Fischetti and Toth ([8]) used MWA problem as a relaxation of Asymmetric Travelling Salesman Problem (ATSP).

Let us formally define the MWA problem. Consider a directed graph $G = (V, E)$ in which $V = \{v_1, v_2, \ldots, v_n\}$ is the vertex set and $E = \{(i, j) : i, j \in V\}$ is the edge set. We associate a weight $w(i, j)$ to each edge (i, j) and we distinguish one vertex $r \in V$ as a root. An arborescence A rooted at r is a directed spanning tree rooted at r. So A is a spanning tree of G if we ignore the direction of edges and there is a directed path in A from r to each other vertex $v \in V$. An MWA $A(G)^\star$ of the graph G is an arborescence with the minimum total cost. Without loss of generality, we can remove any edge entering in the root r. Consider a subset of vertices $S \subseteq V$. Let δ_S^{in} be the set of edges entering S: $\delta_S^{in} = \{(i, j) \in E : (i \in V \setminus S) \wedge (j \in S)\}$. For a vertex $k \in V$, δ_k^{in} is the

© Springer International Publishing AG 2017
D. Salvagnin and M. Lombardi (Eds.): CPAIOR 2017, LNCS 10335, pp. 185–201, 2017.
DOI: 10.1007/978-3-319-59776-8_15

set of edges that enter in k. Let V' be the set of vertices without the root r: $V' = V \setminus \{r\}$. The MWA problem can be formulated as follows [10]:

$$w(A(G)^\star) = \min \sum_{(i,j) \in E} w(i,j) \cdot x_{i,j} \tag{1}$$

$$\text{(MWA)} \quad \sum_{(i,j) \in \delta_j^{in}} x_{i,j} = 1, \forall j \in V' \tag{2}$$

$$\sum_{(i,j) \in \delta_S^{in}} x_{i,j} \geq 1, \forall S \subseteq V' : |S| \geq 2 \tag{3}$$

$$x_{i,j} \in \{0,1\}, \forall (i,j) \in E \tag{4}$$

in which $x_{i,j} = 1$ if the edge (i,j) is in the optimal arborescence $A(G)^\star$ and $x_{i,j} = 0$ otherwise. The first group of constraints imposes that exactly one edge enters in each vertex $j \in V'$ and the constraints (3) enforce the existence of a path from the root r to all other vertices. Without loss of generality [9], we assume that $w(i,i) = \infty, \forall i \in V$ and $w(i,j) > 0, \forall (i,j) \in E$. Then the constraints (4) can be relaxed to $x_{i,j} \geq 0, \forall i,j (5)$ and the constraints (2) become redundant (see [6]).

Here, we address the Constrained Aborescence Problem (CAP) - that requires one to find an arborescence that satisfies other side constraints and is of minimum cost - and show how one can handle them in CP. After some theoretical preliminaries, we define the `MinArborescence` constraint and show how to filter the decision variables. Then we describe how Linear Programming (LP) reduced costs can be improved. Finally, we show some experimental results and conclude.

2   Background

Algorithms to compute an MWA $A(G)^\star$ of a given graph G were proposed independently by Chu and Liu ([3]), Edmonds ([6]) and Bock ([2]). A basic implementation of that algorithm is in $O(|V||E|)$. The associated algorithm is often called Edmonds' Algorithm. An $O(\min\{|V|^2, |E| \log |V|\})$ implementation of the Edmonds' algorithm is proposed by [23]. More sophisticated implementations exist (see for example [12,19]). Fischetti and Toth [9] proposed an $O(|V|^2)$ implementation to compute an MWA and also the associated linear programming reduced costs. We rely on this algorithm for filtering the `MinArborescence` constraint.

An MWA has two important properties that are used to construct it [16].

Proposition 1. *A subgraph $A = (V, F)$ of the graph G is an arborescence rooted at r if and only if A has no cycle, and for each vertex $v \neq r$, there is exactly one edge in F that enters v.*

Proposition 2. *For each $v \neq r$, select the cheapest edge entering v (breaking ties arbitrarily), and let $F^\star$ be this set of $n - 1$ edges. If $(V, F^\star)$ is an arborescence, then it is a minimum cost arborescence, otherwise $w(V, F^\star)$ is a lower bound on the minimum cost arborescence.*

The LP dual problem $\mathcal{D}_{\text{MWA}}$ of MWA is [9]:

$$\max \sum_{S \subseteq V'} u_S$$

$$(\mathcal{D}_{\text{MWA}}) \quad w(i,j) - \sum_{(i,j) \in \delta_S^{in}, \forall S \subseteq V'} u_S \geq 0, \forall (i,j) \in E$$

$$u_S \geq 0, \forall S \subseteq V'$$

in which u_S is the dual variable associated to the subset of edges $S \subseteq V'$.

Let $rc(i,j)$ be the LP reduced cost associated to the edge (i,j). The necessary and sufficient conditions for the optimality of MWA (with primal solution $x_{i,j}^\star$) and $\mathcal{D}_{\text{MWA}}$ (with dual solution $u_S^\star$) are [9]:

1. primal solution $x_{i,j}^\star$ satisfies the constraints (3) and (5)
2. $u_S^\star \geq 0$ for each $S \subseteq V'$
3. reduced cost $rc(i,j) = w(i,j) - \sum_{(i,j) \in \delta_S^{in}, \forall S \subseteq V'} u_S^\star \geq 0$ for each $(i,j) \in E$
4. $rc(i,j) = 0$ for each $(i,j) \in E$ such that $x_{i,j}^\star > 0$
5. $\sum_{(i,j) \in \delta_S^{in}} x_{i,j}^\star = 1$ for each $S \subseteq V'$ such that $u_S^\star > 0$

Algorithm 1 [9] describes the global behavior of algorithms for computing an MWA $A(G)^\star$ for a graph G rooted at a given vertex r. Note that the optimality condition 5. implies that: for each $S \subseteq V'$, $u_S^\star > 0 \implies \sum_{(i,j) \in \delta_S^{in}} x_{i,j}^\star = 1$ and $\sum_{(i,j) \in \delta_S^{in}} x_{i,j}^\star > 1 \implies u_S^\star = 0$ (because $u_S^\star \geq 0$). The different values of dual variables $u_S, \forall S \subseteq V'$ are obtained during the execution of Edmonds algorithm. Actually, for each vertex $k \in V : u_k^\star = \arg\min_{(v,k) \in \delta_k^{in}} w(v,k)$ (line 8 of Algorithm 1). If a subset $S \subseteq V' : |S| \geq 2$ is a strong component[1], then $u_S^\star$ is the minimum reduced cost of edges in δ_S^{in}. Only these subsets can have $u_S^\star > 0$. All other subsets ($S \subseteq V' : |S| \geq 2$ and S is not a strong component) have $u_S^\star = 0$. So there are $O(|V|)$ subsets $S \subseteq V' : |S| \geq 2$ that can have $u_S^\star > 0$. A straightforward algorithm can compute $rc(i,j) = w(i,j) - \sum_{(i,j) \in \delta_S^{in}, \forall S \subseteq V'} u_S^\star, \forall (i,j)$ in $O(|V|^3)$ by considering, for each edge, the $O(|V|)$ subsets that are directed cycles. Fischetti and Toth [9] proposed an $O(|V|^2)$ algorithm to compute $rc(i,j), \forall (i,j) \in E$.

Consider the graph $G_1 = (V_1, E_1)$ in Fig. 1 with the vertex 0 as root. Figures 2, 3 and 4 show the different steps needed to construct $A(G_1)^\star$.

Related works in CP. Many works focused on filtering the Minimum Spanning Tree (MST) constraints on undirected weighted graphs (see for examples [4,5,21,22]). About directed graphs, Lorca ([17]) proposed some constraints on trees and forests. In particular, the `tree` constraint [1,17] is about anti-arborescence on directed graph. By considering a set of vertices (called resource),

---

[1] A strong component of a graph G is a maximal (with respect to set inclusion) vertex set $S \subseteq V$ such that (i) $|S| = 1$ or (ii) for each pair of distinct vertices i and j in S, at least one path exists in G from vertex i to vertex j [9].

Algorithm 1. Computation of a minimum weight arborescence $A(G)^\star$ rooted at vertex r

---

Input: $G = (V, E)$; $r \in V$; $w(e), \forall e \in E$

    // all primal and dual variables $x^\star_{i,j}$ and $u^\star_S$ are assumed to be zero

1 **foreach** *each edge* $(i, j) \in E$ **do**

2     $\lfloor$ $rc(i, j) \leftarrow w(i, j)$

3 $A_0 \leftarrow \emptyset$; $h \leftarrow 0$

    // Phase 1:

4 **while** $G_0 = (V, A_0)$ *is not r-connected* **do**

        // A graph is r-connected iff there is a path from the vertex r
           to each other vertex v in V.

5      $h \leftarrow h + 1$

6   Find any strong component S_h of G_0 such that $r \notin S_h$ and $A_0 \cap \delta^{in}_{S_h} = \emptyset$

         // If $|S_h| > 1$, then S_h is a directed cycle

7   Determine the edge (i_h, j_h) in $\delta^{in}_{S_h}$ such that $rc(i_h, j_h) \leq rc(e), \forall e \in \delta^{in}_{S_h}$

8      $u^\star_{S_h} \leftarrow rc(i_h, j_h)$ // dual variable associated to S_h

9      $x^\star_{i_h, j_h} \leftarrow 1$; $A_0 \leftarrow A_0 \cup \{i_h, j_h\}$

10   **foreach** *each edge* $(i, j) \in \delta^{in}_{S_h}$ **do**

11      $\lfloor$ $rc(i, j) \leftarrow rc(i, j) - u^\star_{S_h}$

    // Phase 2:

12 $t \leftarrow h$

13 **while** $t \geq 1$ **do**

        // Extend A_0 to an arborescence by letting all but one edge of
           each strong component S

14   **if** $x^\star_{i_t, j_t} = 1$ **then**

15   **foreach** *each* $q < t$ *such that* $j_t \in S_q$ **do**

16            $x^\star_{i_q, j_q} \leftarrow 0$

17            $A_0 \leftarrow A_0 \setminus \{(i_q, j_q)\}$

18      $t \leftarrow t - 1$

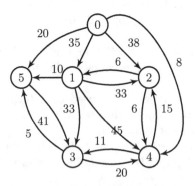

Fig. 1. Initial graph G_1

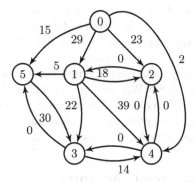

Fig. 2. Computation of $A(G_1)^\star$: Phase 1, after selection of all single vertices. $A_0 = \{(2,1),(4,2),(4,3),(2,4),(3,5)\}$.

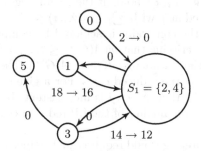

Fig. 3. Computation of $A(G_1)^\star$: Phase 1, after the detection of the size 2 strong component $\{2,4\}$. $A_0 = \{(2,1),(4,2),(4,3),(2,4),(3,5),(0,4)\}$.

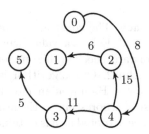

Fig. 4. Computation of $A(G_1)^\star$: Phase 2. $A_0 = \{(2,1),(4,2),(4,3),(3,5),(0,4)\}$. $(2,4)$ is removed.

the `tree` constraint partitions the vertices of a graph into a set of disjoint anti-arborescences such that each anti-arborescence points to a resource vertex. The author proposed a GAC filtering algorithm in $O(|E||V|)$ that is in $O(|V|^3)$. Further, Lorca and Fages [18] propose an $O(|E| + |V|)$ filtering algorithm for this constraint.

Here, we focus on an optimization oriented constraint for MWA mentioned in [11]. The authors consider MWA as a relaxation of the Traveling Salesman Problem (TSP) and use LP reduced costs to filter inconsistent values. In this paper, we formally define the optimization constraint for MWA and propose an algorithm to improve the LP reduced costs of MWA.

3   The `MinArborescence` Constraint

To define the `MinArborescence` constraint, we use the predecessor variable representation of a graph. The arborescence is modeled with one variable X_i for each vertex i of G representing its predecessor. The initial domain of a variable X_i is thus the neighbors of i in G: $j \in D(X_i) \equiv (j, i) \in E$. The constraint `MinArborescence(X, w, r, K)` holds if the set of edges $\{(X_i, i) : i \neq r\}$ is a valid arborescence rooted at r with $\sum_{i \neq r} w(X_i, i) \leq K$.

The consistency of the constraint is achieved by computing an exact MWA $A(G)^\star$ rooted at r and verifying that $w(A(G)^\star) \leq K$. The value $w(A(G)^\star)$ is an exact lower bound for the variable K: $K \geq w(A(G)^\star)$. The filtering of the edges can be achieved based on the reduced costs. For a given edge $(i, j) \notin A(G)^\star$, if $w(A(G)^\star) + rc(i, j) > K$, then $X_i \leftarrow j$ is inconsistent. In Sect. 4, we propose a procedure to strengthen the quality of the LP reduced costs in $O(|V|^2)$ in some cases.

Note that, a GAC filtering would require exact reduced costs, that to the best of our knowledge can only be obtained by recomputing an MWA from scratch in $O(|E||V|^2)$ which is in $O(|V|^4)$.

The basic decomposition of the `MinArborescence` constraint does not scale well due to the exponential number of constraints in Eq. (3). We propose a light constraint (called the `Arborescence` constraint) to have a scalable baseline model for experiments.

Decomposing `MinArborescence`. The constraint `Arborescence(X, r)` holds if the set of edges $\{(X_i, i) : \forall i \in V'\}$ is a valid arborescence rooted at r. We introduce an incremental forward checking like incremental filtering procedure for this constraint in Algorithm 2. Our algorithm is inspired by the filtering described in [20] for enforcing a Hamiltonian circuit. During the search, the bound variables form a forest of arborescences. Eventually when all the variables are bound, a unique arborescence rooted at r is obtained. The filtering maintains for each node:

1. a reversible integer value $localRoot[i]$ defined as $localRoot[i] = i$ if X_i is not bound, otherwise it is recursively defined as $localRoot[X_i]$, and
2. a reversible set $leafNodes[i]$ that contains the leaf nodes of i in the forest formed by the bound variables.

The filtering prevents cycles by removing from any successor variable X_v all the values corresponding to its leaf nodes (i.e. the ones in the set $leafNodes[v]$). Algorithm 2 registers the filtering procedure to the bind events such that $\texttt{bind(i)}$ is called whenever the variable X_i is bind. This $\texttt{bind}$ procedure then finds the root lr of the (sub) arborescence at line 15. Notice that j is not necessarily the root as vertex j might well have been the leaf node of another arborescence that is now being connected by the binding of X_i to value j. This root lr then inherits all the leaf nodes of i. None of these leaf nodes is allowed to become a successor of lr, otherwise a cycle would be created.

Algorithm 2. Class Arborescence

```
1  X: array of |V| variables;
2  localRoot: array of |V| reversible int;
3  leafNodes: array of |V| reversible set of int;

4  Method init()
5  │   X_r ← r ;                              // self loop on r
6  │   foreach each vertex v ∈ V' do
7  │   │   leafNodes[v].insert(v) ;
8  │   │   X_v.removeValue(v) ;               // no self loop
9  │   │   if X_v.isBound then
10 │   │   └   bind(v);
11 │   │   else
12 │   │   └   X_v.registerOnBindChanges() ;

13 Method bind(i: int)
14 │   j ← X_i ;                              // edge (j,i) ∈ arborescence
15 │   lr ← localRoot[j] ;                    // local root of j
16 │   foreach each v ∈ leafNodes[i] do
17 │   │   localRoot[v] ← lr ;
18 │   │   leafNodes[lr].insert(v) ;
19 │   └   X_lr.removeValue(v) ;
```

The $\texttt{MinArborescence(X, w, r, K)}$ constraint can be decomposed as the $\texttt{Arborescence(X, r)}$ constraint plus $\sum_{i \neq r} w(X_i, i) \leq K$, a sum over element constraints.

4   Improved Reduced Costs

Let $A(G)^{\star}_{i \rightarrow j}$ be an MWA of the graph G when the edge (i, j) is forced to be in it. We know that the LP reduced cost $rc(i, j)$ gives a lower bound on the associated cost increase: $w(A(G)^{\star}_{i \rightarrow j}) \geq w(A(G)^{\star}) + rc(i, j)$. However, this lower bound on $w(A(G)^{\star}_{i \rightarrow j})$ can be improved in some cases.

Let us use the following notation to characterize an MWA $A(G)^{\star}$.
Let $pred[v], \forall v \in V'$, be the vertex in V such that the edge $(pred[v], v) \in A(G)^{\star}$: $x^{\star}_{pred[v],v} = 1$. For example, in the graph G_1, $pred[1] = 2$ and $pred[5] = 3$. Let $parent[S]$ be the smallest cycle strictly containing the subset $S \subseteq V'$: $parent[S]$ is the smallest subset $> |S|$ such that $\sum_{(i,j) \in \delta^{in}_{parent[S]}} x^{\star}_{i,j} = 1$ and $S \subset parent[S]$. In other words, $parent[S]$ is the first directed cycle that includes the subset S found during the execution of the Edmonds' algorithm. We assume that $parent[S] = \emptyset$ if there is no such cycle and $parent[\emptyset] = \emptyset$. In the graph G_1, $parent[1] = parent[3] = parent[5] = \emptyset$ and $parent[2] = parent[4] = \{2, 4\}$. Here, $parent[parent[k]] = \emptyset, \forall k \in V'$.

The next three properties give some information to improve the LP reduced costs when all vertices involved do not have a parent.

Property 1. Assume that there is a path $\mathcal{P} = (j, \dots, i)$ from the vertex j to vertex i in $A(G)^{\star}$ such that $\forall k \in \mathcal{P} : parent[k] = \emptyset$. If the edge (i, j) is forced to be in $A(G)^{\star}$, then the cycle $c = (k : k \in \mathcal{P})$ will be created during the execution of Edmonds' algorithm.

Proof. $parent[k] = \emptyset$ means that $pred[k]$ is such that $w(pred[k], k) = \min\{w(v, k) : (v, k) \in \delta^{in}_k\}$ and then $pred[k]$ will be first selected by Edmonds' algorithm $\forall k \in \mathcal{P} \setminus \{j\}$. On the other hand, if the edge (i, j) is forced into the MWA it implies that all other edges entering j are removed. Consequently, the cycle $c = (k : k \in \mathcal{P})$ will be created. □

Let us use the following notations to evaluate the improved reduced costs. Let $min1[k] = \arg\min_{(v,k) \in \delta^{in}_k} w(v, k)$ be the minimum cost edge entering the vertex k. If there is more than one edge with the smallest weight, we choose one of them arbitrarily. Also, let $min2[k] = \arg\min_{(v,k) \in \delta^{in}_k \wedge (v,k) \neq min1} w(v, k)$ be the second minimum cost edge entering k. For each vertex $k \in V$, let $bestTwoDiff[k]$ be the difference between the best two minimum costs of edges entering the vertex k: $\forall k \in V, bestTwoDiff[k] = w(min2[k]) - w(min1[k])$. For instance, in the graph G_1: $min1[5] = (3, 5)$, $min2[5] = (1, 5)$ and $bestTwoDiff[5] = 10 - 5 = 5$.

Property 2. Consider the cycle $c = (i, \dots, j)$ obtained by forcing the edge (i, j) such that $parent[k] = \emptyset, \forall k \in c$ (see Property 1). The minimum cost increase if the cycle is broken/connected by the vertex $k' \in c$ is $bestTwoDiff[k']$.

Proof. For a given $k' \in c$, $parent[k'] = \emptyset$ implies that the edge $(pred[k'], k') = min1[k']$. Then the cheapest way to break the cycle by k' is to use the edge with the second minimum cost $min2[k']$. Hence the minimum cost increase if the cycle is broken by the vertex k' is $w(min2[k']) - w(min1[k']) = bestTwoDiff[k']$. □

When $parent[j] = \emptyset$, the LP reduced costs $rc(i,j)$ are simple and can be easily interpreted.

Property 3. Consider a vertex $j \in V'$ such that $parent[j] = \emptyset$. For all $i \in V \setminus \{j\}$ with $(i,j) \notin A(G)^*$: $rc(i,j) = w(i,j) - w(pred[j],j)$.

Proof. We know that if $parent[j] = \emptyset$, then for each $S \subseteq V'$ with $j \in S$ and $|S| \geq 2$, $u_S^* = 0$ (because none of them is a directed cycle). Then $rc(i,j) = w(i,j) - u_j^*$. On the other hand, since $parent[j] = \emptyset$, the edge $min1[j]$ is the one used to connect j in $A(G)^*$. So $u_j^* = w(min1[j]) = w(pred[j],j)$ and $rc(i,j) = w(i,j) - w(pred[j],j)$. □

By considering an MWA $A(G)^*$, the interpretation of $rc(i,j)$ when $parent[j] = \emptyset$ is that the edge (i,j) is forced into $A(G)^*$ and the edge $(pred[j],j)$ that is used to connect j is removed. Intuitively, if this process induces a new cycle, this latter has to be (re)connected to the rest of the arborescence from a vertex in the cycle different from j. Proposition 3 established below, gives a first improved reduced cost expression when a new cycle c is created by forcing (i,j) into the arborescence and $\forall k \in c, parent[k] = \emptyset$. Note that such new cycle will be created only if there is already a path in $A(G)^*$ from the vertex j to the vertex i.

Proposition 3. *Assume that there is a path $\mathcal{P} = (j, \ldots, i)$ from the vertex j to vertex i in $A(G)^*$ such that $\forall k \in \mathcal{P} : parent[k] = \emptyset$. We have: $w(A(G)^*_{i \to j}) \geq w(A(G)^*) + rc(i,j) + \min_{k \in \mathcal{P} \setminus \{j\}}\{bestTwoDiff[k]\}$.*

Proof. Without loss of generality, we assume that the cycle $c = (k : k \in \mathcal{P})$ is the first one created by the Edmonds' algorithm (see Property 1). After this step, the new set of vertices is $V' = \{c\} \cup \{v \in V : v \notin c\}$. In $A(G)^*$, the edges assigned to vertices in c do not influence the choice of edges for each vertex $v \in V : v \notin c$ (because $parent[k] = \emptyset, \forall k \in c$). So $w(A(G)^*_{i \to j}) \geq \sum_{k \in V \wedge k \notin c} w(pred[k],k) + w(c)$ in which $w(c)$ is the minimum sum of costs of edges when exactly one edge is assigned to each vertex in c without cycle. The cheapest way to connect all vertices in c such that each one has exactly one entering edge is to use all cheapest entering edges $(min1[k], \forall k \in c)$. The cycle obtained must be broken in the cheapest way. To do so, the vertex used: (1) must be different from j (because (i,j) is already there) and (2) have to induce the minimal cost increase. Then, from Property 2, a lower bound on the minimum cost increase is $\min_{k \in \mathcal{P} \setminus \{j\}}\{bestTwoDiff[k]\}$. In addition, we have to add the cost of the forced edge (i,j) and remove the cost of the edge in $A(G)^*$ that enters in j: $w(c) \geq \sum_{k \in c} w(pred[k],k) + \min_{k \in c \setminus \{j\}}\{bestTwoDiff[k]\} + w(i,j) - w(pred[j],j)$. We know that $\sum_{k \in c} w(pred[k],k) + \sum_{k \in V \wedge k \notin c} w(pred[k],k) = w(A(G)^*)$ and $rc(i,j) = w(i,j) - w(pred[j],j)$ (see Property 3).
Thus $w(A(G)^*_{i \to j}) \geq w(A(G)^*) + rc(i,j) + \min_{k \in \mathcal{P} \setminus \{j\}}\{bestTwoDiff[k]\}$. □

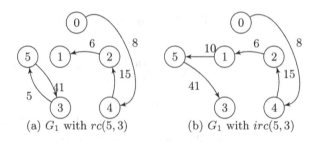

(a) G_1 with $rc(5,3)$ (b) G_1 with $irc(5,3)$

Fig. 5. G_1 with $rc(5,3)$ and $irc(5,3)$

Example 1. Consider the graph G_1 presented in Fig. 1 and its MWA (Fig. 4). We want to force $(5,3)$ to be into the MWA:

- $rc(5,3) = w(5,3) - w(4,3) = 41 - 11 = 30$. This operation leads to the graph shown in Fig. 5(a). We can see that the new cycle created $c = (5,3)$ must be broken from a vertex different from 3.
- $irc(5,3) = rc(5,3) + bestTwoDiff[5] = 30 + 5 = 35$. The corresponding graph is shown in Fig. 5(b), that actually is the new MWA.

Property 2 and Proposition 3 can be generalized into Property 4 and Proposition 4 below to include some vertices that have one parent.

Property 4. Consider an ordered set of vertices $(k_1, k_2, \ldots, k_n)$ in $A(G)^\star$ such that $c = (k_1, k_2, \ldots, k_n)$ is a directed cycle connected (broken) by the vertex $k^\star$ and $parent[c] = \emptyset$. The minimum cost increase if the cycle is broken by another vertex $k' \in c \setminus \{k^\star\}$ is $\geq bestTwoDiff[k'] - u_c^\star$.

Proof. We know that $parent[c] = \emptyset$ implies $u_c^\star = \min\{w(min2[k]) - w(min1[k]) : k \in c\} = w(min2[k^\star]) - w(min1[k^\star])$. So if the cycle is now connected by another vertex than $k^\star$, the edge $min1[k^\star]$ can be used instead of $min2[k^\star]$ and decreases the cost by $w(min2[k^\star]) - w(min1[k^\star]) = u_c^\star$. On the other hand, $\forall k_i \in c \setminus \{k^\star\}$, $(pred[k_i], k_i) = min1[k_i]$. The cheapest way to use k' to connect c is to use $min2[k']$. Hence, a lower bound on the total cost induced is $min2[k'] - min1[k'] - u_c^\star = bestTwoDiff[k'] - u_c^\star$. □

Now the improved reduced costs can be formulated as follows.

Proposition 4. *Assume that there is a path $\mathcal{P} = (j, \ldots, i)$ from the vertex j to vertex i in $A(G)^\star$ such that $\forall k \in \mathcal{P} : parent[parent[k]] = \emptyset$. Then $w(A(G)_{i \to j}^\star) \geq w(A(G)^\star) + rc(i, j) + \min_{k \in \mathcal{P} \setminus \{j\}} \{bestTwoDiff[k] - u_{parent[k]}^\star\}$.*

Proof. Note that if $\forall k \in \mathcal{P} : parent[k] = \emptyset$ (that implies that $u_{parent[k]}^\star = 0$), the formula is the same as the one of Proposition 3. Let Z denote the set of vertices in $\mathcal{P}$ and in all other cycles linked to $\mathcal{P}$. Formally, $Z = \{v \in V : v \in \mathcal{P}\} \cup \{k \in V : \exists v \in \mathcal{P} \wedge parent[k] = parent[v]\}$. We know that, in $A(G)^\star$, the edges assigned to vertices in Z do not influence the choice of edges for each vertex

$k \in V \setminus Z$. So $w(A(G)^\star_{i\to j}) \geq \sum_{k\in V\setminus Z} w(pred[k], k) + w(Z)$ in which $w(Z)$ is the minimum sum of the costs of edges when we assign exactly one edge to each vertex in Z without cycle. The reasoning is close to the proof of Proposition 3. The differences here are:

1. $\exists k \in \mathcal{P} \setminus \{j\}$: $parent[k] \neq \emptyset$. Assume that we want to break the cycle by one vertex $v^\star$ in $\mathcal{P} \setminus \{j\}$. If the vertex $v^\star$ used is such that $parent[v^\star] = \emptyset$, then the minimum cost to pay is $\geq bestTwoDiff[v^\star]$ (here $u^\star_{parent[v^\star]} = 0$ because $parent[v^\star] = \emptyset$). If $v^\star$ is such that $parent[v^\star] \neq \emptyset \wedge parent[parent[v^\star]] = \emptyset$, then from Property 4, the cost to pay is $\geq bestTwoDiff[v^\star] - u^\star_{parent[v^\star]}$. By considering all vertices in $\mathcal{P} \setminus \{j\}$, the cost to pay is then $\geq \min_{k\in\mathcal{P}\setminus\{j\}}\{bestTwoDiff[k] - u^\star_{parent[k]}\}$.

2. the vertex j may have one parent. Let $connect$ be the vertex that is used to connect the cycle $parent[j]$ in $A(G)^\star$.
   Case 1: $parent[i] \neq parent[j]$. If we force the edge (i, j), then the cycle $parent[j]$ should not be created because it is as if all edges entering j but (i, j) are removed. First, assume that $j \neq connect$. The edge in $parent[j]$ not in $A(G)^\star$ should be used and the cost won is $u^\star_{parent[j]}$ (as in the proof of Property 4). Thus, a lower bound on the cost to break the cycle $parent[j]$ by j is: $w(i, j) - w(pred[j], j) - u^\star_{parent[j]}$. This lower bound is equal to $rc(i, j)$ because $w(pred[j], j) = w(min1[j])$. Now assume that $j = connect$. In this case $(pred[j], j) = min2[j]$ (because $parent[parent[j]] = \emptyset$). Using the edge (i, j) instead of $(pred[j], j)$ induces the cost $w(i, j) - w(min2[j]) = w(i, j) - w(min1[j]) - w(min2[j]) + w(min1[j]) = w(i, j) - w(min1[j]) - u^\star_{parent[j]} = rc(i, j)$.
   Case 2: $parent[i] = parent[j] \neq \emptyset$. This means that the edge (i, j) is the one of the cycle that is not in the MWA. In this case $rc(i, j) = 0$, and the new cycle created should be broken as described above (1.).

Hence, a lower bound on $w(Z)$ is

$$\sum_{k\in Z} w(pred[k], k) + \min_{k\in\mathcal{P}\setminus\{j\}}\{bestTwoDiff[k] - u^\star_{parent[k]}\} + rc(i, j)$$

and $w(A(G)^\star_{i\to j}) \geq w(A(G)^\star) + \min_{k\in\mathcal{P}\setminus\{j\}}\{bestTwoDiff[k] - u^\star_{parent[k]}\} + rc(i, j)$. □

Note that $parent[parent[k]] = \emptyset$ if k is not in a cycle or k is in a cycle that is not contained in a larger cycle. The formula of Proposition 4 is available only if $\forall k \in \mathcal{P}$: $parent[parent[k]] = \emptyset$. Let $irc(i, j)$ denote the improved reduced cost of the edge (i, j): $irc(i, j) = rc(i, j) + \max\{\min_{k\in\mathcal{P}\wedge k\neq j}\{bestTwoDiff[k] - u^\star_{parent[k]}\}, 0\}$ if the assumption of Proposition 4 is true and $irc(i, j) = rc(i, j)$ otherwise.

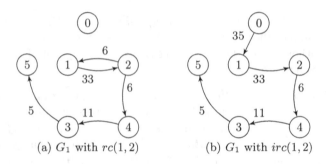

(a) G_1 with $rc(1,2)$ (b) G_1 with $irc(1,2)$

Fig. 6. G_1 with $rc(1,2)$ and $irc(1,2)$

Example 2. Consider the graph G_1 in Fig. 1 and its MWA $A(G_1)^*$ in Fig. 4. For the construction of $A(G_1)^*$, the cycle $c_1 = \{2,4\}$ is created. We want to force into the MWA the edge:

1. $(1,2)$: $rc(1,2) = w(1,2) - u_2^* - u_{c_1}^* = w(1,2) - w(4,2) - (w(0,4) - w(2,4))$. $rc(1,2) = 16$. The corresponding graph is presented in Fig. 6(a). Of course, the new cycle $(1,2)$ created must be broken from the vertex 1. $irc(1,2) = rc(1,2) + (w(0,1) - w(2,1)) = 16 + 29 = 45$. Actually, that is the exact reduced cost since the new graph obtained is an arborescence (see Fig. 6(b)).
2. $(1,4)$: $rc(1,4) = w(1,4) - u_4^* - u_{c_1}^* = w(1,4) - w(2,4) - (w(0,4) - w(2,4))$. $rc(1,4) = 37$. The corresponding graph is presented in Fig. 7 (a). But $irc(1,4) = rc(1,4) + \min\{w(0,1) - w(2,1) = 29, w(1,2) - w(4,2) - u_{c_1}^* = 16\}$. So $irc(1,4) = 37 + 16 = 53$. Here (see Fig. 7 (b)), the graph obtained is not an arborescence and $irc(1,4)$ is a lower bound.

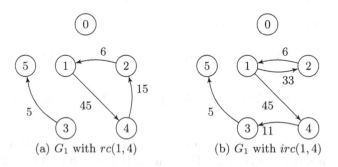

(a) G_1 with $rc(1,4)$ (b) G_1 with $irc(1,4)$

Fig. 7. G_1 with $rc(1,4)$ and $irc(1,4)$

Algorithm 3 computes $irc(i,j), \forall (i,j)$ in $O(|V|^2)$. First, it initializes each $irc(i,j)$ to $rc(i,j), \forall (i,j) \in E$. Then, for each edge (i,j) involved in the assumption of Proposition 4 (Invariant (a)), Algorithm 3 updates its $irc(i,j)$ according to the formula of Proposition 4.

Algorithm 3. Computation of improved reduced costs $irc(i,j), \forall (i,j) \in E$

Input: $parent[k], \forall k \in V$; $pred[k], \forall k \in V$; $u_{c_i}^\star, \forall c_i \in C$ and
$\quad\quad bestTwoDiff[k], \forall k \in V$ that can be computed in $O(|V|^2)$
Output: $irc(i,j), \forall (i,j) \in E$

1 **foreach** each edge $(i,j) \in E$ **do**
2 $\quad \lfloor\ irc(i,j) \leftarrow rc(i,j)$

3 **foreach** each vertex $i \in V$ **do**
4 $\quad$ **if** $parent[parent[i]] = \emptyset$ **then**
5 $\quad\quad min \leftarrow bestTwoDiff[i] - u_{parent[i]}^\star$
6 $\quad\quad j = pred[i]$
7 $\quad\quad$ **while** $(parent[parent[j]] = \emptyset) \wedge min > 0 \wedge j \neq r$ **do**
$\quad\quad\quad$ // **Invariant (a): there is a path $\mathcal{P}$ from j to i such that**
$\quad\quad\quad\quad \forall k \in \mathcal{P} : parent[parent[k]] = \emptyset$
$\quad\quad\quad$ // **Invariant (b):**
$\quad\quad\quad\quad min = min_{k \in \mathcal{P}\setminus\{j\}}\{bestTwoDiff[k] - u_{parent[k]^\star}\}$
8 $\quad\quad\quad irc(i,j) \leftarrow irc(i,j) + min$
9 $\quad\quad\quad$ **if** $bestTwoDiff[j] - u_{parent[j]^\star} < min$ **then**
10 $\quad\quad\quad\quad \lfloor\ min \leftarrow bestTwoDiff[j] - u_{parent[j]}^\star$
11 $\quad\quad\quad j = pred[j]$

---

5 Experimental Results

As a first experiment, we evaluate the proportion of reduced costs affected by Proposition 4. Therefore we randomly generated two classes of 100 instances $w(i,j) \in [1,100]$ with different values of the number of vertices:

- $class1$: for each $i \in V$, $parent[parent[i]] = \emptyset$;
- $class2$: many vertices $i \in V$ are such that $parent[parent[i]] \neq \emptyset$.

The $class1$ was obtained by filtering out the random instances not satisfying the property. Let $exactRC(i,j)$ be the exact reduced cost associated to the edge $(i,j) \in E$. Table 1 shows, for each class of instances (with respectively $|V| = 20$, $|V| = 50$ and $|V| = 100$), the proportion of instances of each class and for each group of instances: (1) the proportion of edges $(i,j) \in E$ such that $rc(i,j) < exactRC(i,j)$; (2) the proportion of edges that have $irc(i,j) > rc(i,j)$; and (3) the proportion of edges such that $irc(i,j) = exactRC(i,j)$. Note that, for this benchmark, at least 37% of 300 instances are $class1$ instances and at least 45% of LP reduced costs (with $rc(i,j) < exactRC(i,j)$) are improved for these instances. Of course, the results are less interesting for the $class2$ instances.

To test the `MinArborescence` constraint, experiments were conducted on an NP-Hard variant of CAP: the Resource constrained Minimum Weight Arborescence Problem (RMWA) [10,14]. The RMWA problem is to find an MWA under the resource constraints for each vertex $i \in V$: $\sum_{(i,j)\in\delta_i^+} a_{i,j} \cdot x_{i,j} \leq b_i$ where δ_i^+ is the set of outgoing edges from i, $a_{i,j}$ is the amount of resource

Table 1. Proportion of reduced costs affected by Proposition 4

	$\|V\| = 20$		$\|V\| = 50$		$\|V\| = 100$	
	Class1	Class2	Class1	Class2	Class1	Class2
%: instances of class$k$ ($k \in \{1,2\}$)	48	52	31	69	32	68
%: $rc(i,j) < exactRC(i,j)$	19.3	38.1	9.8	26.7	2.1	16.6
%: $irc(i,j) > rc(i,j)$	12.6	1.9	4.6	0.9	1.2	0.2
%: $irc(i,j) = exactRC(i,j)$	9.9	1.17	3.49	0.79	1.19	0.2

uses by the edge (i,j) and b_i is the resource available at vertex i. RMWA can be modeled in CP with a `MinArborescence` constraint (or one of its decompositions) for the MWA part of problem and the `binaryKnapsack` constraint [7] together with `weightedSum` constraint for the resource constraints. We have randomly generated the different costs/weights as described in [14]: $a_{i,j} \in [10, 25]$, $w(i,j) \in [5, 25]$. To have more available edges to filter, we have used $b_i = 2 \cdot \lfloor \frac{\sum_{(i,j) \in \delta_i^+} a_{i,j}}{|\delta_i^+|} \rfloor$ (instead of $b_i = \lfloor \frac{\sum_{(i,j) \in \delta_i^+} a_{i,j}}{|\delta_i^+|} \rfloor$) and 75% graph density.

In order to avoid the effect of the dynamic first fail heuristic interfering with the filtering, we use the approach described in [25] to evaluate global constraints. This approach consists in recording the search tree with the weakest filtering as a baseline. It is obtained with the decomposition model using the `Arborescence` constraint. This search tree is then replayed with the stronger reduced cost based filtering for `MinArborescence`. The recorded search tree for each instance corresponds to an exploration of 30 s. As an illustration for the results, Table 2 details the computational results for `MinArborescence` constraint with filtering based on improved reduced costs (MinArbo_IRC), reduced costs (MinArbo_RC), the decomposition with `Arborescence` constraint (Arbo) and Arbo+filtering only based on lower bound on MWA (Arbo+LB) on 4 (arbitrarily chosen) out of the 100 randomly instances with $|V| = 50$. We also report the average results for the 100 instances. On average, the search space is divided by ≈ 460 with the reduced costs based filtering `MinArborescence` constraint (wrt Arbo) and by ≈ 81 with Arbo+LB. This demonstrates the benefits brought by the `MinArborescence` global constraints described in this paper.

To further differentiate the filtering of MinArbo_IRC, we now use MinArbo_RC as a baseline filtering for recording the search tree on another set of 100 randomly generated instances of class1 with $|V| = 50$. Figure 8 shows the corresponding performance profiles wrt the number of nodes visited and the time used respectively. For $\approx 30\%$ of instances the search space is divided by at least 1.5 and for $\approx 7\%$ the search space is divided by at least 4. On the other hand, the average gain for MinArbo_IRC (wrt MinArbo_RC) is 1.7 wrt the number of nodes visited and 1.4 wrt time. Unfortunately, as was expected, the average gain is limited as only $\approx 5\%$ of LP reduced costs can be improved.

Table 2. Results: MinArbo_IRC, MinArbo_RC, Arbo+LB and Arbo

Instance	MinArbo_IRC		MinArbo_RC		Arbo+LB		Arbo	
	Nodes	Time(s)	Nodes	Time	Nodes	Time	Nodes	Time
1	20259	0	20259	0	40061	1	5879717	28
2	12552	0	12552	0	16794	0	6033706	27
3	13094	0	13094	0	121290	2	6383651	28
4	62607	0	62607	0	283854	6	7316899	29
Average	**14385**	**0**	**14385**	**0**	**81239**	**1.4**	**6646748**	**28**

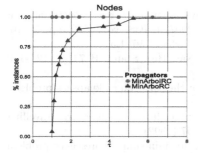
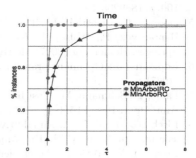

Fig. 8. Performance profiles wrt number of nodes visited and time

The implementations and tests have been realized within the OscaR open source solver [24]. Our source-code and the instances are available at [15]. Our CP model is able to solve and prove optimality of RMWA instances with up to $|V| = 50$. Similar instances can be solved using the Lagrangian decomposition approach of [14]. The branch and cut algorithm of [10] reports results solving instances with up to $|V| = 500$. We believe this lack of performance of CP wrt to the branch and cut approach is due to the $|V|$ independent knapsack constraints inducing a weaker pruning. We do hope this first result will trigger more research in the future to make CP more competitive on this challenging problem.

6 Conclusion

We have defined the `MinArborescence` constraint based on the reduced costs to filter the edges for a constrained arborescence problem. We have proposed an algorithm to improve the LP reduced costs of the minimum weighted arborescence in some cases. Finally, we have demonstrated experimentally the interest of improved reduced costs in some particular graphs and the efficiency of the cost-based filtering on the resource constrained arborescence problem. As future works, we would like to: (1) think about a global constraint wrt resource constraints (2) study the incremental aspects of the `MinArborescence` constraint and (3) propose specialized search heuristics.

References

1. Beldiceanu, N., Flener, P., Lorca, X.: The *tree* Constraint. In: Barták, R., Milano, M. (eds.) CPAIOR 2005. LNCS, vol. 3524, pp. 64–78. Springer, Heidelberg (2005). doi:10.1007/11493853_7
2. Bock, F.: An algorithm to construct a minimum directed spanning tree in a directed network. Dev. Oper. Res. **1**, 29–44 (1971)
3. Chu, Y.J., Liu, T.H.: On the shortest arborescence of a directed graph. Sci. Sin. Ser. A **14**, 1396–1400 (1965)
4. Dooms, G., Katriel, I.: The *minimum spanning tree* constraint. In: Benhamou, F. (ed.) CP 2006. LNCS, vol. 4204, pp. 152–166. Springer, Heidelberg (2006). doi:10.1007/11889205_13
5. Dooms, G., Katriel, I.: The "not-too-heavy spanning tree" constraint. In: Hentenryck, P., Wolsey, L. (eds.) CPAIOR 2007. LNCS, vol. 4510, pp. 59–70. Springer, Heidelberg (2007). doi:10.1007/978-3-540-72397-4_5
6. Edmonds, J.: Optimum branchings. J. Res. Nat. Bur. Stand. B **71**(4), 125–130 (1967)
7. Fahle, T., Sellmann, M.: Cost based filtering for the constrained knapsack problem. Ann. Oper. Res. **115**(1–4), 73–93 (2002)
8. Fischetti, M., Toth, P.: An additive bounding procedure for asymmetric travelling salesman problem. Math. Program. **53**, 173–197 (1992)
9. Fischetti, M., Toth, P.: An efficient algorithm for min-sum arborescence problem on complete digraphs. Manage. Sci. **9**(3), 1520–1536 (1993)
10. Fischetti, M., Vigo, D.: A branch-and-cut algorithm for the resource-constrained minimum-weight arborescence problem. Network **29**, 55–67 (1997)
11. Focacci, F., Lodi, A., Milano, M., Vigo, D.: Solving TSP through the integration of OR and CP techniques. Electron. Notes Discrete Math. **1**, 13–25 (1999)
12. Gabow, H.N., Galil, Z., Spencer, T.H., Tarjan, R.E.: Efficient algorithms for finding minimum spanning trees in undirected and directed graphs. Combinatorica **6**(3), 109–122 (1986)
13. Graham, R.L., Hell, P.: On the history of the minimum spanning tree problem. Hist. Comput. **7**, 13–25 (1985)
14. Guignard, M., Rosenwein, M.B.: An application of lagrangean decomposition to the resource-constrained minimum weighted arborescence problem. Network **20**, 345–359 (1990)
15. Houndji, V.R., Schaus, P.: Cp4cap: Constraint programming for constrained arborescence problem. https://bitbucket.org/ratheilesse/cp4cap
16. Kleinberg, J., Tardos, E.: Minimum-cost arborescences: a multi-phase greedy algorithm. In: Algorithm Design, Tsinghua University Press (2005)
17. Lorca, X.: Contraintes de Partitionnement de Graphe. Ph. D. thesis, Université de Nantes (2010)
18. Fages, J.-G., Lorca, X.: Revisiting the `tree` constraint. In: Lee, J. (ed.) CP 2011. LNCS, vol. 6876, pp. 271–285. Springer, Heidelberg (2011). doi:10.1007/978-3-642-23786-7_22
19. Mendelson, R., Tarjan, R.E., Thorup, M., Zwick, U.: Melding priority queues. In: Hagerup, T., Katajainen, J. (eds.) SWAT 2004. LNCS, vol. 3111, pp. 223–235. Springer, Heidelberg (2004). doi:10.1007/978-3-540-27810-8_20
20. Pesant, G., Gendreau, M., Potvin, J.-Y., Rousseau, J.-M.: An exact constraint logic programming algorithm for the traveling salesman problem with time windows. Transp. Sci. **32**(1), 12–29 (1998)

21. Régin, J.-C.: Simpler and incremental consistency checking and arc consistency filtering algorithms for the weighted spanning tree constraint. In: Perron, L., Trick, M.A. (eds.) CPAIOR 2008. LNCS, vol. 5015, pp. 233–247. Springer, Heidelberg (2008). doi:10.1007/978-3-540-68155-7_19
22. Régin, J.-C., Rousseau, L.-M., Rueher, M., van Hoeve, W.-J.: The weighted spanning tree constraint revisited. In: Lodi, A., Milano, M., Toth, P. (eds.) CPAIOR 2010. LNCS, vol. 6140, pp. 287–291. Springer, Heidelberg (2010). doi:10.1007/978-3-642-13520-0_31
23. Tarjan, R.E.: Finding optimum branchings. Networks **7**(3), 25–35 (1977)
24. OscaR Team. Oscar: Scala in or (2012). https://bitbucket.org/oscarlib/oscar
25. Van Cauwelaert, S., Lombardi, M., Schaus, P.: Understanding the potential of propagators. In: Michel, L. (ed.) CPAIOR 2015. LNCS, vol. 9075, pp. 427–436. Springer, Cham (2015). doi:10.1007/978-3-319-18008-3_29

Learning When to Use a Decomposition

Markus Kruber[1]([✉]) [ID], Marco E. Lübbecke[1] [ID], and Axel Parmentier[2] [ID]

[1] Chair of Operations Research, RWTH Aachen University,
Kackertstrasse 7, 52072 Aachen, Germany
{kruber,luebbecke}@or.rwth-aachen.de
[2] CERMICS, École des Ponts Paristech, Université Paris Est,
6 et 8 Avenue Blaise Pascal, 77420 Champs sur Marne, France
axel.parmentier@enpc.fr

Abstract. Applying a Dantzig-Wolfe decomposition to a mixed-integer program (MIP) aims at exploiting an embedded model structure and can lead to significantly stronger reformulations of the MIP. Recently, automating the process and embedding it in standard MIP solvers have been proposed, with the detection of a decomposable model structure as key element. If the detected structure reflects the (usually unknown) actual structure of the MIP well, the solver may be much faster on the reformulated model than on the original. Otherwise, the solver may completely fail. We propose a supervised learning approach to decide whether or not a reformulation should be applied, and which decomposition to choose when several are possible. Preliminary experiments with a MIP solver equipped with this knowledge show a significant performance improvement on structured instances, with little deterioration on others.

Keywords: Mixed-integer programming · Branch-and-price · Column generation · Automatic Dantzig-Wolfe decomposition · Supervised learning

1 Setting and Approach

Dantzig-Wolfe (DW) reformulation of a mixed-integer program (MIP) became an indispensable tool in the computational mathematical programming bag of tricks. On the one hand, it may be *the* key to solving specially structured MIPs. On the other hand, successfully applying the technique may require a solid background, experience, and a non-negligible implementation effort.

In order to make DW reformulation more accessible also to non-specialists, general solvers were developed that make use of the method. One such solver is GCG [4], an extension to the well-established MIP solver SCIP [1]. Several *detectors* first look for possible DW reformulations of the original MIP model. Different types of reformulations are used by practitioners, and even if most MIPs can be forced into each of these types [2], their relevance highly depends on the model structure. For instance, staircase forms suit well to temporal knapsack problems [3] while bordered block diagonal forms work well on vehicle routing problems. As in general solvers, we do not know *a priori* the structure of

© Springer International Publishing AG 2017
D. Salvagnin and M. Lombardi (Eds.): CPAIOR 2017, LNCS 10335, pp. 202–210, 2017.
DOI: 10.1007/978-3-319-59776-8_16

the original model, and many *decompositions* of each type are detected. These decompositions are then evaluated: if they are of "good quality," the MIP is reformulated according to a "best suited" decomposition.

If one finds and selects a decomposition (DEC) that captures a structure underlying the original model, the reformulated model may be solved much faster than the original one. However, from the many different decompositions, we may select one that does not reflect the actual underlying model structure; and in this case the solver may completely fail. We currently do not have any consistently reliable *a priori* measure to distinguish between these cases, except heuristic proxies, see e.g., [2]. What is more, the latter case is not uncommon, and GCG, presented with an arbitrary MIP, successfully detects a decomposition that leads to an improved performance over SCIP on the original model only in a small fraction of cases. This is to be expected, but will not render GCG (in its present form) a competitive general MIP solver. Our work thus aims at providing a mean to decide whether it pays to DW reformulates a given MIP model or not. Figure 1 illustrates how the result may look like.

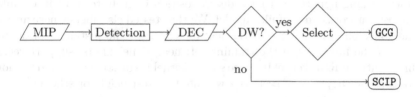

Fig. 1. Multiple detectors DW solver GCG with "SCIP exit strategy"

Literature. Decomposable model structure may be detected directly from the MIP, see e.g., [2,12] and the references therein. Machine learning techniques have been recently used in computational mathematical optimization, e.g., automated MIP solver configuration [13], load balancing in parallel branch-and-bound [10], or variable selection in branching decisions [6,9]. We are not aware of works that try to learn MIP model structure to be exploited in decompositions.

Our Supervised Learning Approach. We would like to learn an answer to the question: Given a MIP $\mathcal{P}$, a DW decomposition $\mathcal{D}$, and a time limit τ, will GCG using $\mathcal{D}$ optimally solve $\mathcal{P}$ faster than SCIP (or have a smaller optimality gap at time τ)? We define a mapping ϕ that transforms a tuple $(\mathcal{P}, \mathcal{D}, \tau)$ into a vector of sufficient statistics or *features* $\phi(\mathcal{P}, \mathcal{D}, \tau) \in \mathbb{R}^d$. Thanks to this mapping ϕ, the question above becomes a standard binary classification problem on $\mathbb{R}^d$. We can therefore train a standard classifier $f : \mathbb{R}^d \rightarrow \{0, 1\}$ to solve this problem. Given an instance $(\mathcal{P}, \mathcal{D}, \tau)$, the quantity $f \circ \phi(\mathcal{P}, \mathcal{D}, \tau)$ is equal to one iff the predicted answer to the question above is positive. Practically, we have built a database of SCIP and GCG runs for tuples $(\mathcal{P}, \mathcal{D}, \tau)$, a mapping ϕ, and have trained classifiers f from the scikit-learn library [11] on the instances $\phi(\mathcal{P}, \mathcal{D}, \tau)$. Answers to the probabilistic versions $g : \mathbb{R}^d \rightarrow [0, 1]$ of these classifiers can be interpreted as the probability that GCG using $\mathcal{D}$ outperforms SCIP if the time limit is τ.

GCG starts by detecting decompositions $\mathcal{D}_1, \ldots, \mathcal{D}_k$ for $\mathcal{P}$. This detection takes time τ_{det}. We then decide how to make use of the remaining time $\tau - \tau_{\text{det}}$:

$$\text{Continue GCG if } \max_{i=1,\ldots,k} g \circ \phi(\mathcal{P}, \mathcal{D}_i, \tau - \tau_{\text{det}}) \geq \alpha. \text{ Otherwise run SCIP.} \qquad (1)$$

The threshold α reflects our level of conservatism towards solving $\mathcal{P}$ using a DW reformulation. If we decide to continue the run with GCG, we

$$\text{use decomposition } \mathcal{D} \text{ with maximum } g \circ \phi(\mathcal{P}, \mathcal{D}, \tau - \tau_{\text{dec}}), \qquad (2)$$

that is, one with largest predicted probability that GCG beats SCIP.

There are three key elements for such an approach to perform well: First, the features must catch relevant information, see Sect. 2.2. Second, for training a classifier we need to present it data, i.e., tuples $(\mathcal{P}, \mathcal{D}, \tau)$ in the learning phase that are similar to those we expect to see later when using the classifier. Our learning dataset contains instances from a wide range of families of structured and non-structured models. Third, an appropriate binary classifier must be used. It is our working hypothesis that a decomposition is likely to work if a similar decomposition works on a similar model. We thus tested classifiers whose answer depends only on the distance of the feature vector of the instance considered to those of the instances in the training set: nearest neighbors, support vector machines with an RBF kernel because such a kernel is stationary [5], and random forests because they can be seen as a weighted nearest neighbor scheme [8].

2   Decompositions and Features

2.1   Bird's View on Dantzig-Wolfe Reformulation

We would like to solve what we call the *original* MIP

$$\min \left\{ c^t x, \ Ax \geq b, \ x \in \mathbb{Z}_+^n \times \mathbb{Q}_+^q \right\}. \qquad (3)$$

Today, the classical approach to solve (3) is branch-and-cut, implemented e.g., in the SCIP solver [1]. Without going into much detail, the data in (3) can sometimes be re-arranged such that a particular *structure* in the model becomes visible. Among several others, one such structure is the so-called *arrowhead* or double-bordered block diagonal form. It consists of partitioning the variables $x = [x^1, \ldots, x^\kappa, x^\ell]^t$ and right hand sides $b = [b^1, \ldots, b^\kappa, b^\ell]^t$ to obtain

$$
\begin{aligned}
\min \ & c^t x \\
\text{s.t.} \ & \begin{bmatrix} D^1 & & & F^1 \\ & D^2 & & F^2 \\ & & \ddots & \vdots \\ & & & D^\kappa & F^\kappa \\ A^1 & A^2 & \cdots & A^\kappa & G \end{bmatrix} \cdot \begin{bmatrix} x^1 \\ x^2 \\ \vdots \\ x^\kappa \\ x^\ell \end{bmatrix} \geq \begin{bmatrix} b^1 \\ b^2 \\ \vdots \\ b^\kappa \\ b^\ell \end{bmatrix} \\
& x \in \mathbb{Z}_+^n \times \mathbb{Q}_+^q.
\end{aligned} \qquad (4)
$$

with sub-matrices of appropriate dimensions. Such a re-arrangement is called a *decomposition* of (3) and finding it is called *detection*. Matrices D^i are called *blocks*, variables x^ℓ are *linking variables* and $(A^1 \cdots A^\kappa \, G)x \geq b^\ell$ are *linking constraints*. Such forms are interesting because a Dantzig-Wolfe reformulation can be applied, leading to a solution of the resulting MIP by branch-and-price. Its characteristic is that the linear relaxations in each node of the search tree are solved by column generation, an alternation between solving *pricing problems*, i.e., integer programs over $D^i x^i \geq b^i$, and a linear program (the so-called *master problem*) involving exponentially many variables and constraints systematically derived from the linking constrains. See [4] for details (which are not necessary for the following). The algorithmic burden is considerable but may pay off if the pricing problems capture a well solvable sub-structure of the model (3). In such a case we call the decomposition *good*. When we know some good decomposition of a MIP model we call the model *structured*, if we do not know a good one the model is "non-structured." Automatic detection of decompositions, DW reformulation, and branch-and-price is implemented in the GCG solver [4].

2.2 Features Considered

We now give an idea of the feature map ϕ that turns a MIP, decomposition, and time limit into a vector $\phi(\mathcal{P}, \mathcal{D}, \tau) \in \mathbb{R}^d$ that we give as input to supervised learning classifiers. We define a large number of features (more than 100) to catch as much information as possible, and then use a regularization approach to avoid overfitting. We only sketch the main types, without being exhaustive.

Our first features are *instance statistics*, like the number of variables, constraints, and non-zeros, the proportion of binary, integer, and continuous variables, or of certain types of constraints, such as knapsack or set covering. We collect *decomposition based statistics* like the number κ of blocks, or the proportion of non-zeros or variable/constraint types per block. As the dimension d of ϕ is fixed and κ varies across decompositions, we consider block statistics via their average, variance, and quantiles. A small number of linking constraints and variables is empirically considered good on "non-structured" MIPs [2].

Richer decomposition based features come from *adjacencies*, e.g., from the bipartite graph with a vertex for each row i of A and each column j of A, and an edge (i, j) iff $a_{ij} \neq 0$. Blocks can be seen as clusters of vertices in this graph. We can then build features inspired from graph clustering.

Many features can be obtained from the *detectors themselves*. The simplest is the *detector indicator feature* $\mathbb{1}_t(\mathcal{D})$, which is a binary feature equal to one iff decomposition $\mathcal{D}$ was found by detector t ("$\mathcal{D}$ is of type t"). Some detectors use metrics to evaluate the quality of their decomposition, for instance, whether blocks are "equal." These metrics can occur in ϕ. Some features mentioned above play different roles in different types of decompositions. Type specific behavior can be captured by products of detector indicators $\mathbb{1}_t$ with other features.

Finally, as the detection time varies from one instance to another, an important feature is the time remaining $\tau - \tau_{\text{dec}}$ after detection. Functions of this time feature can be used within products with other features.

3   Preliminary Computational Results

3.1   Experimental Setup

As Dantzig-Wolfe decomposition leverages embedded structure in MIP instances, we want to learn from a wide range of model structures. At the same time, using a non-appropriate decomposition (i.e., assuming a "wrong" structure) almost certainly leads to poor solver performance. It is therefore most important to detect the "absence" of structure (or structure currently not exploitable by GCG). We have therefore built a dataset of 300 "structured" (instances for which an intuitive decomposition is known, e.g. from literature) and 100 "non-structured" instances. We underline the fact that the information whether an instance contains structure or not is *not* part of the input features. We considered the following families of structured instances: vertex coloring (clr), set covering (stcv), capacitated p-median (cpmp), survivable fixed telecommunication network design (sdlb), cutting stock (ctst), generalized assignment (gap), network design (ntlb), lot sizing (ltsz), bin packing (bp), resource allocation (rap), stable set (stbl), and capacitated vehicle routing (cvrp) problems. The instances assumed "non-structured" are randomly chosen from MIPLIB 2010 [7]. GCG detected decompositions for each instance, leading to a total of 1619 decompositions. We launched SCIP and GCG with a time limit of two hours on each. Table 1 provides the number of instances per family and their difficulty, for which we use as proxy the solution status of SCIP after 2 h. All experiments were performed on a i7-2600 3.4 GHz PC, 8 MB cache, and 16 GB RAM, running OpenSuSE Linux 13.1. We used the binary classifiers of the **python scikit-learn** library [11], SCIP in version 3.2.1 [1], and GCG in version 2.1.1 [4].

Table 1. Number of instances listed per problem class and solution status of SCIP

	All	clr	stcv	cpmp	sdlb	ctst	gap	ntlb	ltsz	bp	rap	stbl	cvrp	miplib
Instances	400	25	25	25	25	25	25	25	25	25	25	25	25	100
Opt. sol.	65.5%	19	3	18	10	25	23	25	25	6	12	22	6	68
Feas. sol.	31.5%	6	21	7	11	–	2	–	–	19	12	3	19	26
No sol.	3.0%	–	1	–	4	–	–	–	–	–	1	–	–	6

3.2   Overall Computational Summary of Our Experiment

We have randomly split our dataset into a training set containing 3/5 of the instances and a test set containing the remaining. Table 2 aggregates the overall results of our experiments when using a k-nearest neighbors (KNN) classifier. We report statistics for four solvers: the standard branch-and-cut MIP solver SCIP; GCG that tries to detect a decomposition and perform a DW reformulation accordingly; columns SL correspond to GCG with the learned classifiers (1) and (2) built in (i.e., the proposal of this paper); and finally, columns OPT list the results we would obtain if we always selected the best solver. The rows show, for the sets of all, structured, and non-structured instances, the number of instances that could not be solved to optimality, the total time needed on the respective entire test set, and the geometric mean of the CPU times per instance. As defined in (1) and (2) the detection time is included in the timelimit for GCG, SL and OPT.

Table 2. Aggregated statistics on a test set of 131 instances, 99 structured and 32 non-structured. Overall GCG achieves a better performance than SCIP in 34 cases.

Instances	All				Structured				Non-structured			
Solver	SCIP	GCG	SL	OPT	SCIP	GCG	SL	OPT	SCIP	GCG	SL	OPT
No opt. sol.	52	66	44	39	39	37	31	26	13	29	14	13
CPU time (h)	111.3	142.6	93.1	85.7	83.5	82.2	65.9	58.5	27.8	56.8	29.2	27.2
Geo. mean (s)	127.1	370.4	78.6	67.8	73.4	146.9	39.2	32.2	672.9	5145.0	766.0	646.5

The OPT columns show that there is potential in applying a DW reformulation on structured instances. However, we see that always using the current default GCG as a standalone solver is not an option. It needs to be complemented with an option to run SCIP in the case that a DW reformulation does not appear to be promising. This is what we do with our supervised learning approach. The resulting solver SL performs better than SCIP on structured instances with only little performance deterioration on "non-structured" instances. This trend should be confirmed on a larger test set. The performance of SL relies on the quality of the decisions taken by classifiers (1) and (2), that we now explain in detail.

3.3 Deciding Whether to Continue Running GCG or to Start SCIP

Table 3 shows the ability of classifier (1) to choose between SCIP and GCG given the decompositions provided by GCG's detectors and a time limit. The first row lists the share of instances where GCG and SCIP are respectively the best options. The 2×2 cells below give the *confusion matrix*, based on the prediction of the respective classifiers on the left side (rows) and on the true classes on the top (columns). The diagonal of each cell corresponds to cases where the classifier has predicted the right answer. The top right corner of each cell corresponds to false negatives (GCG is better, but we predict SCIP) and the bottom left to false positives (SCIP is better but we predict GCG). As GCG may perform very poorly when not a "right" structure is detected we try to keep the false positives low, even if this implies accepting more false negatives and thus not exploiting the full potential of GCG. The confidence threshold α in (1) was set to 0.5 here.

Even if the size of the test set is too small to arrive at final conclusions, we identify some trends. The support vector machine with radial basis function (RBF) classifier shows a highly risk-averse behaviour with predicting GCG only in 7.6% of all cases. Only half of its predictions are correct, which is a poor performance. KNN and random forrests (RF) classifiers are willing to take more risk and predict that GCG is the best option in 20.6% to 29.0% of all instances.

Table 3. Accuracy of solver selection, i.e., classifier (1)

		All instances		Structured		Non-structured	
		SCIP	GCG	SCIP	GCG	SCIP	GCG
Classifier	Pred	74.0%	26.0%	68.7%	31.3%	90.6%	9.4%
RBF	SCIP	73.3%	19.1%	66.7%	23.2%	93.8%	6.3%
Unbal.	GCG	3.8%	3.8%	5.1%	5.1%	0.0%	0.0%
KNN	SCIP	69.5%	9.9%	64.6%	11.1%	84.4%	6.3%
distance	GCG	6.9%	13.7%	7.1%	17.2%	6.3%	3.1%
RF	SCIP	63.4%	11.5%	55.6%	13.1%	87.5%	6.3%
Unbal.	GCG	10.7%	14.5%	13.1%	18.2%	3.1%	3.1%
RF	SCIP	60.3%	10.7%	50.5%	11.1%	90.6%	9.4%
Bal	GCG	13.7%	15.3%	18.2%	20.2%	0.0%	0.0%

With a true predictions over all GCG predictions ratio of around 2/3, KNN shows the best precision. This explains that our SL scheme with KNN catches roughly 2/3 of the improvement potential of OPT with respect to SCIP in Table 2.

3.4 Selecting a Best Decomposition in GCG

Table 4 shows the percentage of instances on which classifier (2) predicts the right decomposition on the entire test set, on the subset of instances for which GCG is selected by classifier (1), and on the subset of instances such that GCG on the best decomposition actually outperforms SCIP. The classifier predicts the right answer only on about half of the instances. However, in practice, this classifier is called only if GCG has been selected by classifier (1). And on these instances, the right answer is predicted on around 80% of the cases. This difference of performance can be explained by the fact that there is no information on the relative performance of GCG on instances where SCIP performs better than GCG on all decompositions in the training dataset. A direction to include this information and improve the performance is to use a one-versus-one approach, where binary classifiers take two decompositions in input and predict which one is better, instead of using our "one-versus-SCIP" approach.

Table 4. Accuracy of best decomposition selection, i.e., classifier (2)

Classifier	All instances	GCG predicted by (1)	GCG better
RBF	42.7%	80.0%	76.7%
KNN	58.8%	88.9%	77.4%
RF unbalanced	51.1%	72.7%	76.5%
RF balanced	64.9%	71.1%	79.4%

4   Discussion and Future Work

It is our understanding that DW reformulation will be successful in terms of solver performance only on a small fraction of MIPs. It is therefore crucial to reliably decide whether a reformulation will pay off on a given input, *or not*. Several factors influence the success of the reformulation, among them the strength of the relaxation, an acceptable computational burden of repeatedly solving many subproblems, etc. An experienced modeler may have a sense for such factors, and this work aims equipping the GCG solver with a sort of such a sense. Our preliminary computational experience suggests that a supervised learning approach be promising. There are three immediate next steps.

1. We are currently completely re-designing the detection loop in GCG, leading to *many* more potential decompositions, and thus a much richer training set for our learning algorithms. 2. Detection can be very time consuming. One should predict *before* detection if the decomposition detected will be worth being used. This may save detection time in particular on "non-structured" instances where GCG is currently unable to find a good decomposition. 3. The quality of the features given as input to a machine learning classifier determines its performance. A direction to improve feature quality is to use run time, i.e., *a posteriori* information about a decomposition. One could run column generation on each decomposition for a limited amount of time in order to collect solution time and integrality gap of pricing problems, degeneracy of and dual bound provided by the master problem, etc. Figure 2 illustrates such a "strong detection" solution scheme. Reporting about this ongoing work is postponed to the full version of this short paper and will be implemented in a near future release of GCG.

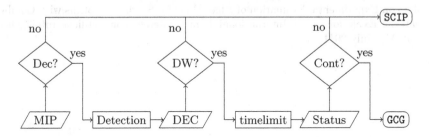

Fig. 2. Three decisions taken along a "strong detection" solution scheme

References

1. Achterberg, T.: SCIP: solving constraint integer programs. Math. Program. Comput. **1**(1), 1–41 (2009)
2. Bergner, M., Caprara, A., Ceselli, A., Furini, F., Lübbecke, M.E., Malaguti, E., Traversi, E.: Automatic Dantzig-Wolfe reformulation of mixed integer programs. Math. Program. **149**(1–2), 391–424 (2015)
3. Caprara, A., Furini, F., Malaguti, E.: Uncommon Dantzig-Wolfe reformulation for the temporal knapsack problem. INFORMS J. Comput. **25**(3), 560–571 (2013)

4. Gamrath, G., Lübbecke, M.E.: Experiments with a generic Dantzig-Wolfe decomposition for integer programs. In: Festa, P. (ed.) SEA 2010. LNCS, vol. 6049, pp. 239–252. Springer, Heidelberg (2010). doi:10.1007/978-3-642-13193-6_21
5. Genton, M.G.: Classes of kernels for machine learning: a statistics perspective. J. Mach. Learn. Res. **2**, 299–312 (2001)
6. Khalil, E., Le Bodic, P., Song, L., Nemhauser, G., Dilkina, B.: Learning to branch in mixed integer programming. In: Proceedings of the 30th AAAI Conference on Artificial Intelligence (2016)
7. Koch, T., Achterberg, T., Andersen, E., Bastert, O., Berthold, T., Bixby, R.E., Danna, E., Gamrath, G., Gleixner, A.M., Heinz, S., Lodi, A., Mittelmann, H., Ralphs, T., Salvagnin, D., Steffy, D.E., Wolter, K.: MIPLIB 2010. Math. Program. Comput. **3**(2), 103–163 (2011)
8. Lin, Y., Jeon, Y.: Random forests and adaptive nearest neighbors. J. Am. Stat. Assoc. **101**(474), 578–590 (2006)
9. Marcos Alvarez, A., Louveaux, Q., Wehenkel, L.: A machine learning-based approximation of strong branching. INFORMS J. Comput. **29**(1), 185–195 (2014)
10. Marcos Alvarez, A., Wehenkel, L., Louveaux, Q.: Machine learning to balance the load in parallel branch-and-bound (2015)
11. Pedregosa, F., Varoquaux, G., Gramfort, A., Michel, V., Thirion, B., Grisel, O., Blondel, M., Prettenhofer, P., Weiss, R., Dubourg, V., Vanderplas, J., Passos, A., Cournapeau, D., Brucher, M., Perrot, M., Duchesnay, E.: Scikit-learn: machine learning in python. J. Mach. Learn. Res. **12**, 2825–2830 (2011)
12. Wang, J., Ralphs, T.: Computational experience with hypergraph-based methods for automatic decomposition in discrete optimization. In: Gomes, C., Sellmann, M. (eds.) CPAIOR 2013. LNCS, vol. 7874, pp. 394–402. Springer, Heidelberg (2013). doi:10.1007/978-3-642-38171-3_31
13. Xu, L., Hutter, F., Hoos, H.H., Leyton-Brown, K.: Hydra-MIP: automated algorithm configuration and selection for mixed integer programming. In: RCRA Workshop on Experimental Evaluation of Algorithms for Solving Problems with Combinatorial Explosion at the International Joint Conference on Artificial Intelligence (IJCAI), July 2011

Experiments with Conflict Analysis in Mixed Integer Programming

Jakob Witzig[1]([✉]), Timo Berthold[2], and Stefan Heinz[2]

[1] Zuse Institute Berlin, Takustr. 7, 14195 Berlin, Germany
witzig@zib.de
[2] Fair Isaac Germany GmbH, Takustr. 7, 14195 Berlin, Germany
{timoberthold,stefanheinz}@fico.com

Abstract. The analysis of infeasible subproblems plays an important role in solving mixed integer programs (MIPs) and is implemented in most major MIP solvers. There are two fundamentally different concepts to generate valid global constraints from infeasible subproblems. The first is to analyze the sequence of implications, obtained by domain propagation, that led to infeasibility. The result of the analysis is one or more sets of contradicting variable bounds from which so-called conflict constraints can be generated. This concept has its origin in solving satisfiability problems and is similarly used in constraint programming. The second concept is to analyze infeasible linear programming (LP) relaxations. The dual LP solution provides a set of multipliers that can be used to generate a single new globally valid linear constraint. The main contribution of this short paper is an empirical evaluation of two ways to combine both approaches. Experiments are carried out on general MIP instances from standard public test sets such as MIPLIB2010; the presented algorithms have been implemented within the non-commercial MIP solver SCIP. Moreover, we present a pool-based approach to manage conflicts which addresses the way a MIP solver traverses the search tree better than aging strategies known from SAT solving.

1 Introduction: MIP and Conflict Analysis

In this paper we consider *mixed integer programs (MIPs)* of the form

$$c^\star = \min\{c^t x \mid Ax \geq b, \ \ell \leq x \leq u, \ x \in \mathbb{Z}^k \times \mathbb{R}^{n-k}\}, \tag{1}$$

with objective coefficient vector $c \in \mathbb{R}^n$, constraint coefficient matrix $A \in \mathbb{R}^{m \times n}$, constraint left-hand side $b \in \mathbb{R}^m$, and variable bounds $\ell, u \in \overline{\mathbb{R}}^n$, where $\overline{\mathbb{R}} := \mathbb{R} \cup \{\pm\infty\}$. Furthermore, let $\mathcal{N} = \{1, \ldots, n\}$ be the index set of all variables.

When omitting the integrality requirements, we obtain the *linear program (LP)*

$$c_{LP}^\star = \min\{c^t x \mid Ax \geq b, \ \ell \leq x \leq u, \ x \in \mathbb{R}^n\}. \tag{2}$$

The linear program (2) is called the *LP relaxation* of (1). The LP relaxation provides a lower bound on the optimal solution value of the MIP (1), i.e., $c_{LP}^\star \leq c^\star$.

© Springer International Publishing AG 2017
D. Salvagnin and M. Lombardi (Eds.): CPAIOR 2017, LNCS 10335, pp. 211–220, 2017.
DOI: 10.1007/978-3-319-59776-8_17

In LP-based branch-and-bound [11,18], the most commonly used method to solve MIPs, the LP relaxation is used for bounding. Branch-and-bound is a divide-and-conquer method that splits the search space sequentially into smaller subproblems, which are (hopefully) easier to solve. During this procedure, we may encounter infeasible subproblems. Infeasibility can be detected by contradicting implications, e.g., derived by domain propagation, or by an infeasible LP relaxation. Modern MIP solvers try to 'learn' from infeasible subproblems, e.g., by *conflict analysis*. Conflict analysis for MIP has its origin in solving satisfiability problems (SAT) and goes back to [21]. Similar ideas are used in constraint programming, e.g., see [14,15,25]. The first suggestions of using conflict analysis techniques in MIP were by [2,12,24]. Further publications suggested the use conflict information for variable selection in branching to tentatively generate conflicts before branching [3,16] and to analyze infeasibility detected in primal heuristics [6,7].

Today, conflict analysis is widely established in solving MIPs. The principal idea of conflict analysis, in MIP terminology, can be sketched as follows.

Given an infeasible node of the branch-and-bound tree defined by the subproblem

$$\min\{c^t x \mid Ax \geq b,\ \ell' \leq x \leq u',\ x \in \mathbb{Z}^k \times \mathbb{R}^{n-k}\} \tag{3}$$

with local bounds $\ell \leq \ell' \leq u' \leq u$. In LP-based branch-and-bound, the infeasibility of a subproblem is typically detected by an infeasible LP relaxation (see next section) or by contradicting implications.

In the latter case, a *conflict graph* gets constructed which represents the logic of how the set of branching decisions led to the detection of infeasibility. More precisely, the conflict graph is a directed acyclic graph in which the vertices represent bound changes of variables and the arcs (v, w) correspond to bound changes implied by propagation, i.e., the bound change corresponding to w is based (besides others) on the bound change represented by v. In addition to these inner vertices which represent the bound changes from domain propagation, the graph features source vertices for the bound changes that correspond to branching decisions and an artificial sink vertex representing the infeasibility. Then, each cut that separates the branching decisions from the artificial infeasibility vertex gives rise to a valid *conflict constraint*. A conflict constraint consists of a set of variables with associated bounds, requiring that in each feasible solution at least one of the variables has to take a value outside these bounds. Note that in general, this is not a linear constraint and that by using different cuts in the graph, several different conflict constraints might be derived from a single infeasibility. A variant of conflict analysis close to the one described above is implemented in SCIP, the solver in which we will conduct our computational experiments. Also, a similar implementation is available in the FICO Xpress-Optimizer.

This short paper consists of two parts which are independent but complement each other in practice. The first part of this paper (Sect. 2) focuses on a MIP technique to analyze infeasibility based on LP theory. We discuss the interaction, differences, and commonalities between conflict analysis and the so-called *dual ray analysis*. Although both techniques have been known before, e.g., [2,22], this will be, to the best of our knowledge, the first published direct comparison of the two. In the second part (Sect. 3), we present a new approach to drop conflicts that do not lead to variable bound reductions frequently. This new concept is an alternative to the *aging scheme* known from SAT. Finally, we present computational experiments comparing the techniques described in Sects. 2 and 3. Applying a combined approach of conflict and dual ray analysis is the main novelty of this paper.

2   Analyzing Dual Unbounded Solutions

The idea of conflict analysis is tightly linked to domain propagation: conflict analysis studies a sequence of variable bound implications made by domain propagation routines. Besides domain propagation, there is another important subroutine in MIP solving which might prove infeasibility of a subproblem: the LP relaxation. The proof of LP infeasibility comes in the form of a so-called "dual ray", which is a list of multipliers on the model constraints and the variable bounds. Those give rise to a globally valid constraint that can be used similarly to a conflict constraint. In this section, we discuss the analysis of the LP infeasibility proof in more detail.

2.1   Analysis of Infeasible LPs: Theoretical Background

Consider a node of the branch-and-bound tree and the corresponding subproblem of type (3) with local bounds $\ell \leq \ell' \leq u' \leq u$. The *dual LP* of the corresponding LP relaxation of (3) is given by

$$\max\{y^t b + \underline{r}^t \ell' + \overline{r}^t u' \mid A^t y + \underline{r} + \overline{r} = c, \ y, \ \underline{r} \in \mathbb{R}^n_{\geq 0}, \ \overline{r} \in \mathbb{R}^n_{\leq 0}\}, \qquad (4)$$

where $A_{\cdot i}$ is the i-th column of A, $\underline{r}_i = \max\{0, c_i - y^t A_{\cdot i}\}$, and $\overline{r}_i = \min\{0, c_i - y^t A_{\cdot i}\}$. By LP theory each unbounded ray $(\gamma, \underline{r}, \overline{r})$ of (4) proves infeasibility of (3). A ray is called unbounded if multiplying the ray with an arbitrary scalar $\alpha > 0$ will not change the feasibility. Note, in this case it holds

$$\underline{r}_i = \max\{0, -y^t A_{\cdot i}\} \quad \text{and} \quad \overline{r}_i = \min\{0, -y^t A_{\cdot i}\}.$$

Moreover, the Lemma of Farkas states that exactly one of the following two systems is satisfiable

$$(F_1) \quad \begin{matrix} Ax \geq b \\ \ell' \leq x \leq u' \end{matrix} \Bigg\} \lor \begin{cases} \gamma^t A + \underline{r} + \overline{r} = 0 \\ \gamma^t b + \underline{r}^t \ell' + \overline{r}^t u' > 0 \end{cases} (F_2)$$

It follows immediately, that if F_1 is infeasible, there exists an unbounded ray $(\gamma, \underline{r}, \overline{r})$ of (4) satisfying F_2. An infeasibility proof of (3) is given by a single constraint

$$\gamma^t A x \geq \gamma^t b, \qquad (5)$$

which is an aggregation of all rows A_j. for $j = 1, \dots, m$ with weight $\gamma_j > 0$. Constraint (5) is globally valid but violated in the local bounds $[\ell', u']$ of subproblem (3). In the following, this constraint will be called a *proof-constraint*.

2.2 Conflict Analysis of Infeasible LPs

The analysis of an infeasible LP relaxation, as it is implemented in SCIP, is a hybrid of the theoretical considerations made in Sect. 2.1 and the analysis of the conflict graph known from SAT. To use the concept of a conflict graph, all variables with a non-zero coefficient in the proof-constraint are converted to vertices of the conflict graph representing bound changes; global bound changes are omitted. Those vertices, called the *initial reason*, are then connected to the artificial sink representing the infeasibility. This neat idea was introduced in [1]. From thereon, conflict analysis can be applied as described in Sect. 1.

In practice, the proof-constraint is often quite dense, and therefore, it might be worthwhile to search for a sparser infeasibility proof. This can be done by a heuristic that relaxes some of the local bounds $[\ell', u']$ that appear in the proof-constraint. Of course, the relaxed local bounds $[\ell'', u'']$ with $\ell < \ell'' \leq \ell' \leq u' \leq u'' < u$ still need to fulfill $\gamma^t b + \underline{r}^t \ell'' + \overline{r}^t u'' > 0$. The more the bounds can be relaxed by this approach, the smaller becomes the initial reason. Consequently, the derived conflict constraints become stronger. Note again that these constraints do not need to be linear, if general integer or continuous variables are present.

2.3 Dual Ray Analysis of Infeasible LPs

The proof-constraint is globally valid but infeasible within the local bounds. It follows immediately by the Lemma of Farkas that the *maximal activity*

$$\Delta_{\max}(\gamma^t A, \ell', u') := \sum_{i \in \mathcal{N}:\, \gamma^t A_i > 0} (\gamma^t A_i) u_i' + \sum_{i \in \mathcal{N}:\, \gamma^t A_i < 0} (\gamma^t A_i) \ell_i'$$

of $\gamma^t A x$ w.r.t. variable bounds $[\ell', u']$ is strictly less than the corresponding left-hand side $\gamma^t b$.

Instead of creating an "artificial" initial reason, the proof-constraint might also be used directly for domain propagation in the remainder of the search. It is a conical combination of global constraints, i.e., it is itself a valid (but redundant) global constraint. In contrast to the method described in Sect. 2.2, using a dual unbounded ray as a set of weights to aggregate model constraints yields exactly one linear constraint.

The proof-constraint along with an activity argument can be used to deduce local lower and upper variable bounds [2]. Consider a subproblem with local bounds $[\ell', u']$. For any $i \in \mathcal{N}$ with a non-zero coefficient in the proof-constraint the *maximal activity residual* is given by

$$\Delta_{\max}^i(\gamma^t A, \ell', u') := \sum_{j \in \mathcal{N} \setminus i:\, \gamma^t A_j > 0} (\gamma^t A_j) u_j' + \sum_{j \in \mathcal{N} \setminus i:\, \gamma^t A_j < 0} (\gamma^t A_j) \ell_j',$$

i.e., the maximal activity over all variables but x_i. Hence, valid local bounds are given by

$$\frac{\gamma^t b - \Delta^i_{\max}(\gamma^t A, \ell', u')}{a_i} \left\{ \begin{array}{c} \leq \\ \geq \end{array} \right\} x_i \left\{ \begin{array}{l} \text{if } a_i > 0 \\ \text{if } a_i < 0 \end{array} \right..$$

This is the bound tightening procedure [9] which is widely used in all major MIP solvers for all kinds of linear constraints.

Just like the dual ray might be heuristically reduced to get a short initial reason for conflict analysis, it might be worthwhile to alter the proof-constraint itself before using it for propagation. This can include the application of presolving steps such as coefficient tightening to the constraint, projecting out continuous variables or applying mixed-integer rounding to get an alternative globally valid constraint that might be more powerful in propagation.

Finally, instead of generating a valid constraint from the dual ray, one could equivalently use the ray itself to simply check for infeasibility [22,23] or to estimate the objective change during branch-and-bound and to derive branching decisions therefrom. While in Sect. 2.2, we described a way to reduce LP infeasibility analysis to conflict analysis based on domain propagation, one could also try to generate a dual ray by solving the LP relaxation after having detected infeasibility by propagation.

3   Managing of Conflicts in a MIP Solver

Maintaining and propagating large numbers of conflict constraints might have a negative impact on solver performance and create a big burden memory-wise. For instances with a high throughput of branch-and-bound nodes, a solver like SCIP might easily create hundreds of thousands of conflicts within an hour of running time. In order to avoid a slowdown or memory short-coming, an aging mechanism is used within SCIP. Once again, aging is a concept inspired by SAT and CP solving. Every time a conflict constraint is considered for domain propagation an age counter (individually for each constraint) is increased if no deduction was found. If a deduction is found, the age will be reset to 0. If the age reaches a predefined threshold the conflict constraint is permanently deleted.

In SAT and CP, this mechanism is a well-established method to drop conflict constraints that are not frequently propagated. In the case of MIP solving, there are two main differences concerning the branch-and-bound search. First, domain propagation is most often not the most expensive part of node processing. Second, SAT and CP solvers often use a pure depth-first-search (DFS) node selection, while state-of-the-art MIP solvers use some hybrid between DFS and best-estimate-search or best-first-search (see, e.g., [2,5,20]). Therefore, it frequently happens that the node processed next is picked from a different part of the tree.

In the following, we describe a pool-based approach to manage conflict constraints. Here, a pool refers to a fixed-size array that allows direct access to a particular element and is independent of the model itself. The *conflict pool* is used to manage all conflict constraints, independently whether they were derived from domain propagation or an infeasible LP relaxation. The number of constraints that can be stored within the conflict pool at the same is limited. In our implementation the maximal size of the conflict pool depends on the number of variables and constraints of the presolved problem. However, the pool provides space for at least 1 000 and at most 50 000 conflict constraints at the same time. The conflict pool allows a central management of conflict constraints independently from the model constraints, i.e., they can be propagated, checked or deleted separately, without the need to traverse through all constraints.

To drop conflict constraints that don't frequently lead to deductions we implemented an update-routine that checks the conflict pool regularly, e.g., any time we create the first new conflict at a node. Moreover, we still use the concept of aging to determine the conflict constraints that are rarely used in propagation. Within this update procedure the oldest conflict constraints are removed.

Beside the regular checks, the conflict pool is updated every time a new improving incumbent solution is found. Conflict constraints might depend on a (previous) best known solution, e.g., when the conflict was created from an LP whose infeasibility proof contained the objective cutoff. Such conflicts become weaker whenever a new incumbent is found and the chance that they lead to deductions becomes smaller the more the incumbent improves. Due to this, for each conflict constraint involving an incumbent solution we store the corresponding objective value. If this value is sufficiently worse than the new objective value, the conflict constraint will be permanently deleted. In our computational experiments (cf. Sect. 4) we use a threshold of 5%.

4   Computational Experiments

In our computational experiments, we compare combinations of the techniques presented in this paper: conflict analysis and dual ray analysis. To the best of our knowledge, most major MIP solvers either use conflict analysis of infeasible LPs and domain propagation (e.g., SCIP, FICO Xpress-Optimizer) or they employ dual ray analysis (e.g., Gurobi, SAS). We will refer to the former as the `conflict` setting and to the latter as the `dualray` setting. We compare those to a setting that uses conflict analysis and dual ray analysis simultaneously, the `combined` setting. Finally, we consider an extension of the `combined` setting that uses a pool for conflict management, the setting `combined+pool`.

All experiments were performed with the non-commercial MIP solver SCIP [13] (git hash 60f49ab, based on SCIP 3.2.1.2), using SoPlex 2.2.1.3 as LP solver. The experiments were run on a cluster of identical machines, each with an Intel Xeon Quad-Core with 3.2 GHz and 48 GB of RAM; a time limit of 3600 s was set.

We used two test sets: the MIPLIB2010 [17] benchmark test set and a selection of instances taken from the MIPLIB [8], MIPLIB2003 [4], MIPLIB2010, the COR@L [19] collection, the ALU[1], and the MARKSHARE [10] test sets. From these

---

[1] The instances are part of the contributed section of MIPLIB2003.

we selected all instances for which (i) all of the above settings need at least 100 nodes, (ii) at least one setting finishes within the time limit of 3600 s, and (iii) at least one setting analyzes more than 100 infeasible subproblems successfully. We refer to this test set as the CONFLICT set, since it was designed to contain instances for which conflict or dual ray analysis is frequently used.

Aggregated results on the number of generated nodes and needed solving time can be found in Table 1. Detailed results for each instance and test set separately can be found in the appendix of [26]. We use the conflict setting as a base line (since it used to be the SCIP default), for which we give actual means of branch-and-bound nodes and the solving time. For all other settings, we instead give factors w.r.t. the base line. A number greater than one implies that the setting is inferior and a number less than one implies that the setting is superior to the conflict setting.

First of all, we observe that solely using dual ray analysis is inferior to using conflict analysis on both test sets w.r.t. both performance measures. Note that we used a basic implementation of dual ray analysis; a solver that solely relies on it might implement further extensions that decrease this difference in performance, see also Sect. 5. However, the combination of conflict and dual ray analysis showed some significant performance improvements. We observed a speed-up of 3% and 18% on MIPLIB2010 and CONFLICT, respectively. Moreover, the number of generated nodes could be reduced by 5% and 25%, respectively. Finally, on the CONFLICT test set, the combined setting solved one instance more than the conflict setting and five more than the dualray setting. We take those results as an indicator that the two techniques complement each other well. In an additional experiment, we also tested applying conflict analysis solely from domain propagation or solely from infeasible LPs. Both variants were inferior to the conflict setting and are therefore not discussed in detail. Detailed results for each instance and test set separately can be found in the appendix of [26].

To partially explain the different extent of the improvements on both test sets, we would like to point out that in the MIPLIB2010 benchmark set, there are only 31 instances which fulfill the filtering criteria mentioned above for the CONFLICT set. On those, the combined setting is 7.2% faster and needs 15.6% less nodes than the conflict setting.

Looking at individual instances, there are a few cases for which the combined setting is the clear winner, e.g., neos-849702 or bnatt350. For neos-849702 and bnatt350, the dualray setting has a timeout, while the conflict setting is a factor of 6.2 and 1.83 slower, respectively, than the combined setting. At the same time, ns1766074 shows the largest deterioration from using a combined setting, being a factor of 1.63 slower than conflict and a factor of 1.06 slower than dualray. Generally, we observed that for infeasible MIPs (like ns1766074), the conflict setting was preferable over combined.

As can be seen in Table 1, using a conflict pool in addition to an aging system makes hardly any difference w.r.t. the overall performance in combination with the combined setting. However, looking at individual instances, there are cases where using a pool leads to a speed up of 17%

Table 1. Aggregated computational results. Columns marked with # show the number of solved instances. Columns 3 and 4 show the shifted geometric mean of absolute numbers of generated nodes (n, shift = 100) and needed solving time in seconds (t, shift = 10), respectively. All remaining columns show the relative number of generated nodes (n_Q) and needed solving time (t_Q) w.r.t. Columns 3 and 4, respectively.

Test set	conflict			dualray			combined			combined+pool		
	#	n	t	#	n_Q	t_Q	#	n_Q	t_Q	#	n_Q	t_Q
MIPLIB2010	60	14382	686	57	*1.365*	*1.167*	60	0.955	0.977	60	0.957	0.975
CONFLICT	105	16769	143	101	*1.616*	*1.256*	106	**0.755**	**0.827**	106	**0.759**	**0.829**

(e.g., ns1766074) and 20% (e.g., neos-1620807) compared to the combined setting on the MIPLIB2010 and CONFLICT test set, respectively. Moreover, using the presented pool-based approach to manage conflicts in combination with the conflict setting shows a decrease of solving time of 8.5% (MIPLIB2010) and 5.5% (CONFLICT) on instances where more than 50 000 infeasible LPs were analyzed successfully.

5 Conclusion and Outlook

In this short paper we discussed the similarities and differences of conflict analysis and dual ray analysis in solving MIPs. Our computational study indicates that a combination of both approaches can significantly enhance the performance of a state-of-the-art MIP solver and help each other. In our opinion both approaches complement each other well and on instances where the analysis of infeasible subproblems succeeds frequently, the solving time improved by 17.3% and the number of branch-and-bound nodes by 24.5%. In contrast, using a pool-based approach in addition to an aging mechanism to manage conflict constraints showed very little impact.

There are several instances for which using either dual ray analysis or conflict analysis exclusively outperformed the combination of both. Thus, we will plan to investigate a dynamic mechanism to switch between both techniques. Furthermore, applying dual ray analysis for infeasibility deduced by domain propagation as well as using more preprocessing (e.g., mixed integer rounding, projecting out continuous variables, etc.) techniques to modify constraints derived from dual ray analysis appear as promising directions for future research.

Acknowledgments. The work for this article has been conducted within the Research Campus Modal funded by the German Federal Ministry of Education and Research (fund number 05M14ZAM). We thank the anonymous reviewers for their valuable suggestions and helpful comments.

References

1. Achterberg, T.: Conflict analysis in mixed integer programming. Discrete Optim. **4**(1), 4–20 (2007)
2. Achterberg, T.: Constraint integer programming (2007)
3. Achterberg, T., Berthold, T.: Hybrid branching. In: Hoeve, W.-J., Hooker, J.N. (eds.) CPAIOR 2009. LNCS, vol. 5547, pp. 309–311. Springer, Heidelberg (2009). doi:10.1007/978-3-642-01929-6_23
4. Achterberg, T., Koch, T., Martin, A.: MIPLIB 2003. Oper. Res. Lett. **34**(4), 361–372 (2006)
5. Bénichou, M., Gauthier, J.-M., Girodet, P., Hentges, G., Ribière, G., Vincent, O.: Experiments in mixed-integer linear programming. Math. Program. **1**(1), 76–94 (1971)
6. Berthold, T., Gleixner, A.M.: Undercover: a primal MINLP heuristic exploring a largest sub-MIP. Math. Program. **144**(1–2), 315–346 (2014)
7. Berthold, T., Hendel, G.: Shift-and-propagate. J. Heuristics **21**(1), 73–106 (2015)
8. Bixby, R.E., Boyd, E.A., Indovina, R.R.: MIPLIB: a test set of mixed integer programming problems. SIAM News **25**, 16 (1992)
9. Brearley, A., Mitra, G., Williams, H.: Analysis of mathematical programming problems prior to applying the simplex algorithm. Math. Program. **8**, 54–83 (1975)
10. Cornuéjols, G., Dawande, M.: A class of hard small 0-1 programs. In: Bixby, R.E., Boyd, E.A., Ríos-Mercado, R.Z. (eds.) IPCO 1998. LNCS, vol. 1412, pp. 284–293. Springer, Heidelberg (1998). doi:10.1007/3-540-69346-7_22
11. Dakin, R.J.: A tree-search algorithm for mixed integer programming problems. Comput. J. **8**(3), 250–255 (1965)
12. Davey, B., Boland, N., Stuckey, P.J.: Efficient intelligent backtracking using linear programming. INFORMS J. Comput. **14**(4), 373–386 (2002)
13. Gamrath, G., Fischer, T., Gally, T., Gleixner, A.M., Hendel, G., Koch, T., Maher, S.J., Miltenberger, M., Müller, B., Pfetsch, M.E., Puchert, C., Rehfeldt, D., Schenker, S., Schwarz, R., Serrano, F., Shinano, Y., Vigerske, S., Weninger, D., Winkler, M., Witt, J.T., Witzig, J.: The SCIP optimization suite 3.2. Technical Report 15–60, ZIB, Takustr. 7, 14195 Berlin (2016)
14. Ginsberg, M.L.: Dynamic backtracking. J. Artif. Intell. Res. **1**, 25–46 (1993)
15. Jiang, Y., Richards, T., Richards, B.: No-good backmarking with min-conflict repair in constraint satisfaction and optimization. In: PPCP, vol. 94, pp. 2–4. Citeseer (1994)
16. Kılınç Karzan, F., Nemhauser, G.L., Savelsbergh, M.W.P.: Information-based branching schemes for binary linear mixed-integer programs. Math. Program. Comput. **1**(4), 249–293 (2009)
17. Koch, T., Achterberg, T., Andersen, E., Bastert, O., Berthold, T., Bixby, R.E., Danna, E., Gamrath, G., Gleixner, A.M., Heinz, S., Lodi, A., Mittelmann, H., Ralphs, T., Salvagnin, D., Steffy, D.E., Wolter, K.: MIPLIB 2010. Math. Program. Comput. **3**(2), 103–163 (2011)
18. Land, A.H., Doig, A.G.: An automatic method of solving discrete programming problems. Econometrica **28**(3), 497–520 (1960)
19. Linderoth, J.T., Ralphs, T.K.: Noncommercial software for mixed-integer linear programming. Integer Program.: Theor. Pract. **3**, 253–303 (2005)
20. Linderoth, J.T., Savelsbergh, M.W.: A computational study of search strategies for mixed integer programming. INFORMS J. Comput. **11**(2), 173–187 (1999)

21. Marques-Silva, J.P., Sakallah, K.: Grasp: a search algorithm for propositional satisfiability. IEEE Trans. Comput. **48**(5), 506–521 (1999)
22. Pólik, I.: (Re)using dual information in MILP. In: INFORMS Computing Society Conference, Richmond, VA (2015)
23. Pólik, I.: Some more ways to use dual information in MILP. In: International Symposium on Mathematical Programming, Pittsburgh, PA (2015)
24. Sandholm, T., Shields, R.: Nogood learning for mixed integer programming. In: Workshop on Hybrid Methods and Branching Rules in Combinatorial Optimization, Montréal (2006)
25. Stallman, R.M., Sussman, G.J.: Forward reasoning and dependency-directed backtracking in a system for computer-aided circuit analysis. Artif. Intell. **9**(2), 135–196 (1977)
26. Witzig, J., Berthold, T., Heinz, S.: Experiments with conflict analysis in mixed integer programming. Technical report 16–63, ZIB, Takustr. 7, 14195 Berlin (2016)

A First Look at Picking Dual Variables for Maximizing Reduced Cost Fixing

Omid Sanei Bajgiran[1,3]([✉]), Andre A. Cire[1], and Louis-Martin Rousseau[2,3]

[1] Department of Management, University of Toronto Scarborough, Toronto, Canada
omid.saneibajgiran@rotman.utoronto.ca
[2] Department of Mathematics and Industrial Engineering,
Polytechnique Montréal, Montreal, Canada
[3] Interuniversity Research Center on Enterprise Networks,
Logistics and Transportation (CIRRELT), Montreal, Canada

Abstract. Reduced-cost-based filtering in constraint programming and variable fixing in integer programming are techniques which allow to cut out part of the solution space which cannot lead to an optimal solution. These techniques are, however, dependent on the dual values available at the moment of pruning. In this paper, we investigate the value of picking a set of dual values which maximizes the amount of filtering (or fixing) that is possible. We test this new variable-fixing methodology for arbitrary mixed-integer linear programming models. The resulting method can be naturally incorporated into existing solvers. Preliminary results on a large set of benchmark instances suggest that the method can effectively reduce solution times on hard instances with respect to a state-of-the-art commercial solver.

Keywords: Mixed-integer programming · Variable fixing methodology · Reduced-cost based filtering

1 Introduction

A key feature of modern mathematical programming solvers refers to the wide range of techniques that are applied to *simplify* an instance. Typically considered during a preprocessing stage, these techniques aim at fixing variables, eliminating redundant constraints, and identifying structure that can either lead to speed-ups in solution times or provide useful information about the model at hand. Examples of valuable information include, e.g., potential numerical issues or which subset of inequalities and variables may be responsible for the infeasibility [12], if that is the case. These simplification methods alone reduce solution times by half in state-of-the-art solvers such as CPLEX, SCIP, or Gurobi [4], thereby constituting an important tool in the use of mixed-integer linear programming (MILP) in practical real-world problems [11].

In this paper we investigate a new simplification technique that expands upon the well-known *reduced cost fixing method*, first mentioned by Balas and

© Springer International Publishing AG 2017
D. Salvagnin and M. Lombardi (Eds.): CPAIOR 2017, LNCS 10335, pp. 221–228, 2017.
DOI: 10.1007/978-3-319-59776-8_18

Martin [2] and largely used both in the mathematical programming and the constraint programming (CP) communities. The underlying idea of the method is straightforward: Given a linear programming (LP) model and any optimal solution to such a model, the *reduced cost* of a variable indicates the marginal linear change in the objective function when the value of the variable in that solution is increased [5]. In cases where the LP encodes a relaxation of an arbitrary optimization problem, we can therefore filter all values from a variable domain that, based on the reduced cost, incur a new objective function value that is worse than a known solution to the original problem. The result is a tighter variable bound which can then trigger further variable fixing and other simplifications.

This simple but effective technique is widely applied in MILP presolving [4,11,12] and plays a key role in a variety of propagation methods for global constraints in CP [7–9]. It can be easily incorporated into solvers since the reduced costs are directly derived from any optimal set of duals, which in turn can be efficiently obtained by solving an LP once. The technique is also a natural way of exploiting the strengths of MILP within a CP framework, since the dual values incorporate a global bound information that is potentially lost when processing constraints one at a time (a concept that is explored, e.g., in [3,17,19]).

However, in all cases typically only *one* reduced cost per variable is considered, that is, the one obtained after solving the LP relaxation of a MILP. In theory, any set of feasible duals provides valid reduced costs that may lead, in turn, to quite different variable bound tightenings. This question was originally raised by Sellmann [16], who demonstrated that not only distinct dual vectors would result in significantly different filtering behaviors, but that potentially sub-optimal dual vectors could yield much more pruning than the optimal ones.

Our goal in this work is to investigate the potential effect of *picking* the dual vector that maximizes reduced-cost-based filtering. By doing so, we revisit the notion of *relaxed consistency* for reduced costs fixing; that is, we wish to influence the choice of the dual values given by a relaxation so as to maximize the amount of pruning that can be performed. We view the proposed techniques as a first direction towards answering some of the interesting questions raised in the field of *CP-based Lagrangian relaxation* [3,16], in particular related to how to select the dual variables (or, equivalently, the Lagrangian multipliers) to maximize propagation.

The contribution of this paper is to formulate the problem of finding the dual vectors that maximize the number of reductions as an optimization problem defined over the space of optimal (or just feasible) dual values. We compare this approach to achieving full *relaxed consistency*, which can be obtained by solving a large (but polynomial) number of LP problems. The resulting technique can be seamlessly incorporated into existing solvers, and preliminary results over the MIPLIB indicate that it can significantly reduce solution time as well as the size of the branching tree when proving the optimality of a primal bound. We hope to motivate further research on the quality of the duals used within both ILP and CP technology.

For the sake of clarity and without loss of generality, the proposed approaches will be detailed in the context of integer linear programs (ILPs), i.e., where all variables are integers, as opposed to mixed-integer linear programming models. This technique is also applicable in the context of CP, if one can derive a (partial) linear relaxation of the model [15].

The paper is organized as follows. Section 2 introduces the necessary notation and the basic concepts of reduced cost fixing and the related consistency notions. Next, we discuss one alternative to obtain an approximate consistency in Sect. 3. Finally, we present a preliminary numerical study in Sect. 4 and conclude in Sect. 5.

2   Reduced Cost Fixing and Relaxed-Consistency

For the purposes of this paper, consider the problem

$$z_P := \min\{c^T x : Ax \geq b, x \geq 0\} \tag{P}$$

with $A \in \mathbb{R}^{n \times m}$ and $b, c \in \mathbb{R}^n$ for some $n, m \geq 1$. We assume that (P) represents the LP relaxation of an ILP problem P_S with an optimal solution value of $z^* \geq z_P$ and where variables $\{x_i : i \in S\}$ are subject to integrality constraints. The dual of the problem (P) can be written as

$$z_D := \max\{u^T b : u^T A \leq c^T, u \geq 0\} \tag{D}$$

where $u \in \mathbb{R}^m$ is the vector of *dual variables*. We assume for exposition that P_S, (P), and (D) are bounded (the results presented here can be easily generalized when that is not the case).

We have $z_P = z_D$ (strong duality) and for every optimal solution x^* of (P), there exists an optimal solution u^* to (D) such that $u^{*T}(b - Ax^*) = 0$ (complementary slackness). Moreover, for some j such that $x_j^* = 0$, the quantity

$$\bar{c}_j = c_j - u^{*T} A_j \tag{RC}$$

is the *reduced cost* of variable x_j and yields the marginal increase in the objective function if x_j^* moves away from its lower bound. Thus, if a given known feasible solution with value $z^{UB} \geq z^*$ is available to the original ILP, the *reduced cost fixing* technique consists of fixing $x_j^* = 0$ if

$$z_P + \bar{c}_j \geq z^{UB}, \tag{RCF}$$

since any solution with $x_j^* > 0$ can never improve upon the existing upper bound z^{UB}. We remark in passing that the condition (RCF) can be generalized to establish more general bounds on a variable. That is, we can use the reduced cost $\bar{c}_j$ to deduce values l_j and u_j such that either $x_j^* \geq l_j$ or $x_j^* \leq u_j$ in any optimal solution (see, e.g., [10]). In this paper we restrict our attention to the classical case described above. We refer to Wolsey [18] and Nemhauser and Wolsey [13] for the formal proofs of correctness.

The dual variables u^* for the computation of (RC) can be obtained with very little computational effort after finding an optimal solution x^* to (P) (e.g., they are computed simultaneously to x^* when using the Simplex method). In the most of practical known implementations concerning ILP presolving and CP propagation methods, the reduced cost fixing is typically carried out using the single u^* computed after solving every LP relaxation [7,12]. Note, however, that (D) may contain multiple optimal solutions, each potentially yielding a different reduced cost $\bar{c}_j$ that may or may not satisfy condition (RCF).

One therefore does not need to restrict its attention to a unique u^*, and can potentially improve the number of variables that are fixed if the whole dual space is considered instead. This would mean that if there exists a reduced cost $\bar{c}_j$ such that variable x_j can be fixed to 0 with respect to z^{UB}, then there exist a dual vector u such that $u^T b + (c_j - u^T A_j) \geq z^{UB}$. This was first demonstrated by [16] in the context of CP-based Lagrangian Relaxation, which also pointed out that often more filtering occurs when u is not an optimal dual vector and therefore we might have that $u^T b < z_P$.

Achieving the notion of *Relaxed Consistency*, as defined in [6], is thus quite consuming in such a condition. This form of consistency can be casted, in the context of ILP, as follows.

Definition 1. *Let P_S be an ILP model with linear programming relaxation (P) and its corresponding dual (D). The model P_S is relaxed consistent (or relaxed-P-consistent) with respect to an upper bound z^{UB} to P_S if for any dual feasible u^* to (D) and its associated reduced cost $\bar{c}$ vector, condition (RCF) is never satisfied, i.e., $z_P + \bar{c}_j < z^{UB}$ for all $j = 1, \ldots, n$.*

If a model is *relaxed consistent* according to Definition 1, then it is not possible to fix any variable x_j via reduced costs.

Consistent with the theory presented in [6,16], any ILP formulation can be efficiently converted into a relaxed consistent formulation in polynomial time. Given an ILP model P_S and the primal (P) and dual (D) of its associated linear programming relaxation, the set of optimal dual solution coincides with the polyhedral set $\mathcal{D} = \{u \in \mathbb{R}^m : u^T A \geq c, u \geq 0\}$. Thus, a variable x_j can be fixed to zero if the optimal solution $\bar{c}_j^*$ of the problem $\bar{c}_j^* = \max\{c_j - u^T A_j : u \in \mathcal{D}\}$ is such that $z_P + \bar{c}_j^* \geq z^{UB}$. This means that the complexity of the procedure is dominated by the cost of solving $O(n)$ LP models, each of which can be done in weakly polynomial time [18].

3   Dual Picking for Maximum Reduced Cost Fixing

Establishing *relaxed consistency* by solving $O(n)$ LP models is impractical when a model has any reasonably large number of variables. We thus propose a simple alternative model that exploits the underlying concept of searching in the dual space for maximizing filtering. Namely, as opposed to solving an LP for each variable, we will search for the dual variables that together maximize the number of variables that can be fixed. This can be written as the following MILP model:

$$\max \quad \sum_{i=1}^{n} y_i \qquad \text{(DP-RCF)}$$

$$\text{s.t.} \quad u^T A \le c \qquad (1)$$
$$u^T b = z_P \qquad (2)$$
$$u^T b + (c_j - u^T A_j) \ge z^{UB} - (1 - y_j)M \quad \forall j \qquad (3)$$
$$u \ge 0 \qquad (4)$$
$$y \in \{0,1\}^n \qquad (5)$$

In the model (DP-RCF) above, we are searching for the dual variables u, on the optimal dual face, that maximize the number of variables fixed. Specifically, we will define a binary variable y_i in such a way that $y_i = 1$ if and only if we fix it allows to deduce that x_j can be fixed to 0.

To enforce this, let M be a sufficiently large number, and $\mathbf{1}$ an n-dimensional vector containing all ones. Constraints (1), and (4) ensure that u^* is dual feasible. If $y_i = 1$, then inequality (3) reduces to condition (RCF) and the associated x_i should be fixed to 0. Otherwise, the right-hand side of (3) is arbitrary small (in particular to account for arbitrarily small negative reduced costs). Constraint (2) enforces strong duality and the investigation of optimal dual vectors only, it can be omitted in order to explore the whole dual feasible space (as sub-optimal dual vectors can perhaps filter more [16]). Finally, constraint (5) defines the domain of the y variable and the objective maximizes the number of variables fixed.

The model (DP-RCF) does not necessarily achieve *relaxed consistency* as it yields a single reduced cost vector and it is restricted to the optimal dual face. However, our experimental results indicate that the model can be solved quite efficiently and yields interesting bounds. Notice also that any feasible solution to (DP-RCF) corresponds to a valid set of variables to fix, and hence any solutions found during search can be used to our purposes as well.

4   Preliminary Numerical Study

We present a preliminary numerical study of our technique on a subset of the MIPLIB 2010 benchmark [1]. All experiments were performed using IBM ILOG CPLEX 12.6.3 on a single thread of an Intel Core i7 CPU 3.40 GHz with 8.00 GB RAM.

Our goal for these experiments is twofold: We wish first to evaluate the filtering achieved by the dual picking models (DP-RCF) in comparison to the full relaxed-consistent model, and next verify what is the impact of fixing these variables in CPLEX when proving optimality. For the first criteria, we considered the number of fixed variables according to three different approaches:

1. The model (DP-RCF) with a time limit of 10 min, denoted by DP.
2. A modified version of the model (DP-RCF) without constraint (2), i.e., we increased our search space by considering any *feasible* dual solution, also fixing a time limit of 10 min. We denote this approach by DP-M.
3. Solving the $O(n)$ LP models to achieve relaxed consistency in view of Definition 1, with no time limit. This method provides the maximum number of variables that can be fixed thought RCF and thus an upperbound for both DP and DP-M, it is denoted by RCC (*relaxed reduced-cost consistency*).

In all the cases above, we considered the optimal solution value of the instance as the upper bound z^{UB} for the (RCF) condition, which results in the strongest possible filtering for a fixed set of duals.

Next, to assess the impact of fixing variables in the ILP solution process, we ran the default CPLEX for each approach above, specifically setting the fixed variables to zero and providing the optimal solution value of each instance to the solver. We have also ran default CPLEX without any variable fixing and with the optimal value as an input, in order to evaluate the impact of variable fixing in proving the optimality of a particular bound.

As a benchmark we considered all "easy" instances of the MIPLIB that could be solved within our memory or limit of 8 GB and a time limit of 60,000 s. We have also eliminated all instances where *relaxed consistency* could not fix any variable. This resulted in 36 instances.

The results are depicted in Table 1, where Aux stands for the CPU time required to solve the auxiliary model (DP-RCF) and Vars is the number of variables which could be fixed. Moreover, Cons and Vars in the Dimension category indicate the number of constraints and decision variables of each instance, respectively, and DC shows the number of variables that default CPLEX can fix as a result of the final dual solution calculated by ourselves. Omitted rows indicates the auxiliary problem reached its time limit. Due to space restrictions, instances *neos-16...*, *neos-47...*, *neos-93...*, *neos-13...*, *neos-84...*, *rmatr-p5*, *rmatr-p10*, *core253...*, *neos-93...*, and *sat...* represent instances *neos-1601936*, *neos-476283*, *neos-934278*, *neos-1337307*, *neos-849702*, *rmatr100-p5*, *rmatr100-p10*, *core2536-691*, *neos-934278*, and *satellites1-25*, respectively.

We first notice that achieving full *relaxed consistency* is quite time consuming and not practical with respect to the default solution time of CPLEX. When looking at both DP models, it is obvious that restricting the search to the optimal dual face, rather than the whole dual feasible space, yields practically the same amount of filtering while being orders of magnitude faster. In fact, in many cases the (DP-RCF) model can be solved in less than a second.

To determine whether such filtering is worth the extra effort, we compare the solution time of DP against the default solution time of CPLEX. For each instance we compute the speedup as the solution time of CPLEX divided by the sum of both solution and dual picking (i.e., solving (DP-RCF)) time of DP. We then compute the geometric mean of all speedups, which yields an average speed up of 20% when using our dual picking methodology over the default CPLEX.

Table 1. General results

Instance	Dimension		DC	CPLEX Default		DP				DP-M				RCC			
	Cons	Vars	Vars	Time	Nodes	Time	Nodes	Aux	Vars	Time	Nodes	Aux	Vars	Time	Nodes	Aux	Vars
MILP instances																	
30n20b8	578	18380	0	3	260	3	260	0.5	7282					3	90	494	13603
aflow40b	1442	2728	33	187	17461	1961	200477	0.03	499					281	21635	42	532
binkar10_1	1026	2298	165	7	1567	8	2135	0.1	200					5	902	51	281
core253...	2539	15293	2494	38304	239195	2414	8816	600	3046					4464	72010	79173	4818
biella1	1203	7328	219	942	1477	257	1107	52	429					204	1089	18220	1540
gmu-35-40	424	1205	0	68	351765	68	351765	0.02	0	124	≈6e5	9	485	26	94209	19	493
mik-...	151	251	50	1	2205	0.7	1115	0.01	50	0.7	1115	0.5	50	0.7	1115	0.8	50
mzzv11	9499	10240	29	19	57	17	46	26	213					26	329	18679	678
neos13	20852	1827	8	6	485	5	399	1	8	5	399	68	8	6	7	588	179
neos-16...	3131	4446	867	327	4264	137	2259	2	1766					377	7549	7747	1782
neos-47	10015	11915	36	1054	1565	195	1129	2	69					195	1129	38293	69
neos-93...	11495	23123	15843	430	97	74	12	3	15981					132	39	86254	16088
net12	14021	14115	0	162	440	16	50	1	32					150	880	3057	334
ns1208400	4289	2883	2	21	860	21	860	0.08	286	21	860	107	286	32	1465	1286	592
ns1830653	2932	1629	0	208	24606	74	9185	0.2	850	217	31211	952	853	110	5203	65	915
pw-myciel4	8164	1059	0	0.4	17	0.4	117	0.05	0	0.4	117	6	0	1	580	54	160
rmatr-p10	7260	7359	100	20	455	27	345	0.2	100					27	345	3260	100
rmatr-p5	8685	8784	100	3	2	5	6	2	100					5	6	9601	100
rococo...	1293	3117	0	523	13193	4475	54077	0.3	511					531	15979	35	553
roll3000	2295	1166	0	2645	177058	2645	177058	0.08	120					1015	61271	50	149
sat...	5996	9013	99	220	442	25	497	2	2499					11	758	13845	4989
sp98ir	1531	1680	110	244	8607	241	13057	1.5	285					204	18659	86	524
timtab1	171	397	0	2393	≈1e6	2393	≈1e6	0.02	13	2393	≈1e6	4	13	2393	≈1e6	1	13
Binary-Only Problems																	
acc-tight5	3052	1339	1339	97	2262	97	2262	0	1339	97	2262	0	1339	97	2262	0	1339
air04	823	8904	0	7	493	7	46	6	64					5	54	1500	126
bab5	4964	21600	0	8313	49311	2903	12717	369	373					883	7904	18563	2309
eil33-2	32	4516	1004	112	15041	18	6825	6	1035					0.3	305	33	4013
eilB101	100	2818	45	595	26753	318	21666	1	125					84	10309	115	1588
n3div36	4484	22120	4755	57917	≈2.7e6	57017	≈2.7e6	1	4755					3518	83900	6568	6372
neos-13...	5687	2840	14	9	300	6	110	5	77					5	30	1098	150
neos18	11402	3312	0	0.4	1	0.4	1	0.08	84					0.4	1	142	84
neos-84...	1041	1737	1737	105	8223	105	8223	0	1737	105	8223	0	1737	105	8223	0	1737
ns1688347	4197	2685	0	8	260	5	1310	0.25	267					1	1	56	716
opm2-z7-s2	31798	2023	0	230	1000	230	1000	1	0	230	1000	35	0	25	260	2484	459
rmine6	7078	1096	0	69	15233	31	5920	0.25	1	31	5920	10	1	52	10497	116	66
sp98ic	825	10894	6902	29041	149629	40306	466555	62	7098					345	35647	4724	8488

5   Conclusion

In this paper, we revisited the notion of reduced-cost based filtering and variable fixing, which are known to be dependent on the available dual information. We defined the problem of identifying the set of dual values that maximize the number of variables which can be fixed as an optimization problem. We demonstrated that looking for a good set of such dual on the optimal dual face is considerably faster and filter almost as many variables as when considering the full feasible dual space. In many cases fixing more variable lead to a reduced search tree that can be explored faster. However, in a good number of cases, solution time increases when more variables are fixed, which is probably due to the fact that early in the tree the search takes a different path.

Future research will consider dual picking during search, so as to try to fix variables when the relative gap becomes small enough in a subtree, as well as applying the techniques in the context of constraint programming.

References

1. MIPLIB2010. http://miplib.zib.de/miplib2010-benchmark.php
2. Balas, E., Martin, C.H.: Pivot and complement-a heuristic for 0–1 programming. Manage. Sci. **26**(1), 86–96 (1980)
3. Bergman, D., Cire, A.A., Hoeve, W.-J.: Improved constraint propagation via lagrangian decomposition. In: Pesant, G. (ed.) CP 2015. LNCS, vol. 9255, pp. 30–38. Springer, Cham (2015). doi:10.1007/978-3-319-23219-5_3
4. Bixby, E.R., Fenelon, M., Gu, Z., Rothberg, E., Wunderling, R.: MIP: theory and practice — closing the gap. In: Powell, M.J.D., Scholtes, S. (eds.) CSMO 1999. ITIFIP, vol. 46, pp. 19–49. Springer, Boston, MA (2000). doi:10.1007/978-0-387-35514-6_2
5. Chvátal, V.: Linear Programming. Freeman, New York (1983). Reprints: (1999), (2000), (2002)
6. Fahle, T., Sellmann, M.: Cost-based filtering for the constrained knapsack problem. Ann. Oper. Res. **115**, 73–93 (2002)
7. Focacci, F., Lodi, A., Milano, M.: Cost-based domain filtering. In: Jaffar, J. (ed.) CP 1999. LNCS, vol. 1713, pp. 189–203. Springer, Heidelberg (1999). doi:10.1007/978-3-540-48085-3_14
8. Focacci, F., Lodi, A., Milano, M., Vigo, D.: Solving TSP through the integration of OR and CP techniques. Electron. Notes Discrete Math. **1**, 13–25 (1999)
9. Focacci, F., Milano, M., Lodi, A.: Solving TSP with time windows with constraints. In: Proceedings of the 1999 International Conference on Logic programming, Massachusetts Institute of Technology, pp. 515–529 (1999)
10. Klabjan, D.: A new subadditive approach to integer programming. In: Cook, W.J., Schulz, A.S. (eds.) IPCO 2002. LNCS, vol. 2337, pp. 384–400. Springer, Heidelberg (2002). doi:10.1007/3-540-47867-1_27
11. Lodi, A.: Mixed integer programming computation. In: Jünger, M., Liebling, T.M., Naddef, D., Nemhauser, G.L., Pulleyblank, W.R., Reinelt, G., Rinaldi, G., Wolsey, L.A. (eds.) 50 Years of Integer Programming 1958–2008, pp. 619–645. Springer, Heidelberg (2010)
12. Mahajan, A.: Presolving mixed-integer linear programs. In: Wiley Encyclopedia of Operations Research and Management Science (2010)
13. Nemhauser, G.L., Wolsey, L.A.: Integer Programming and Combinatorial Optimization (1988)
14. Chichester, W., Nemhauser, G.L., Savelsbergh, M.W.P., Sigismondi, G.S.: Constraint Classification for Mixed Integer Programming Formulations. COAL Bulletin, vol. 20, pp. 8–12 (1992)
15. Refalo, P.: Linear formulation of constraint programming models and hybrid solvers. In: Dechter, R. (ed.) CP 2000. LNCS, vol. 1894, pp. 369–383. Springer, Heidelberg (2000). doi:10.1007/3-540-45349-0_27
16. Sellmann, M.: Theoretical foundations of CP-based lagrangian relaxation. In: Wallace, M. (ed.) CP 2004. LNCS, vol. 3258, pp. 634–647. Springer, Heidelberg (2004). doi:10.1007/978-3-540-30201-8_46
17. Thorsteinsson, E.S., Ottosson, G.: Linear relaxations and reduced-cost based propagation of continuous variable subscripts. Ann. Oper. Res. **115**(1), 15–29 (2002)
18. Wolsey, L.A.: Integer Programming, vol. 4. Wiley, New York (1998)
19. Yunes, T., Aron, I.D., Hooker, J.N.: An integrated solver for optimization problems. Oper. Res. **58**(2), 342–356 (2010)

Experimental Validation of Volume-Based Comparison for Double-McCormick Relaxations

Emily Speakman, Han Yu, and Jon Lee$^{(\boxtimes)}$

Department of Industrial and Operations Engineering,
University of Michigan, Ann Arbor, MI, USA
{eespeakm,yuha,jonxlee}@umich.edu

Abstract. Volume is a natural geometric measure for comparing polyhedral relaxations of non-convex sets. Speakman and Lee gave volume
formulae for comparing relaxations of trilinear monomials, quantifying
the strength of various natural relaxations. Their work was motivated
by the spatial branch-and-bound algorithm for factorable mathematical-
programming formulations. They mathematically analyzed an important
choice that needs to be made whenever three or more terms are multiplied in a formulation. We experimentally substantiate the relevance of
their main results to the practice of global optimization, by applying it
to different relaxations of difficult box cubic problems (boxcup). In doing
so, we find that, using their volume formulae, we can accurately predict
the quality of a relaxation for boxcups based on the (box) parameters
defining the feasible region. Specifically, we are able to conclude from our
experiments that all of the relevant relaxations are round enough so that
average objective gaps for a pair of boxcup relaxations can be predicted
by appropriately combining the volumes of relaxations of the individual
trilinear monomials.

Keywords: Mccormick inequalities · Mixed-integer non-linear optimization · Global optimization · Trilinear monomials · Spatial branch-
and-bound

1 Introduction

1.1 Measuring Relaxations via Volume in Mathematical Optimization

In the context of mathematical optimization, there is often a natural tradeoff
in the tightness of a *convexification* (i.e., a convex relaxation) and the difficulty of optimizing over it. This idea was emphasized by [11] in the context
of mixed-integer non-linear programming (MINLP) (see also the recent work
[7]). Of course this is also a well-known phenomenon for difficult 0/1 linear-
optimization problems, where very tight relaxations are available via extremely
heavy semidefinite-programming relaxations (e.g., the Lasserre hierarchy), and
the most effective relaxation for branch-and-bound/cut may well not be the

© Springer International Publishing AG 2017
D. Salvagnin and M. Lombardi (Eds.): CPAIOR 2017, LNCS 10335, pp. 229–243, 2017.
DOI: 10.1007/978-3-319-59776-8_19

tightest. Earlier, again in the context of mathematical optimization, [12] introduced the idea of using volume as a measure of the tightness of a convex relaxation (for fixed-charge and vertex packing problem). Most of that mathematical work was asymptotic, seeking to understand the quality of families of relaxations with a growing number of variables, but some of it was also substantiated experimentally in [11].

1.2   Spatial Branch-and-Bound

The workhorse algorithm for global optimization of so-called factorable MINLPs (see [13]) is spatial branch-and-bound (sBB) (see [1,18,20]). sBB decomposes model functions against a library of basic functions (e.g., x_1x_2, $x_1x_2x_3$, $x_1^{x_2}$, $\log(x)$, $\sin(x)$, x^3, $\sqrt{x}$, $\arctan(x)$). We assume that each basic function is a function of no more than 3 variables, and that we have a convex outer-approximation of the graph of each such basic function on box domains. sBB realizes convexifications of model functions by composing convexifications of basic functions. sBB subdivides the domains of variables, and re-convexifies, obtaining stronger bounds on model functions. Prominent software for sBB includes: ANTIGONE [16], BARON [19], Couenne [3], SCIP [26], and αBB [1].

1.3   Using Volume to Guide Decompositions for Spatial Branch-and-Bound

[23] applied the idea of [12], but now in the context of the low-dimensional relaxations of basic functions that arise in sBB. Specifically, [23] considered the basic function $f = x_1x_2x_3$ on box domains; that is the graph

$$\mathcal{G} := \left\{ (f,x_1,x_2,x_3) \in \mathbb{R}^4 \ : \ f = x_1x_2x_3, \ x_i \in [a_i,b_i], i = 1,2,3 \right\},$$

where $0 \le a_i < b_i$ are given constants. It is important to realize that $\mathcal{G}$ is relevant for any model where any three quantities (which could be complicated functions themselves) are multiplied. Furthermore, the case of nonzero lower bounds (i.e., $a_i > 0$) is particularly relevant, especially when the multiplied quantities are complicated functions of model variables. We let $\mathcal{P}_h$ denote the convex hull of $\mathcal{G}$. Though in fact polyhedral, the relaxation $\mathcal{P}_h$ has a rather complicated inequality description. Often lighter relaxations are used by modelers and MINLP software. [23] considered three natural further relaxations of $\mathcal{P}_h$. Thinking of the product as $x_1(x_2x_3)$ or $x_2(x_1x_3)$ or $x_3(x_1x_2)$, and employing the so-called McCormick relaxation twice, leads to three *different* double-McCormick relaxations. [23] derive analytic expressions for the volume of $\mathcal{P}_h$ as well as all three of the natural relaxations. The expressions are formulae in the six constants $0 \le a_i < b_i$, $i = 1,2,3$. In doing so, they quantify the quality of the various relaxations and provide recommendations for which to use.

It is important to be aware that sBB software such as ANTIGONE and BARON exclusively use the relatively complicated inequality description of

$\mathcal{P}_h$, while others such as Couenne, SCIP and αBB use an *arbitrary* double-McCormick relaxation. Much more discussion and mathematical guidance concerning the tradeoffs between these and related (extended-variable) relaxations can be found in [23, Sect. 3].

1.4   Our Contribution

The results of [23] are theoretical. Their utility for guiding modelers and sBB implementers depends on the belief that volume is a good measure of the quality of a relaxation. Morally, this belief is based on the idea that with no prior information on the form of an objective function, the solution of a relaxation should be assumed to occur with a uniform density on the feasible region. Our contribution is to experimentally validate the robustness of this theory in the context of a particular use case, optimizing multilinear cubics over boxes ('boxcup'). There is considerable literature on techniques for optimizing quadratics, much of which is developed and validated in the context of so-called 'boxqp' problems, where we minimize $\sum_{i,j} q_{ij} x_i x_j$ over a box domain in $\mathbb{R}^n$. So our boxcup problems, for which we minimize $\sum_{i,j} q_{ijk} x_i x_j x_k$ over a box domain in $\mathbb{R}^n$, are natural and defined in the same spirit and for the same purpose as the boxqp problems.

A main result of [23] is an ordering of the three natural relaxations of individual trilinear monomials by volume. But their formulae are for $n = 3$. Our experiments validate their theory as applied to our use case. We demonstrate that in the setting of 'boxcup' problems, the average objective discrepancy between relaxations very closely follows the prediction of the theory of [23], when volumes are appropriately combined (summing the 4-th root of the volume, across the chosen relaxations of each trilinear monomials). Moreover and very importantly, we are able to demonstrate that these results are robust against sparsity of the cubic forms.

[12] defined the *idealized radius* of a polytope in $\mathbb{R}^d$ as essentially the d-th root of its volume (up to some constants depending on d). For a polytope that is very much like a ball in shape, we can expect that this quantity is (proportional to) the "average width" of the polytope. The average width arises by looking at 'max minus min', averaged over all normalized linear objectives. So, the implicit prediction of [23] is that the *idealized radius* should (linearly) predict the expected 'max minus min' for normalized linear objectives. We have validated this experimentally, and looked further into the *idealized radial distance* between pairs of relaxations, finding an even higher degree of linear association.

Finally, in the important case $a_1 = a_2 = 0$, $b_1 = b_2 = 1$, [23] found that the two worst relaxations have the same volume, and the greatest difference in volume between $\mathcal{P}_h$ and the (two) worst relaxations occurs when $a_3 = b_3/3$. We present results of experiments that clearly show that these predictions via volume are again borne out on 'boxcup' problems.

All in all, we present convincing experimental evidence that volume is a good predictor for quality of relaxation in the context of sBB. Our results strongly suggest that the theoretical results of [23] are important in devising decompositions of complex functions in the context of factorable formulations and therefore our results help inform both modelers and implementers of sBB.

1.5 Literature Review

Computing the volume of a polytope is well known to be strongly #P-hard (see [4]). But in fixed dimension, or, in celebrated work, by seeking an approximation via a randomized algorithm (see [9]), positive results are available. Our work though is motivated not by algorithms for volume calculation, but rather in certain situations where analytic formulae are available.

Besides [12,23] (and the follow-on [22]), there have been a few papers on analytic formulae for volumes of polytopes that naturally arise in mathematical optimization; see [2,5,10,24,25]. But none of these works has attempted to apply their ideas to the low-dimensional polytopes that naturally arise in sBB, or even to apply their ideas to compare relaxations. One notable exception is [6], which is a mostly-computational precursor to [23], focusing on quadrilinear functions (i.e., $f = x_1 x_2 x_3 x_4$). [6] identifies that the quality of a repeated-McCormick relaxation depends quite a lot on the order of grouping, but there was no firm theoretical result grounding the choice of repeated-McCormick relaxation. In contrast, we have firm theoretical grounding, via [23], and here we go a step further to see that the theory of [23] can be used to rather accurately predict the quality (as measured by objective gap) of an aggregate relaxation built from the different relaxations of individual trilinear monomials.

There are many implementations of sBB. E.g., BARON [19], Couenne [3], SCIP [26], ANTIGONE [16], and αBB [1]. Both BARON and ANTIGONE use the complete linear-inequality description of $\mathcal{P}_h$, while Couenne, SCIP and αBB use an *arbitrary* double-McCormick relaxation. Our results indicate that there are situations where the choice of BARON and ANTIGONE may be too heavy, and certainly even restricting to double-McCormick relaxations, Couenne, SCIP and αBB do not systematically choose the best one.

There is a large literature on convexification of multilinear functions. Most relevant to our work are: the polyhedral nature of the convexification of the graphs of multilinear functions on box domains (see [17]); the McCormick inequalities describing giving the complete linear-inequality description for bilinear functions on a box domain (see [13]); the complete linear-inequality description of $\mathcal{P}_h$ (see [14,15]).

2 Preliminaries

2.1 From Volumes to Gaps

For a convex body $C \subset \mathbb{R}^d$, we denote its *volume* (i.e., Lebesgue measure) by $\mathrm{vol}(C)$. Volume seems like an awkward measure to compare relaxations, when typically we are interested in objective-function gaps. Following [12], the *idealized radius* of a convex body $C \subset \mathbb{R}^d$ is

$$\rho(C) := (\mathrm{vol}(C)/\mathrm{vol}(B_d))^{1/d},$$

where B_d is the (Euclidean) unit ball in $\mathbb{R}^d$. $\rho(C)$ is simply the radius of a ball having the same volume as C. The *idealized radial distance* between convex bodies C_1 and C_2 is simply $|\rho(C_1) - \rho(C_2)|$. If C_1 and C_2 are concentric balls, say with $C_1 \subset C_2$, then the idealized radial distance between them is the (radial) height of C_2 above C_1. The *mean semi-width* of C is simply

$$\frac{1}{2} \int_{\|c\|=1} \left(\max_{x \in C} c'x - \min_{x \in C} c'x \right) d\psi,$$

where ψ is the $(d-1)$-dimensional Lebesgue measure on the boundary of B_d, normalized so that ψ on the entire boundary is unity. If C is itself a ball, then (i) its idealized radius is in fact its radius, and (ii) its width in any unit-norm direction c is constant, and so (iii) its (idealized) radius is equal to its mean semi-width.

Key point: What we can hope is that our relaxations of individual trilinear monomials (in a model with many overlapping trilinear monomials) are round enough so that choosing them to be of small volume (which is proportional to and monotone increasing in its idealized radius raised to the power d) is a good proxy for choosing the overall relaxation by *mean width* (which is the same as mean objective-value range). It is this that we investigate experimentally.

2.2   Double-McCormick Convexifications of Trilinear Momomials

Without loss of generality (see [23]), we can relabel the three variables so that

$$a_1 b_2 b_3 + b_1 a_2 a_3 \le b_1 a_2 b_3 + a_1 b_2 a_3 \le b_1 b_2 a_3 + a_1 a_2 b_3. \tag{Ω}$$

is satisfied by the variable bounds. Given $f := x_1 x_2 x_3$ satisfying (Ω), there are three choices of double-McCormick convexifications depending on the bilinear sub-monomial we convexify first. We could first group x_1 and x_2 and convexify $w = x_1 x_2$; after this, we are left with the monomial $f = w x_3$ which we again convexify using McCormick. Alternatively, we could first group variables x_1 and x_3, or variables x_2 and x_3.

To see how to perform these convexifications in general, we exhibit the double-McCormick convexification that first groups the variables x_i and x_j. Therefore we have $f = x_i x_j x_k$, and we let $w_{ij} = x_i x_j$, so $f = w_{ij} x_k$. Convexifying $w_{ij} = x_i x_j$ using the standard McCormick inequalities, and then convexifying $f = w_{ij} x_k$, again using the standard McCormick inequalities, we obtain the 8 inequalities:

$$
\begin{aligned}
w_{ij} - a_j x_i - a_i x_j + a_i a_j &\ge 0, & f - a_k w_{ij} - a_i a_j x_k + a_i a_j a_k &\ge 0, \\
-w_{ij} + b_j x_i + a_i x_j - a_i b_j &\ge 0, & -f + b_k w_{ij} + a_i a_j x_k - a_i a_j b_k &\ge 0, \\
-w_{ij} + a_j x_i + b_i x_j - b_i a_j &\ge 0, & -f + a_k w_{ij} + b_i b_j x_k - b_i b_j a_k &\ge 0, \\
w_{ij} - b_j x_i - b_i x_j + b_i b_j &\ge 0, & f - b_k w_{ij} - b_i b_j x_k + b_i b_j b_k &\ge 0.
\end{aligned}
$$

Using Fourier-Motzkin elimination (i.e., projection), we then eliminate the variable w_{ij} to obtain the following system in our original variables f, x_i, x_j, x_k.

$$x_i - a_i \geq 0, \quad (1)$$

$$x_j - a_j \geq 0, \quad (2)$$

$$f - a_j a_k x_i - a_i a_k x_j - a_i a_j x_k + 2 a_i a_j a_k \geq 0, \quad (3)$$

$$f - a_j b_k x_i - a_i b_k x_j - b_i b_j x_k + a_i a_j b_k + b_i b_j b_k \geq 0, \quad (4)$$

$$- x_j + b_j \geq 0, \quad (5)$$

$$- x_i + b_i \geq 0, \quad (6)$$

$$f - b_j a_k x_i - b_i a_k x_j - a_i a_j x_k + a_i a_j a_k + b_i b_j a_k \geq 0, \quad (7)$$

$$f - b_j b_k x_i - b_i b_k x_j - b_i b_j x_k + 2 b_i b_j b_k \geq 0, \quad (8)$$

$$- f + b_j b_k x_i + a_i b_k x_j + a_i a_j x_k - a_i a_j b_k - a_i b_j b_k \geq 0, \quad (9)$$

$$- f + a_j b_k x_i + b_i b_k x_j + a_i a_j x_k - a_i a_j b_k - b_i a_j b_k \geq 0, \quad (10)$$

$$- x_k + b_k \geq 0, \quad (11)$$

$$- f + b_j a_k x_i + a_i a_k x_j + b_i b_j x_k - a_i b_j a_k - b_i b_j a_k \geq 0, \quad (12)$$

$$- f + a_j a_k x_i + b_i a_k x_j + b_i b_j x_k - b_i a_j a_k - b_i b_j a_k \geq 0, \quad (13)$$

$$x_k - a_k \geq 0, \quad (14)$$

$$f - a_i a_j x_k \geq 0, \quad (15)$$

$$- f + b_i b_j x_k \geq 0. \quad (16)$$

It is easy to see that the inequalities (15) and (16) are redundant: (15) is $a_j a_k$ (1) $+ a_i a_k$ (2) $+$ (3), and (16) is $b_j a_k$ (6) $+ a_i a_k$ (5) $+$ (12).

We use the following notation in what follows. For $\ell = 1, 2, 3$, *system ℓ* is defined to be the system of inequalities obtained by first grouping the pair of variables x_j and x_k, with j and k different from ℓ. For $\ell = 1, 2, 3$, the polytope $\mathcal{P}_\ell$ is defined to be the solution set of system ℓ, while

$$\mathcal{P}_h := \mathrm{conv}\left(\{(f, x_1, x_2, x_3) : f = x_1 x_2 x_3, \ x_i \in [a_i, b_i], \ i = 1, 2, 3\}\right).$$

2.3 Volumes of Convexifications

The main results of [23] are as follows.

Theorem 1 (Volume of the convex hull; see [23])

$$\mathrm{vol}(\mathcal{P}_h) = (b_1 - a_1)(b_2 - a_2)(b_3 - a_3) \times$$
$$(b_1(5 b_2 b_3 - a_2 b_3 - b_2 a_3 - 3 a_2 a_3) + a_1(5 a_2 a_3 - b_2 a_3 - a_2 b_3 - 3 b_2 b_3))/24.$$

Theorem 2 (Volume of the double McCormick with $x_1(x_2 x_3)$; see [23])

$$\mathrm{vol}(\mathcal{P}_1) = \mathrm{vol}(\mathcal{P}_h) + (b_1 - a_1)(b_2 - a_2)^2(b_3 - a_3)^2 \times$$
$$\frac{3(b_1 b_2 a_3 - a_1 b_2 a_3 + b_1 a_2 b_3 - a_1 a_2 b_3) + 2(a_1 b_2 b_3 - b_1 a_2 a_3)}{24(b_2 b_3 - a_2 a_3)}.$$

Theorem 3 (Volume of the double McCormick with $x_2(x_1 x_3)$; see [23])

$$\text{vol}(\mathcal{P}_2) = \text{vol}(\mathcal{P}_h) + \frac{(b_1 - a_1)(b_2 - a_2)^2(b_3 - a_3)^2 \left(5(a_1 b_1 b_3 - a_1 b_1 a_3) + 3(b_1^2 a_3 - a_1^2 b_3)\right)}{24(b_1 b_3 - a_1 a_3)}.$$

Theorem 4 (Volume of the double McCormick with $x_3(x_1 x_2)$; see [23])

$$\text{vol}(\mathcal{P}_3) = \text{vol}(\mathcal{P}_h) + \frac{(b_1 - a_1)(b_2 - a_2)^2(b_3 - a_3)^2 \left(5(a_1 b_1 b_2 - a_1 b_1 a_2) + 3(b_1^2 a_2 - a_1^2 b_2)\right)}{24(b_1 b_2 - a_1 a_2)}.$$

Corollary 1 (Ranking the relaxations by volume; see [23])

$$\text{vol}(\mathcal{P}_h) \le \text{vol}(\mathcal{P}_3) \le \text{vol}(\mathcal{P}_2) \le \text{vol}(\mathcal{P}_1).$$

Corollary 2 (A worst case; see [23]). *For the special case of $a_1 = a_2 = 0$, $b_1 = b_2 = 1$, and fixed b_3, the greatest difference in volume for $\mathcal{P}_3(= \mathcal{P}_h)$ and $\mathcal{P}_2$ (or $\mathcal{P}_1$) occurs when $a_3 = b_3/3$.*

3   Computational Experiments

3.1   Box Cubic Programs and 4 Relaxations

Our experiments are aimed at the following natural problem which is concerned with optimizing a linear function on trinomials. Let H be a 3-uniform hypergraph on n vertices. Each hyperedge of H is a set of 3 vertices, and we denote the set of hyperedges by $E(H)$. If H is complete, then $|E(H)| = \binom{n}{3}$. We associate with each vertex i a variable $x_i \in [a_i, b_i]$, and with each hyperedge (i, j, k) the trinomial $x_i x_j x_k$ and a coefficient q_{ijk} ($1 \le i < j < k \le n$). We now formulate the associated *boxcup* ('box cubic problem'):

$$\min_{x \in \mathbb{R}^n} \left\{ \sum_{(i,j,k) \in E(H)} q_{ijk}\, x_i\, x_j\, x_k \; : \; x_i \in [a_i, b_i],\; i = 1, 2, \ldots, n \right\}. \tag{BCUP}$$

The name is in analogy with the well-known *boxqp*, where just two terms (rather than three) are multiplied ('box' refers to the feasible region and 'qp' refers to 'quadratic program').

(BCUP) is a difficult nonconvex global-optimization problem. Our goal here is not to solve instances of this problem, but rather to solve a number of different *relaxations* of the problem and see how the results of these experiments correlate with the volume results of [23]. In this way, we seek to determine if the guidance of [23] is relevant to modelers and those implementing sBB.

We have seen how for a single trilinear term $f = x_i x_j x_k$, we can build four distinct relaxations: the convex hull of the feasible points, $\mathcal{P}_h$, and three relaxations arising from double McCormick: $\mathcal{P}_1$, $\mathcal{P}_2$ and $\mathcal{P}_3$. To obtain a relaxation of (BCUP), we choose a relaxation $\mathcal{P}_\ell$, for some $\ell = 1, 2, 3, h$ and apply this same relaxation method to *each* trinomial of (BCUP). We therefore obtain 4 distinct linear relaxations of the form:

$$\min_{(x,f)\in\mathscr{P}_\ell} \left\{ \sum_{(i,j,k)\in E(H)} q_{ijk}\, f_{ijk} \right\}.$$

where $\mathscr{P}_\ell$, $\ell = 1, 2, 3, h$ is the polytope in dimension $|E(H)| + n$ arising from using relaxation $\mathcal{P}_\ell$ on each trinomial. This linear relaxation is a linear inequality system involving the n variables x_i $(i = 1, 2, \ldots, n)$, and the $|E(H)|$ new 'function variables' f_{ijk}. Each such 'function variable' models a product $x_i\, x_j\, x_k$.

For our experiments, we randomly generate box bounds $[a_i, b_i]$ on x_i, for each $i = 1, \ldots, n$ independently, by choosing (uniformly) random pairs of integers $0 \le a_i < b_i \le 10$. With each realization of these bounds, we get relaxation feasible regions $\mathcal{P}_\ell$, for $\ell = 1, 2, 3, h$.

3.2   Three Scenarios for the Hypergraph *H*

We designed our experiments with the idea of gaining some understanding of whether our conclusions would depend on how much the monomials overlap. So we looked at three scenarios for the hypergraph H of (BCUP), all with $|E(H)| = 20$ monomials:

- Our **dense** scenario has H being a *complete* 3-uniform hypergraph on $n = 6$ vertices ($\binom{6}{3} = 20$). We note that each of the $n = 6$ variables appears in $\binom{6-1}{3-1} = 10$ of the 20 monomials, so there is considerable overlap in variables between trinomials.
- Our **sparse** scenario has hyperedges: $\{1, 2, 3\}$, $\{2, 3, 4\}$, $\{3, 4, 5\}$ ... $\{18, 19, 20\}$, $\{19, 20, 1\}$, $\{20, 1, 2\}$. Here we have $n = 20$ variables and each variable is in only 3 of the trinomials.
- Our **very sparse** scenario has $n = 30$ variables and each variable is in only 2 of the trinomials.

For each scenario, we generate 30 *sets* of bounds $[a_i, b_i]$ on x_i $(i = 1, \ldots, n)$. To control the variation in our results, and considering that the scaling of

$$Q := \{q_{ijk} \ : \ \{i, j, k\} \in E(H)\}$$

is arbitrary, we generate 100,000 random Q with $|E(H)|$ entries, uniformly distributed on the unit sphere in $\mathbb{R}^{|E(H)|}$. Then, for each Q, we both minimize and maximize $\sum_{i<j<k} q_{ijk}\, f_{ijk}$ over each $\mathscr{P}_\ell$, $\ell = 1, 2, 3, h$ and each set of bounds.

3.3   Quality of Relaxations

For each Q we take the difference in the optimal values, i.e. the maximum value minus the minimum value; this can be thought of as the width of the polytope in the direction Q. We then average these widths for each $\mathscr{P}_\ell$, $\ell = 1, 2, 3, h$, across the 100,000 realizations of Q (which results in very small standard errors), and we refer to this quantity $\omega(\mathscr{P}_\ell)$ as the *quasi mean width* of the relaxation. It is not quite the geometric mean width, because we do not have objective terms for all variables in (BCUP) (i.e., we have no objective terms $\sum_{i=1}^{n} c_i x_i$).

We seek to investigate how well the volume formulae of [23], comparing the volumes of the polytopes $\mathcal{P}_\ell$ ($\ell = 1, 2, 3, h$), can be used to predict the quality of the relaxations $\mathscr{P}_\ell$ ($\ell = 1, 2, 3, h$) as measured by their quasi mean width.

Figure 1 consists of a plot for each scenario: dense, sparse, and very sparse. Each plot illustrates the difference in quasi mean width between $\mathscr{P}_3$ (using the 'best' double McCormick) and each of the other relaxations. Each point represents a choice of bounds and the instances are sorted by $\omega(\mathscr{P}_1) - \omega(\mathscr{P}_h)$. In all three plots $\omega(\mathscr{P}_h) - \omega(\mathscr{P}_3)$ is non-positive, which is to be expected because $\mathscr{P}_h$ is contained in each of the three double-McCormick relaxations. Furthermore the plots illustrate that the general trend is for $\omega(\mathscr{P}_2) - \omega(\mathscr{P}_3)$ and $\omega(\mathscr{P}_1) - \omega(\mathscr{P}_3)$ to be positive and also for $\omega(\mathscr{P}_1) - \omega(\mathscr{P}_3)$ to be greater than $\omega(\mathscr{P}_2) - \omega(\mathscr{P}_3)$. This agrees with Corollary 1 and gives strong validation for the use of volume to measure the strength of different relaxations. It confirms that given a choice of the double-McCormick relaxations, $\mathscr{P}_3$ is the one to choose.

However, there are a few exceptions to the general trend and these exceptions are most apparent in the very sparse case. In both the dense case and the sparse case we only see a deviation from the trend on a small number of occasions when $\mathscr{P}_2$ is very slightly better than $\mathscr{P}_3$. In each of these cases, the difference seems to be so small that we can really regard $\mathscr{P}_3$ and $\mathscr{P}_2$ as being equivalent from a practical viewpoint. In the very sparse case, the general trend is still followed, but we see a few more cases where $\mathscr{P}_2$ is slightly better than $\mathscr{P}_3$. We also see that in a few instances, $\mathscr{P}_1$ is better than $\mathscr{P}_2$ and occasionally even slightly better than $\mathscr{P}_3$.

However it is important to note that when we consider the sparse and very sparse cases, the differences in quasi mean width between *any* two of the relaxations is much smaller than these differences in the dense case. If we were to take the sparsity of H to the extreme and run our experiments with $n = 60$ and each variable only in one trinomial, the difference in quasi mean width between *any* two of the polytopes will become zero for these boxcup problems. Therefore, it is not surprising that our results diverge from the general trend as H becomes sparser.

Using the common technique of 'performance profiles' (see [8]), we can illustrate the differences in quasi mean width of the three double-McCormick relaxations in another way. We obtained the matlab code "perf.m" which was adapted to create these plots from the link contained in [21]. Figure 2 shows a performance profile for each of the dense, sparse and very sparse scenarios. For each choice of bounds, $\mathscr{P}_h$ gives the least quasi mean width (because it is contained in each of the other relaxations). Our performance profiles display the fraction of instances where the quasi mean width of $\mathscr{P}_\ell$ is within a factor α of the mean width of $\mathscr{P}_h$, for $\ell = 1, 2, 3$. The plots are natural log plots where the horizonal axis is

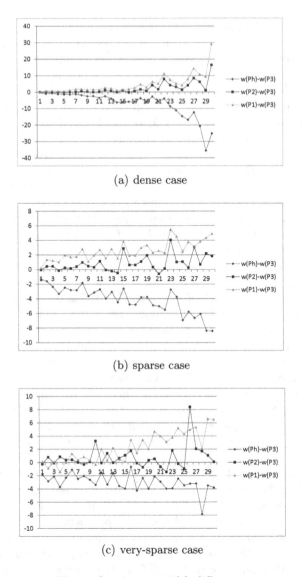

(a) dense case

(b) sparse case

(c) very-sparse case

Fig. 1. Quasi-mean-width differences

$\tau := \ln(\alpha)$. Using this measure, we see that the trend in *all* cases is that $\mathscr{P}_3$ dominates $\mathscr{P}_2$ which in turn dominates $\mathscr{P}_1$. In the very sparse case, we see that $\mathscr{P}_3$ and $\mathscr{P}_2$ are very close for small factors α. In general, all three relaxations are within a small factor of the hull. Displaying the results in this manner gives us a way to see quickly which relaxation performs best for the majority of instances. Again, we see agreement with the prediction of Corollary 1 and confirmation that $\mathscr{P}_3$ is the best double-McCormick relaxation.

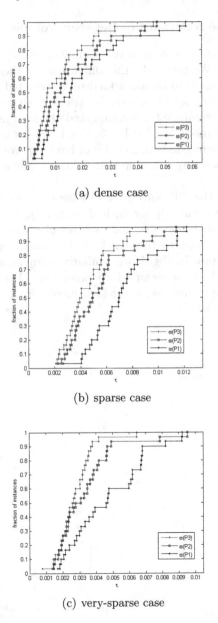

(a) dense case

(b) sparse case

(c) very-sparse case

Fig. 2. Quasi-mean-width performance profiles. Displays the fraction of instances where the quasi mean width of $\mathscr{P}_\ell$ is within a factor $\alpha = e^\tau$ of the mean width of $\mathscr{P}_h$. Note that for small τ, $e^\tau \approx 1 + \tau$.

3.4   Validating the Relationship Between Volume and Mean Width

Using the [23] formulae, we calculate the volume of the relaxation for each individual trinomial, $\mathcal{P}_\ell$. We then take the fourth root of these volumes and sum over all $|E(H)|$ trinomials to obtain a kind of 'aggregated idealized radius' for each relaxation and each set of bounds. Restricting our attention to the dense scenario, in Fig. 3, we compare these aggregated idealized radii with quasi mean width, across all relaxations $\mathcal{P}_\ell, \ell = 1, 2, 3, h$ and each set of bounds (each point in each scatter plot corresponds to a choice of bounds). We see a high R^2 coefficient in all cases, so we may conclude that volume really is a good predictor of relaxation width.

We also compute the difference in width between polytope pairs: $\mathcal{P}_h$ and $\mathcal{P}_3$, $\mathcal{P}_3$ and $\mathcal{P}_2$, $\mathcal{P}_2$ and $\mathcal{P}_1$ for each direction Q. We then average these width differences for each polytope and each set of bounds, across the 100,000 realizations of Q. We refer to this result as the *quasi mean width difference* of the pair of polytopes. In Fig. 4, we similarly compare aggregated idealized radial differences with quasi mean width differences. We see even higher R^2 coefficients, validating volume as an excellent predictor of average objective gap between pairs of relaxations.

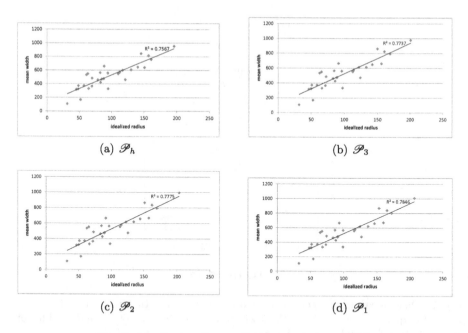

(a) $\mathcal{P}_h$

(b) $\mathcal{P}_3$

(c) $\mathcal{P}_2$

(d) $\mathcal{P}_1$

Fig. 3. Idealized radius predicting quasi mean width

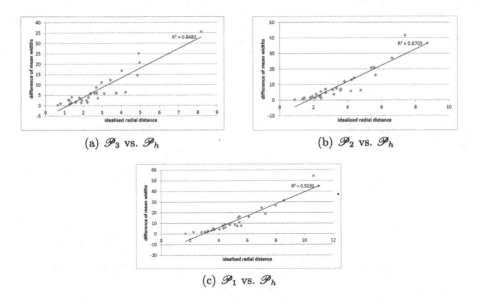

(a) $\mathcal{P}_3$ vs. $\mathcal{P}_h$

(b) $\mathcal{P}_2$ vs. $\mathcal{P}_h$

(c) $\mathcal{P}_1$ vs. $\mathcal{P}_h$

Fig. 4. Idealized radial distance predicting quasi mean width difference

3.5   A Worst Case

Our final set of experiments relate to a 'worst case' as described in [23]. In the important special case of $a_1 = a_2 = 0$ and $b_1 = b_2 = 1$, the two 'bad' double-McCormick relaxations have the same volume and the 'good' double McCormick is exactly the hull. In addition, the greatest difference in volume between the hull and the bad relaxations occurs when $a_3 = b_3/3$.

We compute the same results as we have discussed before (i.e. the differences in quasi mean width between the relaxations) with $n = 6$, but now instead of using random bounds, we fix $a_1 = a_2 = 0$ and $b_1 = b_2 = 1$. We also fix b_3 and run the experiments for $a_3 = 1, 2, \ldots, b_3 - 1$. Here, we only consider the $\binom{5}{3} = 10$ trinomials that have the form $x_j x_k x_6$.

Figure 5 displays a plot of these results for $b_3 = 30, 60, 90, 120$ and 150. From the inequality systems we know that $\mathcal{P}_h$ is exactly $\mathcal{P}_3$, therefore we are interested in the comparison between: $\mathcal{P}_2$ and $\mathcal{P}_3$, and $\mathcal{P}_2$ and $\mathcal{P}_1$. From the plots of these differences, we see exactly what we would expect given the volume formulae. The difference in mean width between $\mathcal{P}_2$ and $\mathcal{P}_1$ is very small; from a practical standpoint it is essentially zero. The difference in mean width between $\mathcal{P}_2$ and $\mathcal{P}_3$ is always positive, indicating again that $\mathcal{P}_3$ is the best choice of double-McCormick relaxation. In addition, we observe that the maximum difference falls close to $a_3 = b_3/3$ in all cases, demonstrating again that volume is a good predictor of how well a relaxation behaves.

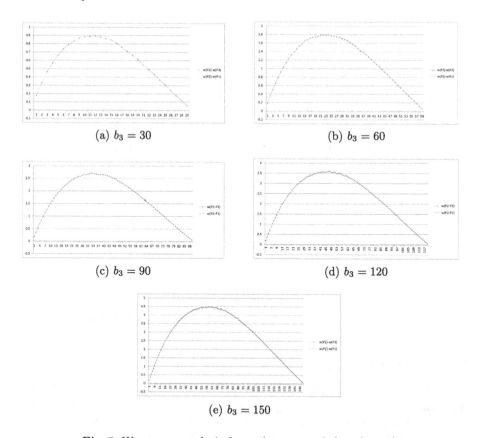

(a) $b_3 = 30$

(b) $b_3 = 60$

(c) $b_3 = 90$

(d) $b_3 = 120$

(e) $b_3 = 150$

Fig. 5. Worst-case analysis for a_3 ($a_1 = a_2 = 0$, $b_1 = b_2 = 1$)

Acknowledgments. Supported in part by ONR grant N00014-14-1-0315.

References

1. Adjiman, C., Dallwig, S., Floudas, C., Neumaier, A.: A global optimization method, αBB, for general twice-differentiable constrained NLPs: I. Theoretical advances. Comput. Chem. Eng. **22**(9), 1137–1158 (1998)
2. Ardila, F., Benedetti, C., Doker, J.: Matroid polytopes and their volumes. Discrete Comput. Geom. **43**(4), 841–854 (2010)
3. Belotti, P., Lee, J., Liberti, L., Margot, F., Wächter, A.: Branching and bounds tightening techniques for non-convex MINLP. Optim. Methods Softw. **24**(4–5), 597–634 (2009)
4. Brightwell, G., Winkler, P.: Counting linear extensions is #P-complete. In: Proceedings of the Twenty-third Annual ACM Symposium on Theory of Computing, STOC 1991, pp. 175–181. ACM, New York (1991)
5. Burggraf, K., De Loera, J., Omar, M.: On volumes of permutation polytopes. In: Bezdek, K., Deza, A., Ye, Y. (eds.) Discrete Geometry and Optimization, vol. 69, pp. 55–77. Springer, Heidelberg (2013)

6. Cafieri, S., Lee, J., Liberti, L.: On convex relaxations of quadrilinear terms. J. Global Optim. **47**, 661–685 (2010)

7. Dey, S.S., Molinaro, M., Wang, Q.: Approximating polyhedra with sparse inequalities. Math. Program. **154**(1), 329–352 (2015)

8. Dolan, D.E., Moré, J.J.: Benchmarking optimization software with performance profiles. Math. Program. **91**(2), 201–213 (2002)

9. Dyer, M., Frieze, A., Kannan, R.: A random polynomial-time algorithm for approximating the volume of convex bodies. J. ACM **38**(1), 1–17 (1991)

10. Ko, C.W., Lee, J., Steingrímsson, E.: The volume of relaxed Boolean-quadric and cut polytopes. Discrete Math. **163**(1–3), 293–298 (1997)

11. Lee, J.: Mixed integer nonlinear programming: some modeling and solution issues. IBM J. Res. Dev. **51**(3/4), 489–497 (2007)

12. Lee, J., Morris, W.: Geometric comparison of combinatorial polytopes. Discrete Appl. Math. **55**, 163–182 (1994)

13. McCormick, G.: Computability of global solutions to factorable nonconvex programs: Part I. Convex underestimating problems. Math. Program. **10**, 147–175 (1976)

14. Meyer, C., Floudas, C.: Trilinear monomials with mixed sign domains: facets of the convex and concave envelopes. J. Global Optim. **29**, 125–155 (2004)

15. Meyer, C., Floudas, C.: Trilinear monomials with positive or negative domains: facets of the convex and concave envelopes. In: Floudas C.A., Pardalos P. (eds) Frontiers in Global Optimization. NOIA, vol. 74, pp. 327–352. Springer, Boston (2004)

16. Misener, R., Floudas, C.A.: ANTIGONE: Algorithms for coNTinuous/Integer Global Optimization of Nonlinear Equations. J. Global Optim. **59**(2), 503–526 (2014). doi:10.1007/s10898-014-0166-2

17. Rikun, A.: A convex envelope formula for multilinear functions. J. Global Optim. **10**, 425–437 (1997)

18. Ryoo, H., Sahinidis, N.: A branch-and-reduce approach to global optimization. J. Global Optim. **8**(2), 107–138 (1996)

19. Sahinidis, N.: BARON 15.6.5: Global Optimization of Mixed-Integer Nonlinear Programs, User's Manual (2015)

20. Smith, E., Pantelides, C.: A symbolic reformulation/spatial branch-and-bound algorithm for the global optimisation of nonconvex MINLPs. Comput. Chem. Eng. **23**, 457–478 (1999)

21. Sofi, A., Mamat, M., Mohid, S., Ibrahim, M., Khalid, N.: Performance profile comparison using matlab. In: Proceedings of International Conference on Information Technology & Society 2015 (2015)

22. Speakman, E., Lee, J.: On sBB branching for trilinear monomials. In: Rocha, A., Costa, M., Fernandes, E. (eds.) GOW 2016, pp. 81–84 (2016). ISBN: 978-989-20-6764-3

23. Speakman, E., Lee, J.: Quantifying double McCormick. To appear in: Math. Oper. Res. (2017)

24. Stanley, R.: Two poset polytopes. Discrete Comput. Geom. **1**(1), 9–23 (1986)

25. Steingrímsson, E.: A decomposition of 2-weak vertex-packing polytopes. Discrete Comput. Geom. **12**(4), 465–479 (1994)

26. Vigerske, S., Gleixner, A.: Scip: Global optimization of mixed-integer nonlinear programs in a branch-and-cut framework. Technical report 16-24, ZIB, Takustr. 7, 14195, Berlin (2016)

Minimum Makespan Vehicle Routing Problem with Compatibility Constraints

Miao Yu⬤, Viswanath Nagarajan⬤, and Siqian Shen$^{(\boxtimes)}$⬤

University of Michigan, Ann Arbor, MI, USA
{miaoyu,viswa,siqian}@umich.edu

Abstract. We study a multiple vehicle routing problem, in which a fleet of vehicles is available to serve different types of services demanded at locations. The goal is to minimize the makespan, i.e. the maximum length of any vehicle route. We formulate it as a mixed-integer linear program and propose a branch-cut-and-price algorithm. We also develop an efficient $O(\log n)$-approximation algorithm for this problem. We conduct numerical studies on Solomon's instances with various demand distributions, network topologies, and fleet sizes. Results show that the approximation algorithm solves all the instances very efficiently and produces solutions with good practical bounds.

Keywords: Vehicle routing · Compatibility constraints · Branch-cut-and-price · Approximation algorithm

1 Introduction

Vehicle routing problems (VRPs), with a goal of finding the optimal routing assignment for a fleet of vehicles to serve demands at various locations, are classical and well studied combinatorial optimization problems. Starting from Dantzig and Ramser [7], many variants of VRPs have been considered, including VRP with time windows, Capacitated VRP, VRP with heterogeneous fleet, VRP with multiple depots, as well as hybrid versions of these variants, all of which are discussed in detail by Golden et al. [15] and Toth and Vigo [24].

In this paper, we consider a minimum makespan VRP with compatibility constraints (VRPCC). We assume that multiple types of services are demanded at various locations of a given network and each type of service can only be served by certain vehicles. The goal of the problem is to minimize the maximum traveling time of all the routes designed for fulfilling the demand, i.e., the makespan for finishing all services. The motivation of studying the minimization of makespan comes from applications that focus on balancing routing assignment for all vehicles and minimizing the time to finish serving all customers. We consider a salient application as deploying shared vehicles for serving patient medical home care service demand, distributed in a geographical network, in which we aim to balance workload of different medical staff teams dispatched

© Springer International Publishing AG 2017
D. Salvagnin and M. Lombardi (Eds.): CPAIOR 2017, LNCS 10335, pp. 244–253, 2017.
DOI: 10.1007/978-3-319-59776-8_20

together with the vehicles. The solution methods investigated in this paper can help scheduling and routing for medical home care delivery.

We review the main VRP literature focusing on different solution methods. To exactly optimize VRPs, branch-and-cut was the dominant approach before 2000s, and the related research (see, e.g., [1,3,19,20,22]) develops effective valid inequalities to improve solution efficiency. Following Fukasawa et al. [13], branch-cut-and-price (BCP) became the best performing exact solution method for (capacitated) VRP: this combines branch-and-cut with column generation. There have also been many approximation algorithms for VRPs with a minimum makespan objective (see, e.g., [4,8,12]) that provide polynomial time algorithms with a provable performance guarantee. To the best of our knowledge, no prior work has provided either exact solution methods or efficient approximation algorithms for VRPCC.

In this paper, we develop the following algorithms for VRPCC: (i) an exact algorithm based on the BCP approach and (ii) an $O(\log n)$-approximation algorithm based on a budgeted covering approach. We provide preliminary numerical experiments for both algorithms. Our results show that the approximation algorithm solves the problem efficiently and yields a practical approximation ratio of at most two (on the test instances).

The remainder of the paper is organized as follow. In Sect. 2, we formally define the problem and present a mixed-integer linear programming formulation. In Sects. 3 and 4, we propose a BCP algorithm and an approximation algorithm for VRPCC, respectively. In Sect. 5, we test our algorithms on Solomon's instances with diverse graph sizes, vehicle fleet sizes, and demand distributions. We present the numerical results for our proposed solution methods. In Sect. 6, we conclude our findings and state future research.

2   Problem Statement and Formulation

Consider an undirected graph $G = (V, E)$, where $V = \{0, 1, \dots, n\}$ is the set of $n+1$ nodes. Node 0 represents the depot and set $V^+ = \{1, 2, \dots, n\}$ corresponds to customer locations. Each edge $(i, j) \in E$, $\forall i, j \in V$ has an associated distance d_{ij} satisfying the triangle inequality such that $d_{ij} \leq d_{ik} + d_{kj}$, $\forall i, k, j \in V$. A fleet K of vehicles, with $K = \{1, 2, \dots, m\}$, initially located at the depot can serve demand from customers. Each vehicle $k \in K$ can only visit a subset $V_k \subset V^+$, based on matches of vehicles and service types. In our problem, a routing decision assigns each vehicle a route such that: (a) the nodes visited by vehicle $k \in K$ are in the set V_k; (b) each node must be visited exactly once; and (c) the maximum distance of all assigned routes is minimized.

We define the vector $x = (x_{ij}^k, (i, j) \in E, k \in K)^T$ where x_{ij}^k takes value 1 if edge $(i, j) \in E$ is used in the route for vehicle k, and 0 otherwise. Consider the binary parameter $u = (u_i^k, k \in K, i \in V^+)^T$ where u_i^k takes value 1 if $i \in V_k$ for vehicle $k \in K$, and 0 otherwise. Let τ represent the maximum distance of all the routes. We formulate a mixed-integer program for the VRPCC as follows.

$$(\textbf{MIP}) \quad \underset{x,\tau}{\text{minimize}} \ \tau \tag{1}$$

$$\text{subject to} \ \sum_{(i,j)\in E} d_{ij} x_{ij}^k \leq \tau \qquad \forall k \in K, \tag{2}$$

$$\sum_{j:(v,j)\in E} x_{vj}^k - \sum_{i:(i,v)\in E} x_{iv}^k = 0 \quad \forall v \in V, \ \forall k \in K, \tag{3}$$

$$\sum_{k\in K} \sum_{i:(i,j)\in E} x_{ij}^k = 1 \qquad \forall j \in V^+, \tag{4}$$

$$\sum_{i:(i,j)\in E} x_{ij}^k \leq u_j^k \qquad \forall j \in V^+, \ \forall k \in K, \tag{5}$$

$$\sum_{(i,j)\in E, i,j\in S} x_{ij}^k \leq |S|-1 \qquad \forall S \subset V^+, \ \forall k \in K, \tag{6}$$

$$x_{ij}^k \in \{0,1\} \qquad \forall (i,j) \in E, \ \forall k \in K, \tag{7}$$

where constraints (2) ensure that τ equals to the maximum distance of all the routes, which is minimized in the objective function (1); constraints (3) enforce flow balance at each node for routing each vehicle; constraints (4) ensure that each node is visited exactly once; constraints (5) ensure that vehicle k can only visit nodes in V_k; constraints (6) are sub-tour elimination constraints.

3   Branch-Cut-and-Price Algorithm

We describe a BCP approach for VRPCC based on a set partition formulation where each decision variable represents a feasible route [18]. Let P^k be the set containing all feasible routes for vehicle $k \in K$. We define the binary decision variable $\lambda = (\lambda_p, \ p \in P^k, \ k \in K)^T$ where λ_p takes value 1 if route $p \in P^k$ is used by vehicle k and 0 otherwise. Denote the binary parameter $a = (a_{ip}, \ i \in V, \ p \in P^k, \ k \in K)^T$ where a_{ip} takes value 1 if $p \in P^k$ visits node $i \in V$, and 0 otherwise. We consider c_p as the cost of route $p \in P^k, \ k \in K$. The set partition formulation is given by

$$(\textbf{SP}) \quad \underset{\lambda,\tau}{\text{minimize}} \ \tau \tag{8}$$

$$\text{subject to} \ \sum_{k\in K} \sum_{p\in P^k} a_{jp} \lambda_p = 1 \qquad \forall j \in V^+, \tag{9}$$

$$\sum_{p\in P^k} c_p \lambda_p \leq \tau \qquad \forall k \in K, \tag{10}$$

$$\lambda_p \in \{0,1\} \qquad \forall p \in P^k, \ \forall k \in K, \tag{11}$$

where constraints (9) enforce that each node is visited exactly once and constraints (10) enforce that τ is the maximum cost of all routes. Due to the exponential size of $P^k, \ k \in K$, we exploit the BCP method to solve SP.

3.1   Column Generation

The idea of column generation is to maintain a subset $\bar{P}^k \subset P^k$ for each $k \in K$. We solve SP with P^k replaced by $\bar{P}^k$ and detect whether any $p \in P^k \backslash \bar{P}^K$ can improve the solution: if yes, we add those favorable routes p to $\bar{P}^k$ and repeat the process; otherwise we claim optimality. We define a restricted master problem (RMP) as linear relaxation of SP with P^k replaced by $\bar{P}^k$ for all $k \in K$.

Clearly, any feasible primal solution to RMP is feasible to the linear relaxation of SP, but this is not necessarily true for their dual solutions. For each vehicle $k \in K$, we want to find if there exists some favorable route $p \in P^k \backslash \bar{P}^k$ that can improve the objective value of RMP. We define α, β as the dual variables corresponding to constraints (9) and (10), respectively, and define the decision variable $y = (y_{ijp}, (i,j) \in E, p \in P^k, k \in K)^T$ where y_{ijp} takes value 1 if $(i,j) \in p$ and 0 otherwise. For each vehicle $k \in K$, with $\hat{\alpha}, \hat{\beta}$ being current optimal dual solution, the problem of pricing out a new route p (if exists) is

$$\textbf{(PP)} \quad z_{\text{PP}}^k(\hat{\alpha}, \hat{\beta}) = \min_y \left\{ \hat{\beta}_k \sum_{(i,j) \in E} d_{ij} y_{ijp}^k - \sum_{(i,j) \in E, j \neq 0} \hat{\alpha}_j y_{ijp}^k \mid p \in P^k \right\} \quad (12)$$

If the optimal objective value of PP is less than 0, then route p can potentially improve the current best solution to RMP. For each vehicle $k \in K$, PP is equivalent to the problem finding the shortest path in $G(V, E)$, where the cost of the edge is replaced by its reduced cost $\bar{c}_{ij} = \hat{\beta}_k d_{ij} - \hat{\alpha}_j$ for each edge $(i,j) \in E$, $j \in V^+$, and $\bar{c}_{i0} = \hat{\beta}_k d_{i,0}$ for each edge $(i,0) \in E$. Since edge costs can be negative in PP, this problem is NP-hard [2]. Solution approaches for PP have been studied in [5,10]. In this paper, we identify and generate the so-called "ng-route" by implementing the following procedure. Suppose that we assign a neighbor set to each node. An *ng-route* is a route where a node $i \in V^+$ can be revisited after a vehicle visits a node whose neighbor set does not contain node i. A label-setting algorithm, as shown in [5], can solve *ng-route* relaxation to PP that would improve the solution time of RMP.

3.2   Cutting Planes

Valid inequalities help improving the quality of the solution produced by RMP. We add subset-row inequalities from [17] as valid cuts when solving RMP. For any set $S = \{i_1, i_2, i_3\}$ containing three vertices $i_1, i_2, i_3 \in V^+$, the corresponding subset-row inequalities ensure that the sum of corresponding variables of all selected routes that visit at least two vertices in S is at most 1. With set I_S containing all the routes in $\bigcup_{k \in K} \bar{P}^k$ that visit at least two nodes in S, subset-row inequalities are in the form of:

$$\sum_{k \in K} \sum_{p \in I_S} \lambda_p \leq 1 \quad \forall S \subseteq V, |S| = 3. \quad (13)$$

We use the algorithm from [21] to solve PP with additional dual variables associated with the added (13).

3.3 Branching Rule

We adopt the following branching rules from [9]: we calculate the sum of flows for each edge as $f_{ij}^k = \sum_{p \in P^k} a_{ijp} \lambda_p$ for all $(i, j) \in E$ and $k \in K$. If f_{ij}^k is fractional, we generate two branches: $f_{ij}^k = 1$, where vehicle k has to use edge (i, j), and $f_{ij}^k = 0$, where we exclude edge (i, j) from any route traveled by vehicle $k \in K$.

4 Approximation Algorithm

Our later computational results show that a straightforward implementation of BCP is time consuming for VRPCC. Here we propose an $O(\log n)$-approximation algorithm, where we recall that n is the number of customer locations. The algorithm is based on the solution approach for the following problem defined on a network $G = (V, E)$.

Problem 1. Maximum Budgeted Cover Problem
Input: A node subset $X \subset V$, a fleet K of vehicles and a budget $B \geq 0$.
Output: A set H of routes, one for each vehicle $k \in K$, where each route has cost less than B.
Objective: Maximize $|H \cap X|$.

A greedy 2-approximation algorithm for this problem follows from [6]. This algorithm was based on the idea of iteratively picking a route that covers the maximum number of remaining nodes for the current node subset X. We propose a variant of this algorithm which is faster and still achieves a 2-approximation.

Our greedy algorithm works with an oracle for the orienteering problem, where $\mathcal{O}(X, B, i)$ outputs a route, with cost less than B, for vehicle $i \in K$, which covers the maximum number of nodes in X. The greedy algorithm, $Greedy(X, B)$, for the maximum budgeted cover problem is described in Algorithm 1.

Algorithm 1. A greedy algorithm for maximum budgeted cover

---

    input : A fleet K of vehicles, a subset $X \subset V$ and a budget of route B
    output: A set H of routes with cost less than B, one route for each vehicle

1 $H \leftarrow \emptyset, X' \leftarrow X$
2 **for** i *in* K **do**
3     $A_i = \mathcal{O}(X', B, i)$
4     $H \leftarrow H \cup \{A_i\}, X' \leftarrow X' \backslash A_i$
5 **end**
6 **return** H

We propose the following lemmas to analyze the above algorithm. The proofs are deferred to a full version.

Lemma 1. *Algorithm 1 is a 2-approximation algorithm for the maximum budgeted cover problem.*

Lemma 2. *Greedy Algorithm needs to be executed at most $\log |X| + 1$ times to cover all nodes in X with sufficient large budget B.*

Next, we propose an approximation algorithm for VRPCC based on Algorithm 1. We set $X = V^+$, where recall that V^+ is the set of customer locations. If budget B is sufficiently large that an optimal solution of maximum budgeted cover problem covers every node in X, then by Lemma 2, we can use Algorithm 1 to produce a feasible solution to VRPCC, where each vehicle carries routes with total cost at most $(\log n + 1)B$. We apply binary search to find the smallest budget B. We detail the algorithmic steps in Algorithm 2.

Algorithm 2. An approximation algorithm for VRPCC

---

   input : A network $G = (V, E)$, a fleet set of vehicles K, an budget B
   output: Routing assignment for each vehicle in $k \in K$

1   Initialize $S_k = \emptyset$ for all $k \in K$, $X \leftarrow V^+$, $Solve \leftarrow$ **true**
2   **while** $X \neq \emptyset$ **do**
3      $H \leftarrow$ output of Algorithm 1 with input K, X, B
4      $n = |X|$, $X \leftarrow X \backslash H$
5   **if** $|X| > \frac{n}{2}$ **then** $Solve \leftarrow$ **false**, go to Step 8
6   Update S_k with $A_k \in H$ for all $k \in K$
7   **end**
8   Apply binary search to find optimal B based on the result of $Solve$, go to Step 1 if B is not optimal
9   Shortcut the routes in S_k, $k \in K$ to produce solution for VRPCC

The approximation factor of Algorithm 2 depends on the oracle for orienteering problem in Algorithm 1. If we use a "bicriteria" approximation for orienteering that covers the optimal number of nodes with cost at most βB then Algorithm 2 yields an approximation factor of $\beta(\log n + 1)$ following the result from Lemma 2. In particular, we use a procedure from [14] which implies $\beta = 3$.

5   Numerical Results

We conduct numerical experiments to evaluate the performance of directly computing the MIP, the BCP algorithm, and the approximation algorithm. We conduct experiments on Solomon's instances [23] adapted to VRPCC settings. The Solomon's instances are classified into three classes according to different distributions of customer locations: random distribution (R), clustered distribution (C), and a mix of both (RC). The customer locations are distributed on a $[0, 100]^2$ square. In the Solomon's instances, there are 100 locations for each instance. We pick the first n locations to test our algorithms, where $n \in \{10, 15, 20, 25, 30\}$. We also test the performance of our algorithms on different fleet sizes $m \in \{3, 5\}$.

For each location $v \in V^+$, we randomly pick two vehicles that are capable of serving the corresponding customer. In practice, the average productivity of a medical home crew ranges from 4–6 visits per day [11]. Therefore, the sizes of our test instances are close to real-world medical home care instances.

For each test instance, we first solve the problem with model MIP and use the result as a benchmark. We use Gurobi 6.5 [16] as the optimization solver. For the BCP algorithm, we use Gurobi as a linear programming solver of RMP at each branched node, and managing the branching process by following the branching rule in Sect. 3.3. For the approximation algorithm, we apply a procedure by Garg [14] that solves orienteering problem for optimal number of nodes but with a budget $5B$ in Algorithm 1. We set the running time limit for all programs as 1,200 s. We use Java and perform numerical test on a Dell desktop with an Intel i7-3770 processor and 16 GB memory.

For MIP and BCP, we report their best upper bounds ("UB") and lower bounds ("LB") achieved, their gaps ("Gap") and computational time ("Time(s)"); for approximation algorithm, we report the objective values of its solutions ("Obj") and computational time ("Time(s)"). Tables 1 and 2 summarize the numerical performances of the three methods. We use "–" to indicate that time limit for the computation of the instance is reached.

Table 1. Numerical results with $m = 3$

Type	n	MIP				BCP				Approx.	
		LB	UB	Gap	Time(s)	LB	UB	Gap	Time(s)	Obj	Time(s)
R	10	87.00	87.00	0.00%	0.15	87.00	87.00	0.00%	4.26	147.00	0.16
	15	100.00	100.00	0.00%	3.77	100.00	100.00	0.00%	288.20	155.00	0.19
	20	114.00	114.00	0.00%	26.94	104.00	143.00	27.27%	–	200.00	0.57
	25	140.00	140.00	0.00%	912.77	122.00	223.00	45.29%	–	230.00	1.10
	30	138.00	153.00	9.80%	–	132.58	252.00	47.39%	–	247.00	1.83
C	10	40.00	40.00	0.00%	1.12	40.00	40.00	0.00%	6.58	48.00	0.18
	15	59.00	79.00	25.32%	–	40.84	80.00	48.95%	–	93.00	0.20
	20	42.00	87.00	51.72%	–	53.00	93.00	43.01%	–	102.00	0.58
	25	44.00	93.00	52.69%	–	57.71	109.00	47.06%	–	111.00	1.20
	30	44.00	94.00	53.19%	–	62.41	136.00	54.11%	–	140.00	3.48
RC	10	87.00	87.00	0.00%	0.63	87.00	87.00	0.00%	5.90	131.00	0.14
	15	119.00	119.00	0.00%	1137.84	71.02	120.00	40.82%	–	171.00	0.27
	20	112.00	132.00	15.15%	–	86.48	132.00	34.48%	–	199.00	0.43
	25	92.00	157.00	41.40%	–	105.27	192.00	45.17%	–	261.00	1.40
	30	137.00	218.00	37.16%	–	126.90	332.00	61.78%	–	282.00	1.88

In Table 1, when $m = 3$, we see that MIP performs well for instances of type R, in which it can solve up to 25-node instances. For instances of types C and RC, we can no longer use MIP to solve instances with more than 15 nodes. This shows that clustered distribution of nodes is more challenging to handle. At the

Table 2. Numerical results with $m = 5$

Type	n	MIP				BCP				Approx.	
		LB	UB	Gap	Time(s)	LB	UB	Gap	Time(s)	Obj	Time(s)
R	10	69.00	69.00	0.00%	0.02	69.00	69.00	0.00%	0.74	119.00	0.04
	15	97.00	97.00	0.00%	0.23	97.00	97.00	0.00%	115.94	162.00	0.09
	20	106.00	106.00	0.00%	3.75	93.00	116.00	19.83%	–	210.00	0.25
	25	120.00	120.00	0.00%	29.45	98.00	142.00	30.99%	–	220.00	0.50
	30	123.00	123.00	0.00%	59.36	103.92	148.00	29.78%	–	193.00	0.78
C	10	40.00	40.00	0.00%	0.04	40.00	40.00	0.00%	0.99	42.00	0.03
	15	78.00	78.00	0.00%	4.71	78.00	78.00	0.00%	465.68	85.00	0.10
	20	82.00	82.00	0.00%	914.80	50.00	85.00	41.18%	–	96.00	0.22
	25	62.00	85.00	27.06%	–	52.38	123.00	57.41%	–	127.00	0.45
	30	61.00	87.00	29.89%	–	55.43	106.00	47.71%	–	110.00	1.00
RC	10	83.00	83.00	0.00%	0.09	83.00	83.00	0.00%	2.08	115.00	0.03
	15	105.00	105.00	0.00%	7.00	105.00	105.00	0.00%	344.88	131.00	0.12
	20	130.00	130.00	0.00%	25.76	94.33	169.00	44.18%	–	197.00	0.28
	25	137.00	139.00	1.44%	–	95.28	202.00	52.83%	–	246.00	0.50
	30	167.00	171.00	2.34%	–	109.27	328.00	66.69%	–	307.00	0.82

same time, BCP does not outperform MIP, and we can only use BCP to solve instances with 15 nodes or fewer. Although BCP algorithm cannot be solved very efficiently yet, it can produce better lower bound for instances that both MIP and BCP cannot solve to optimality, e.g., instances with 20, 25, and 30 nodes of type C. On the other hand, approximation algorithm solves all the instances very quickly. The computation can be finished within 4 s for all instances, and there is no significant difference in computational time when solving instances of type C/RC than type R. Moreover, despite that the theoretical approximation bound of our algorithm could be large when n increases, the practical approximation ratio is within a factor of 2 when comparing the objective values of solutions produced by the approximation algorithm with the best lower bounds produced by MIP and BCP for all instances.

Table 2 summarizes the numerical results for instances with larger fleet size $m = 5$. Results show that the computational time has been significantly improved and more instances can be solved by MIP and BCP. Our approximation algorithm can solve all instances within one second, and the practical approximation ratio is still within a factor of 2 comparing its objective values of solutions with the best lower bounds of MIP and BCP.

To summarize, we recognize that different distributions of demand locations could affect the computational time of our proposed exact solution methods. Despite lacking efficient exact solution approaches for VRPCC, our proposed approximation algorithm can efficiently solve the problem and provides good practical solutions.

6   Conclusions and Future Research

In this paper, we formulated a minimum makespan routing problem with compatibility constraints. We proposed three solution approaches for the problem: MIP, BCP, and an approximation algorithm. Numerical results showed that MIP and BCP could not solve many of these instances to optimality (within our time limit), whereas the approximation algorithm obtained good quality solutions within seconds. Moreover, our approximate solutions (for these instances) are within two times the best lower bounds from MIP or BCP. Future research includes further investigation on BCP to improve its efficiency and designing good implementations to improve the practical approximation factor of the approximation algorithm.

References

1. Achuthan, N.R., Caccetta, L., Hill, S.P.: An improved branch-and-cut algorithm for the capacitated vehicle routing problem. Transp. Sci. **37**(2), 153–169 (2003)
2. Ahuja, R.K., Magnanti, T.L., Orlin, J.B.: Network Flows: Theory, Algorithms, and Applications. Prentice Hall, Upper Saddle River (1993)
3. Applegate, D., Cook, W., Dash, S., Rohe, A.: Solution of a min-max vehicle routing problem. INFORMS J. Comput. **14**(2), 132–143 (2002)
4. Arkin, E.M., Hassin, R., Levin, A.: Approximations for minimum and min-max vehicle routing problems. J. Algorithms **59**(1), 1–18 (2006)
5. Baldacci, R., Mingozzi, A., Roberti, R.: New route relaxation and pricing strategies for the vehicle routing problem. Oper. Res. **59**(5), 1269–1283 (2011)
6. Chekuri, C., Kumar, A.: Maximum coverage problem with group budget constraints and applications. In: Jansen, K., Khanna, S., Rolim, J.D.P., Ron, D. (eds.) APPROX/RANDOM 2004. LNCS, vol. 3122, pp. 72–83. Springer, Heidelberg (2004). doi:10.1007/978-3-540-27821-4_7
7. Dantzig, G.B., Ramser, J.H.: The truck dispatching problem. Manage. Sci. **6**(1), 80–91 (1959)
8. Even, G., Garg, N., Könemann, J., Ravi, R., Sinha, A.: Min-max tree covers of graphs. Oper. Res. Lett. **32**(4), 309–315 (2004)
9. Feillet, D.: A tutorial on column generation and branch-and-price for vehicle routing problems. Q. J. Oper. Res. **8**(4), 407–424 (2010)
10. Feillet, D., Dejax, P., Gendreau, M., Gueguen, C.: An exact algorithm for the elementary shortest path problem with resource constraints: application to some vehicle routing problems. Networks **44**(3), 216–229 (2004)
11. National Association for Home Care & Hospice. Basic Statistics About Home Care, pp. 1–14. National Association for Home Care & Hospice, Washington, DC (2010)
12. Frederickson, G.N., Hecht, M.S., Kim, C.E.: Approximation algorithms for some routing problems. In: 17th Annual Symposium on Foundations of Computer Science, pp. 216–227. IEEE (1976)
13. Fukasawa, R., Longo, H., Lysgaard, J., de Aragão, M.P., Reis, M., Uchoa, E., Werneck, R.F.: Robust branch-and-cut-and-price for the capacitated vehicle routing problem. Math. Program. **106**(3), 491–511 (2006)
14. Garg, N.: A 3-approximation for the minimum tree spanning k vertices. In: Proceedings of the 37th Annual Symposium on Foundations of Computer Science, FOCS 1996, pp. 302–309. IEEE Computer Society, Washington, DC (1996)

15. Golden, B.L., Raghavan, S., Wasil, E.A.: Problem, The Vehicle Routing: Latest Advances and New Challenges. Springer Science & Business Media, New York (2008)
16. Gurobi Optimization, Inc., Gurobi optimizer reference manual (2016). http://www. gurobi.com
17. Jepsen, M., Petersen, B., Spoorendonk, S., Pisinger, D.: Subset-row inequalities applied to the vehicle-routing problem with time windows. Oper. Res. **56**(2), 497–511 (2008)
18. Kallehauge, B., Larsen, J., Madsen, O.B., Solomon, M.M.: Vehicle routing problem with time windows. In: Desaulniers, G., Desrosiers, J., Solomon, M.M. (eds.) Column Generation, pp. 67–98. Springer, New York (2005)
19. Letchford, A.N., Eglese, R.W., Lysgaard, J.: Multistars, partial multistars and the capacitated vehicle routing problem. Math. Program. **94**(1), 21–40 (2002)
20. Lysgaard, J., Letchford, A.N., Eglese, R.W.: A new branch-and-cut algorithm for the capacitated vehicle routing problem. Math. Program. **100**(2), 423–445 (2004)
21. Pecin, D., Pessoa, A., Poggi, M., Uchoa, E.: Improved branch-cut-and-price for capacitated vehicle routing. In: Lee, J., Vygen, J. (eds.) IPCO 2014. LNCS, vol. 8494, pp. 393–403. Springer, Cham (2014). doi:10.1007/978-3-319-07557-0_33
22. Ralphs, T.K., Kopman, L., Pulleyblank, W.R., Trotter, L.E.: On the capacitated vehicle routing problem. Math. Program. **94**(2–3), 343–359 (2003)
23. Solomon, M.M.: Algorithms for the vehicle routing and scheduling problems with time window constraints. Oper. Res. **35**(2), 254–265 (1987)
24. Toth, P., Vigo, D., Routing, V.: Problems, Methods, and Applications. SIAM, Philadelphia (2014)

Solving the Traveling Salesman Problem with Time Windows Through Dynamically Generated Time-Expanded Networks

Natashia Boland[1], Mike Hewitt[2(✉)], Duc Minh Vu[2], and Martin Savelsbergh[1]

[1] Georgia Institute of Technology, Atlanta, USA
[2] Loyola Chicago University, Chicago, USA
mhewitt3@luc.edu

Abstract. The Traveling Salesman Problem with Time Windows is the problem of finding a minimum-cost path visiting each of a set of cities exactly once, where each city must be visited within a specified time window. It has received significant attention because it occurs as a sub-problem in many real-life routing and scheduling problems. We explore an approach in which the strength of a time-expanded integer linear programming (IP) formulation is exploited without ever explicitly creating the complete formulation. The approach works with carefully designed partially time-expanded networks, which are used to produce upper as well as lower bounds, and which are iteratively refined until optimality is reached. Preliminary computational results illustrate the potential of the approach as, for almost all instances tested, optimal solutions can be identified in only a few iterations.

Keywords: Traveling Salesman Problem · Time Windows · Time-expanded networks · Mixed integer programming · Dynamic discretization discovery

1 Introduction

The Traveling Salesman Problem with Time Windows (TSPTW) is defined as follows. Let (N, A) be a complete directed graph with node set $N = \{0, 1, 2, ..., n\}$ representing the set of cities that must be visited by the salesman, starting and ending in designated city 0, and with arc set $A \subseteq N \times N$ representing the roads on which the salesman can travel. Each arc $(i, j) \in A$ has an associated travel time, denoted by τ_{ij}, and an associated cost, denoted by c_{ij}. Furthermore, each city $i \in N$ can only be visited during a specified time interval $[e_i, l_i]$, referred to as the city's time window. The Traveling Salesman Problem with Time Windows seeks a minimum-cost tour that departs city 0 at or after e_0, departs city i ($i \in N \setminus \{0\}$) within its time window $[e_i, l_i]$, and returns to city 0 at or before l_0. Note that arriving at city i ($i \in N \setminus \{0\}$) before e_i is allowed, but implies that the salesman has to wait. Evaluating whether a given city sequence results in a feasible solution can be done in linear time (by departing every city as early

© Springer International Publishing AG 2017
D. Salvagnin and M. Lombardi (Eds.): CPAIOR 2017, LNCS 10335, pp. 254–262, 2017.
DOI: 10.1007/978-3-319-59776-8_21

as possible), but finding a city sequence that results in a feasible solution is NP-hard [8].

There is a large body of work available on exact solution of the TSPTW, from both the operations research and constraint programming communities, e.g., [2,4–6,9,11,12]. The TSPTW is notoriously difficult to solve. Compact integer programming (IP) formulations that use continuous variables to model time have weak linear programming (LP) relaxations. Their solution with current IP solver technology is limited to only small instances. Extended IP formulations, with binary variables indexed by time, have much stronger relaxations, but (tend to) have a huge number of variables. In this paper, we explore an approach in which the strength of an extended IP formulation is exploited without ever explicitly creating the complete formulation. Preliminary work using this approach for a service network design problem arising in less-than-truckload transportation has been very promising: it is able to solve much larger instances than is possible with other techniques [3]. The key to the approach is that it discovers exactly which times are needed to obtain an optimal, continuous-time solution, in an efficient way, by solving a sequence of (small) IPs. The IPs are constructed as a function of a subset of times, with variables indexed by times in the subset. They are carefully designed to be tractable in practice, and to yield a lower bound (it is a cost minimization problem) on the optimal continuous-time value. Once the *right* (very small) subset of times is discovered, the resulting IP model yields the continuous-time optimal value. In this paper, we explore whether such an approach can also be used to solve instances of the TSPTW.

Regarding the existing literature, the method of [4] also dynamically generates a time-expanded network in the context of solving the TSPTW. However, they do so as a preprocessing scheme for a branch-and-cut solution approach. As such, in their approach, the time-expanded network is never changed during the branch-and-cut search, whereas we continue to refine the discretization until the problem is solved. Similarly, the method of [4] uses information from the solution to an LP to heuristically refine the set of time points, whereas we use information from the solution to an IP to carefully refine the set of time points to guarantee convergence to an optimal solution to the TSPTW instance. From a computational perspective, our approach outperforms nearly all existing methods on nearly all instances. While the method of [2] performs better than ours, we believe the performance of our approach can be greatly improved by constraint programming-type ideas.

2 Problem Formulation and Solution Approach

Because of the time dimension, it is natural to use a time-expanded network to model the TSPTW. Assuming integer travel times and time windows, the time-expanded network for the TSPTW has, for each city $i \in N$, a *timed* node, $(i,t) \in \mathcal{N}$, for each $t \in [e_i, l_i] = \{e_i, e_{i+1}, \ldots, l_i\}$. A *timed* arc $a = ((i,t)(j,t')) \in \mathcal{A} \subseteq \mathcal{N} \times \mathcal{N}$ represents travel from city i departing at time t to city j arriving at time $t' = \max\{e_j, t + \tau_{ij}\}$, defined if $t' \leq l_j$. The cost of timed arc $a = ((i,t),(j,t')) \in \mathcal{A}$ is denoted by c_a and taken to be c_{ij}.

An integer programming model of the TSPTW based on the time-expanded network, $(\mathcal{N}, \mathcal{A})$, is as follows. Let x_a for arc $a \in \mathcal{A}$ be a binary variable indicating whether the timed arc is used ($x_a = 1$) or not ($x_a = 0$). Then the following integer program solves the TSPTW:

$$\min \sum_{a \in \mathcal{A}} c_a x_a \tag{1}$$

$$\sum_{a = ((i,t)(j,t')) \in \mathcal{A}} x_a = 1 \qquad \forall i \in N, \tag{2}$$

$$\sum_{a \in \delta^-((i,t))} x_a - \sum_{a \in \delta^+((i,t))} x_a = 0 \qquad \forall (i,t) \in \mathcal{N}, i \neq 0, \tag{3}$$

$$x_a \in \{0,1\} \qquad \text{for all } a \in \mathcal{A}, \tag{4}$$

where $\delta^-()$ and $\delta^+()$ denote the set of timed arcs coming into and going out of a timed node. Constraints (2) force one unit of flow to leave city i for all $i \in N$, whereas constraints (3) are flow balance constraints that ensure that the flow into a timed node equals the flow out of the node at all cities except city 0. Note that this model embeds the assumption, which may be made without loss of generality, that any waiting in the tour occurs prior to the start of a time window; the salesman will either depart a city as soon as he arrives, or at the start of the city's time window, whichever is later.

The TSPTW formulation (1)–(4) defined on the time-expanded network needs no subtour elimination constraints, and we found, in our experiments, that its LP relaxation yields a very strong lower bound. However, it has a huge number of variables because of the time dimension, making solution with IP solvers impractical for all but the smallest instances. In the next section, we show how we iteratively build and update small partially time-expanded networks to solve TSPTW instead of using the whole time-expanded network.

We derive a *partially* time-expanded network, $\mathcal{D}_T = (\mathcal{N}_T, \mathcal{A}_T)$, from a given subset of the timed nodes, $\mathcal{N}_T \subseteq \mathcal{N}$. The arc set, $\mathcal{A}_T \subseteq \mathcal{N}_T \times \mathcal{N}_T$, of a partially time-expanded network consists of arcs of the form $((i,t),(j,t'))$, where $(i,j) \in A$. The cost of arc $a = ((i,t),(j,t')) \in \mathcal{A}_T$ is c_{ij}. We do *not* require that arc $((i,t),(j,t'))$ satisfies $t' = \max\{e_j, t + \tau_{ij}\}$. In fact, the flexibility to introduce timed arcs with travel time different to the actual travel time between the cities at their start and end is an essential feature of our approach. Let TSPTW($\mathcal{D}_T$) denote the IP formulation (1)–(4) defined with respect to $\mathcal{D}_T$ instead of $(\mathcal{N}, \mathcal{A})$. By careful design of $\mathcal{D}_T$, TSPTW($\mathcal{D}_T$) can be guaranteed to yield a lower bound on the value of the TSPTW, or can be guaranteed, if feasible, to yield an upper bound. We first discuss properties of $\mathcal{D}_T$ that ensure a lower bound.

Property 1. *For every $i \in N$, the nodes (i, e_i) and (i, l_i) are in $\mathcal{N}_T$.*

Property 2. *For every node $(i,t) \in \mathcal{N}_T$ and for every $(i,j) \in A$ with $t + \tau_{ij} \leq l_j$, there is an arc of the form $((i,t)(j,t')) \in \mathcal{A}_T$. Furthermore, every arc $((i,t),(j,t')) \in \mathcal{A}_T$ must have either (1) $t + \tau_{ij} < e_j$ and $t' = e_j$ or (2) $e_j \leq t' \leq t + \tau_{ij}$.*

This ensures that the travel time on arc $((i,t),(j,t')) \in \mathcal{A}_T$ is either accurate, or is underestimated: $t' \leq \max\{e_j, t+\tau_{ij}\}$. We say the arc is *too short* if $t' < t+\tau_{ij}$.

Lemma 1. *If $\mathcal{D}_T$ satisfies Properties 1 and 2, then the value of TSPTW($\mathcal{D}_T$) is a lower bound on the value of TSPTW.*

Corollary 1. *If the city sequence specified by an optimal solution to TSPTW($\mathcal{D}_T$) is feasible with actual travel times, then it is an optimal tour for TSPTW. Otherwise, the solution contains at least one arc that is too short.*

In fact, at most one arc in $\mathcal{A}_T$ is required for each $(i,t) \in \mathcal{N}_T$ and each arc $(i,j) \in A$ with $t + \tau_{ij} \leq l_j$, and using the one with the longest possible travel time, consistent with the above two properties, yields the best lower bound.

Property 3. *If arc $((i,t),(j,t')) \in \mathcal{A}_T$, then there is no $(j,t'') \in \mathcal{N}_T$ with $t' < t'' \leq t + \tau_{ij}$. We refer to this property as the longest-feasible-arc property.*

Lemma 2. *For a fixed $\mathcal{N}_T$, the partially time-expanded network $\mathcal{D}_T$ with the longest-feasible-arc property induces an instance of TSPTW($\mathcal{D}_T$) with the largest optimal objective function value.*

Given a partially time-expanded network, $\mathcal{D}_T$, containing an arc, $((i,t),(j,t'))$, that is too short, we can "correct" the resulting travel time underestimation using procedure LENGTHEN-ARC$((i,t),(j,t'))$. In this procedure, a new timed node is added to $\mathcal{N}_T$ to reflect the correct arrival time at j if departing i at time t, and the arcs in $\mathcal{A}_T$ are updated so as to maintain Properties 1–3.

Algorithm 1. LENGTHEN-ARC$((i,t),(j,t'))$

Require: Arc $((i,t),(j,t')) \in \mathcal{A}_T$ with $t' < t + \tau_{ij}$
  1: Add the new node (j,t'') with $t'' = t + \tau_{ij}$ to $\mathcal{N}_T$.
  2: Add new arcs out of node (j,t'') so as to satisfy Properties 2 and 3.
  3: For any too short arc $((h,u),(j,t'))$, if $u + \tau_{hj} \geq t''$, replace this arc by the arc $((h,u),(j,t''))$ (maintain Property 3).

Lemma 3. *If $\mathcal{D}'_T$ is formed from $\mathcal{D}_T$ by lengthening an arc, then the optimal objective function value of TSPTW($\mathcal{D}'_T$) is at least as large as the optimal objective function value of TSPTW($\mathcal{D}_T$). Furthermore, if $\mathcal{D}_T$ satisfies Properties 1–3, then so does $\mathcal{D}'_T$.*

The above properties and lemmas provide the building blocks for an iterative refinement algorithm for solving TSPTW, which is found in Algorithm 2.

By Lemma 1, Corollary 1, Lemma 3, and since the number of arcs is finite, we have the following.

Theorem 1. SOLVE-TSPTW *terminates with an optimal solution.*

Algorithm 2. SOLVE-TSPTW

Require: TSPTW instance (N, A), e, l, τ and c
1: Create a partially time-expanded network $\mathcal{D}_T$ satisfying Properties 1–3.
2: **while** not solved **do**
3:    Solve TSPTW($\mathcal{D}_T$)
4:    Determine whether the tour specified by the solution to TSPTW($\mathcal{D}_T$) is feasible with actual travel times
5:    **if** it is feasible **then**
6:       Stop: the tour is optimal for TSPTW
7:    **end if**
8:    Refine $\mathcal{D}_T$ by correcting an arc in $\mathcal{A}_T$ that is too short: for at least one $a = ((i, t), (j, t')) \in \mathcal{A}_T$ with $t' < t + \tau_{ij}$, run LENGTHEN-ARC(a)
9: **end while**

To improve the overall performance of Algorithm 2, we first use well known preprocessing steps, such as described in [4], to tighten the time windows and to remove unnecessary arcs. We also add valid inequalities to the TSPTW($\mathcal{D}_T$) formulation to eliminate subtours and bad paths (paths that violate time windows if traversed using actual travel times) as they are encountered in the algorithm. We use the simplest form of such inequalities, adapted to the time-expanded setting. Specifically, if $M \subset N$, we eliminate tours on M with the constraint

$$\sum_{(i,j) \in A \cap (M \times M)} \sum_{t,t':((i,t),(j,t')) \in \mathcal{A}_T} x_a \leq |M| - 1.$$

Similarly, if $P \subset A$ is a bad path, we eliminate its use with the inequality

$$\sum_{(i,j) \in P} \sum_{t,t':((i,t),(j,t')) \in \mathcal{A}_T} x_a \leq |P| - 1.$$

We detect node sets that give rise to subtours and bad paths found in integer solutions generated in the course of the algorithm, and add their elimination constraints to the TSPTW($\mathcal{D}_T$) models solved subsequently. We found it helpful to add, *a priori*, all subtour elimination constraints for node sets of size 2.

Finally, we use two heuristics to speed up the search of feasible solutions. Both use $\mathcal{N}_T$ to create a partially time-expanded network, different from $\mathcal{D}_T$, which are designed to generate feasible solutions. Both networks, denoted by $\mathcal{D}_T^1 = (\mathcal{N}_T, \mathcal{A}_T^1)$ and $\mathcal{D}_T^2 = (\mathcal{N}_T, \mathcal{A}_T^2)$, only contain timed arcs that have positive travel time, and so integer solutions on these networks cannot include subtours. The first is more conservative: $\mathcal{A}_T^1$ is created by taking each $(i, t) \in \mathcal{N}_T$ and each $(i, j) \in A$ with $t + \tau_{ij} \leq l_j$, finding the earliest time, t', with $(j, t') \in \mathcal{N}_T$ and $t' \geq t + \tau_{ij}$, and adding arc $((i, t), (j, t'))$ to $\mathcal{A}_T^1$. Note that such an arc must exist, by Property 1 $((i, l_i) \in \mathcal{N}_T$ for all $i \in N)$. Now any integer feasible solution to TSPTW($\mathcal{D}_T^1$) must be a feasible TSPTW tour. The second network may still contain arcs that are too short. Its arcs are created by taking each $(i, t) \in \mathcal{N}_T$ and each $(i, j) \in A$ with $t + \tau_{ij} \leq l_j$, finding the latest time, t', with $(j, t') \in \mathcal{N}_T$

Algorithm 3. SOLVE-TSPTW-ENHANCED

Require: TSPTW instance (N, A), e, l, τ and c, and optimality tolerance ϵ
1: Perform preprocessing, updating A, e and l.
2: Create a partially time-expanded network $\mathcal{D}_\mathcal{T}$.
3: $S = \emptyset$ {S contains feasible solutions to the TSPTW that have been found}
4: **while** not solved **do**
5:    Build and solve TSPTW($\mathcal{D}_\mathcal{T}$), harvest integer solutions, $\bar{S}$, and lower bound, z.
6:    **if** $S = \emptyset$ **then**
7:       Build and solve TSPTW($\mathcal{D}_\mathcal{T}^1$), harvest integer solutions and add to S.
8:    **end if**
9:    Build and solve TSPTW($\mathcal{D}_\mathcal{T}^2$), harvest integer solutions and add to $\bar{S}$.
10:   **for all** $\bar{s} \in \bar{S}$ **do**
11:      **if** $\bar{s}$ can be converted to a feasible solution s **then**
12:         Add s to S.
13:      **end if**
14:      Remove $\bar{s}$ from $\bar{S}$.
15:   **end for**
16:   Compute gap δ between best solution in S and lower bound, z.
17:   **if** $\delta \leq \epsilon$ **then**
18:      Stop: best solution in S is $\epsilon-$optimal for TSPTW.
19:   **end if**
20:   Refine $\mathcal{D}_\mathcal{T}$ by lengthening arcs that are too short.
21: **end while**

---

and $t' \leq t + \tau_{ij}$, and then checking two cases: (1) $t' > t$: in this case, we add $((i, t), (j, t'))$ to $\mathcal{A}_\mathcal{T}^2$; (2) $t' \leq t$: in this case, find the earliest t'' with $(j, t'') \in \mathcal{N}_\mathcal{T}$ and $t'' > t$, and add $((i, t), (j, t''))$ to $\mathcal{A}_\mathcal{T}^2$. Solutions to TSPTW($\mathcal{D}_\mathcal{T}^2$) may have bad paths, so we also add all bad path elimination inequalities to the formulation before solving it. An integer feasible solution to TSPTW($\mathcal{D}_\mathcal{T}^2$) may, or may not, be feasible for the TSPTW.

Preliminary experiments showed that TSPTW($\mathcal{D}_\mathcal{T}^1$) was helpful for finding an initial feasible TSPTW tour sooner, but rarely gave better solutions than could be found with other IP formulations afterwards, and so is disabled after the first feasible TSPTW tour is found. TSPTW($\mathcal{D}_\mathcal{T}^2$) is solved at every iteration.

For every IP solved, (TSPTW($\mathcal{D}_\mathcal{T}$), TSPTW($\mathcal{D}_\mathcal{T}^1$) or TSPTW($\mathcal{D}_\mathcal{T}^2$)) we harvest all integer solutions found by the solver. All are checked for feasibility to the TSPTW, and any subtours and bad paths discovered, in the case that the solution is not feasible, are recorded. This can be done in linear time. Any new best feasible solution is saved. Also, all too short arcs used in one of these solutions, if they led to infeasibility with actual travel times, are lengthened in the iteration in which they are found.

Algorithm 3 summarizes how we solve the TSPTW using partially time-expanded networks.

3   Computational Results

The algorithm is coded in C++ and uses Gurobi 6.5.0 with default settings (and a single thread) to solve IP problems. All computational results are obtained on a workstation with a Intel(R) Xeon (R) CPU E5-4610 v2 2.30 GHz processor running the Ubutu Linux 14.04.3 Operating System. Algorithm performance is evaluated on 5 classes of instances[1], with a total of 337 instances: 32 "easy" and 18 "hard" instances from [1], (AFG-Easy and AFG-Hard), 135 instances from [5], (DDGS), 125 from [10], (SU), and 27 instances from [7], (SPGPR).

The partially time-expanded networks are initialized with $\mathcal{N}_{\mathcal{T}} = \{(i, e_i) : i \in N\} \cup \{(i, l_i) : i \in N\}$. When solving IPs at lines 6, 8 and 10 of Algorithm 3, the IP solver is given an optimality tolerance of ϵ and a time limit of 10 min. At later iterations, as the partially time-expanded network becomes larger, it is possible that the IP solver does not return a solution to TSPTW($\mathcal{D}_{\mathcal{T}}$) within 10 min. If it happens, we let the solver run until it finds an integer solution. An overall time limit of 5 h is set; if Algorithm 3 does not stop at line 18 in that time, the instance is deemed to have "timed out". We assess

1. the average performance of SOLVE-TSPTW in term of running time, optimal gap, and gap versus best solutions in literature,
2. in the case of time-out, the SOLVE-TSPTW solution quality with respect to the best in literature, and
3. the size of the partially time-expanded network the algorithm dynamically discovers versus the size of the full time-space network.

The first two points are addressed in Table 1, where in the top half we report results for all instances and in the bottom half we report results for instances that timed out. Values in columns 2–7 in the top half and columns 3–5 in the bottom half represent averages over all instances in a class. The column with heading "# opt" gives the number of instances that were solved to within a gap of ϵ (the algorithm terminated at line 18) and the column with heading "# $\leq$ best" gives the number of instances for which the best feasible solution found by SOLVE-TSPTW has value within 0.01% of the best known value reported in the literature. The algorithm proved optimality for 313 out of 337 instances and produced 329 solutions of equal or better value than best-known solutions. Except for class SPGPR, the instances that timed out have small optimal gaps, which indicates that the solutions are optimal or close to optimal. In all cases, they are very close to the best solutions reported in the literature. We note that the algorithm produced new best solutions for 41 instances in class SU.

Finally, Table 2 shows that Algorithm 3 requires only a very small number of nodes and arcs to produce high-quality solutions (compared to the full time-expanded network). In summary, the computational results show that a dynamic discretization discovery algorithm using carefully designed partially

---

[1] http://lopez-ibanez.eu/tsptw-instances,   http://homepages.dcc.ufmg.br/~rfsilva/tsptw/.

Table 1. Performance of Algorithm 3

Instance class	# solns	Time (s)	# iter	Gap at 1st iter	Gap at last iter	Gap vs best	# opt	# ≤ best
AFG-Easy	3.97	23	9	0.56%	0.00%	0.00%	32	32
AFG-Hard	9.00	7,122	23	1.02%	0.09%	0.01%	14	15
DDGS	2.03	605	11	4.69%	0.00%	0.01%	132	134
SU	3.34	2,846	14	8.31%	0.06%	-0.06%	112	123
SPGPR	3.78	2,989	9	9.01%	0.73%	0.03%	23	25

Instance class	# timed out	Ave gap	Max gap	Gap vs best
AFG-Hard	4	0.35%	0.64%	0.15%
DDGS	3	0.66%	1.01%	0.06%
SU	13	0.59%	1.16%	-0.16%
SPGPR	4	4.89%	9.01%	0.22%

No instances in class AFG-Easy timed out

Table 2. Time-expanded network size.

Instances	Full time-expanded network		Final partially time-expanded network	
	#Nodes	#Arcs	#Nodes	#Arcs
AFG-Easy	69 k	1,336 k	232	3,549
AFG-Hard	133 k	10,441 k	2,190	50,100
DDGS	3 k	185 k	400	4,773
SU	45 k	7,195 k	1,292	8,512
SPGPR	107,672 k	2,631,490 k	564	14,308

time-expanded networks can produce high-quality solutions to the TSPTW. Further algorithm engineering, e.g., improved cut generation and management and more sophisticated schemes for choosing arcs to lengthen, can enhance its performance. It is easy to modify the algorithm to handle TSPTW variants, e.g., TSPTW with time-dependent travel times.

References

1. Ascheuer, N., Fischetti, M., Grötschel, M.: A polyhedral study of the asymmetric traveling salesman problem with time windows. Networks **36**(2), 69–79 (2000)
2. Baldacci, R., Mingozzi, A., Roberti, R.: New state-space relaxations for solving the traveling salesman problem with time windows. INFORMS J. Comput. **24**(3), 356–371 (2012)
3. Boland, N., Hewitt, M., Marshall, L., Savelsbergh, M.: The continuous time service network design problem. Optimization Online 2015–01-4729 (2015)
4. Dash, S., Günlük, O., Lodi, A., Tramontani, A.: A time bucket formulation for the traveling salesman problem with time windows. INFORMS J. Comput. **24**(1), 132–147 (2012)

5. Dumas, Y., Desrosiers, J., Gélinas, É., Solomon, M.M.: An optimal algorithm for the traveling salesman problem with time windows. Oper. Res. **43**(2), 367–371 (1995)
6. Focacci, F., Lodi, A., Milano, M.: A hybrid exact algorithm for the TSPTW. INFORMS J. Comput. **14**(4), 403–417 (2002)
7. Pesant, G., Gendreau, M., Potvin, J., Rousseau, J.: An exact constraint logic programming algorithm for the traveling salesman problem with time windows. Transp. Sci. **32**(1), 12–29 (1998)
8. Savelsbergh, M.W.P.: Local search in routing problems with time windows. Ann. Oper. Res. **4**(1), 285–305 (1985)
9. Savelsbergh, M.W.P.: The vehicle routing problem with time windows: minimizing route duration. INFORMS J. Comput. **4**(2), 146–154 (1992)
10. da Silva, R.F., Urrutia, S.: A general VNS heuristic for the traveling salesman problem with time windows. Discrete Optim. **7**(4), 203–211 (2010)
11. Wang, X., Regan, A.: Local truckload pickup and delivery with hard time window constraints. Transp. Res. Part B **36**, 97–112 (2002)
12. Wang, X., Regan, A.: On the convergence of a new time window discretization method for the traveling salesman problem with time window constraints. Comput. Ind. Eng. **56**, 161–164 (2009)

A Fast Prize-Collecting Steiner Forest Algorithm for Functional Analyses in Biological Networks

Murodzhon Akhmedov[1,2,3], Alexander LeNail[1], Francesco Bertoni[3],
Ivo Kwee[2,3], Ernest Fraenkel[1], and Roberto Montemanni[2(✉)]

[1] Department of Biological Engineering, Massachusetts Institute of Technology,
Cambridge, MA, USA
{akhmedov,lenail,fraenkel-admin}@mit.edu
[2] Dalle Molle Institute for Artificial Intelligence Research (IDSIA-USI/SUPSI),
Manno, Switzerland
{murodzhon,roberto}@idsia.ch
[3] Institute of Oncology Research (IOR), Bellinzona, Switzerland
{francesco.bertoni,ivo.kwee}@ior.iosi.ch

Abstract. The Prize-collecting Steiner Forest (PCSF) problem is NP-hard, requiring extreme computational effort to find exact solutions for large inputs. We introduce a new heuristic algorithm for PCSF which preserves the quality of solutions obtained by previous heuristic approaches while reducing the runtime by a factor of 10 for larger graphs. By decreasing the draw on computational resources, this algorithm affords systems biologists the opportunity to analyze larger biological networks faster and narrow their analyses to individual patients.

Keywords: Prize-collecting Steiner Forest · Biological networks

1 Introduction

The Prize-collecting Steiner Tree Problem (PCST) is a widely studied problem in the combinatorial optimization literature [2]. In PCST, we are given a graph $G = (V, E)$ such that vertices are assigned prizes $\forall v \in V : p(v) \in \mathbb{R}^+$ and edges are weighted with costs $\forall e \in E : c(e) \in \mathbb{R}^+$. The objective is to find a tree $T = (V_t, E_t)$ which minimizes:

$$f(T) = \sum_{e \in E_t} c(e) + \sum_{v \notin V_t} p(v) \tag{1}$$

Vertices with prize $p(v) = 0$ are referred to as Steiner nodes, and vertices with $p(v) > 0$ are called terminal nodes [7].

In contrast, the Prize-collecting Steiner Forest (PCSF) problem searches for a forest over input graph G. The PCSF is more general form of the PCST, and both problems are proven to be NP-hard [6,21]. The forest solution for the PCSF can be obtained by slightly modifying an input graph, and solving it by PCST algorithms. By adding an artificial *root* node to the input graph with

© Springer International Publishing AG 2017
D. Salvagnin and M. Lombardi (Eds.): CPAIOR 2017, LNCS 10335, pp. 263–276, 2017.
DOI: 10.1007/978-3-319-59776-8_22

edges from $root$ to every other node with weight ω, we construct an augmented graph $G' = (V \cup root, E \cup E_{root})$, where $E_{root} = e_{(root, v \in V)}$ with associated cost function $\forall e \in E_{root} : c'(e) = \omega$. Then, the PCST is solved on G' with a slightly modified objective function [8]:

$$f(T) = \sum_{e \in E_t \setminus E_{root}} c(e) + \sum_{v \notin V_t} p(v) + \sum_{e \in E_t \cap E_{root}} c'(e) \qquad (2)$$

The final forest solution is obtained by removing the $root$ from T as well as the edges emanating from it, such that the single tree solution connected by $root$ becomes a solution of disconnected components. The parameter ω regulates the number of selected outgoing edges from the root, which determines the number of trees in the forest. A forest with distinct sizes and different number of trees can be obtained by running the algorithm for different values of ω.

The PCSF problem from combinatorial optimization maps nicely onto the biological problem of finding differentially enriched sub-networks in the interactome of a cell. An interactome is a graph in which vertices represent biomolecules and edges represent the known physical interactions of those biomolecules. We can assign prizes to vertices based on measurements of differential expression of those cellular quantities in a patient sample and costs to edges from confidence scores for those intra-cellular interactions from experimental observation (high confidence means low edge cost), yielding a viable input to the PCSF problem. Vertices of the interactome which are not observed in patient data are not assigned a prize and become the Steiner nodes.

An algorithm to approximately solve the Prize-Collecting Steiner Forest problem can then be applied against this augmented interactome, resulting in a set of subgraphs corresponding to subsections of the interactome in which functionally related biomolecules may play an important concerted role in the differentially active biological process of interest. Thus, PCSF, when applied in a biological context, can be used to predict neighborhoods of the interactome belonging to the key dysregulated pathways of a disease.

Since the PCSF problem is NP-hard, it requires tremendous amounts of computation to determine exact solutions for large inputs. In biology, large networks are the norm. For example, the human Protein-Protein Interaction (PPI) network has some 15,000 nodes and 175,000 edges [1]. Adding metabolites for a more complete interactome yields inputs of some 36,000 nodes and over 1,000,000 edges. We require efficient algorithms to investigate these huge biological graphs for each patient, which requires solving PCSF many times.

We present a heuristic algorithm for PCSF which outperforms other heuristic approaches from the literature in computational efficiency for larger graphs by a factor of 10, while preserving the quality of the results. Our algorithm, MST-PCSF, builds from the ideas presented in [24,25]. In contrast to [24,25], MST-PCSF searches for a forest structure rather than a tree in the network, which represents the underlying biological problem better. Moreover, MST-PCSF uses more greedy clustering preprocessing step which divides large input graph into smaller clusters, and a heuristic to find approximate solutions for each cluster.

2   Related Work

The authors in [2] initially studied the prize-collecting traveling salesman problem, and proposed a 3-approximation algorithm. A 2-approximation with $O(n^3 \log n)$ running time is proposed in [3] by using the primal-dual method, improved by [4] to $O(n^2 \log n)$ execution time. That runtime was maintained in [5], which improved the approximation factor to $2 - \frac{2}{n}$.

Exact methods were devised in [6,7] using mixed integer linear programming, and a branch-and-cut algorithm based on directed edges was proposed to solve the model. An exact row generation approach was presented in [9] using a new set of valid inequalities.

A relax-and-cut algorithm was studied in [10] to develop effective Lagrangian heuristic. A multi-start local search algorithm with perturbations was integrated with variable neighborhood search in [11]. Seven different variations of PCST were studied in [12], and polynomial algorithms were designed for four of them. Some lower bound and polyhedral analyses were performed in [13–16]. A tabu-search metaheuristic, and a combination of memetic algorithms and integer programming were employed in [17,18] for PCST.

The application of the PCST approach in Biology has led to important results. The specific biological problems were to identify functions of molecules [19], to find protein associations [20], and to reconstruct multiple signaling pathways [21]. We have been developing heuristic and matheuristic PSCT algorithms [24,25] for similar applications, and in this study we aim to extend these ideas to a forest approach in order to make more realistic biological inferences.

3   Algorithm: MST-PCSF, A Fast Heuristic Algorithm for the PCSF Problem

The proposed MST-PCSF heuristic algorithm is composed of two distinct phases. First, we cluster the input graph to transform a global problem into a set of smaller local problems. Second, we bypass the intractability of PCSF by instead solving the Minimum Spanning Tree Problem (MST) on an altered representation of the input graph. These simple ideas dramatically reduce the time needed to obtain high quality solutions to the Prize-collecting Steiner Forest Problem.

3.1   Greedy Clustering Transforms a Global Problem into a Set of Local Problems

The intuition behind the clustering phase of the algorithm is that, in the context of biological interaction networks, we anticipate finding groups of dysregulated terminal nodes within the expression data, which are joined by moderately high confidence edges and Steiner nodes, forming independent high-prize and low-cost patches in the input graph. Simply put, we anticipate the clusters exist quite strongly in the data, and constituent trees from the optimal solution forest will be split between these clusters, but will not span multiple clusters. Clustering therefore is a sensible first step.

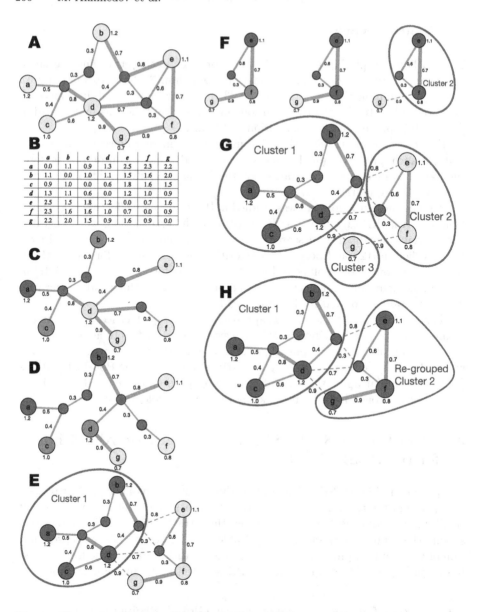

Fig. 1. Clustering phase of MST-PCSF. (**A**) The original network, such that terminal nodes are represented in yellow, Steiner nodes are in brown, edges are in blue and edge thickness relates to cost. (**B**) Compute All-Paths Shortest Path from every terminal to every terminal on the graph, resulting in a matrix of shortest path distances. (**C**) Choose a terminal at random and evaluate which terminals are reachable, assign them to a cluster. (**D**) Iteratively select nodes assigned to the cluster and find additional nodes which satisfy the clustering criterion with respect to other nodes in the cluster. (**E**) Cluster is full when no more terminals satisfy the criterion. Start a new cluster with an unclustered node. (**F**) Repeat steps **B**–**E** until there are no remaining unclustered nodes from the graph in **A**. (**G**) The raw clustering of the graph in **A** may have many singletons and doubletons. (**H**) Merge singleton and doubleton clusters with their nearest cluster, return the final clustering (Color figure online)

PCSF has proven more apt for the problem of highlighting disease-relevant networks [19,20] than PCST because in a tumorous cell, for example, multiple functional pathways may be simultaneously active and dysregulated, but non-overlapping. If we seek to find a tree, we are forced to incorporate spurious edges in our solution. If we seek to find a set of disconnected trees, however, we only select the edges needed to connect the high-prize terminals.

Algorithm 1. MST-PCSF, Clustering phase

Initializalization:
Set $U = \{v : p(v) > 0\}$, the set of 'unassigned terminals'.
Set $A = \varnothing$, the set of 'assigned terminals'.
Vector C such that $|C| = |U|$ and $C_i \leftarrow 0$ for all $i : 1...U$
Matrix $D \in \mathbb{R}^{|U| \times |U|}$ such that $D_{i,j}$ is the weighted distance of the shortest path from v_i to v_j for all pairs $v_i, v_j \in U$
$clusterID \leftarrow 0$

Algorithm
while U is not $\varnothing$ **do**
  $clusterID \leftarrow clusterID + 1$;
  Remove arbitrary vertex v_i from U and insert it into A.
  while A is not $\varnothing$ **do**
    Remove arbitrary vertex v_a from A.
    $C_a = ClusterID$
    for each $v_u \in U$ **do**
      if $(D_{a,u} < p(v_a)$ & $D_{a,u} < p(v_u))$ **then**
        Remove v_u from U and insert it into A.
      end if
    end for
  end while
end while
for each singleton and doubleton cluster $G_k = (V_k, E_k)$ **do**
  Let $G_{min_k} = min \sum_{v_i \in V_k} \sum_{v_j \notin V_k} D_{ij}$, the closest subgraph to G_k.
  Consolidate G_k with the nearest cluster G_{min_k}
end for

Output terminals set A, assignment vector C representing the final clustering.

The pseudocode of the heuristic clustering phase is shown in Algorithm 1. The clustering is performed by considering the pairwise relationship of terminal nodes. Given an input network (Fig. 1A), the algorithm computes all-pairs shortest path distance matrix D_{ij} among the terminal nodes (Fig. 1B). Two terminal nodes i and j are clustered together if they satisfy the strict clustering criterion: $D_{i,j} < p(i)$ & $D_{i,j} < p(j)$, which worked best of those we tested for biological networks. At the first iteration, the algorithm arbitrarily selects a terminal node, evaluates other terminals satisfying the clustering criterion with respect to the selected node, and assigns them into the same cluster (Fig. 1C).

Each terminal node newly added into the cluster is iteratively analyzed according to the clustering criterion, and additional terminal nodes satisfying the criterion are incorporated into the cluster. This process corresponds to the growth phase of the cluster (Fig. 1D). The heuristic obtains the first cluster when there is no more terminal node outside of that cluster which satisfies the clustering criterion (Fig. 1E). Then, the algorithms continues to build new clusters (Fig. 1H) by repeating the same steps in (Fig. 1C–E), and it is terminated when all terminal nodes of the input network are clustered (Fig. 1G).

After clustering the graph, singleton and doubleton clusters are merged with their nearest neighbor clusters (Fig. 1H), since those subgraphs harbor very little biological information on their own.

3.2 Solving MST on an Altered Representation of the Input Graph Bypasses the Complexity of PCSF

In the second phase of the algorithm, we bypass the complexity of PCSF by finding instead the tree covering every terminal node with minimum total edge costs, which is equivalent to solve the MST. The pseudocode of the second phase is shown in Algorithm 2.

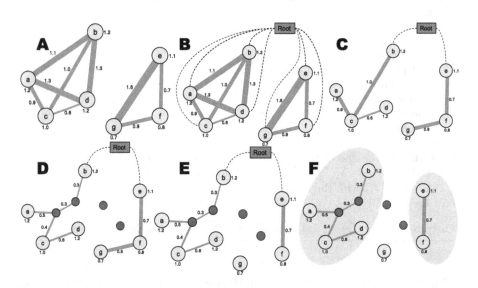

Fig. 2. MST-Phase of MST-PCSF. (A) Construct the complete graph from the terminal nodes in the clustered subgraphs. **(B)** Add an artificial root node with an edge from that root to every other nodes with cost ω. or use the already-existing node specified as the root by the user. **(C)** Determine the MST of this augmented graph by Prim's algorithm. **(D)** Interpolate the Steiner nodes from the original graph. **(E)** Prune all leaves for which the prize of the leaf is less than the cost of the edge. **(F)** Output a forest of subgraphs for further investigation.

The altered representation of the original network is the set of complete subgraphs composed exclusively of terminals from each cluster, in which each edge is weighted with the shortest path distance between the terminal nodes (Fig. 2A). We can solve MST on this representation quickly (Fig. 2C) and then project the solution back into the original graph (Fig. 2D), and finally, disconnecting nodes too expensive to retain in the solution (Fig. 2E).

Algorithm 2. MST-PCSF, MST phase

Initizalization:
For each subgraph G_s, construct the complete subgraph of the terminal nodes in G_s called $G'_s = (V'_s, E'_s)$, such that each edge $\in E'_s$ is weighted with the cost of the shortest path between the terminals it connects in G_s.

Algorithm
Add a vertex called *Root* and edges from *Root* to each terminal with cost ω.
This produces a new graph which is the union of each of the G'_s subgraphs and *Root*.
Find the minimum spanning tree of the new graph to obtain $MST = (V_{mst}, E_{mst})$
Interpolate the omitted Steiner nodes in the edges of MST from G
for each leaf node $v \in MST$ and associated parent node v' **do**
    if $p(v) <$ connection cost **then**
        remove v from MST ;
    end if
end for
Remove *Root* from MST leaving behind a forest of subgraphs we call F.

Output F, an approximate solution to the PSCF problem on graph G.

The algorithmic steps in (Fig. 2A–D) can be repeated several times in order to decrease the tree cost further, by adding the Steiner nodes incorporated in the solution tree (Fig. 2-D) into the cliques constructed in the next iteration (Fig. 2-A), yielding better results at each iteration until convergence. However, we only perform a single iteration of those steps for our results in this work.

4   Results

We compare the performances of MST-PCSF and the message passing algorithm (MSGP), a broadly used heuristic algorithm for PCST and PCSF. The MSGP algorithm has been used to predict unknown protein associations [20], find hidden components of regulatory networks [22], and reconstruct cell-signaling pathways [21]. We test the performances of these algorithms on small benchmark instances from the literature, as well as medium and large networks generated from real biological data. We use the default parameters for MSGP, except the reinforcement parameter g, which is set to 0.001 as in [21]. The computational studies are performed on a server equipped with an AMD Opteron(tm) Processor 6320 and shared memory of 256 GB. A single core is used while running the

Table 1. Performances of MST-PCSF and MSGP for the D instances [21]. The performance of the message passing algorithm [20] and the proposed heuristic are displayed under $MSGP$ and $MST\text{-}PCSF$ for $\omega = \{5, 8\}$. We report the upper bounds obtained from the methods under OBJ column. The running times of the methods are provided in seconds under $t(s)$ column. We provide the average statistics of 10 runs obtained by both methods for each instance.

| Instance | V | E | T | $\omega = 5$ | | | | $\omega = 8$ | | | |
| | | | | MSGP | | MST-PCSF | | MSGP | | MST-PCSF | |
				OBJ	t(s)	OBJ	t(s)	OBJ	t(s)	OBJ	t(s)
D01-A	1001	1255	5	21	0.44	21	0.03	26	0.45	26	0.03
D01-B	1001	1255	5	25	0.44	25	0.03	40	0.44	40	0.03
D02-A	1001	1260	10	46	0.45	46	0.04	58	0.44	58	0.04
D02-B	1001	1260	10	50	0.45	50	0.05	80	0.44	80	0.04
D03-A	1001	1417	83	630	0.66	630	0.52	787	0.7	794	0.51
D03-B	1001	1417	83	782	0.65	782	0.53	1131	0.68	1155	0.52
D04-A	1001	1500	250	928	0.63	937	0.81	1164	0.63	1170	0.79
D04-B	1001	1500	250	1129	0.66	1137	0.84	1574	0.8	1596	0.84
D05-A	1001	1750	500	1774	0.84	1791	1.86	2130	0.82	2195	1.74
D05-B	1001	1750	500	2108	0.77	2123	2.05	2787	0.82	2839	2.06
D06-A	1001	2005	5	21	0.61	21	0.05	26	0.61	26	0.04
D06-B	1001	2005	5	25	0.61	25	0.04	40	0.61	40	0.04
D07-A	1001	2010	10	46	0.67	46	0.06	58	0.67	58	0.06
D07-B	1001	2010	10	50	0.67	50	0.06	80	0.68	80	0.06
D08-A	1001	2167	83	630	0.82	634	0.62	763	0.89	789	0.61
D08-B	1001	2167	83	753	0.84	760	0.64	994	1.54	1023	0.64
D09-A	1001	2250	250	918	1.75	930	0.95	1080	0.95	1126	0.94
D09-B	1001	2250	250	1099	1.28	1116	1.01	1367	1.03	1414	1.01
D10-A	1001	2500	500	1569	1.2	1601	2.3	1705	1.36	1747	2.29
D10-B	1001	2500	500	1812	1.12	1836	2.38	2058	1.25	2078	2.37
D11-A	1001	5005	5	21	1.52	21	0.07	26	4.09	26	0.07
D11-B	1001	5005	5	24	1.56	24	0.07	36	4.09	36	0.07
D12-A	1001	5010	10	42	1.61	42	0.1	50	3.61	50	0.1
D12-B	1001	5010	10	43	1.55	44	0.1	50	2.54	50	0.1
D13-A	1001	5167	83	454	4.7	469	1.06	463	2.46	477	1.07
D13-B	1001	5167	83	501	2.42	516	1.06	510	3.09	526	1.07
D14-A	1001	5250	250	615	3.32	642	1.62	624	3.86	651	1.64
D14-B	1001	5250	250	673	4.16	697	1.62	690	2.09	712	1.64
D15-A	1001	5500	500	1057	3.68	1072	3.58	1076	2.41	1085	3.65
D15-B	1001	5500	500	1124	2.66	1140	3.56	1147	2.69	1154	3.64
D16-A	1001	25005	5	18	31.16	20	0.27	22	24.65	22	0.26
D16-B	1001	25005	5	19	23.26	20	0.28	22	26.69	23	0.27
D17-A	1001	25010	10	28	28.83	29	0.37	31	20.98	33	0.37
D17-B	1001	25010	10	28	29.41	29	0.37	31	31.95	32	0.38
D18-A	1001	25167	83	225	17.53	247	3.81	231	30.31	250	3.81
D18-B	1001	25167	83	233	20.83	256	3.81	238	24.62	259	3.81
D19-A	1001	25250	250	314	17.69	346	5.67	321	33.38	348	5.69
D19-B	1001	25250	250	320	31.16	353	5.68	323	30.83	356	5.67
D20-A	1001	25500	500	542	31.36	547	11.56	545	28.63	549	11.53
D20-B	1001	25500	500	543	21.46	548	11.6	545	34.55	551	11.52
mean				531	7.39	541	1.78	623	8.33	638	1.78

Table 2. The comparison results of the methods for the Breast Cancer network instances generated based on phosphoproteomic data in [23]. The performance of the message passing algorithm [20] and the proposed heuristic are displayed under $MSGP$ and $MST\text{-}PCSF$ for $\omega = \{1, 2\}$. The average statistics of 10 runs provided by both algorithms are reported for each instance.

Instance	V	E	T	$\omega = 1$				$\omega = 2$			
				MSGP		MST-PCSF		MSGP		MST-PCSF	
				OBJ	t(s)	OBJ	t(s)	OBJ	t(s)	OBJ	t(s)
A2-A0CM	36892	1016411	122	35.71	1726	35.64	130	36.69	1189	36.89	128
A2-A0D2	36892	1016411	226	66.38	1838	66.50	234	67.53	1987	67.50	231
A2-A0EV	36892	1016411	69	18.29	1482	18.35	78	19.55	1844	19.60	77
A2-A0EY	36892	1016411	118	40.56	2107	40.70	127	41.77	2273	41.91	128
A2-A0SW	36892	1016411	60	14.45	1655	14.50	69	15.45	1637	15.50	68
A2-A0T6	36892	1016411	92	30.92	1697	31.05	100	32.18	1737	32.31	101
A2-A0YC	36892	1016411	182	48.04	1766	48.17	188	49.04	1761	49.17	190
A2-A0YD	36892	1016411	55	17.92	1593	17.92	64	18.92	1656	18.92	63
A2-A0YF	36892	1016411	165	48.31	1783	48.27	174	49.54	847	49.65	177
A2-A0YM	36892	1016411	236	61.89	2080	62.07	245	62.91	1707	63.19	243
A7-A0CE	36892	1016411	142	38.35	1968	38.42	150	39.61	1719	39.68	149
A7-A13F	36892	1016411	139	43.42	1049	43.57	148	44.42	1471	44.57	145
A8-A06Z	36892	1016411	112	36.24	1394	36.42	119	37.76	1524	37.94	121
A8-A079	36892	1016411	77	25.77	1609	25.96	85	26.77	1691	26.96	87
A8-A09G	36892	1016411	186	46.65	1985	46.61	193	47.61	2229	47.73	196
AN-A04A	36892	1016411	208	63.40	1854	63.41	215	64.58	1916	64.41	216
AN-A0FK	36892	1016411	81	21.82	1582	21.89	89	22.99	1404	23.06	88
AN-A0FL	36892	1016411	126	35.63	1677	35.87	136	36.75	1782	36.99	134
AO-A0JC	36892	1016411	76	22.32	1734	22.39	85	23.32	1843	23.39	83
AO-A0JE	36892	1016411	116	31.60	2362	31.70	124	32.88	2033	32.97	124
AO-A0JM	36892	1016411	216	59.34	2563	59.61	225	60.79	2248	61.09	226
AO-A126	36892	1016411	51	16.26	1708	16.30	59	17.26	1835	17.30	59
AO-A12B	36892	1016411	148	43.94	2101	43.94	157	45.88	1613	45.72	155
AO-A12D	36892	1016411	211	54.94	1671	55.09	221	56.00	1857	56.18	218
AO-A12E	36892	1016411	146	39.68	2178	39.78	154	40.98	2238	40.81	155
AO-A12F	36892	1016411	167	44.88	2150	45.16	176	46.02	2054	46.30	176
AR-A0TR	36892	1016411	120	32.08	1920	32.28	127	33.08	1771	33.28	128
AR-A0TT	36892	1016411	164	43.05	1755	43.33	174	44.22	1826	44.50	172
AR-A0TV	36892	1016411	201	57.08	2453	57.27	210	58.20	3565	58.44	209
AR-A0TX	36892	1016411	169	46.36	2194	46.47	177	47.73	2171	47.74	176
AR-A0U4	36892	1016411	154	41.16	2111	41.39	163	42.48	2065	42.70	160
AR-A1AP	36892	1016411	109	32.49	1568	32.50	117	33.52	1754	33.54	117
AR-A1AS	36892	1016411	181	48.69	2344	48.91	188	49.93	2245	50.09	191
AR-A1AW	36892	1016411	116	31.48	1616	31.72	128	32.65	1849	32.89	127
BH-A0AV	36892	1016411	137	40.06	1983	40.20	144	41.84	1685	41.98	146
BH-A0BV	36892	1016411	220	57.43	1755	57.58	229	58.52	1466	58.59	228
BH-A0DG	36892	1016411	209	62.49	1757	62.58	219	63.56	2020	63.70	218
BH-A0E9	36892	1016411	107	35.57	1808	35.45	114	36.35	1339	36.45	115
BH-A18N	36892	1016411	176	47.57	1608	47.58	183	48.73	2310	48.87	181
BH-A18U	36892	1016411	72	23.87	1627	23.91	81	24.87	1428	24.91	79
C8-A12T	36892	1016411	87	20.25	1646	20.31	95	21.51	1745	21.57	96
C8-A12U	36892	1016411	87	27.03	1561	27.15	95	28.02	1576	28.15	94
C8-A12V	36892	1016411	94	26.38	2330	26.54	101	27.38	2267	27.54	101
C8-A130	36892	1016411	84	26.88	1229	27.06	93	27.88	1363	28.06	92
C8-A131	36892	1016411	187	53.48	1753	53.77	198	54.57	2177	54.80	197
C8-A138	36892	1016411	191	51.18	1930	51.14	199	52.13	1792	52.19	199
D8-A142	36892	1016411	135	41.35	1539	41.20	143	42.03	1142	42.20	142
E2-A154	36892	1016411	145	42.36	1770	42.50	152	43.47	1715	43.61	152
E2-A158	36892	1016411	169	49.12	2135	49.18	175	50.37	2117	50.38	176
E2-A15A	36892	1016411	176	51.55	1599	51.80	182	52.58	1601	52.83	183
mean				39.91	1819	40.02	149	41.06	1822	41.18	148

Table 3. Performances of MST-PCSF and MSGP for the Glioblastoma network instances generated from phosphoproteomic data from [21]. The first four columns provide instance name, number of total nodes, edges, and terminals for each network. The performance of the message passing algorithm [20] and the proposed heuristic are displayed under $MSGP$ and $MST\text{-}PCSF$ for $\omega = \{1, 2\}$. We report the upper bounds obtained from the methods under OBJ column. The running times of the methods are provided in seconds under $t(s)$ column. We report the average statistics of 10 runs obtained by both algorithms for each instance.

Instance	V	E	T	$\omega = 1$				$\omega = 2$			
				MSGP		MST-PCSF		MSGP		MST-PCSF	
				OBJ	t(s)	OBJ	t(s)	OBJ	t(s)	OBJ	t(s)
GBM6	15357	175792	108	26.35	305	26.41	22	28.61	262	28.66	22
GBM8	15357	175792	115	25.45	246	25.50	23	27.45	211	27.50	23
GBM10	15357	175792	132	32.85	247	32.91	26	36.96	206	37.02	26
GBM12	15357	175792	135	32.50	240	32.58	27	35.03	253	35.11	27
GBM15	15357	175792	131	30.49	248	30.58	26	33.62	256	33.70	26
GBM26	15357	175792	122	28.42	274	28.46	24	31.02	269	31.04	24
GBM39	15357	175792	123	29.51	170	29.56	24	31.68	170	31.72	24
GBM59	15357	175792	123	26.54	258	26.53	24	29.56	275	29.55	24
GBM-pool	15357	175792	161	42.90	299	42.98	31	50.90	267	50.97	31
mean				30.56	254	30.61	25	33.87	241	33.92	25

experiments. We run the algorithms for values of $\omega = \{1, 2\}$ for medium and large instances, and we used $\omega = \{5, 8\}$ for small instances due to their higher edge costs. In order to have a fair and robust comparison baseline, we report the average statistics of 10 runs provided by both algorithms for each instance.

The first of these domains is a set of benchmark instances for PCST algorithms from the literature [11] called the D instance set, which is composed of smaller networks of roughly 1,000 nodes and 25,000 edges. However, since the algorithms we compare solve PCSF, we modify the D instances by adding a *root* node beforehand, and specify to use that node as their root. The computational performances of the algorithms are displayed in Table 1. For these small inputs, MSGP provides a slightly better upper bounds for both values of ω. On the other hand, MST-PCSF outperforms MSGP in average running time.

The second domain is phosphoproteomic data from the Glioblastoma patients in [21]. These graphs are obtained by using differentially expressed genes as terminals by mapping them onto the human PPI network. As our PPI, we use STRING (version13), in which the network edges have a experimentally-derived confidence score $s(e)$ [1]. The low confidence edges with $s(e) < 0.5$ are removed to improve the reliability of the findings. We convert edge confidence into edge cost: $c(e) = max(0.01, 1 - s(e))$. The performance statistics of the methods are reported in Table 3. For these medium sized networks, the upper bounds provided

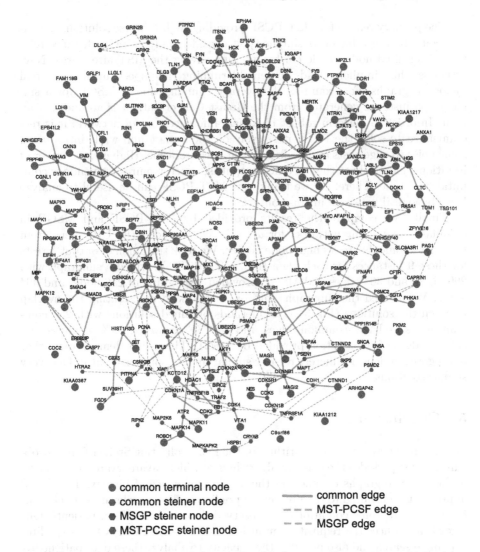

● common terminal node
● common steiner node
● MSGP steiner node
● MST-PCSF steiner node

───── common edge
----- MST-PCSF edge
----- MSGP edge

Fig. 3. A visual representation of the relationship between the solutions obtained by the MST-PCSF and MSGP algorithms on Gliablastoma patient data.

by MST-PCSF and MSGP are comparable for both values of ω. However, MST-PCSF has an order of magnitude speed up in running time on average.

Our largest graphs are generated by mapping the phosphoproteomic data from the Breast Cancer patients in [23] onto an integrated interactome of proteins and metabolites, resulting in networks with 36,897 nodes and 1,016,288 edges. Here as well, each network represents a single patient. The results provided by both algorithms are displayed in Table 2. For these large network instances, MST-PCSF provided very similar quality results compared to MSGP. In addition, MST-PCSF is approximately ten times faster in execution time on average.

The primary reason for MST-PCSF providing slightly worse solutions in general could be the value of its simplicity: it considers only pairwise distance relation of terminal nodes while clustering, and solving the MST afterwards. Nevertheless, this simplicity leads to tenfold running time speed up in large real biological graphs. For example, the average running time of MSGP for a single graph instance generated from the Breast Cancer patient is around half an hour. In order to perform robust biological inferences out of network, it is recommended to solve the PCSF iteratively on the same graph while introducing some noise to edge costs and node prizes at each iteration. Higher the number of iterations, more robust the inference output is, which requires tremendous computational time. Therefore, we think that MST-PCSF can be useful to analyze large biological networks in this context.

Finally, we compare the results provided by MST-PCSF and MSGP in Glioblastoma patient networks for $\omega = 1$. We excluded the UBC gene from the interactome due to its high node degree. We took the union of output forests for these 9 instances for each algorithm. MST-PCSF provided a subgraph with 269 nodes and 301 edges and MSGP provided a subgraph of 286 nodes and 364 edges. We merged the subgraphs into one network, demonstrating the overlap between the solutions (Fig. 3). The methods obtained solutions with 255 common nodes and 251 common edges. MST-PCSF recovered 89% of nodes and 83% of edges of the solution provided by MSGP. This result demonstrates that our algorithm does not merely recover similar quality solutions, but in fact, very similar solutions.

5   Conclusion

We present a new heuristic algorithm for the Prize-collecting Steiner Forest problem which supersedes the existing algorithms of which we are aware, particularly on larger-scale graphs common in the application-space of biology. The PCSF approach is well suited to the problem of predicting disease-relevant subnetworks from an interactome conditional on observed data gathered from patients. Our algorithm reduces the requisite computing time to solve PCSF which expedites existing research and also provides the capacity to analyze these data patient-by-patient, an operation which previously has been prohibitively computationally expensive.

Our algorithm accelerates the pace of relevant subnetwork imputation, which we hope will be a boon for all who apply the Prize-collecting Steiner Forest approach, in biology and elsewhere.

Acknowledgments. M.A. was supported by the Swiss National Science Foundation through the project 205321-147138/1: Steiner trees for functional analysis in cancer system biology. M.A. (partially) and A.L. were supported by the National Institute of Health through the project U54-NS-091046 and U01-CA-184898.

References

1. Szklarczyk, D., Franceschini, A., Kuhn, M., Simonovic, M., Roth, A., Minguez, P., Doerks, T., Stark, M., Muller, J., Bork, P., Jensen, L.J., van Mering, C.: STRING 8 - a global view on proteins and their functional interactions in 630 organisms. Nucleic Acids Res. **37**, D412–D416 (2011)
2. Bienstock, D., Goemans, M.X., Simchi-Levi, D., Williamson, D.: A note on the prize-collecting traveling salesman problem. Math. Program. **59**, 413–420 (1993)
3. Goemans, M.X., Williamson, D.P.: The primal-dual method for approximation algorithms and its application to network design problems. In: Approximation Algorithms for NP-Hard Problems, pp. 144–191 (1996)
4. Johnson, D.S., Minkoff, M., Phillips, S.: The prize-collecting Steiner tree problem: theory and practice. In: Proceedings of 11th ACM-SIAM Symposium on Discrete Algorithms, pp. 760–769 (2000)
5. Feofiloff, P., Fernandes, C.G., Ferreira, C.E., Pina, J.C.: Primal-dual approximation algorithms for the prize-collecting Steiner tree problem. Inf. Process. Lett. **103**(5), 195–202 (2007)
6. Ljubic, I., Weiskircher, R., Pferschy, U., Klau, G., Mutzel, P., Fischetti, M.: Solving the prize-collecting Steiner tree problem to optimality. In: Seventh Workshop on Algorithm Engineering and Experiments, pp. 68–76 (2005)
7. Ljubic, I., Weiskircher, R., Pferschy, U., Klau, G., Mutzel, P., Fischetti, M.: An algorithmic framework for the exact solution of the prize-collecting Steiner tree problem. Math. Progam. **105**(2), 427–449 (2006)
8. Bechet, M.B., Bradde, S., Braunstein, A., Flaxman, A., Foini, L., Zecchina, R.: Clustering with shallow trees. J. Stat. Mech. **2009**, 12010 (2009)
9. Haouari, M., Layeb, S.B., Sherali, H.D.: Algorithmic expedients for the prize collecting Steiner tree problem. Discrete Optim. **7**, 32–47 (2010)
10. Cunha, A.S., Lucena, A., Maculan, N., Resende, M.G.C.: A relax-and-cut algorithm for the prize-collecting Steiner problem in graphs. Discrete Appl. Math. **157**, 1198–1217 (2009)
11. Canuto, S.A., Resende, M.G.C., Ribeiro, C.C.: Local search with perturbation for the Prize-collecting Steiner tree problem in graphs. Networks **38**, 50–58 (2001)
12. Chapovska, O., Punnen, A.P.: Variations of the prize-collecting Steiner tree problem. Networks **47**(4), 199–205 (2006)
13. Beasley, J.E.: An SST-based algorithm for the Steiner problem in graphs. Networks **19**, 1–16 (1989)
14. Duin, C.W., Volgenant, A.: Some generalizations of the Steiner problem in graphs. Networks **17**(2), 353–364 (1987)
15. Fischetti, M.: Facets of two Steiner arborescence polyhedra. Math. Program. **51**, 401–419 (1991)
16. Lucena, A., Resende, M.G.C.: Strong lower bounds for the prize-collecting Steiner problem in graphs. Discrete Appl. Math. **141**, 277–294 (2004)
17. Fu, Z.H., Hao, J.K.: Knowledge guided tabu search for the prize collecting Steiner tree problem in graphs. In: 11th DIMACS Challenge Workshop (2014)
18. Klau, G.W., Ljubić, I., Moser, A., Mutzel, P., Neuner, P., Pferschy, U., Raidl, G., Weiskircher, R.: Combining a memetic algorithm with integer programming to solve the prize-collecting Steiner tree problem. In: Deb, K. (ed.) GECCO 2004. LNCS, vol. 3102, pp. 1304–1315. Springer, Heidelberg (2004). doi:10.1007/978-3-540-24854-5_125

19. Dittrich, M.T., Klau, G.W., Rosenwald, A., Dandekar, T., Mueller, T.: Identifying functional modules in proteinprotein interaction networks: an integrated exact approach. Bioinformatics **26**, 223–231 (2008)
20. Bechet, M.B., Borgs, C., Braunstein, A., Chayesb, J., Dagkessamanskaia, A., Franois, J.M., Zecchina, R.: Finding undetected protein associations in cell signaling by belief propagation. PNAS **108**, 882–887 (2010)
21. Tuncbag, N., Braunstein, A., Pagnani, A., Huang, S.C., Chayes, J., Borgs, C., Zecchina, R., Fraenkel, E.: Simultaneous reconstruction of multiple signaling pathways via the prize-collecting Steiner forest problem. J. Comput. Biol. **20**(2), 124–136 (2013)
22. Tuncbag, N., McCallum, S., Huang, S.C., Fraenkel, E.: SteinerNet: a web server for integrating omic data to discover hidden components of response pathways. Nucleic Acids Res. **40**, 1–5 (2012)
23. Mertins, P., Mani, D.R., Ruggles, K.V., Gillette, M.A., Clauser, K.R., Wang, P., Wang, X., Qiao, J.W., Cao, S., Petralia, F., Kawaler, E., Mundt, F., Krug, K., Tu, Z., Lei, J.T., Gatza, M.L., Wilkerson, M., Perou, C.M., Yellapantula, V., Huang, K., Lin, C., McLellan, M.D., Yan, P., Davies, S.R., Townsend, R.R., Skates, S.J., Wang, J., Zhang, B., Kinsinger, C.R., Mesri, M., Rodriguez, H., Ding, L., Paulovich, A.G., Feny, D., Ellis, M.J., Carr, S.A., NCI CPTAC: Proteogenomics connects somatic mutations to signalling in breast cancer. Nature **534**, 55–62 (2016)
24. Akhmedov, M., Kwee, I., Montemanni, R.: A divide and conquer matheuristic algorithm for the prize-collecting Steiner tree problem. Comput. Oper. Res. **70**, 18–25 (2016)
25. Akhmedov, M., Kwee, I., Montemanni, R.: A fast heuristic for the prize-collecting Steiner tree problem. Lect. Not. Manag. Sci. **6**, 207–216 (2014)
26. Prim, R.C.: Shortest connection networks and some generalizations. Bell Syst. Techn. J. **36**(6), 1389–1401 (1957)

Scenario-Based Learning for Stochastic Combinatorial Optimisation

David Hemmi[1,2]($\boxtimes$), Guido Tack[1,2]($\boxtimes$), and Mark Wallace[1]($\boxtimes$)

[1] Faculty of IT, Monash University, Melbourne, Australia
{david.hemmi,guido.tack,mark.wallace}@monash.edu
[2] Data61/CSIRO, Melbourne, Australia

Abstract. Combinatorial optimisation problems often contain uncertainty that has to be taken into account to produce realistic solutions. This uncertainty is usually captured in *scenarios*, which describe different potential sets of problem parameters based on random distributions or historical data. While efficient algorithmic techniques exist for specific problem classes such as linear programs, there are very few approaches that can handle general Constraint Programming formulations with uncertainty. This paper presents a generic method for solving stochastic combinatorial optimisation problems by combining a *scenario-based decomposition* approach with *Lazy Clause Generation* and strong *scenario-independent nogoods* over the first stage variables. The algorithm can be implemented based on existing solving technology, is easy to parallelise, and is shown experimentally to scale well with the number of scenarios.

1 Introduction

Reasoning under uncertainty is important in combinatorial optimisation, since uncertainty is inherent to many problems in an industrial setting. An example of this especially hard class of optimisation problems is the generalised assignment problem [2], where jobs with an uncertain duration have to be assigned to computers in a network, or the assignment of tasks to entities when developing a project plan. Other examples include the stochastic Steiner tree problem with uncertainty in the set of terminals [12]. At first, edges in the graph can be purchased at a low price. Once the set of terminals is revealed, the Steiner tree must be completed by purchasing edges at an increased price. Various forms of the stochastic set covering problem [11] and stochastic vehicle routing problems [21] are of similar form.

The focus of this paper is on stochastic optimisation problems with a combinatorial structure. The random variables describing the uncertainty have finite support. Each random variable has an underlying discrete probability distribution. A scenario describes the stochastic problem when all the random variables are fixed. Each scenario has a probability of occurrence. Stochastic problems are composed of a first and subsequent stages, the simplest being a two stage problem. The first stage denotes the problem before information about the random variables is revealed and first stage decisions are taken with respect to all scenarios. Second stage decisions are made once the random variables are fixed.

© Springer International Publishing AG 2017
D. Salvagnin and M. Lombardi (Eds.): CPAIOR 2017, LNCS 10335, pp. 277–292, 2017.
DOI: 10.1007/978-3-319-59776-8_23

In constraint programming (CP), modeling frameworks that allow solver agnostic modelling have been developed. Examples include MiniZinc [15] and Essence [10]. Problems described in one of these languages can be solved with a range of CP, MIP or SAT solvers without the user having to tailor the model for specific solvers. Stochastic MiniZinc [17] is an extension of the MiniZinc modelling language that supports uncertainty. Stochastic MiniZinc enables modellers to express combinatorial stochastic problems at a high level of abstraction, independent of the stochastic solving approach. Stochastic MiniZinc uses the standard language set of MiniZinc and admits models augmented with annotations describing the first and second stage. Stochastic MiniZinc automatically transforms the model into a structure that can be solved by standard MiniZinc solvers. At present, stochastic MiniZinc translates a two-stage model into the deterministic equivalent and uses a standard CP, MIP or SAT solver for the search. However, standard solver technology cannot exploit the special model structure of scenario based stochastic optimisation problems. As a result, the search performance is poor and it is desirable to develop a solver that can handle this class of problems without requiring the modeler to have expert knowledge in stochastic optimisation.

The literature on stochastic optimisation problems containing integer variables is diverse, see ([1,4,6,13,18]). Previous works have been concerned with problems that have a specific mathematical structure, for example linear models with integrality requirements. Furthermore, these methods are not available as standalone solvers and it is not possible to easily transform a model defined in a CP framework into a form that can be solved by one of these algorithms, in particular when a CP model contains non-linear constraints that would require (potentially inefficient) linearisation. In addition, methods to solve stochastic constraint satisfaction problems, inspired by the stochastic satisfiability problem, have been introduced, see ([3,14]) for an overview.

This paper proposes an algorithm to solve combinatorial stochastic optimisation problems. The work is based on Lazy Clause Generation (LGC) and translates the *scenario decomposition algorithm for 0–1 stochastic programs* introduced by Ahmed [1] into the CP setting. Our main contributions are the following: first, a search strategy that in combination with Ahmed's algorithm produces scenario-independent nogoods; secondly, an effective use of nogoods in the sub-problems when using decomposition algorithms; thirdly, a scenario bundling method to further improve the performance. Also, the benchmark instances we are using have been made publicly available on the CSPLib.

In contrast to the methods in the literature, the proposed algorithm can be used directly as a back-end solver to address problems formulated in a CP framework. No sophisticated reformulation of the model nor problem specific decompositions are required. The paper concludes with an experiment section.

2   Background and Related Work

This section introduces the basic notation, and then discusses relevant literature.

2.1   Basic Definitions

We will base our definition of stochastic problems on the deterministic case.

Definition 1. *A (deterministic) **constraint optimisation problem** (COP) is a four-tuple P defined as follows:*

$$P = <V, D, C, f>$$

*where V is a set of decision variables, D is a function mapping each element of V to a domain of potential values, and C is a set of constraints. A constraint $c \in C$ acts on variables $x_i, ..., x_j$, termed $scope(c)$ and specifies mutually-compatible variable assignments σ from the Cartesian product $D(x_i) \times ... \times D(x_j)$. The quality of a solution is measured using the objective function f. We write $\sigma(x)$ for the value of x in assignment σ; $\sigma|_X$ for σ restricted to the set of variables X; and $\sigma \in D$ to mean $\forall x : \sigma(x) \in D(x)$. We define the set of **solutions** of a COP as the set of assignments to decision variables from the domain D that satisfies all constraints in C:*

$$sol(P) = \{\sigma \in D \mid \forall c \in C : \sigma|_{scope(c)} \in c\}$$

*Finally, an **optimal solution** is one that minimises the objective function:*

$$\min_{\sigma \in sol(P)} f(\sigma)$$

On the basis of a COP we can define a Stochastic Constraint Optimisation Problem (SCOP). We restrict ourselves to two-stage problems, although the definitions and algorithms can be generalised to multi-stage problems. In a two-stage SCOP, a set of decisions, the *first stage* decisions, is taken before the values for the random variables are known. Once the random variables are fixed, further decisions are made for the subsequent stages. To enable reasoning before the random variables are revealed, we assume that their values can be characterised by a finite set of *scenarios*, i.e., concrete instantiations of the random variables based on historical data or forecasting. Our task is then to make first stage decisions that are optimal *on average* over all scenarios.

The idea behind our definition of an SCOP is to regard each scenario as an individual COP (fixing all random variables), with the additional restriction that all scenario COPs share a common set of first stage variables V. The goal is to find solutions to the scenarios that agree on the first stage variables, and for which the weighted average over all scenario objective values is minimal.

Definition 2. *A **stochastic constraint optimisation problem** (SCOP) is a tuple as follows:*

$$P \quad = < V, P_1, \ldots, P_k, p_1, \ldots, p_k >$$
$$\text{with } P_i = < V_i, D_i, C_i, f_i > \qquad \forall i \in 1..k : V \subseteq V_i$$

where each P_i is a COP, each p_i is the weight (e.g. the probability) of scenario i, and the set V is the set of first stage variables shared by all P_i.

*The set of **solutions** of an SCOP is defined as tuples of solutions of the P_i that agree on the variables in V:*

$$sol(P) = \{\sigma | \sigma = < \sigma_i, .., \sigma_k >, \sigma_i \in sol(P_i),$$
$$\sigma_i(x) = \sigma_j(x)$$
$$\forall i, j \in 1..k, x \in V\}$$

*An **optimal solution** to an SCOP minimises the weighted sum of the individual objectives:*

$$\min_{\sigma \in sol(P)} \textstyle\sum_{i=1}^{k} p_i f_i(\sigma_i)$$

Multiple methods to solve stochastic programs have been developed in the past. The most direct option is to formulate the SCOP as one large (deterministic) COP, called the *deterministic equivalent* (DE). Starting from Definition 2, the DE is constructed simply by taking the union of all the P_i (assuming an appropriate renaming of the second stage variables), and using the weighted sum as the objective function. The number of variables and constraints in the DE increases linearly with the number of scenarios. However, in case of a combinatorial second stage, the solving time may increase exponentially. As a result the DE approach lacks scalability.

An alternative to the DE is to *decompose* the SCOP, in one of two ways. Firstly, the problem can be relaxed by time stages. In the *vertical* decomposition a master problem describes the first stage and contains an approximation of the second stage. A set of sub-problems captures the second stages for each scenario. Complete assignments to the master problem are evaluated against the sub-problems. The L-shaped method [4], which is similar to Benders decomposition, is an example for an algorithm that works on the basis of the *vertical* decomposition. However, the L-Shaped method is limited to linear, continuous SCOPs, with extensions available for problems with integrality requirements. Secondly, the problem can be decomposed *horizontally*, by scenarios. The rest of this paper is based on this scenario decomposition, which is introduced in the next section.

2.2 Scenario-Based Decomposition

A straightforward decomposition for SCOPs is the scenario decomposition, where the P_i of an SCOP are treated as individual problems. A feasible solution to the stochastic program requires the shared variables to agree across all scenarios.

To ensure feasibility, an additional *consistency constraint* $\sigma_i(x) = \sigma_j(x) \forall i, j \in 1 \ldots k, x \in V$ is required. Note how this constraint was built into our definition of SCOP, but now has become an external requirement that can be relaxed. To make this more formal, we introduce the notion of a *pre-solution*, which represents tuples of individual scenario solutions that may violate the consistency constraint.

Definition 3. *A **pre-solution** to a SCOP P is a tuple*

$$pre\_sol(P) = \{\sigma | \sigma =< \sigma_i, .., \sigma_k >, \sigma_i \in sol(P_i), \forall i \in 1..k\}$$

We can redefine the set of solutions as exactly those pre-solutions that agree on the first stage variables:

$$sol(P) = \{\sigma | \sigma \in pre\_sol(P) \wedge \sigma_i(x) = \sigma_j(x) \ \forall i, j \in 1 \ldots k, x \in V\}$$

Algorithms based on the scenario decomposition generate pre-solutions and iteratively enforce convergence (i.e., consistency) on the first stage variables V, using different methods as described below. Any converged pre-solution is a feasible solution of the SCOP, and yields an upper bound on the stochastic objective value. As pre-solutions relax the consistency constraints, any pre-solution that minimises the individual objective functions represents a lower bound to the stochastic objective:

$$\min_{\sigma \in pre\_sol(P)} \textstyle\sum_{i=1}^{k} p_i f_i(\sigma_i) = \textstyle\sum_{i=1}^{k} p_i \min_{\sigma \in sol(P_i)} f_i(\sigma) \leq \min_{\sigma \in sol(P)} \textstyle\sum_{i=1}^{k} p_i f_i(\sigma_i)$$

A number of algorithms have been developed on the basis of iteratively solving the scenarios. In the following, we will introduce the most relevant methods. Progressive Hedging (PH) finds convergence over the shared variables by penalising the scenario objective functions. The added penalty terms represent the Euclidean distance between the averaged first stage variables and each scenario. Gradually, by repetitively solving the scenarios and updating the penalty terms, convergence over the shared variables is achieved. Originally PH was introduced by Rockafellar and Wets [18] for convex stochastic programs, for which convergence on the optimal solution can be guaranteed. However, PH was successfully used as a heuristic for stochastic MIPs, where optimality guarantees do not hold due to non-convexity, such as the stochastic inventory routing problem in [13] or resource allocation problems in [22] to name just a few.

An alternative scenario-based decomposition method was proposed by Carø and Schultz [6], called *Dual Decomposition* (DD), a branch and bound algorithm for stochastic MIPs. To generate lower bounds, the DD uses the dual of the stochastic problem, obtained by relaxing the consistency constraints using Lagrangian multipliers. To generate upper bounds, the solutions to the subproblems are averaged and made feasible using a rounding heuristic. The feasible region is successively partitioned by branching over the shared variables. The efficiency of the DD strongly depends on the update scheme used for the Lagrangian multipliers.

The algorithms introduced so far, including PH, DD and the L-Shaped method, only work for linear programs, with some extensions being available for certain classes of integer linear programs. As a result these approaches may require a solid understanding of the problem and the algorithm, in order to reformulate a given problem to be solved or to adapt the algorithm.

2.3 Evaluate and Cut

In contrast to the specialised approaches, Ahmed [1] proposes a scenario decomposition algorithm for SCOPs with binary first stage variables, which does not rely on the constraints being linear or the problem convex. Ahmed's algorithm solves each scenario COP independently to optimality. This yields a pre-solution to the SCOP and according to the discussion above a lower bound. In addition, each scenario first stage solution is evaluated against all other scenarios. The result is a candidate solution to the SCOP and therefore an upper bound. The evaluated candidates are added to the scenarios as *nogoods* (or *cuts*), forcing them to return different solutions in every iteration. As long as the first stage variables have finite domains, this process is guaranteed to find the optimal SCOP solution and terminates, either when the lower bound exceeds the upper bound or no additional candidate solutions are found. The scenarios can be evaluated separately, allowing highly parallelised implementations.

Algorithm 1 implements Ahmed's [1] ideas using our SCOP notation. We call this algorithm EVALUATEANDCUT, as it evaluates each candidate against all scenarios and adds nogoods that cut the candidates from the rest of the search. In line 7, each scenario i is solved independently, resulting in a variable assignment to the first and second stage σ, and a value obj of the objective function f_i. The lower bound is computed incrementally as the weighted sum (line 8). Furthermore, a set S of all first stage scenario solutions is constructed (line 9, recall that $\sigma|_V$ means the restriction of σ to the first stage variables V). Each candidate solution $\sigma|_V$ in S is evaluated against all scenarios P_i (line 14). This yields a feasible solution to the SCOP, and the tentative upper bound t_{UB} is updated. Once the candidate $\sigma|_V$ is evaluated, a *nogood* excluding $\sigma|_V$ from the search is added to the scenario problems P_i (line 16). We call this the **candidate nogood**. If the tentative upper bound is better than the global upper bound UB, a new incumbent is found (lines 17–19).

Each candidate $\sigma|_V$ is an optimal solution to one scenario problem P_i; excluding it from all scenarios in line 16 implies a monotonically increasing lower bound. The algorithm terminates when the lower bound exceeds the upper bound. Note, the call to solve in line 7 will return infinity if any scenario becomes infeasible. Line 21 will be used in our extensions of the algorithm presented in Sect. 3 and can be ignored for now.

Ahmed [1] reports on an implementation of this algorithm using MIP technology, and evaluates it on benchmarks with a binary first and linear second stage. The paper claims that this algorithm can find high quality upper bounds quickly. For linear SCOPs, improvements to the algorithm are introduced in [20], taking

---

Algorithm 1. A scenario decomposition algorithm for 0–1 stochastic programs

---

1: **procedure** EVALUATEANDCUT
2:     Initialise: $UB = \infty$, $LB = -\infty$, sol = NULL
3:     **while** $LB < UB$ **do**
4:         $LB = 0$, $S = \varnothing$
5:         % *Obtain lower bound and find candidate solutions*
6:         **for** i in 1..k **do**
7:             $< \sigma, obj> = \text{SOLVE}(P_i)$
8:             $LB += p_i * obj$
9:             $S \cup = \sigma|_V$
10:        % *Check candidate solutions and obtain upper bound*
11:        **for** $\sigma_V \in S$ **do**
12:            $t_{UB} = 0$
13:            **for** i in 1..k **do**
14:                $<\_, obj> = \text{solve}(P_i[C \cup = \{\sigma_V\}])$
15:                $t_{UB} += p_i * obj$
16:                $P_k = P_k[C \cup = \{\neg \sigma_V\}]$
17:            **if** $t_{UB} < UB$ **then**
18:                sol $= \sigma_V$
19:                $UB = t_{UB}$
20:        % *Evaluate partial first stage assignments*
21:        DIVE(P,UB,S,sol)
22:    **return** sol

---

advantage of a linear relaxation to improve the lower bounds, and additional optimality cuts based on the dual of the second stage problem.

Discussion: The EVALUATEANDCUT algorithm has the distinct advantage over other decomposition approaches that it can be applied to arbitrary SCOPs. This makes it an ideal candidate as a backend for solver-independent stochastic modelling languages such as Stochastic MiniZinc. However, in the worst case the algorithm requires $O(k^2)$ checks for k scenarios in each iteration. This is not prohibitive if evaluating the candidates is cheap. However, for problems that have many scenarios and a combinatorial second stage, this quadratic behaviour may dominate the solving time. In the following section, we propose a technique to decrease the time to solve and evaluate the individual candidates. Furthermore, we propose a method to reduce the number of iterations required to solve the SCOP.

3   Scenario-Based Algorithms for CP

This section introduces three modifications to EVALUATEANDCUT that improve its performance: using *Lazy Clause Generation* solvers for the scenario subproblem; using *dives* to limit the number of candidate verifications and iterations required; and *scenario bundling* as a hybrid between the DE and scenario-based decomposition.

Traditional CP solvers use a combination of propagation and search. Propagators reduce the variable domains until no further reduction is possible or a constraint is violated. Backtracking search, usually based on variable and value selection heuristics, explores the options left after propagation has finished. In contrast to traditional CP solvers, Lazy Clause Generation (LCG) [16] solvers *learn* during the search. Every time a constraint is violated the LCG solver analyses the cause of failure and adds a constraint to the model that prevents the same failure from happening during the rest of the search. The added constraints are called *nogoods*, and they can be seen as the CP equivalent of cutting planes in integer linear programming – they narrow the feasible search space.

Chu and Stuckey showed how nogoods can be reused across multiple instances of the same model if the instances are structurally similar [7]. This technique is called *inter-instance learning*. The nogoods learned during the solving of one instance can substantially prune the search space of future instances. Empirical results published in [7] indicate that for certain problem classes, a high similarity between models yields a high resusability of nogoods and therefore increased performance.

The EVALUATEANDCUT algorithm repeatedly solves very similar instances. In each iteration, every scenario is solved again, with the only difference being the added nogoods or the projection onto the first stage variables to evaluate candidates (see line 16 in Algorithm 1). As a consequence, inter-instance learning can be applied to speed up the search within the same scenario. We call this concept *vertical learning*. No changes are required to Algorithm 1, except that we assume the calls to SOLVE in line 14 to be incremental and remember the nogoods learned for each P_i in the previous iteration. To the best of our knowledge, this is the first use of inter-instance learning in a vertical fashion.

3.1   Search Over Partial Assignments

As introduced earlier, the EVALUATEANDCUT algorithm quickly finds high quality solutions, by checking the currently best solution for an individual scenario (the *candidate*) against all other scenarios. However, in order to prove optimality, EVALUATEANDCUT relies on the lower bound computed from the pre-solution found in each iteration. The quality of the lower bound, and the number of iterations required for it to reach the upper bound and thus prove optimality, crucially depends on the candidate nogoods added in each iteration.

Compared to nogoods as computed during LCG search, candidate nogoods are rather weak: They only cut off a single, complete first stage assignment. Furthermore, in the absence of Lagrangian relaxation or similar methods, candidate nogoods are the *only* information about the global lower bound that is available to each scenario problem P_i.

Stronger nogoods: To illustrate the candidate nogoods produced by EVALUATEANDCUT, consider a set of shared variables of size 5, and a first stage candidate solution $x = [3, 6, 1, 8, 3]$. The resulting candidate nogood added to the constraint set of each scenario subproblem P_i is:

$$P_i = P_i[C \cup = \{x_1 \neq 3 \vee x_2 \neq 6 \vee x_3 \neq 1 \vee x_4 \neq 8 \vee x_5 \neq 3\}]$$

The added constraint cuts off exactly one solution. A nogood composed of only a *subset* of shared variables would be much stronger. For example, assume that we can prove that even the *partial* assignment $x_1 = 3 \wedge x_2 = 6 \wedge x_3 = 1$ cannot be completed to an optimal solution to the stochastic problem. We could add the following nogood:

$$P_i = P_i[C \cup = \{x_1 \neq 3 \vee x_2 \neq 6 \vee x_3 \neq 1\}]$$

In contrast to the original candidate nogood that pruned exactly one solution, the new, shorter nogood can cut a much larger part of the search space. We call this stronger kind of nogood a **partial candidate nogood**.

Diving: We now develop a method for finding partial candidate nogoods based on *diving*. The main idea is to iteratively fix first stage variables across all scenarios and compute a pre-solution, until the lower bound exceeds the global upper bound. In that case, the fixed variables can be added as a partial nogood.

The modification to Algorithm 1 consists of a single added call to a procedure DIVE in line 21, which is executed in each iteration after the candidates have been checked. The definition of DIVE is described in Algorithm 2.

Line 5 constructs a constraint c that fixes a subset of the first stage variables, based on the current set of scenario solutions S. This is done according to a heuristic that is introduced later. The loop in lines 7–10 is very similar to the computation of a pre-solution in Algorithm 1, except that *all* scenarios are forced to agree on the selected subset of shared variables by adding the constraint c. As in Algorithm 1, we compute a lower bound based on the pre-solution. However, since we have arbitrarily forced the scenario sub-problems to agree using c, this lower bound is not valid globally – there can still be better overall solutions with different values for the variables $\sigma|_V$.

In essence, the algorithm is adding specific consistency constraints one by one. Adding such constraint can lead to one of three states:

1. The lower bound does not exceed the upper bound, but all scenarios agree on a common first stage assignment (even if c does not constrain all first stage variables). This means that a new incumbent solution is found (lines 11–17) and the constraint c can be added as a partial candidate nogood.
2. The lower bound meets or exceeds the upper bound. In this case, the partial consistency constraint c cannot be extended to any global solution that is better than the incumbent (lines 18–22). We can therefore add c as a partial candidate nogood.
3. The lower bound is smaller than the upper bound, and the scenarios have not converged on the first stage variables. In this case, an additional constraint is added to the partial consistency constraint c.

Algorithm 2. Searching for partial candidate nogoods using diving

```
 1: procedure DIVE(P,UB,S,sol)
 2:     t_LB = -∞
 3:     while t_LB < UB do
 4:         t_LB = 0
 5:         c = SELECTFIXED(S,V) % Select first stage variables to fix
 6:         S = ∅
 7:         for i in 1..k do
 8:             <σ,obj> = SOLVE(P_i[C ∪ = {c}])
 9:             t_LB += p_i * obj
10:             S ∪ = σ|_V
11:         if t_LB < UB ∧ S = {σ_V} then
12:             UB = t_LB
13:             sol = σ_V
14:             % Add partial candidate nogood
15:             for i in 1..k do
16:                 P_i = P_i[C ∪ = {¬c}]
17:             return
18:         if t_LB >= UB then
19:             % Add partial candidate nogood
20:             for i in 1..k do
21:                 P_i = P_i[C ∪ = {¬c}]
22:             return
```

The DIVE procedure terminates because in each iteration, SELECTFIXED fixes at least one additional first stage variable, which means that either case (1) or (2) above must eventually hold.

Note, we do not evaluate the scenario solutions constructed during diving against all scenarios. The rationale is that the additional consistency constraints are likely to yield non-optimal overall solutions (i.e., worse than the current incumbent). The $O(k^2)$ evaluations would therefore be mostly useless.

Diving heuristic: Let us now define a heuristic for choosing the consistency constraints to be added in each iteration during a dive. The goal is to produce short, relevant partial candidate nogoods and achieve convergence across all scenarios quickly. Our heuristic focuses on the first stage variables that have already converged in the current pre-solution. The procedure SELECTANDFIX in Algorithm 3 implements the heuristic. First, it picks all the first stage variables that have already converged, including those fixed in previous steps of this dive (line 5 right side of ∧). Secondly, an additional first stage variable assignment is chosen (line 5 left side of ∧), based on the variable/value combination that occurs most often over all scenarios. Given the current pre-solution S, it first constructs a mapping *Count* from variables to multisets of their assignments (line 3). Thereafter a variable/value combination is picked that occurs most often, but is not converged yet (line 4). In example, if variable x_3 is assigned to the value 4 in three scenario solutions, and to the value 7 in another two, then *Count* would contain the pair $< x_3, \{4, 4, 4, 7, 7\} >$. The value 4 would be assigned to x_3 assuming it is the most prevalent variable/value combination over all first stage variables. Finally, we construct a constraint that assigns x_e to v_e, in addition to assigning all variables that have converged across all scenarios (line 5).

Algorithm 3. Diving heuristic

1: **procedure** SELECTFIXED(S,V)

2:     $Vals = < \{\sigma_i(x) : \sigma_i \in S\} : x \in V >$

3:     $Count = < card(\{i : \sigma_i(x) = val, \sigma_i \in S\}) :< x \in V, val \in Vals_x >>$

4:     $< x_e, v_e > = \underset{\substack{<x,v> \\ val \in Vals_x \\ Count_{<x,v>} < k}}{\arg\max} \ Count_{<x,val>}$

5:     $c = (x_e = v_e \wedge \underset{\substack{x \in V \\ val \in Vals_x \\ Cout_{<x,v>} = k}}{\bigwedge} x = v)$

6:     **return** c

Scenario Bundling: The final extension of Ahmed's algorithm is based on the observation that the sub-problems in any scenario decomposition method can in fact comprise multiple scenarios, as long as we can find the optimal *stochastic* solution for that subset of scenarios in each iteration. We call this *scenario bundling*.

Since EVALUATEANDCUT has $O(k^2)$ behaviour for k scenarios, bundling can have a positive effect on the runtime, as long as the solving time for each bundle is not significantly higher than that for an individual scenario. Furthermore, the bundling of scenarios yields better lower bounds, since each component of a pre-solution is now an optimal stochastic solution for a subset of scenarios, which is guaranteed to be worse than the individual scenario objectives. By bundling scenarios in this way, we can therefore combine the fast convergence of EVALUATEANDCUT for large numbers of scenarios with the good performance of methods such as the DE on low numbers of scenarios. Scenario bundling has been applied to progressive hedging [9] and to EVALUATEANDCUT in [19].

4   Experiments

This section reports on our empirical evaluation of the algorithms discussed above. As a benchmark set we use a stochastic assignment problem with recourse similar to the stochastic generalised assignment problem (SGAP) described in [2]. A set of jobs, each composed of multiple tasks, is to be scheduled on a set of machines. Precedence constraints ensure that the tasks in a job are executed sequentially. Furthermore, tasks may be restricted to a sub-set of machines.

The processing time of the tasks varies across the set of machines, and in the stochastic version of the problem, this is a random variable. In the first stage, tasks must be assigned to machines. An optimal schedule with respect to the random variables is created in the second stage. The objective is to find a task to machine assignment minimizing the expected makespan over all scenarios.

Work on the SGAP with uncertainty on whether a job must be executed is described in [2]. However, to the best of our knowledge, there are no public benchmarks for the SGAP with uncertain processing times. Benchmarks for our experiments are created as described for deterministic flexible job shop instances in [5]. The scenarios are created by multiplying the base task durations by a number drawn from a uniform distribution with mean 1.5, variance 0.3, a lower limit of 0.9 and upper limit of 2.

We modelled the benchmark problems in MiniZinc. Each scenario is described in a separate MiniZinc data file, and compiled separately. This enables the MiniZinc compiler to optimise the COPs individually before solving. The models use a fixed search strategy (preliminary studies using activity based search did not improve the performance). The solver is implemented using Chuffed [8] and Python 2.7. The scenarios are solved using Chuffed, and learned nogoods are kept for subsequent iterations to implement vertical learning. A Python script coordinates the scenarios and dives. Up to twenty scenarios are solved in parallel. The experiments were carried out on a 2.9 GHz Intel Core i5, Desktop with 8 GB running OSX 10.12.1. A timeout of 3600 s was used.

4.1 Results

Table 1 contains the results for 9 representative problem instances with a range of 20 to 400 scenarios. In every instance, the optimal solution is found within the first few iterations and the remaining time is used to prove optimality. No results for the deterministic equivalent are presented as the run time is not competitive once the number of scenarios exceeds 20.

The impact of diving: The first two rows per instance contain the time it takes to solve the problem instances without scenario bundling. No substantial time difference can be reported for finding the optimal solution when dives are enabled. However, using dives improves the overall search performance in every instance. Figure 1 contains plots displaying the impact of dives when solving 100 scenarios without bundling.

The monotonically increasing graphs are the lower bounds. The horizontal black line is the optimal solution. The upper bound progress is not displayed, as the optimal solution is always found within a few seconds. Once the lower bound meets the upper bound, optimality is proven and the search terminates. The lower bound increases quickly at the beginning of the search. Over time the *evaluate and cut* method without diving flattens and the lower bound converges slowly towards upper bound. In strong contrast is the progress of the lower bound when using diving. At first during the initial iterations, the partial candidate nogoods are not showing any effects and the two curves are similar. However,

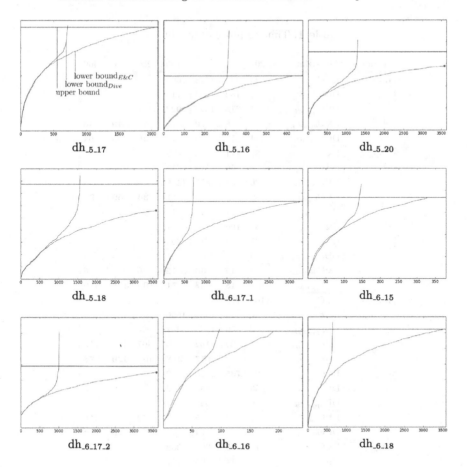

Fig. 1. The impact of dives

after the initial phase, generated strong partial candidate nogoods are paying off and the lower bound jumps drastically.

Using DE to proof optimality: The third row in each instance displays the times it takes to prove optimality using the deterministic equivalent. The upper bounds obtained from EVALUATEANDCUT are used to constrain the stochastic objective function in the DE. Similarly to using the DE directly, it does not scale with an increased number of scenarios. Furthermore, in contrast to EVALUATE-ANDCUT no optimality gap is produced.

Scenario bundling: The last two rows per instance contain the time it takes to solve the instances using scenario bundling. Four scenarios are randomly grouped to form a DE. Each scenario group becomes a sub-problem solved with EVALUATEANDCUT with and without diving enabled. As a result, the run time decreases in every instance. For the benchmark instances, diving is less powerful when using scenario bundles. This can be explained by the decreasing number

Table 1. Time to prove optimality [sec]

Instance	Algorithm	20	40	80	100	200	300	400
dh_5_17	E&C	24	343	1748	2060			
	Dive	12	152	584	716			
	$DE_{upperBound}$	7	215	1342	1923	-	-	-
	$E\&C_{bundle}$	1	**8**	**19**	17	27	**49**	**46**
	$Dive_{bundle}$	1	11	20	17	**26**	45	64
dh_5_16	E&C	36	111	436	638			
	Dive	23	56	212	318			
	$DE_{upperBound}$	20	131	877	1136	-	-	-
	$E\&C_{bundle}$	**2**	**5**	**9**	**11**	**34**	**69**	**91**
	$Dive_{bundle}$	3	7	12	13	41	72	93
dh_5_20	E&C	338	196	-	-			
	Dive	87	880	713	1304			
	$DE_{upperBound}$	43	293	2355	-	-	-	-
	$E\&C_{bundle}$	**3**	**17**	**36**	72	240	423	645
	$Dive_{bundle}$	7	18	43	**71**	**190**	**328**	**510**
dh_5_18	E&C	163	719	-	-			
	Dive	50	179	984	1566			
	$DE_{upperBound}$	**3**	50	1547	3467	-	-	-
	$E\&C_{bundle}$	7	**31**	162	226	407	712	1233
	$Dive_{bundle}$	8	32	**157**	**218**	**401**	**629**	**873**
dh_6_17_1	E&C	94	735	2328	3242			
	Dive	33	152	447	712			
	$DE_{upperBound}$	6	62	647	1073	-	-	-
	$E\&C_{bundle}$	2	3	15	15	59	91	156
	$Dive_{bundle}$	2	3	**14**	15	**52**	**75**	**114**
dh_6_15	E&C	10	35	209	338			
	Dive	5	14	95	147			
	$DE_{upperBound}$	5	69	495	1142	-	-	-
	$E\&C_{bundle}$	1	**3**	**16**	**21**	67	81	**138**
	$Dive_{bundle}$	1	4	19	25	**61**	**66**	183
dh_6_17_2	E&C	253	755	3345	-			
	Dive	58	153	725	1014			
	$DE_{upperBound}$	**0**	31	477	386	2159	-	-
	$E\&C_{bundle}$	7	8	33	49	89	182	303
	$Dive_{bundle}$	6	**7**	**24**	**24**	**48**	**121**	**178**
dh_6_16	E&C	21	69	141	214			
	Dive	13	36	71	97			
	$DE_{upperBound}$	12	111	777	1921	-	-	-
	$E\&C_{bundle}$	11	**24**	**29**	**42**	125	206	336
	$Dive_{bundle}$	11	25	37	54	**76**	**122**	**199**
dh_6_18	E&C	134	672	2140	3548			
	Dive	44	150	454	645			
	$DE_{upperBound}$	9	157	1790	3323	-	-	-
	$E\&C_{bundle}$	**7**	**11**	**19**	**29**	**120**	**219**	338
	$Dive_{bundle}$	9	20	24	31	131	212	**310**

Table 2. Speedup using vertical learning

		20	20	40	40	80	80	100	100	200	300	400
E&C	mean	23.4	29.6	15.8	19.2	9.7	27.4	7.2	26.0	21.8	19.6	17.6
	variance	6.0	17.2	5.0	37.1	2.4	13.8	3.3	12.0	10.7	10.6	8.4
Dive	mean	21.4	32.7	13.7	28.7	8.1	29.9	5.9	23.8	17.1	14.7	12.4
	variance	6.3	23.6	4.0	21.9	2.4	16.6	2.1	11.5	8.9	6.8	5.8

Speedup: 100 - 100 / time no learning * time with learning
Colored: Speedup in [%] when using vertical learning without bundling scenarios
White: Speedup in [%] when using vertical learning and bundling scenarios

of iterations required to find the optimal solution. More time is spend solving the sub-problems and less effort is required to coordinate the scenarios.

Vertical learning: Table 2 shows the speed-up when using vertical learning. Each column displays the speed-up over a scenario group. Overall the solving time decreased by 19.3% with 10.7% variance with an increase of at most 72%.

5   Conclusion

This paper has presented the first application of Ahmed's scenario decomposition algorithm [1] in a CP setting. Furthermore, we have introduced multiple algorithmic innovations that substantially improve the EVALUATEANDCUT algorithm. The most significant improvement is the partial search to create strong candidate nogoods. All our algorithmic innovations can be implemented in a parallel framework.

To further strengthen the EVALUATEANDCUT algorithm we will continue to work on the following ideas. First, the heuristic used to determine the variables to be fixed strongly impacts the performance of the search. A good heuristic is able to produce strong, relevant nogoods. Introducing a master problem that enables a tree search and improved coordination is worthwhile exploring. For the results we have used the standard system to manage nogoods within Chuffed. Further analysing to role of nogoods will help us understand their impact and how we can use nogoods to improve vertical learning and efficiently incorporate inter-instance learning.

References

1. Ahmed, S.: A scenario decomposition algorithm for 0–1 stochastic programs. Oper. Res. Lett. **41**(6), 565–569 (2013)
2. Albareda-Sambola, M., Van Der Vlerk, M.H., Fernández, E.: Exact solutions to a class of stochastic generalized assignment problems. Eur. J. Oper. Res. **173**(2), 465–487 (2006)

3. Balafoutis, T., Stergiou, K.: Algorithms for stochastic CSPs. In: Benhamou, F. (ed.) CP 2006. LNCS, vol. 4204, pp. 44–58. Springer, Heidelberg (2006). doi:10. 1007/11889205_6

4. Birge, J.R., Louveaux, F.: Introduction to Stochastic Programming. Springer Science & Business Media, New York (2011)

5. Brandimarte, P.: Routing and scheduling in a flexible job shop by tabu search. Ann. Oper. Res. **41**(3), 157–183 (1993)

6. CarøE, C.C., Schultz, R.: Dual decomposition in stochastic integer programming. Oper. Res. Lett. **24**(1), 37–45 (1999)

7. Chu, G., Stuckey, P.J.: Inter-instance nogood learning in constraint programming. In: Milano, M. (ed.) CP 2012. LNCS, pp. 238–247. Springer, Heidelberg (2012). doi:10.1007/978-3-642-33558-7_19

8. Chu, G.G.: Improving combinatorial optimization. Ph.D. thesis, The University of Melbourne (2011)

9. Crainic, T.G., Hewitt, M., Rei, W.: Scenario grouping in a progressive hedging-based meta-heuristic for stochastic network design. Comput. Oper. Res. **43**, 90–99 (2014)

10. Frisch, A.M., Harvey, W., Jefferson, C., Martínez-Hernández, B., Miguel, I.: Essence: a constraint language for specifying combinatorial problems. Constraints **13**(3), 268–306 (2008)

11. Goemans, M., Vondrák, J.: Stochastic covering and adaptivity. In: Correa, J.R., Hevia, A., Kiwi, M. (eds.) LATIN 2006. LNCS, vol. 3887, pp. 532–543. Springer, Heidelberg (2006). doi:10.1007/11682462_50

12. Hokama, P., San Felice, M.C., Bracht, E.C., Usberti, F.L.: A heuristic approach for the stochastic steiner tree problem (2014)

13. Hvattum, L.M., Løkketangen, A.: Using scenario trees and progressive hedging for stochastic inventory routing problems. J. Heuristics **15**(6), 527–557 (2009)

14. Manandhar, S., Tarim, A., Walsh, T.: Scenario-based stochastic constraint programming. arXiv preprint arXiv:0905.3763 (2009)

15. Nethercote, N., Stuckey, P.J., Becket, R., Brand, S., Duck, G.J., Tack, G.: MiniZinc: towards a standard CP modelling language. In: Bessière, C. (ed.) CP 2007. LNCS, vol. 4741, pp. 529–543. Springer, Heidelberg (2007). doi:10.1007/ 978-3-540-74970-7_38

16. Ohrimenko, O., Stuckey, P.J., Codish, M.: Propagation via lazy clause generation. Constraints **14**(3), 357–391 (2009)

17. Rendl, A., Tack, G., Stuckey, P.J.: Stochastic MiniZinc. In: O'Sullivan, B. (ed.) CP 2014. LNCS, vol. 8656, pp. 636–645. Springer, Cham (2014). doi:10.1007/ 978-3-319-10428-7_46

18. Rockafellar, R.T., Wets, R.J.B.: Scenarios and policy aggregation in optimization under uncertainty. Math. Oper. Res. **16**(1), 119–147 (1991)

19. Ryan, K., Ahmed, S., Dey, S.S., Rajan, D.: Optimization driven scenario grouping (2016)

20. Ryan, K., Rajan, D., Ahmed, S.: Scenario decomposition for 0–1 stochastic programs: improvements and asynchronous implementation. In: Parallel and Distributed Processing Symposium Workshops, pp. 722–729. IEEE (2016)

21. Toth, P., Vigo, D.: Vehicle Routing: Problems, Methods, and Applications, vol. 18. Siam, Philadelphia (2014)

22. Watson, J.P., Woodruff, D.L.: Progressive hedging innovations for a class of stochastic mixed-integer resource allocation problems. Comput. Manage. Sci. **8**(4), 355–370 (2011)

Optimal Stock Sizing in a Cutting Stock Problem with Stochastic Demands

Alessandro Zanarini[(✉)]

ABB Corporate Research Center, Baden-Dättwil, Switzerland
alessandro.zanarini@ch.abb.com

Abstract. One dimensional cutting stock problems arise in many manufacturing domains such as pulp and paper, textile and wood. In this paper, a new real life variant of the problem occuring in the rubber mold industry is introduced. It integrates both operational and strategical planning optimization: on one side, items need to be cut out of stocks of different lengths while minimizing trim loss, excess of production and the number of required cutting operations. Demands are however stochastic therefore the strategic choice of which mold(s) to build (i.e. which stock lengths will be available) is key for the minimization of the operational costs. A deterministic pattern-based formulation and a two-stage stochastic problem are presented. The models developed are solved with a mixed integer programming solver supported by a constraint programming procedure to generate cutting patterns. The approach shows promising experimental results on a set of realistic industrial instances.

1 Introduction

Classical one-dimensional Cutting Stock Problems (CSP) consist of minimizing the trim loss in the manufacturing process of cutting small items out of a set of larger size stocks. Each item (also referred to as small object) $i = 1, \ldots, I$ has an associated demand d_i, and a length l_i; the stock (also referred to as large object or roll) has a length L out of which one or more items can be cut out. The objective is to utilize the minimum amount of stocks, i.e. minimize the trim loss (problem type 1/V/I/R, according to Dyckhoff's classification [1]). Many different variations of the classical CSP have been proposed since the first formulation was introduced by Kantorovich in 1939 [2], including stock of different lengths, setup costs, open stock minimization, number of pattern minimization, to name just a few (see [3] for a survey).

In this paper, we present a new variation of the problem arising in the production of a set of specific items in the rubber mold industry. In this manufacturing process, a mold is used to create one stock typology of a given length. From the mold, several stocks can be produced and they are then cut by an operator (with the help of a cutting machine) in multiple pieces to meet the individual item demands. Possibly, the left-over of the initial stock may be discarded (trim loss) in case its trim length does not correspond to the length of any required

© Springer International Publishing AG 2017
D. Salvagnin and M. Lombardi (Eds.): CPAIOR 2017, LNCS 10335, pp. 293–301, 2017.
DOI: 10.1007/978-3-319-59776-8_24

items; alternatively, if the remaining length matches one of the item lengths and the demand for that particular item is already met, it is stored in inventory (with an associated cost) as excess of production for future use. Due to process constraints, inventory items cannot be further reworked, i.e. they cannot be re-cut to meet future demands of smaller items. The operational planning optimization consists of minimizing the trim loss, the over-production (i.e., inventory costs) and the number of cuts to be performed by the operator.

The peculiarity of the problem at hand comes from the fact that the mold(s) also needs to be built; that is, the stock length(s) has to be decided up-front in order to meet the future demands. In the unrealistic case of infinite budget and capacity, the trivial optimal solution would be to produce one mold for each item length. In reality, however, the number of molds that can be manufactured is limited. This leads to an integrated strategical and operational real-life planning problem that, to the best of the author's knowledge, has not been explored before in the literature.

The contributions of the paper are: the introduction and formalization of a new real-life cutting stock problem and its solution via an integrated strategical and operational CSP model with either deterministic or stochastic demands. The problems have been solved with a Mixed Integer Programming (MIP) solver supported by a Constraint Programming (CP) procedure to generate different cutting patterns.

The remainder of the paper is organized as follows: Sect. 2 summarizes the relevant literature; Sect. 3 formalizes the strategical and operational planning problem; experimental results are shown in Sect. 4; Sect. 5 draws some conclusions and describes future work.

2   Background

One-dimensional cutting stock problems have been extensively studied in the literature; several approaches have been proposed such as approximation algorithms, heuristics and meta-heuristics, population-based methods, constraint programming, dynamic programming and mathematical programming (please refer to the recent survey in [4] for an ample review and comparison). Among many, the most employed mathematical formulations are either *item-based* (originally introduced in [2]) or *pattern-based* (see [5]). In the former, a binary variable is used to indicate whether item i is cut out of stock j. In the latter, a cutting pattern defines a priori how many items and which item types are cut out of a single stock; then an integer variable i represents the number of stocks that are cut according to a given pattern i. As the number of cutting patterns grows exponentially with the average number of items that can fit into a stock, the problem is typically solved with column generation approaches. Pattern-based branch-and-price and branch-and-cut-and-price algorithms are considered the state-of-the-art complete methods (see [4,6]).

Very few research papers investigate uncertainty of demands arising in cutting stock problems. Kallrath et al. [8] studied a real-life problem arising in the paper industry where they minimize the trim loss and the number of patterns used, while not allowing over production. They solved the problem using a column generation approach that falls back to column enumeration in certain cases (i.e. precomputing all the columns explicitly). The column enumeration is claimed to be easily applicable to MILP problems, easier to implement and maintain and at times more efficient when the pricing problem becomes difficult or when the number of columns is relatively small (see also [9]). In the same paper, they also introduced a two-stage stochastic version of the problem with uncertain demands: the first stage variables represent the patterns employed, and in the second stage the related production is defined. They employ column enumeration in a sample average approximation framework; unfortunately, no experimental results are shown for the stochastic problem. Beraldi et al. studied in [10] a similar problem for which they proposed an ad-hoc approach designed to exploit the specific problem structure. Alem et al. in [7] also considered stochastic CSP, where in the first stage the patterns and production plans are decided and the second stage decision determines over and under production.

The main differences between the problems already studied in the literature and this contribution are: firstly, the main cost drivers are the trim loss and the allowed over production; as a secondary objective the number of cuts to be performed (as opposed to the numbers of patterns utilized); secondly, the molds to be built are the key initial investment and they determine the available stock lengths (as opposed to predefined stock lengths).

In order to clarify the difference in solutions between minimizing the total number of cuts or alternatively the number of patterns, let's consider an example in which the stock length is equal to 16 and two items, of lengths 7 and 8 respectively, need to be produced with a demand of 2 each. All the solutions that are optimal in term of trim loss require 2 stocks. The solution that minimizes the number of patterns employs only one pattern $(8, 7)$; it will be used twice to meet the demand and the production will require a total of 4 cutting operations. Oppositely, the minimization of the number of cuts will make use of one pattern $(8, 8)$ and one pattern $(7, 7)$; the first pattern requires only one cut to produce two items, whereas the second two cuts.

3   Problem Formulation

In this section the problem is formalized: at first the deterministic operational planning problem is shown followed by the integrated strategical and operational planning problem. All the approaches use a pattern-based modelling[1].

---

[1] For comparison, an item-based model was also developed; it showed quickly its limitations even in relatively small instances as soon as included the over production. For brevity, we omit the item-based model and results.

3.1   Operational Planning with Deterministic Demands

In this section, the lengths of the molds is assumed to be already defined and the problem is to find the optimal production plan fulfilling some deterministic demands. We will use the following notation:

Parameters and Variables

I	The total number of different items to be produced; the items are indexed by $i = 1, \ldots, I$
l_i	The length of the item i
d_i	The demand for item i
c_i	The cost for over producing item i
M	The total number of molds; molds are indexed by $m = 1, \ldots, M$
L_m	The length of the stock produced with mold m
P_m	The total number of patterns related to the stock m; the patterns are indexed by $j = 1, \ldots, P_m$
p_{mji}	The number of items i produced with pattern j of mold m
w_{mj}	The amount of trim loss caused by pattern j of mold m
o_{mj}	The number of cutting operations needed for pattern j of mold m
W	The trim loss cost per unit of length
α	The weight on the secondary objective function, i.e. number of cutting costs ($\alpha \ll 1$)
x_{mj}	The integer non-negative variable indicating the number of times pattern j of mold m is used in the final solution

Note that the cost of over production is linked to the length l_i (the longer the higher will be the cost) and to the demand of item i, i.e. the higher is its demand, the lower will be the inventory cost; in general: $W \cdot l_i \geq c_i \geq 0$. The real industrial problem analyzed in this paper exhibits integer item and stock lengths. The following mathematical model formalizes the problem for the optimal production planning:

$$\min \sum_{m=1}^{M} \sum_{j=1}^{P_m} \left(W \cdot w_{mj} \cdot x_{mj} + \alpha \cdot o_{mj} \cdot x_{mj} \right) + \sum_{i=1}^{I} c_i \cdot \left(\sum_{m=1}^{M} \sum_{j=1}^{P_m} p_{mji} \cdot x_{mj} - d_i \right)$$

(1)

$$\sum_{m=1}^{M} \sum_{j=1}^{P_m} p_{mji} \cdot x_{mj} \geq d_i, \qquad i = 1, \ldots, I$$

(2)

$$x_{mj} \in \mathbb{Z}$$

(3)

The objective function in Eq. (1) is composed by three main elements: the waste produced ($w_{mj} \cdot x_{mj}$), the number of cuts required ($o_{mj} \cdot x_{mj}$) and lastly, the over production. Equation (2) constrains the production of item i to meet its demand d_i.

Pattern Enumeration. An early analysis of the industrial instances showed that for relevant mold sizes, the number of feasible cutting patterns is relatively contained (about up to 50). For this reason and for the simplicity of the approach, we decided to enumerate exhaustively all the patterns (see [8,9]). The pattern enumeration is a pure constraint satisfaction problem that has been solved with Constraint Programming.[2] Once all the patterns have been generated, they are fed as input to the mathematical model defined in Sect. 3.1. The parameters p_{mji}, w_{mj} and o_{mj} of the previous section become variables for this sub-problem, respectively:

- $Z_i \in \{0, \ldots, \lfloor L_m/l_i \rfloor\}$ represents the number of units produced for item i
- $Q \in \{0, \ldots L_m\}$ represents the amount of trim loss for the pattern
- $O \in \{0, \ldots L_m - 1\}$ indicates the number of cut operations required

The model follows:

$$L_m = Q + \sum_{i=1}^{I} l_i \cdot Z_i \tag{4}$$

$$Q \neq q \cdot l_i, \qquad q = 1, \ldots, \lfloor L_m/l_i \rfloor, \quad i = 1, \ldots, I \tag{5}$$

$$O = \sum_{i=1}^{I} Z_i - 1 + (Q > 0) \tag{6}$$

Equation (4) constrains the waste and the sum of the item lengths to be equal to the mold size. The set of Eq. (5) avoids that the waste is equal to (or a multiple of) the size of one of the items in demand. This restricts the number of feasible patterns and it is valid as long as $c_i \leq W \cdot l_i$; that is, the inventory cost of an item of length l_i is at most as expensive as throwing away the same quantity.[3] Finally, Eq. (6) defines the number of cuts to be equal to the number of items in the pattern minus 1; an additional cut is required if there is also a final trim (reified constraint).

3.2 Strategical and Operational Planning Under Stochastic Demands

In this industrial context the main practical challenge is that an initial investment is required for the construction of the molds that will be used to meet future stochastic demands. The problem is a two-stage stochastic problem in which the first stage decision variables are the mold to be built, and the second-stage is the actual production plan once the demands are known.

We follow a sample average approximation approach in which a set of scenarios $s \in \Omega$ with corresponding probabilities $\sum_{s \in \Omega} p_s = 1$ is Monte Carlo sampled. The vector of demands d_i^s now depends on the specific scenario realization. We

---

[2] For comparison an equivalent MIP model was also developed, however it performed orders of magnitude slower for enumerating all the solutions.

[3] This was the case in the real industrial context examined.

further introduce the binary variables y_m indicating whether mold m is produced or not. Due to manufacturing and operational constraints the molds considered have to be limited to a maximum length $\mathbb{L}$ $(L_m \leq \mathbb{L})$. The model follows:

$$\min \sum_{s \in \Omega} p_s \left(\sum_{m=1}^{M} \sum_{j=1}^{P_m} (W \cdot w_{mj} \cdot x_{mjs} + \alpha \cdot o_{mj} \cdot x_{mjs}) + \sum_{i=1}^{I} c_i \cdot \left(\sum_{m=1}^{M} \sum_{j=1}^{P_m} p_{mji} \cdot x_{mjs} - d_{is} \right) \right)$$

$$\sum_{m=1}^{M} y_m \leq \mathbb{M} \tag{7}$$

$$x_{mjs} \leq B \cdot y_m, \qquad s \in \Omega, m = 1, \ldots, M, j = 1, \ldots, P_m \tag{8}$$

$$\sum_{j=1}^{P_m} x_{mjs} \geq y_m, \qquad s \in \Omega, m = 1, \ldots, M \tag{9}$$

$$\sum_{m=1}^{M} \sum_{j=1}^{P_m} p_{mji} \cdot x_{mjs} \geq d_{is}, \qquad s \in \Omega, i = 1, \ldots, I \tag{10}$$

$$y_m \in \{0, 1\}, \quad x_{mjs} \in \mathbb{Z} \tag{11}$$

where $\mathbb{M}$ is the maximum number of molds that can be built (Eq. (7)). Equations (8) and (9) indicate that if a mold is not built then none of its pattern can be employed, and vice versa, if only one pattern is used then the associated mold must be built (B is a large number). Finally, Eq. (10) constrains the production to meet the demand for each possible scenario.

For pure cutting stock problems minimizing the number of stocks used, a still open conjecture, the *Modified Integer Round-Up Property* (MIRUP), states that the integral optimal solution is close to the linear relaxation: $z_{opt} - \lceil L_{LP} \rceil \leq 1$ (see [4] and [11]). In order to render the second stage sub-problem computationally tractable, the integrality constraints on x_{mjs} have been lifted (y_m are kept binary).

4   Experimental Results

The MIP and CP model have been developed using Google OR-Tools. The MIP solver employed is the open source CBC. All the experiments have been conducted on a MacBook Pro with an Intel Dual-core i5-5257U and 8 GB of RAM.

Pattern Enumeration. Creating all the possible patterns for all the molds ranging from length 1 to 16 (lengths relevant in the studied industrial context) takes 30 ms for a total combined number of patterns of about 200. For reference, the enumeration of the patterns for molds of lengths 15, 25 and 35 takes respectively 2, 28 and 300 ms, to enumerate resp. 40, 328 and 1995 patterns.

Operational Planning with Deterministic Demands. We generated 2250 synthetic instances with $\mathbb{L} = 20$, and random demands (the ratio between the maximum demand and the minimum demand - $\frac{\max_i d_i}{\min_i d_i}$ - ranged from 3 to 20); in this deterministic version the molds are predefined (M ranged from 1 to 3).

For brevity, we will not report the detailed results but just some observations. Firstly, all the generated instances but one were solved to optimality within one second (with all the integrality constraints); evidently, the item and stock lengths at play in this domain are not challenging for the MIP solver.

Secondly, we compared the optimal objective value of the integer solution with the one from the linear relaxation: the difference between the two is at most 0.03%. This is reassuring for the two-stage stochastic approach in which the integrality constraints on the second-stage decision variables have been lifted.

Thirdly, we used the deterministic model to analyze the variability of the number of cutting operations on a real industrial instance after the trim loss and over production had been fixed to their respective optimal values. Despite being a secondary objective, the best and worst solutions showed as much as almost 10% of difference in the number of cutting operations; in the analyzed industrial context, this translates to about 50 h of reduced time in cutting operations for producing the same quantity of items (about 150 thousands).

Strategical and Operational Planning Under Stochastic Demands. We generated the stochastic instances starting from a real industrial demand vector. We perturbated it using a gaussian distribution in order to create 20 different demand vectors. Each of them represents the mean demand vector, on which another set of gaussian distributions are centered for the scenario generation; their standard deviations is proportional to the demands: $\sigma_i = \frac{d_i}{k}$. We tested different standard deviation with $k \in \{1, 3, 5\}$, scenario set cardinalities, $|\Omega| \in \{10, 20, 50\}$, and number of molds to be constructed, $\mathbb{M} \in \{3, 4\}$; $\mathbb{L}$ and $\max_i\{d_i\}$ are both set to 16, as per the industrial setting. The total number of tests amount to 360 instances.

In order to evaluate the quality of the stochastic approach, we computed the Expected Value of Perfect Information (EVPI) and the Value of the Stochastic Solution (VSS). The EVPI represents how much one could gain with a wait-and-see strategy, i.e. the expected decrease of the objective value in case of a priori knowledge of the stochastic variable realizations; a high EVPI connotes that the stochastic approach is not capable of capturing well the uncertainty on the demands. The VSS indicates the expected increase of the objective value when using a deterministic optimization fed with the expected values of the stochastic variables (see [12] for a general procedure to compute it); a high VSS means that capturing uncertainty with the stochastic approach is actually bringing a benefit; oppositely, a low VSS shows that a deterministic optimization using expected values leads to a solution that is similar to the one computed by the stochastic approach.

The results are presented in Table 1: each row reports aggregated values for 20 instances; the first three columns represent the instance parameters (described at the beginning of the paragraph); four pairs of columns follow reporting averages

Table 1. Aggregated results for 360 instances.

M	k	$|\Omega|$	Time (secs) μ	σ	Obj μ	σ	EVPI μ	σ	VSS μ	σ
3	1	10	2.7	0.2	25769.8	5042.6	10818.1	3484.8	4185.0	8861.5
3	1	20	6.2	0.6	28142.6	5993.8	13942.7	4252.7	8975.3	11289.0
3	1	50	21.2	2.1	26616.4	4261.9	13106.8	3046.9	9861.5	13553.5
3	3	10	2.8	0.5	13600.8	4601.2	2392.0	1861.2	457.8	1602.4
3	3	20	7.0	1.1	13560.2	4650.3	3185.7	2031.3	375.0	1148.8
3	3	50	23.6	2.8	13741.9	4276.0	3415.7	2146.0	977.4	2861.5
3	5	10	2.8	0.5	11991.5	4517.3	1146.3	1363.7	66.3	215.6
3	5	20	7.3	0.9	11777.4	4502.6	1544.0	1474.0	16.0	69.6
3	5	50	26.2	4.4	11793.1	4274.3	1553.6	1494.7	252.1	876.7
4	1	10	2.3	0.2	9628.9	3423.0	5796.7	2226.7	3321.9	5148.8
4	1	20	5.7	0.4	11546.6	3550.4	8101.1	2601.3	11531.5	11090.0
4	1	50	21.4	2.3	11492.7	2635.8	7864.9	1764.7	11942.2	12523.1
4	3	10	2.4	0.2	4383.6	2217.2	1006.9	823.7	231.7	945.6
4	3	20	6.1	0.6	4738.9	2240.3	1626.1	1091.7	490.6	1265.9
4	3	50	22.6	2.1	4669.2	1995.9	1649.8	1003.8	1071.8	2142.2
4	5	10	2.4	0.2	3748.4	2018.3	441.2	517.9	17.1	74.4
4	5	20	6.0	0.6	3857.9	1970.7	745.8	626.0	149.2	363.7
4	5	50	20.9	3.0	3812.3	1826.7	783.7	607.0	164.1	492.4

and standard deviations of respectively the solution time, the objective value, the EVPI and the VSS.

As expected, by decreasing the level of stochasticity (lower standard deviation of the scenario generation, $k = 5$), the VSS drops significantly as well as the EVPI. Oppositely, as the scenario set cardinality grows, the VSS gets bigger in proportion than the EVPI. Finally, increasing $\mathbb{M}$ allows to significantly decrease the value of the objective function, though the VSS, in proportion, increases. From a computational standpoint, for reference, should the integrality constraints be kept, an instance takes more than one minute to solve (with $|\Omega| = 10$). The parameters that impact the most the solution time are $|\Omega|$ and $\mathbb{L}$. Increasing $|\Omega|$ to 100, 200 and 500 leads to a solution time of respectively about 65, 189 and 831 s. Similarly, increasing $\mathbb{L}$ (the real life problem presents $\mathbb{L} = 16$) to 20, 25 and 30 (with $|\Omega| = 10$) results to a solution time of respectively about 17, 162 and 852 s.

5   Conclusion

In this paper, we introduced a new two-stage stochastic problem arising in the production of some rubber elements in the rubber mold industry. The uniqueness comes from deciding the stock lengths before knowing the actual production demand. We believe this problem setup can be relevant for other domains as well where stock purchase orders need to be placed despite uncertainty on the demands. We formalized the problem and developed a solution that was able to solve real-life instances to optimality within an acceptable time. Future work includes the exploration of other optimization techniques to improve scalability, the integration of other real-life constraints and explore a multi-stage stochastic setup for planning the production and inventory for multiple time slots.

Acknowledgements. The author would like to thank Davide Zanarini for bringing to his attention this industrial problem.

References

1. Dyckhoff, H.: A typology of cutting and packing problems. Eur. J. Oper. Res. **44**, 145–159 (1990)
2. Kantorovich, L.V.: Mathematical methods of organizing and planning production. Manag. Sci. **6**(4), 366–422 (1960)
3. Wäscher, G., Hauner, H., Schumann, H.: An improved typology of cutting and packing problems. Eur. J. Oper. Res. **183**, 1109–1130 (2007)
4. Delorme, M., Iori, M., Martello, S.: Bin packing and cutting stock problems: mathematical models and exact algorithms. Eur. J. Oper. Res. **255**(1), 1–20 (2016)
5. Gilmore, P.C., Gomory, R.E.: A linear programming approach to the cutting-stock problem. Oper. Res. **9**(6), 849–859 (1961)
6. Belov, G., Scheithauer, G.: A branch-and-cut-and-price algorithm for one-dimensional stock cutting and two-dimensional two-stage cutting. Eur. J. Oper. Res. **171**(1), 85–106 (2006)
7. Alem, D.J., Munari, P.A., Arenales, M.N., Ferreira, P.A.: On the cutting stock problem under stochastic demand. Ann. Oper. Res. **179**(1), 169–186 (2010)
8. Kallrath, J., Rebennack, S., Kallrath, J., Kusche, R.: Solving real-world cutting stock-problems in the paper industry: mathematical approaches, experience and challenges. Eur. J. Oper. Res. **238**(1), 374–389 (2014)
9. Rebennack, S., Kallrath, J., Pardalos, P.M.: Column enumeration based decomposition techniques for a class of non-convex MINLP problems. J. Global Optim. **43**(2–3), 277–297 (2009)
10. Beraldi, P., Bruni, M.E., Conforti, D.: The stochastic trim-loss problem. Eur. J. Oper. Res. **197**(1), 42–49 (2009)
11. Scheithauer, G., Terno, J.: Theoretical investigations on the modified integer round-up property for the one-dimensional cutting stock problem. Oper. Res. Lett. **20**(2), 93–100 (1997)
12. Escudero, L.F., Garn, A., Merino, M., Prez, G.: The value of the stochastic solution in multistage problems. TOP **15**(1), 48–64 (2007)

Stochastic Task Networks
Trading Performance for Stability

Kiriakos Simon Mountakis[1], Tomas Klos[2], and Cees Witteveen[1(✉)]

[1] Delft University of Technology, Delft, The Netherlands
C.Witteveen@tudelft.nl
[2] Utrecht University, Utrecht, The Netherlands

Abstract. This paper concerns networks of precedence constraints between tasks with random durations, known as stochastic task networks, often used to model uncertainty in real-world applications. In some applications, we must associate tasks with reliable start-times from which realized start-times will (most likely) not deviate too far. We examine a dispatching strategy according to which a task starts as early as precedence constraints allow, but not earlier than its corresponding *planned release-time*. As these release-times are spread farther apart on the time-axis, the randomness of realized start-times diminishes (i.e. *stability* increases). Effectively, task start-times becomes less sensitive to the outcome durations of their network predecessors. With increasing stability, however, performance deteriorates (e.g. expected makespan increases). Assuming a sample of the durations is given, we define an LP for finding release-times that minimize the performance penalty of reaching a desired level of stability. The resulting LP is costly to solve, so, targeting a specific part of the solution-space, we define an associated Simple Temporal Problem (STP) and show how optimal release-times can be constructed from its earliest-start-time solution. Exploiting the special structure of this STP, we present our main result, a dynamic programming algorithm that finds optimal release-times with considerable efficiency gains.

Keywords: Activity network · Stochastic scheduling · Solution robustness

1 Introduction

A *stochastic task network* is a directed acyclic graph $G(V, E)$ with each node in $V = \{1, \ldots, n\}$ representing a task with a random duration and each arc $(i, j) \in E$ representing a precedence-constraint between tasks i and j, specifying that task j cannot start unless task i has finished. Such networks appear in several domains like project scheduling [16], parallel computing [22], or even digital circuit design [4], where there is a need to model a partial order of events with uncertain durations. Postulating that a model of uncertainty is known, task durations are described by a random vector $D = (D_1, \ldots, D_n)$ with a known

© Springer International Publishing AG 2017
D. Salvagnin and M. Lombardi (Eds.): CPAIOR 2017, LNCS 10335, pp. 302–311, 2017.
DOI: 10.1007/978-3-319-59776-8_25

probability distribution. In project scheduling, for example, the duration D_i of task i may turn out to be shorter or longer than a nominal value according to a certain distribution.

A given task network is typically mapped to a *realized schedule* (i.e. an assignment of start-times to tasks) via *earliest-start dispatching*; i.e. observing outcome durations and starting a task immediately when precedence-constraints allow (i.e. not later than the maximum finish-time of its network predecessors). Random durations make the realized start-time of a task (and the overall realized schedule makespan) also random. Since *PERT networks* [17], a large body of literature focused on the problem of determining the makespan distribution [1], eventually shown to be a hard problem [11]. A variety of efficient heuristics have been developed so far (see [4]), among which Monte Carlo sampling remains, perhaps, the most practical.

Consider, for example, the stochastic task network in Fig. 1, detailing the plan of a house construction project, assuming task durations are random variables that follow the uniform distribution within respective intervals. With earliest-start dispatching, the overall duration of the project (i.e. the realized schedule makespan) will range between 12 and 20 days with an expected value of a little over 16 days.

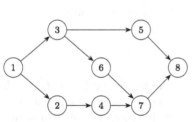

Tasks	Durations (days)
1. Erect walls	2-4
2. Finish walls	3-5
3. Finish roof	2-6
4. Install plumbing	3-5
5. Finish exterior	6-8
6. Install electricity	3-5
7. Paint interior	2-4
8. Finishing touches	1-1

(a) Example construction plan.   (b) Estimated task durations.

Fig. 1. A motivating example.

This paper addresses a problem which, to our knowledge, has not been addressed in existing literature. To motivate our problem, let us return to the earlier example and suppose task 7 ("Paint interior") is assigned to a painting crew charging $100 per day. Assume we are willing to hire them for at least 4 days (the maximum number of days they will need) and for at most 6 days; i.e. we have a budget of $600 for painting. With earliest-start dispatching, 7 may start within 8 to 15 days from the project start (the start-date of task 1). A challenge that arises in this situation is deciding when to hire the painting crew, because to allow for an expected makespan of a little over 16 days (as mentioned earlier), we must book the painting crew from the 8-th day and until the 19-th day, at the excessive cost of $1100. The solution we examine here, is to use a different dispatching strategy, associating task 7 with a *planned release-time*, t_7, before which it may not start even if installing plumbing and electricity are finished earlier than t_7. If we choose that 7 may not start earlier than, e.g., $t_7 = 13$ days from the project start, we only need to book the painting crew on the 13-th day until the 19-th day, for an acceptable cost of $600. However, the price to pay for this stability is an expected makespan increase to a little over 17 days.

Now suppose that after assessing our budget carefully it turns out that each task may deviate at most, say w days, from its respective planned release-time. The emerging question addressed in this paper is:

Which planned release-times reach the desired level of stability[1] while minimizing the incurred performance penalty?

This problem does not involve resource-constraints. However, task networks are often used in the area of resource-constrained scheduling under uncertainty (see [2,12]) to represent solutions (e.g. the *earliest-start policy* [13], the *partial-order schedule* [5,10,20]). Thus, our work is expected to be useful in dealing with associated problems, such as distributing slack in a resource-feasible schedule to make it insensitive to durational variability [8].

Organization. A formal problem statement and its LP formulation are presented in Sect. 2. As the resulting LP can be quite costly to solve, Sect. 3 presents our main result, an efficient dynamic programming algorithm. Section 4 concludes the paper and outlines issues to be addressed in future work.

2   Problem Definition

We are given a task network $G(V, E)$ and a stochastic vector $D = (D_1, \ldots, D_n)$ describing task durations. Let Q index the space of all possible realization scenarios for D such that d_{ip} denotes the realized duration of task i in scenario $p \in Q$. We assume to know the probability distribution of D; i.e. the probability $\mathbb{P}[D = (d_{1p}, \ldots, d_{np})]$ for all $p \in Q$. To limit the unpredictability of the realized schedule, we want to associate tasks with respective *planned release-times* $t = (t_1, \ldots, t_n)$ such that the realized schedule is formed by starting a task as early as permitted by precedence-constraints, but not earlier than its release-time. That is, the start-time s_{jp} of task j in scenario p will be determined as:

$$s_{jp} = \max[\max_{(i,j)\in E} (s_{ip} + d_{ip}), t_j] \tag{1}$$

Given a sample $\mathcal{P} \subseteq Q$ of size m of the stochastic durations vector, this paper is devoted to the following problem:

---

[1] As in Bidot et al. [3], stability here refers to the extent that a predictive schedule (planned release-times in our case) is expected to remain close to the realized schedule.

$$\min_{t \geq 0} \quad F := \sum_{j \in V, p \in \mathcal{P}} s_{jp} \qquad (P)$$

$$\text{subject to} \quad s_{jp} = \max[\max_{(i,j) \in E}(s_i + d_i), t_j] \qquad \forall j \in V, p \in \mathcal{P} \qquad (2)$$

$$s_{jp} - t_j \leq w \qquad \forall j \in V, p \in \mathcal{P} \qquad (3)$$

This problem tries to optimize a trade-off between stability and performance: release-times are sparsely spread in time in order to form a stable schedule, i.e. such that in every considered scenario a realized start-time will stay within w time-units from the corresponding release-time.

Since the whole space of possible duration realizations, $\mathcal{Q}$, may be too large, or even infinite, we only consider a manageable sample $\mathcal{P} \subseteq \mathcal{Q}$ during optimization.[2] At the same time, we want to ensure a minimal performance penalty $F - F^*$ where F^* denotes the throughput of earliest-start dispatching with no release-times.[3]

Instead of minimizing a standard performance criterion like expected makespan, we choose to maximize *expected throughput*, $\frac{1}{m} \frac{n}{\sum_{j,p} s_{jp}}$, which equals the average rate at which tasks finish over all scenarios. It can be shown that a schedule of maximum throughput is one of minimum makespan and/or tardiness (in case tasks are associated with deadlines). We maximize throughput indirectly by minimizing its inverse, with the constant $\frac{m}{n}$ omitted for simplicity.

LP Formulation. The resulting problem is not easy to handle due to the equality constraint, but using a standard trick it can be rewritten as the following linear program (LP):

$$\min_{s,t \geq 0} \quad F := \sum_{j \in V, p \in \mathcal{P}} s_{jp} \qquad (P)$$

$$\text{subject to} \quad s_{jp} \geq s_{ip} + d_{ip} \qquad (i,j) \in E, p \in \mathcal{P} \qquad (4)$$

$$s_{jp} \geq t_j \qquad j \in V, p \in \mathcal{P} \qquad (5)$$

$$s_{jp} - t_j \leq w \qquad j \in V, p \in \mathcal{P} \qquad (6)$$

Note that the solution-space of the resulting LP encompasses that of the original formulation. However, it is easy to show that both problems have the same set of optimal solutions, because a solution (s,t) for the LP cannot be optimal unless it satisfies (2).

Currently, the best (interior-point) LP solvers have a complexity of $O(N^3 M)$ where N is the number of variables and M the input complexity [21]. Thus, letting $\delta \leq n$ denote the max in-degree in $G(V, E)$, the cost of solving (P) as an LP with nm variables and $O(n\delta m)$ constraints can be bounded by $O(n^4 m^4 \delta) \subseteq$

---

[2] Knowing the distribution of D, we assume to be able to draw $\mathcal{P}$.

[3] The reader can easily recognize the similarity of the proposed LP with a so-called Sample Average Approximation (SAA) of a stochastic optimization problem [14].

$O(n^5m^4)$, which can be daunting even for small instances. Fortunately, as shown in the following section, we manage to obtain the substantially tighter bound of $O(n^2m)$ for solving (P), by exploiting its simple structure to devise a dynamic programming algorithm.

3   Fast Computation of Planned Release-Times

We first show that a fixed relationship between variables s_{jp} and t_j can be assumed while looking for an optimal (s,t). Based on this, a problem (P') is defined which can be solved instead of (P).

A Tighter Formulation. Begin by rewriting (6) as $t_j \geq \max_p s_{jp} - w, \forall j \in V$. Now, let Λ denote the set of all feasible (s,t) for problem (P) and let $\Lambda^* \subseteq \Lambda$ be that part of the solution-space that only contains (s,t) for which $t_j = \max_p s_{jp} - w$ for all j.

Lemma 1. *For every feasible $(s,t) \in \Lambda \setminus \Lambda^*$ there exists $(s',t') \in \Lambda^*$ with equal objective value.*

Proof. Consider feasible (s,t) with $t_j = \max_p s_{j^*p} - w + c$ with $c > 0$ for some j^*. Construct t' by letting $t'_j = t_j$ for all $j \neq j^*$ and $t'_{j^*} = t_{j^*} - c = \max_p s_{j^*p} - w$. Trivially, if (s,t) is feasible, so is (s,t'), with the same objective value. Keeping s fixed, we may repeat this construction to enforce that $t_j = \max_p s_{jp} - w$ for all j and have $(s,t') \in \Lambda^*$.

The previous result allows us to consider the following problem, obtained by substituting $\max_{p' \in \mathcal{P}} s_{jp'} - w$ for t_j in (P):

$$\min_{s \geq 0} \quad \sum_p s_{np} \qquad (P')$$

$$\text{subject to} \quad s_{jp} \geq s_{ip} + d_{ip} \qquad\qquad (i,j) \in E, p \in \mathcal{P} \qquad (7)$$

$$s_{jp} \geq \max_{p' \in \mathcal{P}} s_{jp'} - w \qquad\qquad j \in V, p \in \mathcal{P} \qquad (8)$$

$$s_{jp} - (\max_{p' \in \mathcal{P}} s_{jp'} - w) \leq w \qquad\qquad j \in V, p \in \mathcal{P} \qquad (9)$$

Clearly, $(s,t) \in \Lambda^*$ iff s is feasible for (P').[4] In other words, the solution-space of P' comprises only those s that can be paired with t by letting $t_j = \max_p s_{jp} - w$ to form a feasible (s,t) for (P). By Lemma 1, if s is optimal for (P'), then (s,t) is optimal for (P). Also, if (P) has a solution (i.e. if $G(V,E)$ is acyclic), then (P') also has a solution.

---

[4] Since $(s,t) \in \Lambda^*$ implies $\max_{p'} s_{jp'} - w = t_j$ for all j.

The Resulting STP. Formulation (P') is useful because it can be cast as a certain type of Temporal Constraint Satisfaction Problem (TCSP) [9]. We start by noting that (9) is always true and can be omitted. Moreover, (8) can be rewritten as (11), to obtain the following reformulation:

$$\min_{s \geq 0} \sum_p s_{np} \qquad (P')$$

$$\text{subject to} \quad s_{ip} - s_{jp} \leq -d_{ip} \qquad (i,j) \in E, p \in \mathcal{P} \qquad (10)$$

$$s_{jp} - s_{jp'} \leq w \qquad (p,p') \in \mathcal{P}^2, j \in V \qquad (11)$$

Constraints (10) and (11) effectively represent the solution-space of a Simple Temporal Problem (STP) [9] with temporal variables $\{s_{jp} : j \in V, p \in \mathcal{P}\}$. The structure of the resulting STP (specifically, of its *distance graph* [9]) is demonstrated in Fig. 2.

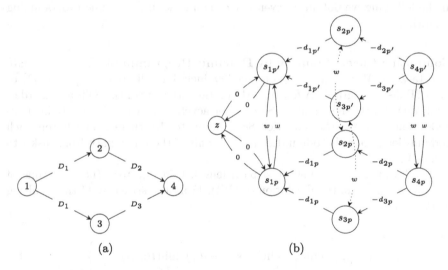

(a)   (b)

Fig. 2. Example task network (a) and resulting STP (b) for a sample $P = \{p, p''\}$.

The *earliest start time* (est) solution of any given STP (assuming it is consistent) assigns to each variable the smallest value it may take over the set of feasible solutions. Therefore, the est solution of the resulting STP optimally solves (P'), leading us to the following observation.

Observation 1. *By Lemma 1, an optimal solution (s,t) for (P) can be formed by finding the earliest start time solution s of the resulting STP and pairing it with t formed by letting $t_j = \max_{p \in \mathcal{P}} s_{jp} - w$ for all j.*

Algorithm 1

1: $l(s_{1p}) \leftarrow 0$ for all $p \in \mathcal{P}$
2: **for** each tier j in a topological sort of $G(V, E)$ **do**
3:     $k_{jp} \leftarrow \min\{l(s_{ip}) - d_{ip} : (i, j) \in E\}$ for all $p \in \mathcal{P}$
4:     $p^* \leftarrow \arg\min\{k_{jp} : p \in \mathcal{P}\}$
5:     $l(s_{jp}) \leftarrow \max\{k_{jp}, k_{jp^*} + w\}$ for all $p \in \mathcal{P}$
6: **end for**
7: $s_{jp} \leftarrow l(s_{jp})$ for all $j \in V, p \in \mathcal{P}$
8: $t_j \leftarrow \max_{p \in \mathcal{P}} s_{jp} - w$ for each $j \in V$

The est value of s_{jp} is the length of the shortest-path (in the distance graph) from (the node corresponding to) s_{jp} to the special-purpose variable z which is fixed to zero. Those values can be found with a single-source shortest-path algorithm (e.g. Bellman-Ford [19]) in time $O(NM)$ where N is the number of nodes and M the number of arcs. In our case, $N = nm$ and $M = O(nm\delta)$, yielding $O(n^3m^2)$; already a better bound than that of solving (P) as an LP. However, in the following we obtain an even better bound with a dynamic programming algorithm.

Computing the est Solution by Dynamic Programming. Let us associate each task $j \in V$ with a corresponding *tier* including all nodes $\{s_{jp} : p \in \mathcal{P}\}$ of the STP distance graph. A few remarks on the structure of the STP are in order. First, due to (11) the resulting STP is not acyclic, but each cycle only includes nodes that belong to the same tier. Second, due to (10) there is a path from each node in tier j to each node in tier i if and only if there is a path from task i to j in $G(V, E)$.

Let $l(s_{jp})$ denote the shortest-path length from s_{jp} to z (i.e. the value of variable s_{jp} in an optimal solution of (P')). From the structure of the resulting STP, we have:

$$l(s_{jp}) = \min \left\{ \min_{(i,j) \in E} (l(s_{ip}) - d_{ip}), \min_{p' \neq p} l(s_{jp'}) + w \right\} \tag{12}$$

The existence of cycles complicates solving subproblem $l(s_{jp})$ as it depends on (and is a dependency of) other subproblems $l(s_{jp'})$ in the same tier. However, we can "break" dependencies between subproblems in the same tier as shown below.

Define $k_{jp} := \min_{(i,j) \in E}(l(s_{ip}) - d_{ip})$ and $p^* := \arg\min_{p \in \mathcal{P}} k_{jp}$.

Lemma 2. $l(s_{jp}) = \min\{k_{jp}, k_{jp^*} + w\}$

Proof. Begin by noting that the shortest-path from s_{jp} to z visits at most one node $s_{jp'}$ from the same tier. As such, for every s_{jp} we have that: either $l(s_{jp}) = k_{jp}$, or $l(s_{jp}) = k_{jp'} + w < k_{jp}$ for some $p' \neq p$.

Now, note that $l(s_{jp^*}) = k_{jp^*}$, since if not (i.e. if $l(s_{jp^*}) \neq k_{jp^*}$), then $l(s_{jp^*}) = k_{jp'} + w < k_{jp^*}$ with $p' \neq p^*$, which contradicts the definition of p^*.

Last, we show that if $l(s_{jp}) \neq k_{jp}$ then $l(s_{jp}) = k_{jp^*} + w$. Suppose not. Since $l(s_{jp}) \neq k_{jp}$ then according to (12), $l(s_{jp}) = l(s_{jp'}) + w$ but with $p' \neq p^*$. Expanding $l(s_{jp'})$ according to (12),

$$\min\{k_{jp'}, \min_{p'' \neq p'} l(s_{jp''}) + w\} + w < l(s_{jp^*}) + w = k_{jp^*} + w$$

and since $k_{jp'} \geq k_{jp^*}$,

$$\min_{p'' \neq p'} l(s_{jp''}) + w < k_{jp^*}$$

$$\Leftrightarrow l(s_{jp^*}) + w < k_{jp^*}$$

which contradicts that $l(s_{jp^*}) = k_{jp^*}$.

The resulting recursion suggests a dynamic programming approach, summarized in Algorithm 1. It involves solving the subproblems of one tier at a time, visiting tiers according to a topological sort of $G(V, E)$ (recall that tiers correspond to tasks $j \in V$). Finding a topological sort takes $O(n\delta)$ [23], recalling that δ denotes the max in-degree of a task in the network. The overall complexity of Algorithm 1 is therefore $O(nm\delta) \subseteq O(n^2 m)$.

4   Conclusion

Given a stochastic task network with n tasks we consider dispatching the tasks as early as possible, subject to (planned) release-times. Assuming a sample with m realizations of the stochastic durations vector is drawn, we defined an LP for finding optimal release-times; i.e. that minimize the performance penalty of reaching a desired level of stability. The resulting LP is costly to solve, so pursuing a more efficient solution method we managed to show that optimal release-times can be expressed as a function of the earliest start time solution of an associated Simple Temporal Problem. Exploiting the structure of this STP, we were able to define a dynamic programming algorithm for finding optimal release-times with considerable efficiency, in time $O(n^2 m)$.

Future Work. Since we optimize according to a manageable sample $\mathcal{P}$, there is a (potentially non-zero) probability $\mathbb{P}_v$ that the realized start-time of a task deviates further than w time-units from its planned release-time. The question of how $\mathbb{P}_v$ (or $\mathbb{E}[\mathbb{P}_v]$ as in [6]) depends on m (the size of $\mathcal{P}$) should be addressed in future work. Furthermore, in an earlier paper [18], an LP similar to (P) was used it in a two-step heuristic for a flavor of the *stochastic resource constrained project scheduling problem* (stochastic RCPSP) [15,24]. Given a resource allocation determined in a first step, in a second step a LP was used to find planned release-times that minimize the total expected deviation of the realized schedule from those release-times. This heuristic was found to outperform the state-of-the-art in the area of *proactive project scheduling*. In future work, we shall investigate using the algorithm presented here in order to stabilize the given resource-allocation, expecting gains in both efficiency and effectiveness. Finally, a potentially related problem, namely PERTCONVG, is studied by Chrétienne and Sourd in [7], which involves finding start-times for a task network so as to minimize the sum of convex cost functions. In fact, their algorithm bears structural similarities to ours, since subproblems are solved in a topological order. It would be worth investigating if their analysis can be extended in order to enable casting the problem studied here as an instance of that problem.

References

1. Adlakha, V., Kulkarni, V.G.: A classified bibliography of research on stochastic pert networks: 1966–1987. INFOR **27**, 272–296 (1989)
2. Beck, C., Davenport, A.: A survey of techniques for scheduling with uncertainty (2002)
3. Bidot, J., Vidal, T., Laborie, P., Beck, J.C.: A theoretic and practical framework for scheduling in a stochastic environment. J. Sched. **12**, 315–344 (2009)
4. Blaauw, D., Chopra, K., Srivastava, A., Scheffer, L.: Statistical timing analysis: from basic principles to state of the art. IEEE Trans. Comput. Aided Des. Integr. Circuits Syst. **27**, 589–607 (2008)
5. Bonfietti, A., Lombardi, M., Milano, M.: Disregarding duration uncertainty in partial order schedules? Yes, we can!. In: Simonis, H. (ed.) CPAIOR 2014. LNCS, vol. 8451, pp. 210–225. Springer, Cham (2014). doi:10.1007/978-3-319-07046-9_15
6. Calafiore, G., Campi, M.C.: Uncertain convex programs: randomized solutions and confidence levels. Math. Program. **102**, 25–46 (2005)
7. Chrétienne, P., Sourd, F.: Pert scheduling with convex cost functions. Theoret. Comput. Sci. **292**, 145–164 (2003)
8. Davenport, A., Gefflot, C., Beck, C.: Slack-based techniques for robust schedules. In: Sixth European Conference on Planning (2014)
9. Dechter, R., Meiri, I., Pearl, J.: Temporal constraint networks. Artif. Intell. **49**, 61–95 (1991)
10. Godard, D., Laborie, P., Nuijten, W.: Randomized large neighborhood search for cumulative scheduling. In: ICAPS. vol. 5 (2005)
11. Hagstrom, J.N.: Computing the probability distribution of project duration in a pert network. Networks **20**, 231–244 (1990)
12. Herroelen, W., Leus, R.: Project scheduling under uncertainty: survey and research potentials. EJOR **165**, 289–306 (2005)
13. Igelmund, G., Radermacher, F.J.: Preselective strategies for the optimization of stochastic project networks under resource constraints. Networks **13**, 1–28 (1983)
14. Kleywegt, A.J., Shapiro, A., Homem-de Mello, T.: The sample average approximation method for stochastic discrete optimization. SIAM J. Optim. **12**, 479–502 (2002)
15. Lamas, P., Demeulemeester, E.: A purely proactive scheduling procedure for the resource-constrained project scheduling problem with stochastic activity durations. J. Sched. **19**, 409–428 (2015)
16. Leus, R.: Resource allocation by means of project networks: dominance results. Networks **58**, 50–58 (2011)
17. Malcolm, D.G., Roseboom, J.H., Clark, C.E., Fazar, W.: Application of a technique for research and development program evaluation. Oper. Res. **7**, 646–669 (1959)

18. Mountakis, S., Klos, T., Witteveen, C., Huisman, B.: Exact and heuristic methods for trading-off makespan and stability in stochastic project scheduling. In: MISTA (2015)

19. Pallottino, S.: Shortest-path methods: complexity, interrelations and new propositions. Networks **14**, 257–267 (1984)

20. Policella, N., Oddi, A., Smith, S.F., Cesta, A.: Generating robust partial order schedules. In: Wallace, M. (ed.) CP 2004. LNCS, vol. 3258, pp. 496–511. Springer, Heidelberg (2004). doi:10.1007/978-3-540-30201-8_37

21. Potra, F.A., Wright, S.J.: Interior-point methods. J. Comput. Appl. Math. **124**, 281–302 (2000)

22. Shestak, V., Smith, J., Maciejewski, A.A., Siegel, H.J.: Stochastic robustness metric and its use for static resource allocations. J. Parallel Distrib. Comput. **68**, 1157–1173 (2008)

23. Tarjan, R.E.: Edge-disjoint spanning trees and depth-first search. Acta Informatica **6**, 171–185 (1976)

24. Van de Vonder, S., Demeulemeester, E., Herroelen, W.: Proactive heuristic procedures for robust project scheduling: an experimental analysis. EJOR **189**, 723–733 (2008)

Rescheduling Railway Traffic on Real Time Situations Using Time-Interval Variables

Quentin Cappart[✉] and Pierre Schaus

Université catholique de Louvain, Louvain-la-Neuve, Belgium
{quentin.cappart,pierre.schaus}@uclouvain.be

Abstract. In the railway domain, the action of directing the traffic in accordance with an established timetable is managed by a software. However, in case of real time perturbations, the initial schedule may become infeasible or suboptimal. Subsequent decisions must then be taken manually by an operator in a very limited time in order to reschedule the traffic and reduce the consequence of the disturbances. They can for instance modify the departure time of a train or redirect it to another route. Unfortunately, this kind of hazardous decisions can have an unpredicted negative snowball effect on the delay of subsequent trains. In this paper, we propose a Constraint Programming model to help the operators to take more informed decisions in real time. We show that the recently introduced time-interval variables are instrumental to model this scheduling problem elegantly. We carried experiments on a large Belgian station with scenarios of different levels of complexity. Our results show that the CP model outperforms the decisions taken by current greedy strategies of operators.

1 Introduction

Since the dawn of the nineteenth century, development of railway systems has taken a huge importance in many countries. Over the years, the number of trains, the number of tracks, the complexity of networks increase and are still increasing. In this context, the need of an efficient and reliable train schedule is crucial. Indeed, a bad schedule can cause train conflicts, unnecessary delays, financial losses, and a passenger satisfaction decrease. While earliest train schedules could be built manually without using optimisation or computer based methods, it is not possible anymore. Plenty of works deal with this problematic of building the most appropriate schedule for general [1–3], or specific purposes [4,5].

Practical schedules must also deal with real time perturbations. Disturbances, technical failures or simply consequences of a too optimistic theoretical schedule can cause delays which can be propagated on other trains. The initial schedule may then become infeasible. Real-time modifications of the initial schedule therefore may be required. Several works already tackle this problem. A recent survey (2014) initiated by Cacchiani et al. [6] recaps the different trends on models and algorithms for real-time railway disturbance management. For instance, Fay et al. [7] propose an expert system using fuzzy rules and Petri Net for the

© Springer International Publishing AG 2017
D. Salvagnin and M. Lombardi (Eds.): CPAIOR 2017, LNCS 10335, pp. 312–327, 2017.
DOI: 10.1007/978-3-319-59776-8_26

modelling. Such a method requires to define the rules, which can differ according to the station. Higgins et al. [8] propose to use local search methods in order to solve conflicts. However, this work does not take into account the alternative routes that trains can have in order to reach their destination, and that the planned route remains not always optimal in case of real time perturbations. For that, D'Ariano et al. [9] model the problem with an alternative graph formulation and propose a branch and bound algorithm to solve it. They suggest to enrich their method with a local search algorithm for rerouting optimization purposes [10]. Besides, some works consider the passenger satisfaction [11–13] in their model. It is also known as the Delay Management Problem. Generally speaking, most of the methods dealing with train scheduling are based on Mixed Integer Programming [14–17]. However, the performance of Mixed Integer Programming models for solving scheduling problems is known to be highly dependant of the granularity of time chosen.

Constraint Based Scheduling [18], or in other words, applying Constraint Programming on scheduling problems seems to be a good alternative over Mixed Integer Programming. According to the survey of Bartak et al. [19], Constraint Programming is particularly well suited for real-life scheduling applications. Furthermore, several works [20–22] show that Constraint Programming can be used for solving scheduling problems on large and realistic instances. By following this trend, Rodriguez [23] proposes a Constraint Programming model for real-time train scheduling at junctions. However, despite the good performances obtained, this model can be improved. Firstly, the modelling do no use the strength of global constraints which can provide a better propagation. Secondly, the search can also be improved with heuristics and the use of Local Search techniques. Finally, the objective function is only defined in function of train delays without considering the passengers or the different categories of trains.

In this paper, we propose a new model for rescheduling the railway traffic in a real time context. The contributions are as follows:

- A Constraint Programming model for rescheduling the traffic through real time disturbances on the railway network. In addition to being a relevant application for railway companies, the proposed model scales on a realistic instance, the station of Courtrai (Belgium).
- The application of state of the art scheduling algorithms and relevant global constraints in order to achieve a better propagation and a faster search. Concretely, the conditional time-intervals introduced by Laborie et al. [24,25] as well as their dedicated global constraints have been used. Furthermore, the exploration of the state space has been carried out using Failure Directed Search together with Large Neighbourhood Search [26].
- The formulation of an objective function aiming at the same time to minimise the total delay and to maximise the overall passenger satisfaction. Furthermore, the objective function also considers the heterogeneity of the traffic and different levels of priority between trains through a defined lexicographical order. For instance a freight train has a lower priority than a passenger train.

The implementation of the model has been performed with IBM ILOG CP Optimizer V12.6.3 [27] which is particularly fitted for designing scheduling models [28–30]. The next section describes the signalling principles and the components considered for the modelling. Section 3 introduces the model and its specificities. Experiments and discussions about its performances are then carried in Sect. 4.

2   Signalling Principles

In the railway domain, the goal of a signalling system is to ensure a safe control of the traffic [31]. Such a task encompasses the regulation of the traffic. In Belgium, it is mainly performed by a software, called the Traffic Management System, that automatically regulates the traffic according to predefined rules. Let us consider the fictive station presented in Fig. 1. Several components are depicted on it:

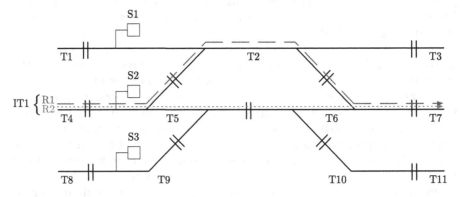

Fig. 1. Track layout of a fictive station with two Routes ($R1$ and $R2$). (Color figure online)

- The **track segments** (e.g. $T1$) are the portions of the railway structure where a train can be detected. They are delimited by the **joints** ($\dashv\vdash$).
- The **signals** (e.g. $S1$) are the devices used to control the train traffic. They are set on a proceed state (green) if a train can safely move into the station or in a stop state (red) otherwise.

Besides these physical components, signalling also involves logical structures:

- The **routes** correspond to the paths that trains can follow inside a station in order to reach a destination. They are expressed in term of track segments and signals. For instance, $R1$ is a route going from $T4$ to $T7$ by following the path $[T4, S2, T5, T2, T6]$. The first track segment of a route is always in front of a signal which is initially at a stop state and which turns green when the Traffic Management System allows the train to proceed. The track segment used for the destination is not a part of the route. This action is called a **route activation**. At this step, all the track segments forming the route after the start signal are reserved and cannot be used by another train. Once a route has been activated, the train can move through it in order to reach its destination. For instance, once a train following route $R1$ has reached $T6$ and is not on the previous track segments anymore, $T5$ and $T2$ can be released in order to allow other trains to use them. Detailed explanations about the route management is provided in [32].

- The **itineraries** correspond to a non physical path from a departure point to an arrival point. An itinerary can be constituted of one or several routes which can be alternative. For instance, as depicted in Fig. 1, two routes ($R1$ and $R2$) are possible in order to accomplish itinerary $IT1$ from $T4$ to $T7$. In normal situations, a route is often preferred than others, but in case of perturbations, the route causing the less conflicts is preferred.

According to an established timetable, the Traffic Management System activates routes in order to ensure the planned traffic. However, in case of real time perturbations, signalling must be handled manually and the actions to perform are decided by human operators. The procedure follows this pattern:

1. The Traffic Management System has predicted a future conflict caused by perturbations on the traffic.
2. The operators controlling the traffic analyse the situation and evaluate the possible actions to do in order to minimise the consequences of the perturbations. Besides the safety, the criterion considered for the decision is the sum of delays per passenger. In other words, the objective is to minimise the sum of delays of trains pondered by their number of passenger. Furthermore, some trains can have a higher priority than others.
3. According to the situation, they perform some actions (stopping a train, changing its route, etc.) or do nothing.

In this work, we are interested by the actions of operators facing up real time perturbations. In such situations, they must deal with several difficulties:

- The available time for analysing the situation and taking a decision can be very short according to the criticality of the situation (less than one minute).
- For large stations with a dense traffic, particularly during the peak hours, the effects of an action are often difficult to predict.
- The number of parameters that must be considered (type of trains, number of passengers, etc.) complicates the decision.

Such reasons can lead railway operators to take decisions that will not improve the situation, or worse, will degrade it. Most of the time, the decision taken is to give the priority either to the first train arriving at a signal, or to the first one that must leave the station [9]. However, it is not always the best decision. It is why we advocate the use of a Decision Support Tool based on optimisation in order to assist operators in their decisions. The requirement for this software is then to provide a good solution to this rescheduling problem within a short and parametrisable computation time.

3 Modelling

This section presents how we model the problem. Basically, the goal is to schedule adequately trains in order to bring them to their destination. The decision is then to chose, for each train, which route must be activated and at what time. Each track segment can host at most one train at a time. Furthermore, they can also be reserved only for one train. An inherent component of scheduling problems are the activities. Roughly speaking, a classical activity A is modelled with three variables, a start date $s(A)$, a duration $d(A)$ and an end date $e(A)$. The activity is *fixed* if the three variables are reduced to a singleton. Our model contains three kinds of activities that are linked together:

- The **itinerary activities** define the time interval when a train follows a particular itinerary. Each train has one and only one itinerary activity. We define $A^{t,it}$ as the activity for itinerary it of train t.
- The **route activities** define the interval when a train follows a particular route of an itinerary. We define $A^{t,it,r}$ as the activity for route r of Itinerary it related to train t.
- The **train activities** correspond to the movements of a train through the station in order to complete a route. Such activities use the track segments as resources. We define $A_i^{t,it,r}$ as the ith train activity of route r of itinerary it and related to train t. Inside a same route, there are as many train activities as the number of elements on the route path. The element can be a track segment or a signal. For instance, by following the example of Fig. 1, $A_1^{t,IT1,R1}$ is a train activity related to track segment $T4$, $A_2^{t,IT1,R1}$ to signal $S2$, $A_3^{t,IT1,R1}$ to $T5$, etc.

One particularity of our problem is that some activities are optional. In other words, they may or may not be executed in the final schedule. For instance, let us assume that a train has to accomplish an itinerary from $T4$ to $T7$. To do so, it can follow either $R1$, or $R2$. If $R1$ is chosen, the activity related to $R2$ will not be executed. For that, we model optional activities with the *conditional time-interval variables* introduced by Laborie et al. [24,25] that implicitly encapsulate the notion of optionality. It also allows an efficient propagation through dedicated global constraints (**alternative** and **span** for instance) as well as an efficient search. Roughly speaking, when an activity is fixed, it can be either *executed*, or *non executed*. If the activity is executed, it behaves as a classical activity that executes on its time interval, otherwise, it is not considered by any constraint. This functionality is modelled with a new variable $x(A) \in \{0,1\}$ for each activity A such that $x(A) = 1$ if the activity is executed and $x(A) = 0$ otherwise.

Figure 2 presents how the different activities are organised for the case study shown in Fig. 1 for a train t. Activity $A^{t,IT1}$ is mandatory, it models the fact that the train has to reach its destination. **Alternative** constraint ensures that the train can follow only one route to do so, $R1$ or $R2$. The other route activity is not executed. Furthermore, the start date and the end date of the chosen route is synchronised with the itinerary activity. **Span** constraint enforces the route activity to be synchronised with the start date of the first track segment activity and with the end date of the last track segment activity. For instance, $A^{t,IT1,R1}$ is synchronised with the start date of $A_1^{t,IT1,R1}$ and the end date of $A_5^{t,IT1,R1}$. The detailed explanations of alternative and span constraints are provided thereafter.

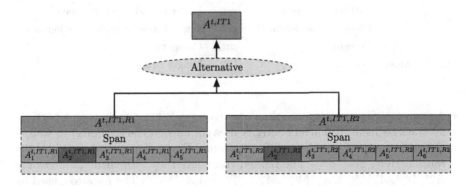

Fig. 2. Breakdown structure of the model using alternative and span constraints.

As we will see, conditional time-interval variables facilitate the construction of our model. Let us now define the different components of the model.

Parameters. Two entities are involved in our model: the trains and the track segments. Table 1 recaps the parameters considered. The *speed, number of passengers,* and *length* are straightforward to understand. The *estimated arrival time* of a train is a prediction of its arrival time at the station. The *earliest start time* defines a lower bound on the starting time of a train. In other words, a train cannot start its itinerary before this time. Indeed, a train cannot leave its platform before the time announced to the passengers. The *planned completion time* is the time announced on the initial schedule. It defines when the train is supposed to arrive at a platform. It is used in the objective function in order to compute the delays generated. The *category* defines the nature of the train. More explanations about the category is provided in Sect. 3.

Decision Variables. As previously said, the problem is to chose, for each train, which route must be activated and at what time. Such a problem can be seen as a slightly variant of a job shop scheduling problem [33] where the machines represent the track segments and the jobs represent train activities. The difference are as follows:

- Some activities are optional. In other words, they may or not be executed in the final schedule.
- The end of a train activity must be synchronised by the start of the next one.

Table 1. Parameters related to a train t or a track segment ts.

Entity	Parameter	Name	Meaning
Train	Speed	spt_t	Speed of t
	Passengers	p_t	Number of passengers of t
	Estimated arrival time	eat_t	When t arrives to the station
	Earliest start time	est_t	Lower bound on start time of t
	Planned completion time	pct_t	When t must complete its journey
	Category	cat_t	Category of t
Track segment	Length	lgt_{ts}	Length of ts

- A train activity can use more than one resource. When a train is on a particular track segment, its current activity uses the current track segment as well as the next ones that are reserved for the route. For instance, let us consider the route activity $A^{t,IT1,R1}$. The related train activities with their resources are $A_1^{t,IT1,R1}$, $A_2^{t,IT1,R1}$, $A_3^{t,IT1,R1}$, $A_4^{t,IT1,R1}$ and $A_5^{t,IT1,R1}$. The resources used by the activities are $(T4)$, $(T4, S2)$, $(T5, T2, T6)$, $(T2, T6)$ and $(T6)$ respectively. Let us remember from Sect. 2 that only the track segments located after the start signal are reserved through the route activation. Concerning the initial track segment T_4, it is released after the train has passed the start signal.

From a Constraint Based Scheduling approach, the problem is to assign a unique value to each train activity. The decision variables and their domain are, for all trains t, itineraries it, routes r and indexes i:

$$s(A_i^{t,it,r}) \begin{cases} \in [eat_t, horizon] & if\ t\ on\ track\ segment\ ts \\ \in [est_t, horizon] & if\ t\ in\ front\ of\ a\ signal \end{cases} \quad (1)$$

$$d(A_i^{t,it,r}) \begin{cases} = lgt_{ts}/spd_t & if\ t\ on\ track\ segment\ ts \\ \in [0, horizon] & if\ t\ in\ front\ of\ a\ signal \end{cases} \quad (2)$$

$$e(A_i^{t,it,r}) = s(A_i^{t,it,r}) + d(A_i^{t,it,r}) \quad (3)$$

$$x(A_i^{t,it,r}) \in \{0,1\} \quad (4)$$

The domain is determined in order to be as restricted as possible without removing a solution. Eq. (1) indicates that an activity cannot begin before the *estimated arrival time* of t. The upper bound of the start date is defined by the time horizon considered. More details about the horizon chosen is provided in Sect. 4. Eq. (2) models the time required to achieve the activity. If t is on a track segment, the duration is simply the length of the track segment divided the speed of t. Otherwise, the time that t will have to wait is unknown. Eq. (3) is an implicit constraint of consistency. Finally, Eq. (4) states that the activity is optional. Concerning route and itinerary activities, they are linked to train activities through constraints.

Constraints. This section describes the different constraints considered. Most of them are expressed in term of a train, an itinerary and a route. Let us express T as the set of trains, IT_t as the set of possible itineraries for t, R_{it} the set of possible routes for $it \in IT_t$ and N_r as the number of train activities of a route $r \in R_{it}$. Furthermore, let us state TS as the set of all the track segments in the station.

Precedence. This constraint (Eq. (5)) ensures that train activities must be executed in a particular order. It links the end of a train activity $A_i^{t,it,r}$ with the start of $A_{i+1}^{t,it,r}$.

$$e(A_i^{t,it,r}) = s(A_{i+1}^{t,it,r}) \qquad \forall t \in T, \forall it \in IT_t, \forall r \in R_{it}, \forall i \in [1, N_r[\qquad (5)$$

All precedence constraints are aggregated into a temporal network in order to have a better propagation [24,34]. Instead of having a bunch of independent constraints, they are considered as a global constraint.

Execution Consistency. As previously said, some activities are alternative. If a route is not chosen for a train, none of activities $A_i^{t,it,r}$ will be executed. Otherwise, all of them must be executed. Eq. (6) states that all the train activities related to the same environment must have the same execution status.

$$x(A_1^{t,it,r}) \equiv x(A_i^{t,it,r}) \qquad \forall t \in T, \forall it \in IT_t, \forall r \in R_{it}, \forall i \in]1, N_r] \qquad (6)$$

Alternative. Introduced by Laborie and Rogerie [24], this constraint models an exclusive alternative between a bunch of activities. It is expressed in Eq. (7).

$$\texttt{alternative}\left(A^{t,it}, \left\{A^{t,it,r} \middle| r \in R_{it}\right\}\right) \qquad \forall t \in T, \forall it \in IT_t \qquad (7)$$

It means that when $A^{t,it}$ is executed, then exactly one of the route activities must be executed. Furthermore, the start date and the end date of $A^{t,it}$ must be synchronised with the start and end date of the executed route activity. if $A^{t,it}$ is not executed, none of the other activities can be executed. In our model, $A^{t,it}$ is an mandatory activity. It models the fact that each train must reach its destination through an itinerary but for that, it must follow exactly one route.

Span. Also introduced in [24], this constraint states that an executed activity must span over a bunch of other executed activities by synchronising its start date with the earliest start date of other executed activities and its end date with the latest end date. It is expressed in Eq. 8.

$$\texttt{span}\left(A^{t,it,r}, \left\{A_i^{t,it,r} \middle| i \in [1, N_r]\right\}\right) \qquad \forall t \in T, \forall it \in IT_t, \forall r \in R_{it} \qquad (8)$$

It models the fact that the time taken by a train to complete a route is equal to the time required for crossing each of its components. If the route is not chosen, then none activity will be executed. A representation of **span** constraint is shown in Fig. 2.

Unary Resource. An important constraint is that trains cannot move or reserve a track segment that is already used for another train. It is a `unary resource` constraint (Eq. (9)). Each track segment can then be reserved only once at a time. Let us state ACT_{ts} as the set of all the train activities using track segment ts.

$$\texttt{noOverlap}\Big(\big\{A\big|A \in ACT_{ts}\big\}\Big) \qquad \forall ts \in TS \qquad (9)$$

The semantic of this constraint, as well as comparisons with existing frameworks, is presented in [25] which extends state of the art filtering methods in order to handle conditional time-interval variables.

Train Order Consistency. This last constraint ensures that trains cannot overtake other trains if they are on the same track segment. An illustration of this scenario is presented in Fig. 3. Even if train t_2 has a higher priority than t_1, it cannot begin its activities before t_1 because t_1 has an earlier estimated arrival time. In other words, on each track segment in front of a signal, the start date of the first activity of each train is sorted by their estimated arrival time.

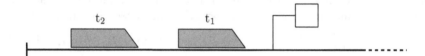

Fig. 3. Two trains waiting on the same track segment.

Let us state $TSB \subset TS$ as the set of all first track segments, N_{tsb} as the number of trains beginning their itinerary on track segment tsb, and (AB_i^{tsb}) as the sequence of the first activities of trains t_i beginning on tsb with $i \in [1, N_{tsb}]$. The sequence is ordered by the estimated arrival time of trains t_i. Then, `train order consistency` constraint can be expressed through Eq. (10).

$$s(AB_{i-1}^{tsb}) < s(AB_i^{tsb}) \qquad \forall tsb \in TSB, \forall i \in]1, N_{tsb}] \qquad (10)$$

These constraints are also considered in the temporal network.

Objective Function. The criterion frequently used for the objective function is the sum of train delays [23]. Let us state jct_t as the *journey completion time* of a train t. It corresponds to the end date of the last train activity of t. The delay d_t of t is expressed in Eq. 11.

$$d_t = \max\Big(0, jct_t - pct_t\Big) \qquad (11)$$

The `max` function is used to nullify the situation where t is in advance on its schedule. Eq. 12 presents a first objective function.

$$\min\left(\sum_{t \in T} d_t\right) \tag{12}$$

However, in real circumstances, railway operators must also consider other parameters such as the number of passengers and the priority of trains. Section 3 introduced the *category* parameter, here are the different categories considered:

1. Maintenance or special vehicles (C_1).
2. Passenger trains with a correspondence (C_2).
3. Simple passenger trains (C_3).
4. Freight trains (C_4).

Such categories are sorted with a decreasing order according to their priority. Special vehicles have then the highest priority and freight trains the lowest. Furthermore, if a passenger train has more passengers than another one, the cost of the delay will be more important. The objective function is then threefold:

- Scheduling trains according to their priority. For instance, a maintenance vehicle must be scheduled before a passenger train, if possible.
- Minimising the sum of delays.
- Maximising the overall passenger satisfaction. The passenger satisfaction decreases if its train is late.

A second objective function can then be expressed (Eq. 13).

$$\texttt{lex\_min}\left(\sum_{t \in C_1} d_t, \sum_{t \in C_2} p_t \times d_t, \sum_{t \in C_3} p_t \times d_t, \sum_{t \in C_4} d_t\right) \tag{13}$$

Where p_t corresponds to the number of passengers of train t, as defined in Table 1. This equation gives a lexicographical ordering of trains according to their category from $C1$ to $C4$. For passenger categories (C_2 and C_3) the delay is expressed by passengers. In this way, the more is the number of passenger, the greater will be the penalty for delays. The objective is then to minimise this expression with regard to its lexicographical ordering.

Search Phase. The exploration of the state space is performed with the algorithm of Vilim et al. [26] which combines a Failure-Directed Search with Large Neighborhood Search. The implementation proposed on CP Optimizer V12.6.3 [27] is particularly fitted to deal with conditional time-interval variables, precedence constraints, and optional resources. The search is performed on execution and on start date variables. Concerning the variable ordering, trains are sorted according to their category. For instance, activities related to passenger trains will be assigned before activities of freight trains. The value ordering is let by default.

4   Experiments

This section evaluates the performance of the Constraint Programming model through different experiments. Concretely, we compare our solution with the solutions obtained with classical dispatching methods on a realistic station: Courtrai. Its track layout is presented in Fig. 4. It contains 23 track segments, 14 signals, 68 itineraries and 84 possible routes. Three systematic strategies are commonly used in practice:

- **First Come First Served** (FCFS) strategy gives the priority to the trains according to their estimated arrival time.
- **Highest Delay First Served** (HDFS) strategy gives the priority to the train that has the earliest planned completion time.
- **Highest Priority First Served** (HPFS) strategy gives the priority to a train belonging to the category with the highest priority. The planned completion time is then used as a tie breaker for trains of a same category.

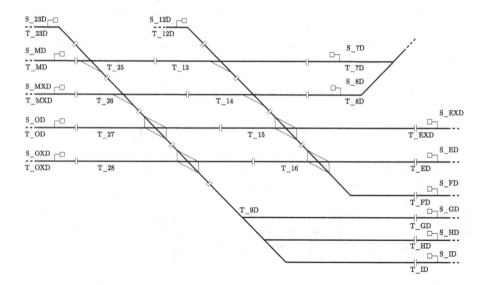

Fig. 4. Track layout of Courtrai.

Furthermore, three meta-parameters must be considered for the experiments, the *time horizon*, the *decision time* and *the number of trains*. The time horizon defines an upper bound on the estimated arrival time of trains. According to D'Ariano et al. [9], the practical time horizon for railway managers is usually less than one hour. In our experiments, we considered three time horizons (30 min, one hour and two hours). The decision time is the time that railway operators have at disposal for taking a decision. It is highly dependant to the criticality of the current situation. However, according to Rodriguez [23], the system must be able to provide an acceptable solution within 3 min for practical uses. Concerning the number of trains, we considered scenarios having 5, 10, 15, 20, 25 and 30 trains.

Experiments have be done under the assumption that we allow a feasible schedule. This responsibility is beyond the scope of the station. Two situations are considered: a homogeneous and a heterogeneous traffic.

All the experiments have been realised on a MacBook Pro with a 2.6 GHz Intel Core i5 processor and with a RAM of 16 Go 1600 MHz DDR3 using a 64-Bit HotSpot(TM) JVM 1.8 on El Capitan 10.11.15. The model has been designed with CP Optimizer V12.6.3 and the optimisation is performed with four workers.

Homogeneous Traffic. In this first situation, the number of passengers and the category are not considered. Each train has then the same priority and Eq. 12 is the objective function used. Table 2 recaps the experiments performed. Each scenario is repeated one hundred times with a random schedule. The different values of the schedule are generated randomly with uniform distributions. For instance, let us consider a schedule from 1 pm to 3 pm with 10 trains. For each train, we randomly choose its itinerary among the set of possible itineraries, its departure time and its expected arrival time in the interval [1 pm, 3 pm]. The delay indicated for each strategy corresponds to the arithmetic mean among all the tests. For each scenario, the average, the minimum and the maximum improvement ratio of the CP solution in comparison to the best solution obtained with classical methods is also indicated. POS indicates the number of tests where the CP approach has improved the solution while OPT indicates the number of tests where CP has reached the optimum.

As we can see, CP improves the solution for almost all the tests. The average improvement ratio is above 20% in all the scenarios. Optimum is often reached (more than 75% of the instances) when 10 trains or less are considered.

Heterogeneous Traffic. In this second situation, we use Eq. 13 for the objective function. The number of passengers and the category are then considered. As for the heterogeneous case, such values are chosen randomly with a uniform distribution. Among the classical approaches, only HPFS deals with an heterogeneous traffic. Table 3 recaps the experiments performed. Optimisations are performed sequentially for each category. The time is allocated according to the priority of categories. The allocation is then not done a priori but dynamically according to the time taken by the successive optimisations. Unlike the previous experiments, we do not compute the improvement ratio, but the number of experiences where CP has improved the solution obtained with HPFS. Our choice was motivated by the subjective aspect of defining an improvement ratio for a heterogeneous traffic. For instance there is no clear preference between decreasing the delay of one train of category C1 and of 10 trains with lower priorities. This kind of questions usually requires the consideration of signalling operators. We consider that CP has improved the solution when the sum of delay per category is lexicographically lower than the result provided by HDFS. For the three

Table 2. Comparison between CP and classical scheduling approaches for an homogeneous traffic with a decision time of 3 min.

Horizon	# Trains	Average delay (min.)			Improvement ratio (%)			POS	OPT
		FCFS	HDFS	CP	Mean	Min	Max	$(x/100)$	$(x/100)$
2 h	5	192.06	191.10	148.91	22.08	-7.69	100.00	99	97
	10	800.40	797.82	575.04	27.92	5.13	66.08	100	85
	15	1917.95	1896.64	1341.53	29.27	1.16	58.549	100	28
	20	3457.45	3397.03	2414.45	28.92	10.26	53.29	100	0
	25	5581.10	5632.69	3993.04	28.45	8.58	48.37	100	0
	30	8004.84	8018.76	5714.69	28.61	14.35	43.61	100	0
1 h	5	225.36	228.75	72.06	23.65	0.71	100.00	100	96
	10	911.01	906.32	664.18	26.72	3.96	62.41	100	83
	15	2128.34	2104.95	1494.67	28.99	5.93	51.29	100	17
	20	3675.72	3680.00	2612.6	28.92	8.25	55.86	100	1
	25	5971.54	6004.81	4246.55	28.89	9.41	53.67	100	0
	30	8595.56	8576.92	6085.51	29.05	7.89	45.51	100	0
30 min	5	231.16	229.06	173.95	24.06	1.49	100.00	100	94
	10	950.30	929.71	692.18	25.55	3.12	64.32	100	83
	15	2145.36	2161.20	1535.86	28.41	7.927	51.61	100	14
	20	3858.67	3888.29	2728.0	29.30	6.63	52.50	100	0
	25	6137.02	6135.59	4320.14	29.59	7.75	53.38	100	0
	30	8863.49	8775.84	6357.08	27.56	8.72	49.08	100	0

horizons considered and for 100 tests per scenario, CP improves the solution for almost all the tests, even when the optimum is not reached.

Scalability. This experiment deals with the scalability of the CP model in function with the decision time. We observed that setting the decision time to 10 min instead of 3 min do not increase significantly the performances. The gain of the improvement ratio is less than 1% for 60 random instances on an homogeneous traffic of 5, 10, 15, 20, 25 or 30 trains (10 instances per configuration). We can then conclude that even if the CP approach gives a feasible and competitive solution within 3 min, the quality of the solution do not increase significantly with time.

Criticality. In some cases, railway operators do not have an available decision time of 180 s, they have to react almost instantly because of the criticality of the situation. For this reason, we analysed how the CP model performs with a decision time lower than 10 s. To do so, we recorded the number of experiments where the CP approach has improved the solution in comparison to FCFS and HDFS strategies. For the scenarios depicted in Table 2, we observed that CP provides a same or better solution in more than 99% of the cases and can then also be used to deal with critical situations.

Table 3. Comparison between CP and HPFS approach for an heterogeneous traffic with a decision time of 3 min.

Horizon	# Trains	POS ($x/100$)	OPT ($x/100$)
2 h	5	100	99
	10	98	87
	15	100	64
	20	100	51
	25	100	13
	30	100	5
1 h	5	100	99
	10	99	89
	15	100	75
	20	100	51
	25	100	22
	30	100	8
30 min	5	100	99
	10	99	87
	15	100	75
	20	100	52
	25	100	26
	30	100	7

Reproducibility. A shortcoming in the literature about this field of research is the lack of reproducibility. To overcome this lack, we decided to provide information about our instances and the tests performed[1]. The information provided are enough to build a model, perform experiments and to compare them with ours.

5   Conclusion

Nowadays, railway operators must deal with the problem of rescheduling the railway traffic in case of real time perturbations in the network. However, the systematic and greedy strategies (FCFS, HDFS and HPFS) currently used for this purpose often give a suboptimal decision. In this paper, we presented a CP model for rescheduling the railway traffic on real time situations. The modelling is based on the recently introduced time-interval variables. Such a structure allows to design the model elegantly with variables and global constraints especially dedicated for scheduling. Finally, an objective function taking into account the heterogeneity of the traffic is presented. Experiments have shown that a dispatching better than the classical approaches is obtained in less than 3 min in

---

[1] Available at https://bitbucket.org/qcappart/qcappart_opendata.

almost all the situations that can occur in a large station, even when the optimum is not reached.

Two aspects are considered for our future works: improving the model and its scalability. Firstly, we plan to modify the model and analyse the consequences. For instance, we will use a `xor` instead of `alternative` constraint. Secondly, we will consider experiments on larger areas covering several stations.

Acknowledgments. This research is financed by the Walloon Region as part of the Logistics in Wallonia competitiveness pole. Experiments have been performed with ILOG CP Optimizer through the academic initiative of IBM.

References

1. Higgins, A., Kozan, E., Ferreira, L.: Optimal scheduling of trains on a single line track. Transp. Res. Part B Methodol. **30**, 147–161 (1996)
2. Ghoseiri, K., Szidarovszky, F., Asgharpour, M.J.: A multi-objective train scheduling model and solution. Transp. Res. Part B Methodol. **38**, 927–952 (2004)
3. Zhou, X., Zhong, M.: Bicriteria train scheduling for high-speed passenger railroad planning applications. Eur. J. Oper. Res. **167**, 752–771 (2005)
4. Harabor, D., Stuckey, P.J.: Rail capacity modelling with constraint programming. In: Quimper, C.-G. (ed.) CPAIOR 2016. LNCS, vol. 9676, pp. 170–186. Springer, Cham (2016). doi:10.1007/978-3-319-33954-2_13
5. Senthooran, I., Wallace, M., Koninck, L.: Freight train threading with different algorithms. In: Michel, L. (ed.) CPAIOR 2015. LNCS, vol. 9075, pp. 393–409. Springer, Cham (2015). doi:10.1007/978-3-319-18008-3_27
6. Cacchiani, V., Huisman, D., Kidd, M., Kroon, L., Toth, P., Veelenturf, L., Wagenaar, J.: An overview of recovery models and algorithms for real-time railway rescheduling. Transp. Res. Part B Methodol. **63**, 15–37 (2014)
7. Fay, A.: A fuzzy knowledge-based system for railway traffic control. Eng. Appl. Artif. Intell. **13**, 719–729 (2000)
8. Higgins, A., Kozan, E., Ferreira, L.: Heuristic techniques for single line train scheduling. J. Heuristics **3**, 43–62 (1997)
9. Dariano, A., Pacciarelli, D., Pranzo, M.: A branch and bound algorithm for scheduling trains in a railway network. Eur. J. Oper. Res. **183**, 643–657 (2007)
10. D'Ariano, A., Corman, F., Pacciarelli, D., Pranzo, M.: Reordering and local rerouting strategies to manage train traffic in real time. Transp. Sci. **42**, 405–419 (2008)
11. Schachtebeck, M., Schöbel, A.: To wait or not to wait and who goes first? delay management with priority decisions. Transp. Sci. **44**, 307–321 (2010)
12. Dollevoet, T., Huisman, D., Schmidt, M., Schöbel, A.: Delay management with rerouting of passengers. Transp. Sci. **46**, 74–89 (2012)
13. Dollevoet, T., Huisman, D., Kroon, L., Schmidt, M., Schöbel, A.: Delay management including capacities of stations. Transp. Sci. **49**, 185–203 (2014)
14. Corman, F., Goverde, R.M.P., D'Ariano, A.: Rescheduling dense train traffic over complex station interlocking areas. In: Ahuja, R.K., Möhring, R.H., Zaroliagis, C.D. (eds.) Robust and Online Large-Scale Optimization. LNCS, vol. 5868, pp. 369–386. Springer, Heidelberg (2009). doi:10.1007/978-3-642-05465-5_16
15. Foglietta, S., Leo, G., Mannino, C., Perticaroli, P., Piacentini, M.: An optimized, automatic TMS in operations in Roma Tiburtina and monfalcone stations. WIT Trans. Built Environ. **135**, 635–647 (2014)

16. Araya, S., Abe, K., Fukumori, K.: An optimal rescheduling for online train traffic control in disturbed situations. In: The 22nd IEEE Conference on Decision and Control, 1983, pp. 489–494. IEEE (1983)
17. Lamorgese, L., Mannino, C.: An exact decomposition approach for the real-time train dispatching problem. Oper. Res. **63**, 48–64 (2015)
18. Baptiste, P., Le Pape, C., Nuijten, W.: Constraint-Based Scheduling: Applying Constraint Programming to Scheduling Problems, vol. 39. Springer Science & Business Media, US (2012)
19. Barták, R., Salido, M.A., Rossi, F.: New trends in constraint satisfaction, planning, and scheduling: a survey. Knowl. Eng. Rev. **25**, 249–279 (2010)
20. Kelareva, E., Brand, S., Kilby, P., Thiébaux, S., Wallace, M., et al.: CP and MIP methods for ship scheduling with time-varying draft. In: ICAPS (2012)
21. Kelareva, E., Tierney, K., Kilby, P.: CP methods for scheduling and routing with time-dependent task costs. In: Gomes, C., Sellmann, M. (eds.) CPAIOR 2013. LNCS, vol. 7874, pp. 111–127. Springer, Heidelberg (2013). doi:10.1007/978-3-642-38171-3_8
22. Ku, W.Y., Beck, J.C.: Revisiting off-the-shelf mixed integer programming and constraint programming models for job shop scheduling. Dept Mech. Ind. Eng., Univ. Toronto, Toronto, ON, Canada, Technical report. MIE-OR-TR2014-01 (2014)
23. Rodriguez, J.: A constraint programming model for real-time train scheduling at junctions. Transp. Res. Part B Methodol. **41**, 231–245 (2007)
24. Laborie, P., Rogerie, J.: Reasoning with conditional time-intervals. In: FLAIRS Conference, pp. 555–560 (2008)
25. Laborie, P., Rogerie, J., Shaw, P., Vilím, P.: Reasoning with conditional time-intervals. Part II: an algebraical model for resources. In: FLAIRS Conference (2009)
26. Vilím, P., Laborie, P., Shaw, P.: Failure-directed search for constraint-based scheduling. In: Michel, L. (ed.) CPAIOR 2015. LNCS, vol. 9075, pp. 437–453. Springer, Cham (2015). doi:10.1007/978-3-319-18008-3_30
27. Laborie, P.: IBM ILOG CP optimizer for detailed scheduling illustrated on three problems. In: Hoeve, W.-J., Hooker, J.N. (eds.) CPAIOR 2009. LNCS, vol. 5547, pp. 148–162. Springer, Heidelberg (2009). doi:10.1007/978-3-642-01929-6_12
28. Vilím, P.: Max energy filtering algorithm for discrete cumulative resources. In: Hoeve, W.-J., Hooker, J.N. (eds.) CPAIOR 2009. LNCS, vol. 5547, pp. 294–308. Springer, Heidelberg (2009). doi:10.1007/978-3-642-01929-6_22
29. Salido, M.A., Escamilla, J., Barber, F., Giret, A., Tang, D., Dai, M.: Energy-aware parameters in job-shop scheduling problems. In: GREEN-COPLAS 2013: IJCAI 2013 Workshop on Constraint Reasoning, Planning and Scheduling Problems for a Sustainable Future, pp. 44–53(2013)
30. Hait, A., Artigues, C.: A hybrid CP/MILP method for scheduling with energy costs. Eur. J. Ind. Eng. **5**, 471–489 (2011)
31. Theeg, G.: Railway Signalling & Interlocking: International Compendium. Eurailpress (2009)
32. Cappart, Q., Schaus, P.: A dedicated algorithm for verification of interlocking systems. In: Skavhaug, A., Guiochet, J., Bitsch, F. (eds.) SAFECOMP 2016. LNCS, vol. 9922, pp. 76–87. Springer, Cham (2016). doi:10.1007/978-3-319-45477-1_7
33. Yamada, T., Nakano, R.: Job shop scheduling. IEE Control Engineering Series, p. 134 (1997)
34. Dechter, R., Meiri, I., Pearl, J.: Temporal constraint networks. Artif. Intell. **49**, 61–95 (1991)

Dynamic Temporal Decoupling

Kiriakos Simon Mountakis[1], Tomas Klos[2], and Cees Witteveen[1(✉)]

[1] Delft University of Technology, Delft, The Netherlands
C.Witteveen@tudelft.nl
[2] Utrecht University, Utrecht, The Netherlands

Abstract. Temporal decoupling is a method to distribute a temporal constraint problem over a number of actors, such that each actor can solve its own part of the problem. It then ensures that the partial solutions provided can be always merged to obtain a complete solution. This paper discusses static and dynamic decoupling methods offering maximal flexibility in solving the partial problems. Extending previous work, we present an exact $O(n^3)$ flexibility-maximizing static decoupling method. Then we discuss an exact $O(n^3)$ method for updating a given decoupling, whenever an actor communicates a commitment to a particular set of choices for some temporal variable. This updating method ensures that: (i) the flexibility of the decoupling never decreases and (ii) every commitment once made is respected in the updated decoupling. To ensure an efficient updating process, we introduce a fast heuristic to construct a new decoupling given an existing decoupling in nearly linear time. We present some experimental results showing that, in most cases, updating an existing decoupling in case new commitments for variables have been made, significantly increases the flexibility of making commitments for the remaining variables.

Keywords: Simple Temporal Problem · Decoupling · Flexibility

1 Introduction

In quite a number of cases a joint task has to be performed by several actors, each controlling only a part of the set of task constraints. For example, consider making a joint appointment with a number of people, constructing a building that involves a number of (sub)contractors, or finding a suitable multimodal transportation plan involving different transportation companies. In all these cases, some task constraints are under the control of just one actor (agent), while others require more than one agent setting the right values to the variables to satisfy them. Let us call the first type of constraints *intra-agent* constraints and the second type *inter-agent* constraints.

If there is enough time, solving such multi-actor constraint problems might involve consultation, negotiation or other agreement technologies. Sometimes, however, we have to deal with problems where communication between agents during problem solving is not possible or unwanted. For example, if the agents are in competition, there are legal or privacy issues preventing communication, or there is simply no time for communication.

© Springer International Publishing AG 2017
D. Salvagnin and M. Lombardi (Eds.): CPAIOR 2017, LNCS 10335, pp. 328–343, 2017.
DOI: 10.1007/978-3-319-59776-8_27

In this paper we use an approach to solve multi-actor constraint problems in the latter contexts: instead of using agreement technologies, we try to avoid them by providing *decoupling techniques*. Intuitively, a decoupling technique modifies a multi-actor constraint system such that each of the agents is able to select a solution for its own set of (intra-agent) constraints and a simple merging of all individual agent solutions always satisfies the total set of constraints. Usually, this is accomplished by tightening intra-agent constraints such that inter-agent constraints are implied. Examples of real-world applications of such decoupling techniques can be found in e.g. [2,8].

Quite some research focuses on finding suitable decoupling techniques (e.g., [1,2,7]) for Simple Temporal Problems (STPs) [3]. An STP $S = (T, C)$ is a constraint formalism where a set of temporal variables $T = \{t_0, t_1, \ldots, t_n\}$ are subject to binary difference constraints C and solutions can be found in low polynomial ($O(n^3)$) time. Since STPs deal with temporal variables, a decoupling technique applied to STPs is called *temporal decoupling*. The quality of a temporal decoupling technique depends on the degree to which it tightens intra-agent constraints: the more it restricts the flexibility of each individual agent in solving their own part of the constraints, the less it is preferred. Therefore, we need a flexibility metric to evaluate the quality of a temporal decoupling.

Originally, flexibility was measured by summing the differences between the highest possible (latest) and the lowest possible (earliest) values of all variables in the constraint system after decoupling ([6,7]). This so-called *naive flexibility metric* has been criticized, however, because it does not take into account the dependencies between the variables and, in general, seriously overestimates the "real" amount of flexibility. An alternative metric, the *concurrent flexibility metric* has been proposed in [15]. This metric accounts for dependencies between the variables and is based on the notion of an *interval schedule* for an STP S: For each variable $t_i \in T$ an interval schedule defines an interval $[l_i, u_i]$, such that choosing a value within $[l_i, u_i]$ for each variable t_i, always constitutes a solution for S. The sum $\sum_{i=1}^{n}(u_i - l_i)$ of the widths of these intervals determines the flexibility of S. The concurrent flexibility of S then is defined as the maximum flexibility we can obtain for an interval schedule of S and can be computed in $O(n^3)$ (see [9]).

As shown in [15], the concurrent flexibility metric can also be used to obtain an optimal (i.e. maximum flexible) temporal decoupling of an STP. This decoupling is a *total decoupling*, that is, a decoupling where the n variables are evenly distributed over n independent agents and thus every agent controls exactly one variable. It has been shown that this total decoupling is optimal for every partitioning of the set of temporal variables if one considers the concurrent flexibility metric as the flexibility metric to be used. In this paper, we concentrate on such (optimal) total decouplings of an STP.

In all existing approaches, a single temporal decoupling is computed in advance and is not changed, even if later on some agents announce their commitment to a specific value (or range of values) for a variable they control. Intuitively, however, we can use such additional information for the benefit of all agents, by possibly increasing the flexibility of variables they are not yet committed to. Specifically, when an agent announces a commitment to a sub-range of values within the given interval schedule (that represents the current decoupling), we are interested in updating the decoupling such that the individual flexibility of no agent is affected negatively.

More precisely, the overall aim of this paper is to show that a decoupling update method with the following nice properties do exist: first of all, it never affects the current flexibility of the agents negatively, and, secondly, it never decreases (and possibly increases) the individual flexibility of the variables not yet committed to. We will also show that updating a temporal decoupling as the result of a commitment for a single variable can be done in almost linear (amortized) time ($O(n \log n)$), which compares favourably with the cost of computing a new optimal temporal decoupling ($O(n^3)$).

Organisation. In Sect. 2 we discuss existing relevant work on STPs and temporal decoupling (TD). Then, in Sect. 3, extending some existing work, we briefly show how a total TD can be computed in $O(n^3)$, using a minimum matching approach. In Sect. 4, we first provide an exact approach to update an existing decoupling after commitments to variables have been made. We conclude, however, that this adaptation is computationally quite costly. Therefore, in Sect. 5 we offer an alternative, heuristic method, that is capable to adapt a given temporal decoupling in almost linear time per variable commitment. To show the benefits of adapting the temporal decoupling, in Sect. 6 we present the results of some experiments using STP benchmarks sets with the heuristic decoupling update and compare the results with an exact, but computationally more intensive updating method. We end with stating some conclusions and suggestions for further work.

2   Preliminaries

Simple Temporal Problems. A Simple Temporal Problem (STP) (also known as a Simple Temporal Network (STN)) is a pair $S = (T, C)$, where $T = \{t_0, t_1, \ldots, t_n\}$ is a set of temporal variables (events) and C is a finite set of binary difference constraints $t_j - t_i \leq c_{ij}$, for some real number c_{ij}.[1] The problem is to find a solution, that is a sequence $(s_0, s_1, s_2, \ldots, s_n)$ of values such that, if each $t_i \in T$ takes the value s_i, all constraints in C are satisfied. If such a solution exists, we say that the STN is *consistent*.[2] In order to express absolute time constraints, the time point variable $t_0 \in T$, also denoted by z, is used. It represents a fixed reference point on the timeline, and is always assigned the value 0.

---

[1] If both $t_j - t_i \leq c_{ij}$ and $t_i - t_j \leq c_{ji}$ are specified, we sometimes use the more compact notations $-c_{ji} \leq t_j - t_i \leq c_{ij}$ or $t_j - t_i \in [-c_{ji}, c_{ij}]$.

[2] W.l.g., in the remainder of the paper we simply assume an STN to be consistent. Consistency of an STN can be determined in low-order polynomial time [4].

Example 1. Consider two trains 1 and 2 arriving at a station. Train 1 has to arrive between 12:05 and 12:15, train 2 between 12:08 and 12:20. People traveling with these trains need to change trains at the station. Typically, one needs at least 3 min for changing trains. Train 1 will stay during 5 min at the station and train 2 stays 7 min before it departs. Let t_i denote the arrival time for train $i = 1, 2$. Let $t_0 = z = 0$ represent noon (12:00). To ensure that all passengers have an opportunity to change trains, we state the following STP $S = (T, C)$ where: $T = \{z, t_1, t_2\}$, and $C = \{5 \leq t_1 - z \leq 15,\ 8 \leq t_2 - z \leq 20,\ -2 \leq t_2 - t_1 \leq 4\}$. As the reader may verify, two possible solutions[3] for this STP are $s = (0, 10, 10)$ and $s' = (0, 15, 17)$. That is, there is a solution when both trains arrive at 12:10, and there is also a solution where train 1 arrives at 12:15, while train 2 arrives at 12:17. ∎

An STP $S = (T, C)$ can also be specified as a directed weighted graph $G_S = (T_S, E_S, l_S)$ where the vertices T_S represent variables in T and for every constraint $t_j - t_i \leq c_{ij}$ in C, there is a directed edge $(t_i, t_j) \in E_S$ labeled by its weight $l_S(t_i, t_j) = c_{ij}$. The weight c_{ij} on the arc (t_i, t_j) can be interpreted as the length of the path from t_i to t_j. Using a shortest path interpretation of the STP $S = (T, C)$, there is an efficient method to find all shortest paths between pairs (t_i, t_j) using e.g. Floyd and Warshall's all-pairs shortest paths algorithm [5]. A shortest path between time variables t_i and t_j then corresponds to a *tightest constraint* between t_i and t_j. These tightest constraints can be collected in an $n \times n$ *minimum distance matrix* $D = [d_{ij}]$, containing for every pair of time-point variables t_i and t_j the length d_{ij} of the *shortest path* from t_i to t_j in the distance graph. In particular, the first row and the first column of the distance matrix D contain useful information about the possible schedules for S: The sequence $lst = (d_{00}, d_{01}, \ldots, d_{0n})$ specifies the latest starting time solution for each time point variable t_i. Analogously, $est = (-d_{00}, -d_{10}, \ldots, -d_{no})$ specifies the *earliest starting time* solution.

Example 2. Continuing Example 1, the minimum distance matrix D of S equals

$$D = \begin{bmatrix} 0 & 15 & 19 \\ -5 & 0 & 4 \\ -8 & 2 & 0 \end{bmatrix}$$

The earliest time solution therefore is $est = (0, 5, 8)$, and the latest time solution $lst = (0, 15, 19)$. ∎

Given S, the matrix D can be computed in low-order polynomial ($O(n^3)$) time, where $n = |T|$, see [3].[4] Hence, using the STP-machinery we can find earliest and latest time solutions quite efficiently.

---

[3] Of course, there are many more.

[4] An improvement by Planken et al. [12] has shown that a schedule can be found in $O(n^2 w_d)$-time where w_d is the graph width induced by a vertex ordering.

Temporal Decoupling. In order to find a solution for an STP $S = (T, C)$ all variables $t_i \in T$ should be assigned a suitable value. Sometimes these variables are controlled by different agents. That is, $T - \{z\} = \{t_1, \ldots, t_n\}$ is partitioned into k non-empty and non-overlapping subsets $T_1, \ldots, T_k$ of T, each T_j corresponding to the set of variables controlled by agent $a_j \in A = \{a_1, a_2, \ldots, a_k\}$. Such a partitioning of $T - \{z\}$ will be denoted by $[T_i]_{i=1}^k$. In such a case, the set of constraints C is split in a set $C_{intra} = \bigcup_{i=1}^k C_i$ of intra-agent constraints and a set $C_{inter} = C - C_{intra}$ of inter-agent constraints. Here, a constraint $t_i - t_j \leq c_{ji}$ is an intra-constraint occurring in C_h if there exists a subset T_h such that $t_i, t_j \in T_h$, else, it is an inter-agent constraint. Given the partitioning $[T_j]_{j=1}^k$, every agent a_i completely controls the (sub) STP $S_i = (T_i \cup \{z\}, C_i)$, where C_i is its set of intra-agent constraints, and determines a solution s_i for it, independently from the others. In general, however, it is not the case that, using these sub STPs, merging partial solutions s_i will always constitute a total solution to S:

Example 3. Continuing Example 1, let train i be controlled by agent a_i for $i = 1, 2$ and assume that we have computed the set of tightest constraints based on C. Then $S_1 = (\{z, t_1\}, \{5 \leq t_1 - z \leq 15\})$ and $S_2 = (\{z, t_2\}, \{8 \leq t_2 - z \leq 19\})$. Agent a_1 is free to choose a time t_1 between 5 and 15 and suppose she chooses 10. Agent a_2 controls t_2 and, therefore, can select a value between 8 and 19. Suppose he chooses 16. Now clearly, both intra-agent constraints are satisfied, but the inter-agent constraint $t_2 - t_1 \leq 4$ is not, since $16 - 10 > 4$. Hence, the partial solutions provided by the agents are not conflict-free. ∎

The *temporal decoupling* problem for $S = (T, C)$, given the partitioning $[T_j]_{j=1}^k$, is to find a suitable set $C'_{intra} = \bigcup_{i=1}^k C'_i$ of (modified) intra-agent constraints such that if s'_i is an arbitrary solution to $S'_i = (T_i \cup \{z\}, C'_i)$, it always can be merged with other partial solutions to a total solution s' for S.[5]

We are interested, however, not in arbitrary decouplings, but an *optimal decoupling* of the k agents, allowing each agent to choose an assignment for its variables independently of others, while maintaining *maximum flexibility*. This optimal decoupling problem has been solved in [14] for the total decoupling case, that is the case where $k = n$ and each agent a_i controls a single time point variable t_i. In this case, the decoupling results in a specification of a lower bound l_i and an upper bound u_i for every time point variable $t_i \in T$, such that t_i can take any value $v_i \in [l_i, u_i]$ without violating any of the intra- or inter-agent constraints. This means that if agent a_i controls T_i then her set of intra-agent constraints is $C_i = \{l_j \leq t_j \leq u_j \mid t_j \in T_i\}$.

The total flexibility the agents have, due to this decoupling, is determined by the sum of the differences $(u_i - l_i)$. Therefore, the decoupling bounds $l = (l_i)_{i=1}^n$ and $u = (u_i)_{i=1}^n$ are chosen in such a way that the flexibility $\sum_i (u_i - l_i)$ is maximized. It can be shown (see [14]) that such a pair (l, u) can be obtained as a maximizer of the following LP:

---

[5] In other words, the set $\bigcup_{i=1}^k C'_i$ logically implies the set C of original constraints.

$$\max_{l,u} \sum_i (u_i - l_i) \qquad\qquad \textbf{(TD}(D)\textbf{)}$$

$$\text{s.t.} \quad l_0 = u_0 = 0 \qquad\qquad\qquad\qquad\qquad (1)$$
$$l_i \le u_i \qquad\qquad \forall i \in T \qquad\qquad (2)$$
$$u_j - l_i \le d_{ij} \qquad\qquad \forall i \ne j \in T \qquad (3)$$

where D is the minimum distance matrix associated with S.

Example 4. Consider the matrix D in Example 2 and the scenario sketched in Example 3. Then the LP whose maximizers determine a maximum decoupling is the following:

$$\max_{l,u} \quad \sum_i (u_i - l_i)$$
$$\text{s.t.} \quad u_0 = l_0 = 0,$$
$$l_1 \le u_1, \quad l_2 \le u_2$$
$$u_1 - l_0 \le 15, \quad u_2 - l_0 \le 19$$
$$u_0 - l_1 \le -5, \quad u_2 - l_1 \le 4$$
$$u_0 - l_2 \le -8, \quad u_1 - l_2 \le 2$$

Solving this LP, we obtain $\sum_{i=0}^{2}(u_i - l_i) = 6$ with maximizers $l = (0, 15, 13)$ and $u = (0, 15, 19)$. This implies that in this decoupling (l, u) agent a_1 is forced to arrive at 12:15, while agent a_2 can choose to arrive between 12:13 and 12:19. ∎

Remark. Note that the total decoupling solution (l, u) for S also is a solution for a decoupling based on an arbitrary partitioning $[T_i]_{i=1}^k$ of S. Observe that (l, u) is a decoupling, if for any value v_i chosen by a_h and any value w_j chosen by $a_{h'}$ for every $1 \le h \ne h' \le k$, we have $v_i - w_j \le d_{ji}$ and $w_j - v_i \le d_{ij}$. Since (l, u) is a total decoupling, it satisfies the conditions of the LP $\textbf{TD}(D)$. Hence, $u_i - l_j \le d_{ij}$ and $u_j - l_i \le d_{ji}$. Since $v_i \in [l_i, u_i]$ and $w_j \in [l_j, u_j]$, we then immediately have $v_i - w_j \le u_i - l_j \le d_{ji}$ and $w_j - v_i \le u_j - l_i \le d_{ji}$. Therefore, whatever choices are made by the individual agents satisfying their local constraints, these choices always will satisfy the original constraints, too. ∎

Remark. In [14] the (l, u) bounds found by solving the LP $\textbf{TD}(D)$ are used to compute the concurrent flexibility $flex(S) = \sum_i^n (u_i - l_i)$ of an STP S. Taking the concurrent flexibility as our flexibility metric, the (l, u) bounds for decoupling are always optimal, whatever partitioning $[T_i]_{i=1}^n$ is used: first, observe that due to a decoupling, the flexibility of an STN can never increase. Secondly, if the (l, u) bounds for a total decoupling are used, by definition, the sum $\sum_{i=1}^k flex(S_i)$ of the (concurrent) flexibilities of the decoupled subsystems equals the flexibility $flex(S)$ of the original system. Hence, the total decoupling bounds (l, u) are optimal, whatever partitioning $[T_i]_{i=1}^n$ used. ∎

In this paper, we will consider concurrent flexibility as our flexibility metric. Hence, a total decoupling is always an optimal decoupling for any partitioning of variables. Therefore, in the sequel, we concentrate on total decouplings.

3   Total Decoupling by Minimum Matching

In the introduction we mentioned that an (optimal) total decoupling can be achieved in $O(n^3)$ time. In the previous section, we presented an LP to compute such a decoupling. If the STP has n variables, the LP to be solved has $2n$ variables and n^2 constraints. Modern interior-point methods solve LPs with n variables and m constraints in roughly $O(n^3 m)$ [13]. Thus, the complexity of solving total decoupling as an LP might be as high as $O(n^5)$. In a previous paper ([9]), we have shown that computing the concurrent flexibility of an STP can be reduced to a *minimum matching* problem (see [10]) using the distance matrix D to construct a weighted cost matrix D^* for the latter problem.

This reduction, however, does not allow us to directly compute the corresponding flexibility maximizers (l, u). In this section we therefore show that there is a full $O(n^3)$ alternative method for the LP-based flexibility method to compute the concurrent flexibility and the corresponding maximizers (l, u), thereby showing that an optimal total decoupling can be computed in $O(n^3)$.

Flexibility and Minimum Matching. Given an STN $S = (T, C)$ with minimum distance matrix D, let $flex(S)$ be its concurrent flexibility, realised by the maximizers (l, u). Hence, $flex(S) = f(l, u) = \sum_{i=1}^{n}(u_i - l_i)$. Unfolding this sum –as was noticed in [9]– we obtain

$$f(l, u) = \sum_{i \in T}(u_i - l_i) = \sum_{i \in T}(u_{\pi_i} - l_i) \tag{4}$$

for every permutation π of T.[6] Since (l, u) is a total decoupling, we have

$$u_j - l_i \le d_{ij} \qquad \qquad \forall i \ne j \in T \tag{5}$$
$$u_j - l_j = (u_j - z) + (z - l_j) \le d_{0j} + d_{j0} \qquad \forall j \in T \tag{6}$$

Hence, defining the modified distance matrix (also called *weight matrix*) $D^* = [d_{ij}^*]_{n \times n}$ by

$$d_{ij}^* = \begin{cases} d_{ij}, & 1 \le i \ne j \le n \\ d_{0i} + d_{i0}, & 1 \le i = j \le n \end{cases}$$

we obtain the following inequality:

$$f(l, u) \le \min\{\sum_{i \in T} d_{i\pi_i}^* : \pi \in \Pi(T)\} \tag{7}$$

---

[6] To avoid cumbersome notation, we will often use $i \in T$ as a shorthand for $t_i \in T$.

where $\Pi(T)$ is the space of permutations of T. Equation (7) states that the maximum flexibility of an STN is upper bounded by the value of a minimum matching in a bipartite graph with weight matrix D^*. The solution of such a matching problem consists in finding for each row i in D^* a unique column π_i such that the sum $\sum_{i \in T} d^*_{i\pi_i}$ is minimized. As we showed in [9], by applying LP-duality theory, Eq. (7) can be replaced by an equality: $f(l, u) = \min_{\pi \in \Pi(T)} \sum_{i \in T} d^*_{i\pi_i}$, so the concurrent flexibility $flex(S) = f$ can be computed by a minimum matching algorithm as e.g. the Hungarian algorithm, running in $O(n^3)$ time.

Finding a Maximizer (l, u) Using Minimum Matching. With the further help of LP-duality theory i.c., complementary slackness conditions [10], the following result is immediate:

Observation 1. *If π is a minimum matching with value m for the weight matrix D^*, then there exists a maximizer (l, u) for the concurrent flexibility $flex(S)$ of S, such that $flex(S) = m$ and for all $i \in T$, $u_{\pi_i} - l_i = d^*_{i\pi_i}$.*

Now observe that the inequalities stated in the LP-specification $\mathbf{TD}(D)$ and the inequalities $u_{\pi_i} - l_i = d^*_{i\pi_i}$ in Observation 1 all are binary difference constraints. Hence, the STP $S' = (T', C')$, where

$$T' = L \cup U = \{l_i \mid i \in T\} \cup \{u_i \mid i \in T\},$$
$$C' = \{u_i - l_j \le d^*_{ij} \mid i, j \in T\} \cup \{l_i - u_{\pi_i} \le -d^*_{i\pi_i} \mid i \in T\} \cup \{l_i - u_i \le 0 \mid i \in T\}$$

is an STP[7] and every solution $s' = (l_1, \dots l_n, u_1, \dots, u_n)$ of S' in fact is a maximizer (l, u) for the original STP S, since the flexibility associated with such a solution (l, u) satisfies $flex(S) \ge f(l, u) = \sum_{i \in T}(u_i - l_i) \ge \sum_{i \in T} d^*_{i\pi_i} = flex(S)$. Hence, this pair (l, u) is a maximizer realizing $flex(S)$.

In particular, the earliest and latest solutions of S' have this property. Hence, since (i) D^* can be obtained in $O(n^3)$ time, (ii) a minimum matching based on D^* can be computed in $O(n^3)$, and (iii) the STN S' together with a solution $s = (l, u)$ for it can be computed in $O(n^3)$, we obtain the following result:

Corollary 1. *Given an STN $S = (T, C)$ with distance matrix D, an optimal total decoupling (l, u) for S can be found in $O(n^3)$.*

4   Dynamic Decoupling by Updating

A temporal decoupling allows agents to make independent choices or commitments to variables they control. As pointed out in the introduction, we want to adapt an existing (total) temporal decoupling (l, u) whenever an agent makes a new commitment to one or more temporal variables (s)he controls. To show that such a commitment could affect the flexibility other agents have in making their possible commitments, consider our leading example again:

---

[7] We should note that this STP has two external reference variables $u_0 = l_0 = 0$.

Example 5. In Example 3 we obtained a temporal decoupling $(l, u) = ((0, 15, 13), (0, 15, 19))$. Here, agent 1 was forced to arrive at 12:15, but agent 2 could choose to arrive between 12:13 and 12:19. Suppose agent 2 commits to arrive at 12:13. As a result, agent 1 is able to arrive at any time in the interval $[9, 15]$. Then, by adapting the decoupling to the updated decoupling $(l', u') = ((0, 9, 13), (0, 15, 13))$, the flexibility of agent 1 could increase from 0 to 6, taking into account the new commitment agent 1 has made. If the existing commitment (l, u), however, is not updated, agent 1 still has to choose 12:15 as its time of arrival.   ■

In order to state this *dynamic decoupling* or *decouple updating* problem more precisely, we assume that at any moment in time the set T consists of a subset T_c of variables committed to and a set of not committed to, or *free*, variables T_f. The commitments for variables $t_i \in T_c$ are given as bounds $[l_i^c, u_i^c]$, specifying that for $i \in T_c$, t_i is committed[8] to choose a value in the interval $[l_i^c, u_i^c]$. The total set of commitments in T_c is given by the bounds (l^c, u^c). We assume that these committed bounds do not violate decoupling conditions.

Whenever (l, u) is a total decoupling for $S = (T, C)$, where $T = T_c \cup T_f$, we always assume that (l, u) respects the commitments, i.e. $[l_i, u_i] = [l_i^c, u_i^c]$ for every $t_i \in T_c$, and (l, u) is an optimum decoupling for the free variables in T_f, given these commitments. Suppose now an agent makes a new commitment for a variable $t_i \in T_f$. In that case, such a commitment $[v_i, w_i]$ should satisfy the existing decoupling, that is $l_i \leq v_i \leq w_i \leq u_i$, but as a result, the new decoupling where $[l_i, u_i] = [v_i, w_i]$, $T_c' = T_c \cup \{t_i\}$, and $T_f' = T_f - \{t_i\}$ might no longer be an optimal decoupling w.r.t. T_f' (e.g. see Example 5). In that case we need to update (l, u) and to find a better decoupling (l', u').

The decoupling updating problem then can be stated as follows:

Given a (possibly non-optimal) total decoupling (l, u) for an STP $S = (T, C)$, with $T = T_c \cup T_f$, find a new total decoupling (l', u') for S such that

1. no individual flexibility of free variables is negatively affected, i.e., for all $j \in T_f$, $[l_j, u_j] \sqsubseteq [l_j', u_j']$[9];
2. all commitments are respected, that is, for all $j \in T_c$, $[l_j'^c, u_j'^c] = [l_j^c, u_j^c]$;
3. the flexibility realized by (l', u'), given the commitments (l'^c, u'^c), is maximum.

Based on the earlier shown total decoupling problem **TD**(D), this decoupling update problem can also be stated as the following LP:

$$\max \quad \sum_j (u_j' - l_j') \qquad \qquad (\textbf{DTD}(D, T_c, T_f, (l, u))$$

---

[8] Note that this is slightly more general concept than a strict commitment of a variable t_i to one value v_i.

[9] That is, the interval $[l_j', u_j']$ contains $[l_j, u_j]$.

$$\text{s.t.} \quad u_0' = l_0' = 0 \tag{8}$$

$$u_j' - l_i' \leq d_{ij} \qquad \forall i \neq j \in T \tag{9}$$

$$u_j \leq u_j', \ l_j' \leq l_j \qquad \forall j \in T_f \tag{10}$$

$$l_j' = l_j, \ u_j' = u_j \qquad \forall j \in T_c \tag{11}$$

Here, (l, u) is the existing total decoupling.

In fact, by transforming **DTD**-instances into **TD**-instances, we can show that the dynamic decoupling (decoupling updating) problem can be reduced to a minimum-matching problem and can be solved in $O(n^3)$, too.[10]

5   A Fast Heuristic for Updating

Although the dynamic total decoupling problem can be reduced to the static temporal decoupling problem, in practice, the computational complexity involved might be too high. While an initial decoupling might be computed off-line, an update of a decoupling requires an on-line adaptation process. Since the costs of solving such a dynamic matching problem are at least $O(n^2)$ per update, we would like to alleviate this computational investment. Therefore, in this section we discuss a fast heuristic for the decoupling updating problem. Using this heuristic, we show that an updated decoupling can be found in (amortized) $O(n \log n)$ per update step if $O(n)$ updates are assumed to take place.

The following proposition is a simple result we base our heuristic on:

Proposition 1. *If (l, u) is a total decoupling for S with associated weight matrix D^*, then, for all $j \in T$, $l_j = \max_{k \in T}(u_k - d_{kj}^*)$ and $u_j = \min_{k \in T}(l_k + d_{jk}^*)$.*

Proof. Since (l, u) is a maximizer of the LP **TD**(D), $u_i - l_j \leq d_{ij}^*$, for every $i, j \in T$. Hence, for each $i, j \in T$, we have $l_j \geq \max_{k \in T}(u_k - d_{kj}^*)$ and $u_i \leq \min_{k \in T}(l_k + d_{ik}^*)$. Now, on the contrary, assume that for some $i \in T$, $l_i > \min_{k \in T}(u_k - d_{ki}^*)$. Then the bounds (l', u) where $l' = l$, except for $l_i' = \min_{k \in T}(u_k - d_{ki}^*)$, would satisfy the constraints $u_i - l_j' \leq d_{ij}^*$ for every $i, j \in T$, as well as $l_j' \leq u_j$ for every $j \in T$. Hence, (l', u) is a decoupling as well. But then (l', u) satisfies the conditions of the LP **TD**(D) but $\sum_j(u_j - l_j') > \sum_j(u_j - l_j)$, contradicting the fact that (l, u) is a maximizer of this LP. Hence, such an $i \in T$ cannot exist. The proof for $u_i < \min_{k \in T}(l_k + d_{ik})$ goes along the same line and the proposition follows. ∎

The converse, however, of this proposition is not true: not every solution (l, u) satisfying the two equalities is a maximum decoupling. It can only be shown that in such a case (l, u) is a *maximal* total decoupling. That means, if (l, u) satisfies

---

[10] As has been observed by a reviewer, there exists an $O(n^2)$ algorithm for the dynamic variant of the minimum-matching problem. It is likely that our dynamic decoupling problem can be solved by such a dynamic minimum matching algorithm in $O(n^2)$ time, too.

the equations above, there does not exist a decoupling (l', u') containing (l, u) that has a higher flexibility.

It turns out that these *maximal total decouplings* and their updates can be computed very efficiently: given a (non-maximal) total decoupling (l, u), and a set $T = T_c \cup T_f$ of committed and free variables, there exists a very efficient algorithm to compute a new total decoupling $[l', u']$ such that

1. (l', u') preserves the existing commitments: for all $j \in T_c$, $[l'_j, u'_j] = [l_j, u_j]$;
2. $[l', u']$ respects the existing bounds of the free variables: for all $j \in T_f$, $[l_j, u_j] \sqsubseteq [l'_j, u'_j]$;
3. (l', u') satisfies the conditions stated in Proposition 1 above w.r.t. the free variables, i.e. (l', u') is a maximal decoupling with respect to variables in T_f.

The following surprisingly simple algorithm finds a maximal flexible total decoupling for a set $T_f \cup T_c$ of free and committed temporal variables that satisfies these conditions. The algorithm iteratively updates the existing (l, u)-decoupling bounds for the variables in T_f until all free variables satisfy the equations stated in Proposition 1.

Algorithm 1. Finding an update (l', u') of an existing total decoupling (l, u)

Require: (l, u) is a total decoupling for S; $D^* = [d^*_{ij}]$ is the weight matrix; $T = T_f \cup T_c$;
1: $l' = l$; $u' = u$;
2: **for** $i = 1$ to $|T_f|$ **do**
3:     $\min_i := \max_{k \in T}(u'_k - d^*_{ik})$;
4:     $\max_i := \min_{k \in T}(l'_k + d^*_{kj})$;
5:     **if** $l'_i > \min_i$ **then**
6:         $l'_i \leftarrow \min_i$
7:     **if** $u'_i < \max_i$ **then**
8:         $u'_i \leftarrow \max_i$
9: **return** (l', u');

It is easy to see that the algorithm preserves the existing commitments for variables in T_c, since only bounds of free variables in T_f are updated.

It is also easy to see that the existing bounds $[l_j, u_j]$ of free variables $j \in T_f$ are respected: For every j it holds that either $u_j < \max_j$ ($l'_j > \min_j$, respectively) or $u_j = \max_j$ ($l'_j > \min_j$, respectively). Hence, $u'_j \geq u_j$ and $l'_j \leq l_j$.

To show that the obtained decoupling (l', u') is maximal with respect to the free variables, we state two key observations: first of all, in every step i it holds that $l'_i \geq \min_i$ and $u'_i \leq \max_i$, because (l, u) is a decoupling and the $(\min_i, \max_i)$ bounds are not violated during updating. Secondly, once the bounds (l'_i, u'_i) for a variable $i \in T_f$ have been updated to $\min_i$ and $\max_i$, (l'_i, u'_i) will never need to be updated again, since $\min_i$ depends on values u'_k that might increase or stay the same; hence $\min_i$ cannot decrease in subsequent steps. Likewise, $\max_i$ depends on values l'_k that might decrease or stay the same. Therefore, $\max_i$ cannot increase in later steps. Therefore, it is sufficient to update the bounds for the variables only once.

As a result, at the end all free variables have been updated and achieved their minimal/maximal bound. Then a maximal total decoupling has been obtained, since all variables in T_f will satisfy the equations stated in Proposition 1.

Example 6. Continuing our example, notice that the weight matrix D^* obtained via the minimum distance matrix D (see Example 2) equals

$$D^* = \begin{bmatrix} 0 & 15 & 19 \\ -5 & 10 & 4 \\ -8 & 2 & 6 \end{bmatrix}$$

Given the decoupling $(l, u) = ((0, 15, 13), (0, 15, 19))$ with $T_f = \{t_1, t_2\}$, let agent 2 commit to $t_2 = 13$. We would like to compute a new decoupling maximizing the flexibility for t_1. Note that $\min_1 = \max\{5, 5, 9\} = 9$ and $\max_1 = \min\{15, 25, 15\} = 15$. Hence, the heuristic finds an updated decoupling $(l', u') = ((0, 9, 13), (0, 15, 13))$.

Complexity of the Heuristic. As the reader quickly observes, a naive implementation of Algorithm 1 would require $O(n^2)$-time to obtain an updated decoupling: To update a single variable $i \in T_f$, we have to compute $\min_i$ and $\max_i$. This costs $O(n)$ per variable and there may be n variables to update.

Fortunately, if there are $O(n)$ updating steps, the computational cost per step can be significantly reduced: First compute, for each $i \in T$, a priority queue $Q_{\min(i)}$ of the entries $u_k - d^*_{ik}$, $k \in T$, and a priority queue $Q_{\max(i)}$ of the entries $l_k + d^*_{ki}$, $k \in T$. The initialisation of these priority queues will cost $O(n \cdot n \log n) = O(n^2 \cdot \log n)$.

After a new commitment for a variable j, say $[v_j, w_j]$, we have to compute a decoupling update. It suffices to update the priority queues $Q_{\min(i)}$ and $Q_{\max(i)}$ for every $i \in T_f$. In this case, the element $u_j - d^*_{ij}$ in the queue $Q_{\min(i)}$ has to be changed to $w_j - d^*_{ij}$ and the element l_j in queue $Q_{\max(i)}$ has to be changed to $v_j + d^*_{ji}$. Maintaining the priority order in the priority queues will cost $O(\log n)$ per variable. Hence, the total cost for computing a new decoupling are $O(n \log n)$. If there are $O(n)$ updates, the total cost of performing these $O(n)$ updates are $O(n^2 \cdot \log n) + O(n) \cdot O(n \cdot \log n) = O(n^2 \cdot \log n)$. Hence, the amortized time costs per update are $O(n \cdot \log n)$.

In the next section we will verify the quality of such maximal flexible decouplings experimentally, comparing them with maximum flexible total decouplings and a static decoupling.

6   Experimental Evaluation

We discuss an experimental evaluation of the dynamic total decoupling method. These experiments have been designed (i) to verify whether updating a decoupling indeed significantly increases the decoupling flexibility compared to using a static decoupling and (ii) to verify whether the heuristic significantly reduces the computational investments in updating as expected.

Material. We applied our updating method to a dataset of STP benchmark instances (see [11]). This dataset contains a series of sets of STP-instances with STPs of varying structure and number of variables. Table 1 contains the main characteristics of this benchmark set. The size of the STP-instances in this dataset varies from instances with 41 variables with 614 constraints to instances with 4082 nodes and 14110 constraints. In total, these sets contain 1122 STP-instances. We used MATLAB (R2016b) on an iMac 3.2 Ghz, Intel Core i5, with 24 Gb memory to perform the experiments.

Table 1. STP Benchmarksets used in the experiments.

Benchmark set	NR instances	Min vars	AV vars	Max vars
bdh	300	41	207	401
Diamonds	130	111	857	2751
Chordalgraphs	400	200	200	200
NeilYorkeExamples	180	108	1333	4082
sf-fixedNodes	112	150	150	150

Method. For each instance in the benchmark set we first computed a maximum decoupling (l, u) with its flexibility $flex = \sum_{j=1}^{n}(u_j - l_j)$, using the minimum matching method. Then, according to an arbitrary ordering $< t_1, t_2, \ldots t_n >$ of the variables in T, each variable t_i is iteratively committed to a fixed value v_i, where v_i is randomly chosen in an interval (l_i, u_i) belonging to the current temporal decoupling (l, u). After each such a commitment, we compute a new updated decoupling (l^i, u^i) where $T_c = \{t_1, \ldots t_i\}$ and $T_f = \{t_{i+1}, \ldots t_n\}$. The total flexibility of the new decoupling (l^i, u^i) is now dependent upon the $n-i$ free variables in T_f and will be denoted by f_h^i. We initially set $f_h^0 = flex$. In order to account for the decreasing number of free variables after successive updates, we compute after each update the heuristic update flexibility per free variable in T_f: $f_h^i/(n-i)$. To compare these flexibility measures with the static case, for the latter we take the total flexibility of the free variables $flex^i = \sum_{j=i+1}^{n}(u_j - l_j)$ and then compute the flexibility per free time variable without updating: $flex^i/(n-i)$.

As a summary result for each benchmark instance k, we compute

1. the average over all updates of the static flexibility per free variable: $av\_stat = \sum_{i=1}^{n} flex^i/((n-i) \cdot n)$
2. the average over all updates of the heuristic update flexibility per free variable: $av\_h = \sum_{i=1}^{n} f_h^i/((n-i) \cdot n)$

Note that the ratio $rel\_flex_h = av\_h/av\_stat$ indicates the impact of the heuristic updating method for a particular instance: a value close to 1 indicates almost no added flexibility (per time variable) by updating, but a value of 2 indicates that the flexibility (per time variable) doubled by updating the decoupling.

Finally, we collected the rel_flex_h results per instance for each benchmark set and measured their minimum, mean and maximum values per benchmark set.

Results. The rel_flex_h results are grouped by benchmark set and their minimum, mean, and maximum per benchmark set are listed in Table 2.

Table 2. Statistics of flexibility ratio's rel_flex_h of decoupling updates vss static decoupling per benchmark set.

Benchmark set	Min	Mean	Max
bdh-agent-problems	1.00	1.00	1.002
Diamonds	1.34	1.95	2.39
Chordalgraphs	1.08	1.31	1.65
NeilYorkeExamples	1.20	1.74	2.39
sf-fixedNodes	1.21	1.38	1.78

As can be seen, except for bdh-agent-problems, dynamic updating of a temporal decoupling increases the mean flexibility per variable rather significantly: For example, in the diamonds and NeilYorke benchmark sets, updating almost doubled the flexibility per free time variable.[11]

We conclude that, based on this set of benchmarks, one might expect a significant increase in flexibility if a decoupling is updated after changes in commitments have been detected, compared to the case where no updating is provided.

To verify whether the heuristic was able to reduce the computation time significantly, unfortunately, we can only present partial results. The reason is that for the more difficult instances in these benchmark sets, computing the updates with an exact minimum matching method was simply too time-consuming.[12] We therefore selected from each benchmark set the easy[13] instances and collected them in the easy-<benchmark> variants of the original benchmark sets. We then measured the mean computation time per benchmark set for both the exact and heuristic updating method and also the mean performance ratio rel_flex/rel_flex_h of the exact versus the heuristic update method. The latter metric indicates how much the exact method outperforms the heuristic method in providing flexibility per time variable. The results are collected in Table 3.

---

[11] The reason that in the bdh-agent problems the updating did not increase, is probably due to the fact that in these instances, as we observed, the flexibility was concentrated in a very few time point variables. Eliminating the flexibilities by commitments of these variables did not affect the flexibility of the remaining variables that much.

[12] Some of the more difficult benchmark problems even did not finish within 36 h.

[13] A benchmark problem instance in [11] was considered to be "easy" if the exact update method finished in 15 min. For the bdh-agent set we collected the instances until easybdh 8_10_50_350_49, for the diamonds set all instances up to diamonds-38-5.0, for the chordal graphs all instances until chordal-fixedNodes-150,5,1072707262, for the NeilYorkeExamples all instances until ny-419,10, and for the sf-fixedNodes the instances until sf-fixedNodes-150,5,1072707262.

Table 3. Comparing the time (sec.) and performance ratio of the exact and heuristic updating methods (easy variants of benchmark instances)

Benchmark set	CPU heuristic	CPU exact	Performance ratio
easy-bdh-agent-problems	0.21	2.75	1.00
easy-diamonds	0.38	67.65	1.05
easy-chordalgraphs	0.78	12.6	1.06
easy-NeilYorkeExamples	0.61	135.62	1.01
easy-sf-fixedNodes	0.13	4.71	1.03

From these results we conclude that even for the easy cases, the heuristic clearly outperforms the exact update method, being more than 10 times and sometimes more than 200 times faster. We also observe that the heuristic method does not significantly lose on flexibility: The performance ratio's obtained are very close to 1.

7 Conclusions and Discussion

We have shown that adapting a decoupling after a new variable commitment has been made in most cases significantly increases the flexibility of the free, non-committed, variables. We also showed that this updating method did not induce any disadvantage to the actors involved: every commitment is preserved and the current decoupling bounds are never violated. We introduced a simple heuristic that replaces the rather costly computation of a decoupling with maximum flexibility by a computation of a decoupling with maximal flexibility. This heuristic reduces the computation time per update from $O(n^3)$ to $O(n \cdot \log n)$ (amortized) time. As we showed experimentally, the computational investments for the heuristic are significantly smaller, while we observed almost no loss of flexibility compared to the exact method.

In the future, we want to extend this work to computing flexible schedules for STPs: Note that the update heuristic also can be used to find a maximal flexible schedule given a solution s of an STP. Such a solution is nothing more than a non-optimal decoupling (s, s) for which, by applying our heuristic, an $O(n \log n)$ procedure exists to find an optimal flexible schedule. Furthermore, we are planning to construct dynamic decoupling methods in a distributed way, like existing approaches to static decoupling methods have done.

References

1. Boerkoel, J., Durfee, E.: Distributed reasoning for multiagent simple temporal problems. J. Artif. Intell. Res. (JAIR) **47**, 95–156 (2013)
2. Brambilla, A.: Artificial Intelligence in Space Systems: Coordination Through Problem Decoupling in Multi Agent Planning for Space Systems. Lambert Academic Publishing, Germany (2010)

3. Dechter, R.: Constraint Processing. Morgan Kaufmann, USA (2003)
4. Dechter, R., Meiri, I., Pearl, J.: Temporal constraint networks. Artif. Intell. **49**, 61–95 (1991)
5. Floyd, R.: Algorithm 97: shortest path. Commun. ACM **5**(6), 345 (1962)
6. Hunsberger, L.: Algorithms for a temporal decoupling problem in multi-agent planning. In: Proceedings of the Association for the Advancement of Artificial Intelligence (AAAI-02), pp. 468–475 (2002)
7. Hunsberger, L.: Group decision making and temporal reasoning. Ph.D. thesis, Harvard University, Cambridge, Massachusetts (2002)
8. van Leeuwen, P., Witteveen, C.: Temporal decoupling and determining resource needs of autonomous agents in the airport turnaround process. In: Proceedings of the 2009 IEEE/WIC/ACM International Joint Conference on Web Intelligence and Intelligent Agent Technology, WI-IAT 2009, vol. 02, pp. 185–192. IEEE Computer Society, Washington, DC, USA (2009)
9. Mountakis, S., Klos, T., Witteveen, C.: Temporal flexibility revisited: maximizing flexibility by computing bipartite matchings. In: Proceedings Twenty-Fifth International Conference on Automated Planning and Scheduling (2015). http://www.aaai.org/ocs/index.php/ICAPS/ICAPS15/paper/view/10610
10. Papadimitriou, C.H., Steiglitz, K.: Combinatorial Optimization: Algorithms and Complexity. Prentice-Hall Inc., Upper Saddle River (1982)
11. Planken, L.: Algorithms for simple temporal reasoning. Ph.D. thesis, Delft University of Technology (2013)
12. Planken, L., de Weerdt, M., van der Krogt, R.: Computing all-pairs shortest paths by leveraging low treewidth. J. Artif. Intell. Res. **43**(1), 353–388 (2012). http://dl.acm.org/citation.cfm?id=2387915.2387925
13. Potra, F.A., Wright, S.J.: Interior-point methods. J. Comput. Appl. Math. **124**(1–2), 281–302 (2000). http://www.sciencedirect.com/science/article/pii/S0377042700004337. Numerical Analysis 2000. Vol. IV: Optimization and Nonlinear Equations
14. Wilson, M., Klos, T., Witteveen, C., Huisman, B.: Flexibility and decoupling in the simple temporal problem. In: Rossi, F. (ed.) Proceedings International Joint Conference on Artificial Intelligence (IJCAI), pp. 2422–2428. AAAI Press, Menlo Park, CA (2013). http://ijcai.org/papers13/Papers/IJCAI13-356.pdf
15. Wilson, M., Klos, T., Witteveen, C., Huisman, B.: Flexibility and decoupling in simple temporal networks. Artif. Intell. **214**, 26–44 (2014)

A Multi-stage Simulated Annealing Algorithm for the Torpedo Scheduling Problem

Lucas Kletzander[(✉)] and Nysret Musliu

TU Wien, Karlsplatz 13, 1040 Wien, Austria
{lkletzan,musliu}@dbai.tuwien.ac.at

Abstract. In production plants complex chains of processes need to be scheduled in an efficient way to minimize time and cost and maximize productivity. The torpedo scheduling problem that deals with optimizing the transport of hot metal in a steel production plant was proposed as the problem for the 2016 ACP (Association for Constraint Programming) challenge. This paper presents a new approach utilizing a multi-stage simulated annealing process adapted for the provided lexicographic evaluation function. It relies on two rounds of simulated annealing each using a specific objective function tailored for the corresponding part of the evaluation goals with an emphasis on efficient moves. The proposed algorithm was ranked first (ex aequo) in the 2016 ACP challenge and found the best known solutions for all provided instances.

Keywords: Torpedo scheduling · Simulated annealing · Lexicographic evaluation function

1 Introduction

Production plants have a wide range of complex chains of processes that raise the need for optimization. To be competitive, cost needs to be minimized and efficiency and productivity need to be maximized. A selected scheduling problem in steel production was chosen for the 2016 ACP (Association for Constraint Programming) challenge [12]. The aim was to provide the best solutions for the torpedo scheduling problem where the transport of hot metal across various zones needs to optimized. The goal is to use as few transport vehicles (torpedoes) as possible and keep the time spent for a chemical process called desulfurization low while satisfying all deadlines and moving through the zones respecting capacity and duration constraints.

So far we know one other approach to the same problem by Geiger [4] ranking third in the competition. He used a branch and bound procedure to traverse feasible transport assignments and solved a resource-constrained scheduling problem based on variable neighborhood search to minimize desulfurization times. Other optimization problems in steel production have been researched in the past years. The molten iron allocation problem, modeled as a parallel machine scheduling problem [13] and the molten iron scheduling problem modeled as a flow shop

© Springer International Publishing AG 2017
D. Salvagnin and M. Lombardi (Eds.): CPAIOR 2017, LNCS 10335, pp. 344–358, 2017.
DOI: 10.1007/978-3-319-59776-8_28

problem [6,9] deal with assignments of torpedoes to machines. Various torpedo scheduling problems are defined for planning the transport of hot metal with the focus on vehicle routing across a network of rails and the objective to minimize transportation times [2,7,10]. Further production stages of steel making are also considered, e.g. steelmaking-continuous casting [14] which comes after the stage considered in this paper.

The approach we used is to utilize a multi-stage simulated annealing process adapted for the provided lexicographic evaluation function. The technique to simulate the physical process of annealing was introduced in [8] and is a widely used technique in many applications [3,11]. Applications of multi-stage algorithms can be found in the domain of vehicle routing problems with time windows. These applications range back to [5] and also include application of simulated annealing [1], however, only for one stage. Several of these problems share a primary goal to minimize a number of vehicles, but pursue very different secondary objectives.

In our approach the first round of simulated annealing focuses on the primary goal of optimizing the number of torpedoes while the second round deals with the secondary goal of reducing the desulfurization time. The algorithm is designed to try a large number of moves in a short time by emphasis on efficient move calculations. Various parameters for the algorithm were determined by empiric evaluation. With this process it was possible to obtain the best solutions found in the competition for all given instances.

2   Problem Definition

The selected problem represents a part of the steel production process in a steel production plant. The *blast furnace (BF)* continually produces hot metal with a certain level of sulfurization that needs to be picked up at certain deadlines. From there the metal is transported via torpedoes which are cars suited for the transport of hot metal. There are two possible routes the torpedoes can take.

The first and standard route is to transition to a buffer zone (*full buffer, FB*) where torpedoes can wait for an arbitrary amount of time. Afterwards they reach the *desulfurization zone (D)* where the sulfurization level of the hot metal can be lowered. The time that needs to be spend at this station is proportional to the desired decrease in sulfurization levels. The delivery zone is at the *converter (C)* where the hot metal is unloaded from the torpedoes. The hot metal is required at certain deadlines at the converter. However, there is a maximum sulfuriziation level set for each deadline that the delivered hot metal must respect. Finally, the torpedo returns to another buffer zone (*empty buffer, EB*) where empty torpedoes wait for their next turn.

The other route can be used in order to get rid of excess hot metal that is not needed at the converter. It consists of a transition to the *emergency pit* where the torpedo disposes the hot metal and then returns to the buffer zone for empty torpedoes (EB). This route takes a fixed amount of time.

2.1   Instance Representation

For each instance of the problem several global parameters specifying the details of the production plant are given. *durBF* specifies the duration of filling a torpedo at the blast furnace. *durConverter* specifies the duration of unloading a torpedo at the converter.

The sulfurization level of hot metal is assumed to have five possible levels from 1 (low) to 5 (high). *durDesulf* specifies the time needed at the desulfurization zone to lower the sulfurization level by 1.

Further the three parameters specifying the amount of available slots in individual zones are *nbSlotsFullBuffer*, *nbSlotsDesulf* and *nbSlotsConverter*. These denote the maximum number of torpedoes that are allowed to stay at the zone at any given time. For the blast furnace and any transition between stations only one torpedo is allowed at the same time. There are no limits for the empty buffer and the emergency pit.

Finally the minimum transition times between any two adjacent zones are specified via the values *tt⟨Zone1⟩To⟨Zone2⟩*.

The goal of the algorithm is to schedule the routes of the torpedoes for an extended period of time. For this purpose, a series of blast furnace deadlines and converter deadlines is included in every instance.

Blast furnace deadlines are described as *BF ⟨id⟩ ⟨time⟩ ⟨sulf⟩* where ⟨id⟩ identifies the individual deadline. ⟨time⟩ denotes the exact point in time when the hot metal will be released, therefore at this point in time a torpedo must be waiting to receive the metal. ⟨sulf⟩ denotes the sulfurization level the metal will have.

Converter deadlines are similarly described as *C ⟨id⟩ ⟨time⟩ ⟨maxSulf⟩* where ⟨id⟩ identifies the individual deadline. ⟨time⟩ denotes the exact point in time when the hot metal has to be provided by a torpedo located at the converter. ⟨maxSulf⟩ denotes the maximum sulfurization level that can be accepted.

The instance files for the competition can be found on the web page.[1]

2.2   Solution Representation

The solution format contains the used number of torpedoes *nbTorpedoes*, followed by a list of torpedo tours. Each tour first contains three IDs. *idTorpedo* specifies the torpedo used for this tour in the range [0, *nbTorpedoes* − 1]. *idBF* specifies the ID of the blast furnace output the torpedo receives on this tour and *idC* denotes the ID of the converter demand the torpedo satisfies on this tour or −1 in case the metal is disposed at the emergency pit.

Next for each tour a set of time points is given, determining the exact sequence of the tour. For each zone *start⟨Zone⟩* and *end⟨Zone⟩* are specified according to the order the torpedo passes through the zones. The time points from *startFB* to *endC* are only specified for regular tours, but are missing for tours using the emergency pit.

2.3   Evaluation Function

The function given for the evaluation of the solution quality considers the necessary number of torpedoes and the time spent in the desulfurization zone. First a solution is only valid if all deadlines at the blast furnace and the converter are met, the maximum sulfurization levels are respected, torpedoes move through the system correctly according to the specified transition and duration times and capacity constraints in the various zones are respected.

Then valid solutions are evaluated according to the lexicographic evaluation function that ranks solutions first based on the number of torpedoes and then on the total desulfurization time *timeDesulf*. More precisely the function

$$cost = nbTorpedoes + timeDesulf/(upperBoundTorpedo \cdot durDesulf) \qquad (1)$$

is used and the goal is to minimize the cost. Here *upperBoundTorpedo* is four times the number of converter deadlines as in the worst case every single torpedo tour to the converter needs to lower the sulfurization level from 5 (maximum) to 1 (minimum), therefore spending $4 \cdot durDesulf$ in this zone each time. This way for any reasonable solution the desulfurization time is normalized to the interval $[0,1)$, the integer value of the cost represents the number of torpedoes and the part after the comma represents the desulfurization time, enforcing the lexicographic order.

3   Solution Approach

To solve the given problem we introduce a two-stage simulated annealing based algorithm. We first formulate solution candidates and a set of equations to efficiently track data determining the solution quality. We then provide the overview of the algorithm and describe the motivation for the design of individual parts and their contributions for improving the solution quality. Emphasis is put on the design of the objective functions to use the power of the two-stage approach.

3.1   Representing a Solution Candidate

We model a possible solution with the concept of a *torpedo run*. Such a run represents one round trip of a torpedo and is tied to exactly one blast furnace deadline and either one converter deadline or the emergency pit. This is very similar to a tour in the required solution representation except that no specific torpedo ID is assigned. It allows to determine the amount of such runs directly from the given instance. The total amount of runs is equal to the number of blast furnace deadlines. The amount of runs targeting the converter is equal to the number of converter deadlines and the rest is targeting the emergency pit.

As a solution candidate we use an array of such torpedo runs that contains the correct number of runs targeting the converter and emergency pit. The algorithm is designed to respect the order of zones, prevent violations of zone durations and transition times and to always meet the blast furnace deadlines. All other violations of the constraints previously described are possible in the algorithm. Instead of preventing them they are monitored and penalized by the objective functions used in the simulated annealing algorithm.

Monitoring Constraint Violations. Constraint violations are tracked by maintaining an array of 10 values called *badness* each representing a certain kind of violation that can be individually weighted in the objective function.

The first set of constraint violations regarding the converter demands can be described as follows given the notation that R_C is the set of torpedo runs targeting the converter and BF and C hold the corresponding blast furnace and converter deadline objects:

$$\sum_{i \in R_C} \max\{startC_i - C[idC_i] . time, 0\} \tag{2}$$

$$diff_i = BF[idBF_i] . sulf - \left\lfloor \frac{endD_i - startD_i}{durD} \right\rfloor - C[idC_i] . maxSulf \tag{3}$$

$$\sum_{i \in R_C} \max\{diff_i, 0\} \tag{4}$$

Equation (2) counts the total converter deadline miss time by summing up the delay across all runs i that miss their deadline, in which case the start time at the converter $startC_i$ will be after the time of the assigned converter deadline.

Equation (3) first calculates the final sulfurization level for torpedo run i using the level at the blast furnace and the amount of time spend in the desulfurization zone. By subtracting the maximum allowed level at the converter it calculates the difference $diff_i$ between the maximum allowed level and the actual level. Values below 0 mean that the hot metal has lower sulfurization level than the allowed maximum indicating a potentially wasteful, but feasible pairing. A value of 0 exactly meets the requirement and values above 0 indicate sulfurization level misses and therefore constraint violations. Equation (4) sums up this difference in sulfurization levels across all runs that miss this requirement.

For each of the remaining capacity constraints the algorithm maintains an array with the size T equal to the amount of time units in the whole planning period. One such array is maintained for each capacitated zone (cBF, cFB, cD and cC) and for the corresponding transitions ($cBFtoFB$, $cFBtoD$, $cDtoC$ and $cCtoE$). Each element of such an array counts the amount of torpedo runs using the respective zone or transition at the given point in time.

To track the violations for each of these arrays X the corresponding badness value is calculated by

$$\sum_{0 \leq t < T} \max\{X[t] - maxOccupation_X, 0\} \tag{5}$$

where $maxOccupation_X$ is either one of the given capacities *nbSlotsFullBuffer*, *nbSlotsDesulf*, *nbSlotsConverter* for the corresponding arrays or 1 in all other cases.

Monitoring Optimization Goals. To keep track of the number of torpedos another array spanning across the whole time span of the planning period is used. The array *occupation* counts for each point in time how many torpedo runs are currently active. Therefore the maximum value of this array represents the current value of the main evaluation goal.

However, for this array tracking the sum of all values is not good enough for use in the objective functions. Therefore the array *occupationCount* counts the number of elements of *occupation* having a certain value by

$$occupationCount[i] = \text{count}\{t : occupation[t] = i\}, \tag{6}$$

e.g. *occupationCount*[1] counts the amount of time points where exactly one torpedo run is active. This concept allows to individually weight specific levels of occupation.

Finally the desulfurization is tracked by *timeDesulf* holding the total amount of desulfurization time for all torpedo runs. Further the array *desulfMismatch* keeps track of the difference between the sulfurization levels of the metal provided at the blast furnace compared to the required levels at the converter. It calculates

$$desulfMismatch[i] = \text{count}\{j : BF[idBF_j] . sulf - C[idC_j] . maxSulf = i\} \tag{7}$$

for $0 \leq i < 5$, e.g. *desulfMismatch*[3] counts the amount of torpedo runs that need at least 3 desulfurization steps in order to be feasible. This again allows individual weights for specific difference values.

3.2   The Simulated Annealing Algorithm

The simulated annealing process is done in two rounds using almost the same parameters, but very different objective functions. The general design of each round uses the usual process of simulated annealing:

```
generateInitialSolution();
for round = 0...1 {
  value = evaluateSolution();
  t = value/10;
  for outer = 0...10000 {
    for inner = 0...innerIterations() {
      chooseMove();
      newValue = evaluateSolution();
      if (shouldAccept()) {
        acceptMove();
      } else {
        abortMove();
      }
    }
    t *= 0.998;
  }
}
```

Heuristic Generation of the Initial Solution Candidate. The concept of *generateInitialSolution*() is to take the ordered lists of blast furnace and converter deadlines and pair them according to this ordering. The torpedo runs are initialized to spend the minimum amount of time in the desulfurization zone that allows them to meet the sulfurization level requirement of the assigned converter. They arrive at the converter just at the time of the deadline or too late in case this pairing is actually not feasible and spend any required waiting time in the full buffer.

As long as emergency pit runs are available, they are set whenever it is possible to put the current blast furnace output to the emergency pit and still transport the next blast furnace output to the current converter deadline in time. As emergency pit runs tend to get scheduled a bit too early by this approach and cause converter deadline misses a few runs later, the algorithm includes backtracking on such deadline misses to remove earlier emergency pit runs again and schedule those later.

At first time and space proportional to the total length of the planning period are required as all the capacity tracking arrays need to be initialized. The effort of constructing the initial solution is proportional to the number of blast furnace deadlines (actually not linear due to the backtracking, but in practice still very close) times the duration of a run as each run needs to be added to the capacity tracking system.

Note that in most cases the initial solution will not be feasible as some degree of constraint violation in regard to missing deadlines and capacity constraints is to be expected. However, it is designed to have a structure that produces a small amount of constraint violations while still keeping the execution very fast.

Parameter Tuning. The parameters for the algorithm were carefully chosen by experimental evaluation of various combinations of parameters to increase the performance. In the following the best values used in the final computation for the competition are presented. Reasons for the provided choice are given as well as problems encountered with different values. Unless otherwise stated, changes in most parameters only resulted in small changes in the results.

Iteration Parameters. For each round the algorithm uses a fixed number of outer iterations that was set to 10000. This value ensures that the algorithm converges to a stable result.

The temperature was set to start with a value of one tenth of the initial value of the objective function. In each outer iteration the temperature is decreased by multiplying with the factor 0.998. This choice, especially combined with the starting temperature and the number of inner iterations was one of the more critical choices for the quality of the results and therefore subject of detailed empirical evaluation. The initial solution is constructed in a way to already have a structure limiting the amount of constraint violations. While a certain increase in such violations is expected in the early stage of the simulated annealing process to prevent getting stuck too close to the initial solution, keeping the temperature high for too long destroys the structure of the initial solution requiring extensive amounts of repair at lower temperatures that make the overall result worse. On the other hand, dropping the temperature too fast leads to getting stuck in local optima.

The number of inner iterations *innerIterations*() depends on the size of the instance, more precisely it is the number of blast furnace deadlines. This choice was made as the number of possible moves also depends on this value. For the second round experiments showed an increase by a factor of 4 to be beneficial.

Efficient Moves. We proposed three moves that are chosen randomly with certain probabilities in *chooseMove*(). The first move is a switch between the assigned converter deadlines for two torpedo runs. It is chosen with a probability of 0.4 in the first round and 0.6 in the second round. The rather high probability is due to the fact that this is the move with the highest impact on the structure of the solution. The selection of the two runs is not randomized, but actually a sensitive choice. The reason is that choosing two runs at very different time points in the whole planning period will likely not be a good move as large deadline misses can be expected. On the other hand, only switching closely adjacent runs will likely end up in local optima too soon. Therefore selecting the distance of the two runs when sorted by their start times randomly between 1 and 10 showed to provide good results.

Additionally runs are locally improved after such a move to reduce the amount of converter deadline misses and sulfurization level misses. First the time spent at the desulfurization station is set to exactly the amount of time needed to pass the required maximum sulfurization level at the converter. Second if the converter deadline is missed the time at the full buffer is reduced just enough to get the deadline, or to 0 if the deadline miss is larger than the full buffer time.

The other moves consist of changing the time spend at the full buffer (probability 0.4 in the first round and 0.2 in the second), the time spend at the desulfurization zone (probability 0.1) or the time spend at the converter (probability 0.1). As these moves are intended to change the internal structure of a run towards a good feasible solution, new values for the respective times are chosen randomly within limits preventing a converter deadline miss if possible.

The key concept in tracking the capacity and goal data is to allow very fast calculation of changes triggered by moves in the algorithm and therefore be able to try a lot of moves in a short amount of time. For the moves it is necessary to first compute the effects of the move and then either accept or reject it. Therefore, every move is reflected by first creating copies of the torpedo runs that are affected by the move. Then all tracking data is updated by removing the original runs and adding the changed copies. In case the move is rejected, the copies are removed again and the originals added back to the tracking data. In case it is accepted, the copies replace the originals in the array of torpedo runs.

The important aspect is that all badness and goal tracking data can be updated incrementally. The sums or counts in (2), (4) and (7) allow easy removal and addition of torpedo runs. For (5) and (6) only the parts of the arrays X and *occupation* affected by the currently changed torpedo runs need to be updated, the effects on the sum and the count can easily be computed incrementally again.

Using this principle it is possible to update all tracked data in time that just depends on the duration of the respective torpedo runs that are removed or added. This duration is usually very small compared to the whole time span of the planning period and is independent of the number of torpedo runs.

Moves are accepted by *shouldAccept*() if their evaluation yields a better or equal result according to *evaluateSolution*() or else with a probability of $\exp\left(\frac{value - newValue}{t}\right)$.

Selection Bias. As the main objective in the first round is to reduce the maximum number of torpedoes used, optimization in areas with already low numbers of torpedoes might not be relevant for the result at all while a single point with a high number determines the value of the result. Therefore, the selection for a move is biased to prefer runs in areas with a high number of torpedoes in use. On the other hand, for the desulfurization time the total sum is relevant, therefore every optimization matters and no selection bias is used in the second round.

3.3   Objective Functions

To use the power of the two-stage approach we proposed specific objective functions for *evaluateSolution*() for each round tailored to the respective goals:

$$\sum_{0 \le i < 10} w_1[i] \cdot badness[i] + \sum_{i \ge 0} occupationCount[i] \cdot i^4 + cost \qquad (8)$$

$$\sum_{0 \le i < 10} w_2[i] \cdot badness[i] + \sum_{i > fixed} 1000000 \cdot occupationCount[i] \cdot i^4 +$$

$$\sum_{0 \le i < 5} 10000 \cdot desulfMismatch[i] \cdot i + timeDesulf \qquad (9)$$

Equation (8) denotes the objective function for the first round and (9) the objective function for the second round. Again, the weights were chosen by experimental evaluation.

Constraint Violations. First, both objective functions take into account the constraint violations maintained by *badness*, however, with different weight vectors w_1 and w_2.

Both objective functions use a weight of 100000 for the total converter deadline miss time as missing such deadlines potentially indicates structural problems of the solution and therefore is considered a priority for optimization.

For optimization of the sulfurization level misses the first round again uses a weight of 100000 as it focuses on finding a feasible solution with the least possible amount of torpedoes. The high value ensures the focus on feasibility. For the second round, however, the weight is only 1000 as there is a special part of the objective function that also deals with the sulfurization level misses in more detail.

Capacity constraint violations are all penalized by a weight of 10000 in the first round. Again the values were chosen rather high to focus on feasibility. For the second run weights for capacity misses at specific zones are only weighted by 10, for transitions by 1000. Transitions need to be weighted higher as their maximum occupation of 1 leaves less margin in general, but the weights for the zones were chosen very low to prevent the algorithm to get stuck in local optima. This actually introduces a small probability for constraint violations still present in the final result, however, higher values focused the process more on these constraints in the first place and only afterwards optimizing the desulfurization times within the limits already set by the constraints while the low values allow a kind of parallel optimization of both desulfurization times and capacity violations at a similar pace.

Number of Torpedoes. The next part of the objective functions uses the *occupationCount* array. For the first round each element *occupationCount*[i] is weighted by i^4. This polynomial weighting strategy ensures a strong optimization towards a small number of currently active torpedo runs at any point in time with a special emphasis on eliminating areas using a high number of torpedoes. This showed to be a successful approach to optimize the first objective.

For the second round the goal of this part of the objective function is completely changed. This part is the reason for choosing to use two separate rounds of optimization in the first place. The key point is the structure of the solution created in the first round. The number of currently active torpedoes is kept as low as possible across the whole time span in order for the optimization to work. However, as only the maximum number of torpedoes counts for the value of the solution and this maximum value might only be reached at a small part of the whole process, this kind of optimization restricts the possibilities to optimize the desulfurization time more than necessary. Section 4.1 will highlight this in the results.

Therefore, the second round memorizes the amount of torpedoes reached in the first round (*fixed*) and sets the weight for every i up to this value to 0. For all i above this value previous weights are increased by a factor of 1000000. This allows free use of any number of torpedoes up to the set limit allowing much more flexibility for the reduction of desulfurization times by utilization of torpedoes that would otherwise be on standby. On the other hand the excessive weights for going beyond this limit ensure that the optimization result from the first round is kept throughout the second round.

Desulfurization. In the first round the objective function is completed by adding the actual evaluation function used for the final solution as given by (1). This adds the optimization of desulfurization times as a low priority goal to the process.

In the second round the desulfurization times are included in more detail to encourage better matching of blast furnace and converter deadlines with respect to their sulfurization levels. To incorporate the difference in initial sulfurization levels compared to the actual converter level demands the array *desulfMismatch* is used. A level miss is weighted by $10000 \cdot i$ where i counts the number of missed levels, therefore linearly penalizing the distance to the required level. Here a range of other methods was tried as well, in particular using polynomial strategies like for the number of torpedoes or also penalizing levels that are lower than required in order to reduce potentially wasteful situations where torpedo runs with low level are used for converter demands with high maximum level. However, none of these strategies gave better results than the one described above.

Finally, the total desulfurization time is added to the objective function as well. Using a weight of 1, this (actually the overall optimization goal of the second round) is a rather low priority target in the optimization process. This is because putting more emphasis on parts like the sulfurization level mismatch works towards producing an optimal structure of the solution earlier. This is important especially regarding the assignment of converter deadlines to the torpedo runs. As such switches can easily produce at least temporary constraint violations it is beneficial to work on an optimized assignment while the temperature is still high and then focus on optimizations within runs by shifting times to reduce desulfurization times locally at a later point in the process.

4   Results

The algorithm was executed on an Intel Core i7-6700K with 4×4.0 GHz. Table 1 shows the main characteristics of the competition instances. The duration and capacity constraints were rather similar for the given instances while the main differences were in the number of blast furnace and converter deadlines and the time span covered. A single run for each instance used only 4 to 10 min on a single CPU core for all instances except 6. Here, even though the instance size is not that different, one run takes almost 30 min due to the structure of the instance creating longer torpedo runs.

4.1   Structure of a Solution

To see the importance of using two rounds of simulated annealing, data collected from one particular computation of instance 1 is presented after the creation of the initial solution and after each round of simulated annealing. The resulting distribution is similar in all instances, therefore it is only described for instance 1 to highlight the way the algorithms transforms the solution.

Table 2 shows the elements of *occupationCount* and *timeDesulf* for instance 1. All values *occupationCount*[i] with $i > 5$ are 0.

The initial solution generated by the heuristic typically uses only few torpedoes more than the final result, in this case 5 torpedoes are used. However,

Table 1. Instance characteristics

Instance	1	2	3	4	5	6
Blast furnace deadlines	850	1500	2200	1000	1800	2500
Converter deadlines	800	1400	2100	1000	1780	2350
Time span	131100	144165	394723	133798	256216	251460

Table 2. Objective values in various stages of the algorithm (instance 1)

Value	[0]	[1]	[2]	[3]	[4]	[5]	*timeDesulf*
Initial	41077	58754	25050	6091	239	11	18333
Round 0	68141	49830	12136	1053	62	0	18512
Round 1	681	17303	55387	46254	11597	0	7776

as to be expected, this solution is not feasible, capacity constraints are slightly violated at the transitions *FBtoD*, *DtoC* and *CtoE* as well as at the desulfurization zone. Figure 1 shows the distribution of the occupation values. The most frequent occupation at this stage is to have 0 or 1 torpedo active.

After the first round of simulated annealing the number of torpedoes was lowered to 4, however, the desulfurization time slightly increased despite the fact that desulfurization is added as a low level optimization goal to this stage. This solution is already feasible and fixes the number of torpedoes used in the final solution. As the figure shows, the distribution for *occupationCount* is shifted as far as possible to the lower indices. The highest number of torpedoes is only used at a very small number of time points. In fact, for almost half the time no torpedoes are active at all.

In the second round of simulated annealing this distribution completely changes as the algorithm now permits free use of any number of torpedoes up to

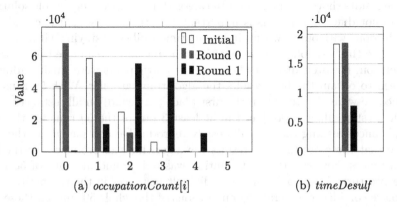

(a) *occupationCount[i]*   (b) *timeDesulf*

Fig. 1. Objective values in various stages of the algorithm (instance 1)

4 in order to optimize the desulfurization time as much as possible. The results show that there is almost no time left without any torpedo on the move while the most frequent occupation shifts to 2 and 3. This allows much more flexibility for optimizing the desulfurization time and results in less than half the time compared to the first round.

4.2   Results for Competition Instances

For the competition 50 runs were performed for each instance and the best result was handed in. There were no restrictions on computation times.

Table 3 presents the results obtained by the 50 runs for each instance. Some runs did not find the best number of torpedoes. In case a previous run already got less torpedoes, the run was aborted after the first round of simulated annealing (Abort 1). Some runs resulted in minor levels of constraint violations and were discarded (Abort 2). The amount of times the best result (Minimum) was produced is also shown (Freq.). Maximum, mean and standard deviation are calculated from runs that were not aborted only.

Table 3. Computation results (50 runs for each instance)

Instance	Abort 1	Abort 2	Minimum	Freq.	Maximum	Mean	Std. dev
1	1	4	4.0890625	3	4.09094907	4.089995	0.000484
2	1	0	4.08607143	16	4.08712662	4.086279	0.000217
3	16	0	3.12928571	5	3.12964286	3.129415	0.000080
4	0	3	3.157	47	3.157	3.157	0
5	6	1	4.08483146	1	4.08609551	4.085374	0.000282
6	1	0	4.075	1	4.07585106	4.075358	0.000172

The results show that in general the algorithm produces very stable solutions that do not differ much. This is important for practical use as the calculation could be done with only few runs in a short time while still staying within a short distance to the best solutions the algorithm might find given more runs. For the competition, however, distances between the participants were small leading to the need to obtain the best results the algorithm can offer. With 50 runs per instance we were able to obtain the first place (ex aequo) on all instances.

The best result was produced in at least 3 out of the 50 runs for the first four instances making those results easily reproducible and increasing the confidence that these are the best results the algorithm can produce. For the last two instances, however, the best solution was only found in 1 out of 50 runs. These two turned out to be the most difficult instances in the competition. To gain more confidence, several recomputations of the whole 50 runs on these two instances were able to reproduce these results, even though only about every second recomputation for instance 5, but did not find any better results.

Constraint violations in the results were only encountered in a minor fraction of the runs (maximum of 4 out of 50). Moreover, for all aborted instances the violations only occurred at most at 2 time units across the whole planning period. Given the fact that penalties for capacity violations were deliberately set low in the second round of simulated annealing to focus further on the optimization of desulfurization times this is considered a good result.

Instance 3, a large instance with the longest time span, yet a very small amount of torpedoes in the solution, showed to be the hardest instance for the first round. Here 32% of runs did not find the best amount of torpedoes compared to at most 12% for any other instance.

Instance 4 was clearly the easiest instance, being one of the smaller instances, but also providing equal amounts of blast furnace and converter deadlines. This eliminates the need for emergency pit runs. In fact, 47 out of 50 runs on this instance produced the best found result.

5   Conclusion

This paper presented an approach using a multi-stage simulated annealing process to solve a scheduling problem in steel production plants. Utilizing efficient moves it optimizes results for the lexicographic evaluation function using two different objective functions each tailored for optimal progress towards the corresponding part of the evaluation in a two-stage simulated annealing process.

The results show that the approach is a valid and competitive method to solve the given problem. As the general idea is generic, it can also be adapted to various other problems utilizing lexicographic evaluation functions. Further the framework to track changing capacity violations in a fast manner showed to improve efficiency of the algorithm.

Future work could include the adaption of the approach to various other problems in this domain to see how well it competes with different approaches. Further the selection of critical parameters for simulated annealing, especially regarding automated parameter selection, could be the goal of research.

Acknowledgments. This work was supported by the Austrian Science Fund (FWF): P24814-N23.

References

1. Bent, R., Van Hentenryck, P.: A two-stage hybrid local search for the vehicle routing problem with time windows. Transp. Sci. **38**(4), 515–530 (2004)
2. Deng, M., Inoue, A., Kawakami, S.: Optimal path planning for material and products transfer in steel works using ACO. In: The 2011 International Conference on Advanced Mechatronic Systems, pp. 47–50. IEEE (2011)
3. Dowsland, K.A., Thompson, J.M.: Simulated annealing. In: Rozenberg, G., Bck, T., Kok, J.N. (eds.) Handbook of Natural Computing, pp. 1623–1655. Springer, Heidelberg (2012)

4. Geiger, M.J.: Optimale Torpedo-Einsatzplanung - Analyse und Lösung eines Ablaufplanungsproblems der Stahlindustrie. In: Entscheidungsunterstützung in Theorie und Praxis - Tagungsband des gemeinsamen Workshops der GOR-Arbeitsgruppen "Entscheidungstheorie und -praxis", "Fuzzy Systeme, Neuronale Netze und Künstliche Intelligenz" sowie "OR im Umweltschutz" am 10. und 11. März 2016 in Magdeburg. Springer (in press)

5. Homberger, J., Gehring, H.: Two evolutionary metaheuristics for the vehicle routing problem with time windows. INFOR: Inf. Syst. Oper. Res. **37**(3), 297–318 (1999)

6. Huang, H., Chai, T., Luo, X., Zheng, B., Wang, H.: Two-stage method and application for molten iron scheduling problem between iron-making plants and steel-making plants. IFAC Proc. Volumes **44**(1), 9476–9481 (2011)

7. Kikuchi, J., Konishi, M., Imai, J.: Transfer planning of molten metals in steel worksby decentralized agent. Memoirs Fac. Eng. Okayama Univ. **42**(1), 60–70 (2008)

8. Kirkpatrick, S., Gelatt, C.D., Vecchi, M.P.: Optimization by simulated annealing. Science **220**(4598), 671–680 (1983)

9. Li, J., Pan, Q., Duan, P.: An improved artificial bee colony algorithm for solving hybrid flexible flowshop with dynamic operation skipping. IEEE Trans. Cybern. **46**(6), 1311–1324 (2016)

10. Liu, Y.Y., Wang, G.S.: The mix integer programming model for torpedo car scheduling in iron and steel industry. In: International Conference on Computer Information Systems and Industrial Applications, pp. 731–734. Atlantis Press (2015)

11. Pham, D., Karaboga, D.: Intelligent Optimisation Techniques: Genetic Algorithms, Tabu Search, Simulated Annealing and Neural Networks. Springer Science & Business Media, London (2012)

12. Schaus, P., Dejemeppe, C., Mouthuy, S., Mouthuy, F.-X., Allouche, D., Zytnicki, M., Pralet, C., Barnier, N.: The torpedo scheduling problem: description (2016). http://cp2016.a4cp.org/program/acp-challenge/problem.html. Accessed: 02 Feb 2017

13. Tang, L., Wang, G., Liu, J.: A branch-and-price algorithm to solve the molten iron allocation problem in iron and steel industry. Comput. Oper. Res. **34**(10), 3001–3015 (2007)

14. Tang, L., Zhao, Y., Liu, J.: An improved differential evolution algorithm for practical dynamic scheduling in steelmaking-continuous casting production. IEEE Trans. Evol. Comput. **18**(2), 209–225 (2014)

Combining CP and ILP in a Tree Decomposition of Bounded Height for the Sum Colouring Problem

Maël Minot[1,3]($\boxtimes$), Samba Ndojh Ndiaye[1,2], and Christine Solnon[1,3]

[1] Université de Lyon - LIRIS, Lyon, France
{mael.minot,samba-ndojh.ndiaye,christine.solnon}@liris.cnrs.fr
[2] Université Lyon 1, LIRIS, UMR5205, 69622 Lyon, France
[3] INSA-Lyon, LIRIS, UMR5205, 69621 Lyon, France

Abstract. The Sum Colouring Problem is an $\mathcal{NP}$-hard problem derived from the well-known graph colouring problem. It consists in finding a proper colouring which minimizes the *sum* of the assigned colours rather than the number of those colours. This problem often arises in scheduling and resource allocation. In this paper, we conduct an in-depth evaluation of ILP and CP's capabilities to solve this problem, with several improvements. Moreover, we propose to combine ILP and CP in a tree decomposition with a bounded height. Finally, those methods are combined in a portfolio approach to take advantage from their complementarity.

1 Introduction

The Sum Colouring Problem (SCP) is an $\mathcal{NP}$-hard problem derived from the well-known graph colouring problem. It consists in finding a proper colouring (*i.e.* an assignment of colours to vertices such that neighbour vertices have different colours) which minimizes the *sum* of the assigned colours rather than the number of those colours. This problem arises in a variety of real-world problems, especially in scheduling and resources allocation [21]. Many incomplete approaches have been proposed [12], whereas only few complete approaches have been proposed: mainly Integer Linear Programming (ILP) in [12], Branch and Bound (B&B), SAT and Constraint Programming (CP) in [22]. In this paper, we propose to conduct a more in-depth evaluation of ILP and CP's capabilities to solve the SCP, with several improvements. Moreover, we use tree decomposition to improve the solution process by decomposing SCPs into independent subproblems which are solved by ILP and CP, with promising results. Finally, we show that a portfolio approach can take advantage of the complementarity of the different approaches.

Section 2 defines the SCP and gives an overview of existing approaches. Sections 3 and 4 introduce improvements for CP and ILP models. Section 5 explains how CP and ILP may be combined by means of a tree decomposition, and Sect. 6 introduces a portfolio approach.

D. Salvagnin and M. Lombardi (Eds.): CPAIOR 2017, LNCS 10335, pp. 359–375, 2017.
DOI: 10.1007/978-3-319-59776-8_29

2   The Sum Colouring Problem

An undirected graph $G = (V, E)$ is defined by a set V of nodes and a set $E \subseteq V \times V$ of edges. Each edge of G is an undirected pair of nodes. We note $\deg(v)$ the degree of a vertex v, *i.e.* $\deg(v) = |\{u \in V, \{u, v\} \in E\}|$, and $\Delta(G)$ the largest degree in the graph, *i.e.* $\Delta(G) = \max\{\deg(v), v \in V\}$.

A legal (or "proper") k-colouring of a graph $G = (V, E)$ is a mapping $c : V \to [1, k]$ such that $\forall \{x, y\} \in E, c(x) \neq c(y)$. Classic graph colouring aims at finding a proper k-colouring that minimizes k, whereas the SCP aims at finding a proper k-colouring that minimizes the sum of assigned colours, *i.e.*, $\sum_{x \in V} c(x)$. The lowest achievable sum for G is called the *chromatic sum* and is denoted $\Sigma(G)$.

Existing Bounds. In [30], it is shown that $\lceil \sqrt{8|E|} \rceil \leq \Sigma(G) \leq \lfloor \frac{3(|E|+1)}{2} \rfloor$. In [21], it is demonstrated that $\Sigma(G) \leq |V| + |E|$, and that an optimal sum colouring will never use strictly more than $\Delta(G) + 1$ colours. Finally, in [24, 33] a lower bound is defined with respect to a clique decomposition of the graph.

Definition 1. *A* clique *is a subset of nodes which are all linked pairwise. A* clique decomposition *of a graph is a partition $\mathcal{C}$ of its vertices such that, for each set $C_i \in \mathcal{C}$, the subgraph induced by C_i is a clique.*

Given a clique decomposition $\mathcal{C}$ of G, we have $\Sigma(G) \geq \sum_{C_i \in \mathcal{C}} |C_i| \cdot (|C_i| + 1)/2$, as all vertices in a same clique must have different colours.

Dominant Colourings. Dominant colourings were discussed in [22]: a k-colouring may be seen as an ordered partition on the vertices of the coloured graph, such that the i-th set S_i contains all vertices using colour i (with $1 \leq i \leq k$). A colouring is dominated when the vertex colour sum can be lowered simply by reassigning the indices of these sets (*i.e.*, swapping colours) without actually altering the partition. The *dominant colouring* of a k-colouring c is the colouring obtained by ordering the partition defined by c by decreasing size of sets, so that S_1 is the largest set, and S_k the smallest. This dominant colouring has the lowest vertex colour sum among all possible colour swappings of c.

Incomplete Approaches. Many incomplete approaches were used to find approximate solutions for the SCP. A review of most of these approaches may be found in [12]. It classifies main contributions in three classes: greedy algorithms [32, 33], local search heuristics [3, 7] and evolutionary algorithms [11, 13, 23]. None of these algorithms are able to reach all best known upper and lower bounds. The percentage of instances on which the best known upper bound is reached ranges from 46% [32, 33] to 90% [11] on tested graphs. Such approaches can prove the optimality of a solution if the lowest upper bound happens to reach the highest lower bound. However, such proofs were only made on 21 instances out of 94, even when using all the bounds found by every methods in [12] simultaneously.

CP. A basic CSP model was proposed in [22]. For each node $u \in V$, this model includes a variable x_u whose domain is $D(x_u) = [1, \Delta(G) + 1]$. There is a disequality constraint for each edge of the graph, i.e., $\forall \{u, v\} \in E, x_u \neq x_v$. The objective is to minimize the sum of all variables, i.e., $\sum_{u \in V} x_u$. This model was evaluated using Choco [14]. The results were not competitive with state-of-the-art approaches: solution times on rather easy instances were long [22].

B&B. In [22], a B&B approach is described. At each node of the search, a lower bound is obtained by computing a clique decomposition $\mathcal{C}$ of the subgraph induced by uncoloured vertices. However, instead of bounding the colour sum for each clique $C_i \in \mathcal{C}$ by $|C_i| \cdot (|C_i|+1)/2$ (as proposed in [24,33]) the authors bound it by the sum of the $|C_i|$ smallest available colours among the vertices of C_i. This new bound is tighter since it takes – to some degree – the availability of colours into account. Besides, each time a proper colouring is found, its corresponding dominant colouring is computed to improve the bound. This approach obtains better results than the basic CP model, but remains limited to small graphs.

SAT. Different SAT encodings for the SCP are described in [22]. They are experimentally compared (using different SAT solvers) with B&B and CP, on randomly generated graphs and on six DIMACS instances. On these instances, the proposed B&B and CP approaches are not competitive with the SAT portfolio ISAC [15], which obtains the best results. In this paper, we introduce new CP models which are competitive with these SAT models, and an ILP model which outperforms them on the six DIMACS instances for which results were given.

ILP. An ILP model was proposed in [31]. It associates a binary variable x_{uk} with every pair $(u, k) \in V \times [1, \Delta(G) + 1]$, so that x_{uk} equals 1 iff node u uses colour k. The objective is to minimize the sum of the integers corresponding to used colours, i.e. $\sum_{u=1}^{|V|} \sum_{k=1}^{\Delta(G)+1} k \cdot x_{uk}$ so that each node $u \in V$ is assigned one colour, i.e., $\sum_{k=1}^{\Delta(G)+1} x_{uk} = 1$, and for each edge $\{u, v\} \in E$, u and v have different colours, i.e., $x_{uk} + x_{vk} \leq 1, \forall k \in [1, \Delta(G) + 1]$. This model was evaluated with CPLEX, showing that it is very efficient for small graphs, but that the memory cost was too high for larger ones.

3 New CP Models for the SCP

The CP model of [22] is very limited, as it only propagates binary difference constraints, and bounds the objective function with the sum of minimal values in domains. In this section, we propose and compare several improvements.

3.1 Initial Domain Reduction

Instead of using the same domain $[1, \Delta(G) + 1]$ for all variables, we propose to tighten domains by using the following property:

Property 1. For every optimal sum colouring c of a graph $G = (V, E)$, we have $\forall v \in V, c(v) \le \deg(v) + 1$.

To prove this property, let us suppose that it does not hold for a given optimal colouring c of a graph G. It follows that there exists a vertex v in V such that $c(v) > \deg(v) + 1$. In such a case, there exists $x \in [1, \deg(v) + 1]$ such that every neighbour of v has a colour different from x (since v only has $\deg(v)$ neighbours). As a consequence, a better colouring than c can be obtained by colouring v with x instead of $c(v)$. Therefore, c is not optimal, which contradicts our initial claim.

Hence, we define $D(x_u) = [1, \deg(u) + 1]$ for each $u \in V$. This is a minor but natural improvement, with a negligible cost.

3.2   Dominant Colourings

As pointed out in [22] and recalled in Sect. 2, colourings found during the search may be dominated, and can be improved simply by swapping colours. This makes the upper bound go down faster at a low computing cost. Besides, these swappings break symmetries by forbidding, thanks to the update of the upper bound, the computation of colourings that are dominated by the ones already found.

3.3   *AllDifferent* Constraints

Instead of using only binary disequality constraints to prevent neighbour vertices from being assigned the same colour, we propose to use *AllDifferent* constraints. This may be done in several different ways. A first possibility is to compute a clique decomposition of the graph, as defined in Definition 1. In this case, we post a global *AllDifferent* constraint for each clique, and a binary disequality constraint for each edge such that no clique contains its two endpoints. A second possibility is to compute a set of maximal cliques such that, for each edge, there exists at least one clique that contains its two endpoints. In this case, we post a global *AllDifferent* constraint for each maximal clique. This introduces redundancies in *AllDifferent* constraints and may prune more values, but at a higher cost.

For both approaches, we may consider different heuristics to build cliques. Several tests (not reported due to lack of space) showed us that a simple greedy construction of maximal cliques yields a good tradeoff between the time spent building the cliques, the time spent to propagate *AllDifferent* constraints, and the reduction of the search space. More precisely, for each vertex of the graph, we build a maximal clique in a greedy way: starting from a clique that contains this vertex, we iteratively choose the vertex with the largest degree among the vertices that can correctly extend the clique, until no such vertex exists. We then post a global *AllDifferent* constraint for each of these maximal cliques.

3.4   Lower Bound

The main drawback of the CP model proposed in [22] is due to the poor lower bound, which is the sum of minimal values in domains. This lower bound does not take into account the disequality constraints. A better lower bound is obtained by using a clique decomposition $\mathcal{C}$, as proposed in the B&B approach of [22]: it is defined by the sum, for each clique C_i of $\mathcal{C}$, of the sum of the $|C_i|$ smallest values in the union of the domains of the variables associated with vertices of C_i. This new bound takes into account some disequality constraints (those between pairs of variables that belong to a same clique). However, the sum of the $|C_i|$ smallest values may be a bad approximation of the chromatic sum of the subgraph induced by C_i when variables of C_i have different domains. Let us consider for example a clique $C_i = \{a, b, c\}$ with $D(a) = \{1, 2, 3\}$ and $D(b) = D(c) = \{7, 8\}$. The sum of the 3 smallest values in $D(a) \cup D(b) \cup D(c)$ is $1 + 2 + 3 = 6$, whereas the chromatic sum of the subgraph induced by $\{a, b, c\}$ is $1 + 7 + 8 = 16$.

Combining *AllDifferent* and Sum Constraints. Better bounds may be computed by considering the global constraint that combines an *AllDifferent* constraint with a sum constraint on the same set of variables [2]. In particular, [2] proposed a bound consistency algorithm for this global constraint. The main idea relies on the notion of "blocks" of variables, defined in such a way that, for a given *AllDifferent* constraint, variables of the same block are interchangeable. Each block also has a set of values. An initial lower bound is computed in $\mathcal{O}(n \log(n))$, where n is the number of variables in the *AllDifferent* contraint. During the search, if a variable of a block is assigned with a value of this block, the lower bound is left unchanged, otherwise, it is updated in $\mathcal{O}(1)$.

Note that the clique decomposition used to compute lower bounds is different from the set of maximal cliques used to propagate disequality constraints as proposed in Sect. 3.3. In the clique decomposition used to compute lower bounds, some disequality constraints are missing (those between vertices that belong to different cliques). Hence, the propagation of the conjunctions of *AllDifferent* and sum global constraints (to compute the lower bound) does not ensure a proper colouring. It must be combined either with binary disequality constraints between neighbour vertices that belong to different cliques, or with *AllDifferent* constraints as defined in Sect. 3.3.

Computation of the Clique Decomposition. We may consider different heuristics to build clique decompositions, and different clique decompositions may lead to different bounds. To build a clique decomposition, we consider a basic greedy approach very similar to the one described in Sect. 3.3 to compute a set of maximal cliques: the only difference is that each time a vertex is added to a clique, it is removed from the graph so that it cannot be selected for another clique. A key point to obtain a good tradeoff between the time spent to compute bounds and the reduction of the search space lies in the frequency of the computation of a clique decomposition. In the B&B approach of [22], a new clique decomposition is computed at each node of the search tree, on the subgraph induced by uncoloured vertices. This allows to compute more accurate bounds, but clique decomposition computations are rather expensive. Hence, we propose to compute a clique decomposition only once, at the root of the search tree. This partition is then used at each node of the tree to compute a new bound.

Triggering of the Bound Computation. Trying too early to prune branches with bound computations often leads to a loss of efficiency: when only a few vertices are coloured, we do not have enough information as to how the colouring will turn out. The computed lower bound is thus too low to be of any use. To prevent unnecessary computations, we set a lower limit for the triggering of the bound computation: if the distance between the sum of currently assigned colours and the current upper bound amounts to more than *gap* %, we refrain from computing the lower bound. In addition to this lower triggering limit, we added an upper one: we refrain from using this bound when *unc* or less vertices are uncoloured. The reason behind this is that when only a few vertices are left uncoloured, it may be faster to explore what remains in this part of the search space rather than using a bound to try to prune this very small branch.

3.5   Experimental Comparison

Experimental Setup and Benchmark. Programs are executed on an Intel® Xeon® CPU E5-2670 at 2.60 GHz processor, with 0480 KB of cache memory and 4 GB of RAM. We consider 126 instances which are classically used for sum colouring, as in [12,31]. Some are from COLOR02/03/04[1], but most of them are DIMACS instances designed for the classical colouring problem[2]. The timeout was set to 24 h. For each instance, the *reference solution* is the best known upper bound, either available in the literature (mainly [12,31]), or previously computed by one of our approaches. It gives an overview of the state of the art. Tables also give detailed results for a set of ten instances that we chose to highlight the peculiarities of each approach.

Configurations. We implemented CP models in Gecode (version 4.2.1) [29]. We cannot report results for all possible combinations of the different improvements described in Sects. 3.1, 3.2, 3.3 and 3.4. We have chosen the following configurations:

- *Base*: Basic model, with binary disequality constraints, a lower bound defined as the sum of smallest values in variable domains, and bound consistency ensured.
- *AllDiff+Bound*: Model with *AllDifferent* constraints (as defined in Sect. 3.3), a lower bound defined by using a clique partition and computing for each clique C_i the sum of the $|C_i|$ smallest values in variables' domains, and bound consistency ensured. Parameters *gap* and *unc* are set to 20% and 5, respectively. Experiments not detailed in this paper showed these values offer the best compromise.
- *AllDiff+Bound+Swap*: Same as *AllDiff+Bound*, but with colour swapping.
- *AllDiff+Bound+Swap+Dom*: Same as *AllDiff+Bound+Swap*, but with domain consistency instead of bound consistency.

---

[1] http://mat.gsia.cmu.edu/COLOR02.
[2] ftp://dimacs.rutgers.edu/pub/challenge/graph/benchmarks/color/.

- *AllDiff+SumBound+Swap*: Same as *AllDiff+Bound+Swap*, but lower bound computation is done by using the bound consistency algorithm of [2]. *gap* and *unc* are respectively set to 50% and 0 (the best setting for this configuration).

For these five configurations, the Branch and Bound (BAB) search engine was selected. As the goal is to minimize the sum of the variables, the value ordering heuristic chooses the smallest value. We have designed and compared different variable ordering heuristics (including well-known ones such as *Activity* and *wDeg*) and the best results are obtained with *minElim*, that chooses the variable that has the smallest value in its domain, and break ties by choosing the variable for which this smallest value would be removed from the fewest domains. *Luby* was used as a restart policy, with a scale of 500.

Results. Table 1 compares CP models on 10 representative instances, and then gives global results for the whole benchmark. *AllDiff+Bound* outperforms *Base* on 6 instances out of the 10, and adding colour swapping (*+Swap*) allows an overall improvement of bounds and generally faster proofs. Replacing bound consistency (in *AllDiff+Bound+Swap*) with domain consistency (in *AllDiff+Bound+Swap+Dom*) pays off on some instances, but degrades the solution process on some others. Finally, using the global constraint that combines a sum and an *AllDifferent* constraint improves the solution process on some instances, but also often degrades it. Actually, we noticed that, in many cases, all variables in the global constraint have very similar domains. Therefore, the bound computed for their sum is very close to the sum of the smallest values in the union of the domains.

As a conclusion, none of the proposed CP model appears to be competitive with state-of-the-art incomplete approaches, as the best model (*AllDiff+Bound+Swap*) is able to reach the reference solution for only 49 instances.

Table 1. Comparison of CP models. The first ten lines detail results for ten representative instances: best upper bound found within the time limit (UB), time needed to find UB (t_{UB}) and to prove optimality (t_{proof}), if optimality has been proven. The last three lines give the average distance between UB and the reference solution (in percentage), and the number of instances for which the reference solution has been found (# Ref. sol.), and optimality has been proven (# Optim. proofs) for the 126 instances.

Name	Ref. sol.	Base			AllDiff+Bound			AllDiff+Bound +Swap			AllDiff+Bound +Swap+Dom			AllDiff+SumBound +Swap		
		UB	t_{UB}	t_{proof}	UB	t_{UB}	t_{proof}	UB	t_{UB}	t_{proof}	UB	t_{UB}	t_{proof}	UB	t_{UB}	t_{proof}
DSJC250.5	3,210	3,598	24,823		3,580	16,179		3,591	840		3,540	51,076		3,577	9,505	
DSJC1000.1	8,991	10,323	84,001		10,339	29,570		10,328	3,340		10,315	66,581		10,295	61,921	
ash331GPIA	1,432	1,437	14,439		1,440	55,543		1,438	30,122		1,437	6,755		1,442	14,258	
le450_5b	1,350	1,518	46,685		1,510	63,361		1,509	8,819		1,487	6,439		1,457	3,827	
3-Insert._3	92	92	0		92	0	18,650	92	0	17,838	92	0	19,542	92	0	
qg.order60	109,800	109,800	404		109,800	592	593	109,800	203	203	109,800	138	139	109,800	218	218
r125.1	257	258	4		257	1,472	4,193	257	1,561	4,797	257	1,570	4,713	258	0	
inithx.i.3	1,986	1,988	17,571		1,988	84,043		1,986	142		1,986	463		1,987	4	
school1	2,674	3,646	77,581		3,646	52,114		3,531	75,519		3,644	48,434		3,648	1,072	
school1_nsh	2,392	3,067	1,329		3,139	42,737		3,031	1,047		3,031	2,604		2,992	13,581	
Average dist. (%)		5.57			5.58			5.32			5.44			5.34		
# Ref. sol.		43			45			49			48			45		
# Optim. proofs		5			11			11			11			8		

Making Proofs with CP. None of the configurations considered above are good at proving optimality, as proofs were only made for 11 instances. This may be due to the variable ordering heuristic *minElim*, which aims at quickly finding good solutions. Hence, we conducted experiments with a new CP configuration, designed to prioritize proof-making. It employs a hybrid restart policy, coupled with a scheduled heuristic change. In this configuration, we use the same setting as for *AllDiff+Bound+Swap+Dom* but we change the variable ordering heuristic during the search: as soon as the search endured 50 consecutive restarts without having improved the global upper bound, *minElim* is replaced by a heuristic that aims at proving optimality (it chooses the variable that has the highest number of uncoloured neighbours, and break ties by first choosing the variable that has the smallest value in its domain, and then the smallest current domain). When the variable ordering heuristic is changed, we also change the restart policy for a geometric policy, with a scale value of 100 and a base of 2.

Using this configuration, we are able to prove optimality for 15 instances instead of 11. Though this is an improvement of 36%, this is still far from the state of the art. For example, using all known bounds computed with heuristic approaches, optimality was proven for 21 of the 94 instances considered in [12].

4   Integer Linear Programming

Two improvements introduced in the previous section for CP may be easily adapted to ILP. Firstly, initial domain reduction is enacted simply by declaring less variables and by removing difference constraints between nodes when the considered colour is not in the domains of both nodes. Secondly, *AllDifferent* constraints are modeled by adding the constraint $\sum_{v \in C_i} x_{vk} \leq 1$, for each clique C_i and each colour k (instead of binary disequalities).

Implementation. We used ILOG CPLEX (version 12.6.2) [4]. To help CPLEX to avoid running out of memory, the two following parameters were added: Depth-First Search was forced as a node selection strategy, and the cuts factor was set to 1.5. Previous experiments showed us that these parameters did not significantly lessen CPLEX's ability to solve the instances we use.

Experimental Results. Table 2 compares results of the initial ILP model as proposed in [31], denoted *ILP*, with the ILP model that includes the two improvements, denoted *ILP+*. The most notable improvement is due to domain reduction, since reducing domains for ILP also removes variables and constraints. Overall, improvements allowed us to increase the number of optimality proofs from 61 to 65, and the number of times the reference solution has been found from 66 to 73. Besides, the number of memory outs goes down from 28 to 23.

When comparing ILP+ to CP, we note that they perform very differently. For four instances (DSJC* and school*), ILP+ ran out of memory. Therefore, the best solution found is far from the reference solution, and from the best solution found with CP models. For `qg.order60`, the best solution found by ILP is far from optimality, whereas all CP models are able to reach the optimum, with two of them even proving optimality. However, for the five remaining instances, ILP either finds better solutions (`ash331GPIA`, `le450_5b`), proves optimality quicker (`3-Insertions_3`, `r125.1`), or proves optimality while CP cannot (`inithx.i.3`). When considering global results (on the whole benchmark), ILP+ is able to find reference solutions and to prove optimality much more often than CP, but the average distance to reference solutions is much larger, mostly because of the times it ran out of memory.

5 Combining CP and ILP

Experiments reported previously showed us that CP and ILP have complementary abilities: ILP is very efficient to solve small instances, but it runs out of memory for 23 instances; it never occurs with CP, but its solutions are often far from the optimum, despite a rather good average distance. We therefore propose to decompose the problem into smaller independent subproblems, and to combine CP and ILP to solve these subproblems. The goal is to identify a subset $K_r \subseteq V$ of nodes, for which we compute all proper colourings with CP. From each of these colourings, we derive an independent subproblem. Given the optimal sum colouring of each of these independent subproblems, we deduce the optimal solution of the original problem in a straightforward way. If the subproblems are small enough, they can be solved with ILP. Furthermore, when instances are well structured, we may choose the subset K_r so that, for each colouring of K_r, we obtain several independent subproblems (instead of a single one), even easier to solve with ILP. This idea is reminiscent of the approach called *Backtracking on Tree Decomposition* (BTD) [9].

In Sect. 5.1, we recall the basic principles of BTD and show how to use it to solve the SCP. In Sect. 5.2, we introduce a new decomposition, as well as a way to use it. This approach is called BFD, and can be employed to combine CP and ILP. Decomposition methods are experimentally compared in Sect. 5.3.

5.1 Tree Decomposition

Definition 2. *[27] A tree decomposition of a graph (V, E) is a couple (K, T) where $T = (I, F)$ is a tree, and $K : I \to \mathcal{P}(V)$ is a function which associates a subset of variables $K_i \subseteq V$ (called a cluster) with every node $i \in I$ such that the following conditions are satisfied: (i) $\cup_{i \in I} K_i = V$; (ii) for each edge $(v_j, v_k) \in E$, there exists a node $i \in I$ such that $\{v_j, v_k\} \subseteq K_i$; and (iii) for all $i, j, k \in I$, if k is in a path from i to j in T, then $K_i \cap K_j \subseteq K_k$. The width of a tree decomposition is $\max_{i \in I} |K_i| - 1$. Intersections of neighbour clusters are called separators.*

BTD is a generic approach that exploits a tree decomposition of the constraint (hyper)graph of a CSP to identify independent subproblems which are solved separately. More precisely, given a tree decomposition (K, T) and a root node $r \in I$, BTD first assigns the variables of the root cluster K_r. Then, BTD recursively solves, for each child i of r, the independent subproblem that contains all variables occurring in the clusters associated with the subtree rooted in i. To avoid the repeated exploration of same parts of the search space, BTD records *structural (no)goods*, that allow to reduce time complexity to $\mathcal{O}(nd^{w+1})$ with n the number of variables, d the maximum domain size and w the width of T. The space complexity is $\mathcal{O}(nsd^s)$ with s the size of the largest separator.

BTD may be used to solve optimization problems provided that the objective function is decomposable, *i.e.*, once all variables of a root cluster r are assigned, the optimal solution that extends this partial assignment may be obtained by computing separately, for each child of r, the optimal solution of the subproblem associated with this child [5,9]. In this case, we have to record *structural valued goods*, *i.e.*, pairs composed by an assignment of the variables of a separator and the optimal solution of the subproblem associated with this child for this assigment. To avoid solving to optimality a subproblem if it is obvious its optimal solution cannot be extended to a global solution, an upper bound is added to the subproblem. As soon as it is proved that the optimal solution of the subproblem cannot be lower than the upper bound, the solving of the subproblem is stopped.

BTD may be used to solve the SCP as its objective function is decomposable: given a tree decomposition of the graph to colour, once the nodes of a root cluster r are coloured, the best sum colouring that extends this partial colouring may be obtained by searching separately the best sum colouring of each child of r. Note that this is not the case, for example, of the classical colouring problem, as we cannot colour children separately when the goal is to minimize the number of used colours (as we must know the colours used by other clusters).

However, experiments on our 126 instances have shown us that many instances are poorly structured: when computing a tree decomposition with *MinFill* [16], 65 instances are not decomposed at all (*i.e.*, there is only one cluster, which contains all variables), and 103 instances have at least one cluster that contains more than 90% of the variables. For these instances, BTD behaves poorly.

5.2   Backtracking with Flower Decomposition (BFD)

Even when the tree decomposition only contains one cluster, we may decompose it into two subsets (K_r and $V \setminus K_r$): we use CP to enumerate all proper colouring of K_r, and then, for each of these colourings, we use ILP to find the optimal sum colouring of $V \setminus K_r$ (given the colours assigned to K_r).

The idea of BFD is to exploit instance structure so that, for each colouring of K_r, we may split $V \setminus K_r$ into several subsets that may be solved to optimality with ILP independently. In other words, we propose to compute a tree decomposition with a height of 1, composed of a root cluster K_r, and a set of leaf clusters which are all children of K_r. The key point is to choose the nodes of K_r so that we obtain leaves that are small enough to be solved to optimality with ILP. To this end, we introduce a parameter l, that enforces a limit on the size of the leaves. More precisely, for each leaf cluster K_i, we ensure that $|K_i \setminus K_r| \leq l \times |V|$. Besides this hard constraint on the leaf sizes, we also aim at favoring small roots (as we have to enumerate all its proper colourings).

This flower decomposition is built as follows. We first build a tree decomposition $(K, T = (I, F))$ of the graph $G = (V, E)$ with *MinFill*. Let S be the set of all separators, *i.e.*, $S = \{K_{i_1} \cap K_{i_2} | \{i_1, i_2\} \in F\}$. Given a subset $S' \subseteq S$, we define a flower decomposition whose root is the cluster $K'_r = \cup_{s \in S'} s$. The other clusters of the flower (the leaves) are defined by the connected components of the subgraph of G induced by $V \setminus K'_r$. Each leaf cluster K'_i is then extended by adding to it any vertex of K'_r adjacent to a vertex of K'_i. A similar process was employed in [10], where it was also demonstrated that it results in a correct tree decomposition. Of course, the quality of the obtained flower decomposition depends on the initial subset S'. The goal is to find the subset S' such that the resulting flower decomposition satisfies the size limit l on the leaf clusters while minimizing the size of K'_r. We use Gecode to solve this problem. As it is $\mathcal{NP}$-hard, we limit the CPU time for computing it to 15 min, and use the best flower decomposition computed within this time limit. If Gecode has not found any flower decomposition that satisfies the size limit l on the leaf clusters, we build a flower decomposition which only contains two clusters and such that the root cluster contains the $|V| \times (1 - l)$ vertices with largest degrees.

As with BTD, we also enforce a limit on the size of cluster separators. In the context of BTD, this is done by merging clusters whose separators contain too many variables. In our case, however, we keep the clusters as they are: the only effect of the limit is that no valued good is recorded on the separators which exceed the limit, since doing so would be very likely to consume a large amount of memory. Moreover, if a separator corresponds to the full root cluster, there is no need to record any valued good on it, since affectations on such a separator cannot be produced more than once.

5.3   Experimental Evaluation

We compare two approaches, denoted BTD and BFD. BTD refers to the classical BTD approach. In this case, the tree decomposition is built using the *MinFill* algorithm and CP is used to solve subproblems: leaf clusters are solved with the Gecode configuration *AllDiff+Bound+Swap* (that uses restarts), whereas non-leaf clusters are solved with the same configuration, but without restarts, as we need to enumerate all solutions.

BFD refers to our new approach, based on a flower decomposition. CP (*All-Diff+Bound+Swap* without restarts) is used to enumerate the solutions of the non-leaf clusters and ILP+ to solve the subproblems induced by the leaves for each assignment of the separators. The maximal size for the separators is set to 30. We report results with two values for the "l" parameter (that limits the size of leaf clusters): 75% (denoted BFD 75) and 90% (denoted BFD 90).

Table 2 reports experimental results of BTD, BFD 90 and BFD 75. When looking at the detailed results on our ten representative instances, we note that they have complementary results: BTD is better than BFD 90 and BFD 75 on DSJC250.5 and school1, BFD 90 is better on ash331GPIA, 3-Insertions_3, inithx.i.3, school1_nsh and r125.1, and BFD 75 is better on DSJC1000.1, le450_5b, qg.order60 and r125.1. For two of these instances (r125.1 and

370   M. Minot et al.

Table 2. Comparison of ILP, ILP+, BTD, BFD 90 and BFD 75. The first ten lines detail results for as many representative instances. "#M#" in t_{proof} means a memory out occurred. The last four lines give, for the 126 instances: the average distance from UB to the reference solution (in percentage of the reference solution); the number of instances for which the reference solution was found (# Ref. sol.); the number of instances for which optimality was proved (# Optim. proofs); the number of times search was aborted due to a lack of memory (# Out of memory).

Name	Ref. sol.	ILP			ILP+			BTD			BFD 90			BFD 75		
		UB	t_{UB}	t_{proof}	UB	t_{UB}	t_{proof}	UB	t_{UB}	t_{proof}	UB	t_{UB}	t_{proof}	UB	t_{UB}	t_{proof}
DSJC250.5	3,210	4,587	795		4,587	2,457	#M#	**4,377**	1		4,855	86,400		15,918	0	#M#
DSJC1000.1	8,991	12,892	1	#M#	12,892	8,138	#M#	12,879	62		50,629	0	#M#	**12,467 86,400**		
ash331GPIA	1,432	1,458	7,066		**1,432 29,870**			1,767	13		1,448	86,400		1,528	1,894	
3-Insert._3	92	92	0	1	**92**	**0**	**0**	92	25	1,796	92	261	331	92	857	1,007
1e450_5b	1,350	1,450	53,963		**1,398 22,554**			2,227	16,237		1,914	86,400		1,883	86,400	
qg.order60	109,800	216,000	0	#M#	116,520	86,393		110,453	4,193		115,259	86,400		**110,198 86,400**		
r125.1	257	**257**	**0**	**0**	257	0	0	257	0	1	257	0	0	257	0	0
inithx.i.3	1,986	2,523	1	#M#	**1,986**	**9**	**20**	2,010	5		1,986	106	686	6,560	10,907	
school1	2,674	5,723	7,962	#M#	5,769	1	#M#	**5,556 82,466**			19,480	0	#M#	19,480	0	#M#
school1_nsh	2,392	5,069	3,573		4,980	7,990	#M#	4,740	10,747		**2,539 86,400**			3,996	86,400	
Average dist. (%)		73.36			63.89			24.42			83.40			85.05		
# Ref. sol.		66			73			17			46			48		
# Optim. proofs		61			65			13			39			26		
# Out of memory		28			23			6			18			16		

school1_nsh), the best results, over all considered approaches, are actually obtained by BFD 90.

When looking at global results over the whole benchmark, BTD is able to find the reference solution for only 17 instances (instead of 46 and 48 for BFD 90 and BFD 75), and it proves optimality for 13 instances only (instead of 39 and 26). Actually, as pointed out previously, most instances have no structure at all, or only a very poor structure, and BTD is generally outperformed on them by the CP approaches introduced in Sect. 3.

BFD 90 and BFD 75 are able to find reference solutions and to prove optimality for much more instances than BTD. Actually, even if the instance is not structured at all (i.e., there is only one cluster in the tree decomposition, which happens 40 times for our 126 instances), BFD is still able to build a flower decomposition with one root cluster (that contains $|V| \cdot (1-l)$ nodes) and one leaf (that contains the remaining $|V| \cdot l$ variables). In this case, BFD often behaves much better than BTD. However, BFD also suffers from a relatively high average distance to reference solutions. Actually, on some instances, BFD spends a lot of time to enumerate colourings for the root cluster which cannot be extended to good solutions. However, for each of these colourings, BFD wastes a lot of time solving to optimality useless subproblems.

Comparing BFD 90 and BFD 75 proves that allowing larger leaf clusters increases the memory needs but also eases the computation of upper bounds, as it makes the root cluster smaller (less enumeration) and gives ILP a more global view of the problem, preventing it in some cases to spend too much time solving a useless subproblem to optimality. When comparing BFD with the CP approaches of Sect. 3, we note an increase in the number of proofs (39 and 26 instead of 11), but the average distance to the reference solution is larger. Compared with

ILP+, BFD finds the reference solution less often, as with optimality proofs, but there also are less failures due to memory. Actually, BFD and ILP+ have complementary performance: BFD 90 (resp. BFD 75) performs strictly better than CPLEX on 21 (resp. 28) instances, and strictly worse on 91 (resp. 85) instances.

6 Portfolio Approach

We have introduced different approaches for the SCP in the previous sections. Some of them are dominated, in the sense that, for each instance, there is always another approach that performs better on this instance (it finds the same solution quicker, or a better solution). This is the case of the CP configurations *Base* and *AllDiff+Bound*, as well as ILP, BTD and BFD 75. The five other approaches (namely, the three remaining CP configurations, BFD 90 and ILP+) are complementary, and we propose to combine them in a portfolio approach.

More precisely, given a solver portfolio [6,8], the per-instance algorithm selection problem [26] consists in selecting the solver of the portfolio which is expected to perform best on a given instance. Algorithm selection systems usually build machine learning models to forecast which solver should be used in a particular context. Using the predictions, one or more solvers from the portfolio may be selected to be run sequentially or in parallel. In our SCP context, solver performance is highly constrained by memory bandwidth, in particular for ILP. Therefore, we cannot simply run our different solvers in parallel, and we consider the case where exactly one solver is selected.

One of the most prominent and successful systems that employs this approach is SATzilla [34], which defined the state of the art in SAT solving for a number of years. Other application areas include constraint solving [25], the travelling salesperson problem [19], subgraph isomorphism [20] and AI planning [28]. The reader is referred to a survey [18] for additional information on algorithm selection.

The selection process is composed of two steps: given an SCP instance to be solved, we first extract features from instances; then, we run algorithm selection to choose a solver. Finally we run the selected solver on the instance.

Feature Extraction. Given a graph $G = (V, E)$ for which we are looking for the chromatic sum, we compute the following features (a "*" denoting the use of the minimum, maximum, mean and standard deviation): number of nodes $|V|$ and edges $|E|$, degrees of the vertices in V^*, size of connected components in G^*, number of constraints and variables in the ILP+ model, number of *AllDifferent* constraints (arity of more than 2) in the *AllDiff+Bound* CP model, arity of these *AllDifferent* constraints*. We also added features computed from the largest connected component G' of G: density, theoretical upper and lower bounds of $\Sigma(G')$, number of clusters in the tree decomposition computed with *MinFill*. Moreover, this tree decomposition is used to compute a flower decomposition (with $l = 90$), which gives additional features: size of the root cluster, Cartesian product of the sizes of the domains in the root cluster, number of clusters, distance between the theoretical upper and lower bounds of the root cluster, density of the root cluster, density of leaf clusters*, number of proper variables in clusters*, separator density*, separator sizes*, the distance between theoretical upper and lower bounds on leaf clusters*, the number of binary variables* and constraints* in the ILP+ model associated with leaf clusters.

Selection Model. We use LLAMA [17] to build our solver selection model. LLAMA supports the most common algorithm selection approaches used in the literature. We performed a set of preliminary experiments to determine the approach that works best here, *i.e.*, a pairwise regression approach with random forest regression. This approach trains a model that predicts the performance difference between every pair of solvers in the portfolio, similarly to what is done in [34]: if the first solver is better than the second, the difference is positive, otherwise negative. The solver with the highest cumulative performance difference (*i.e.*, the most positive difference over all other solvers) is chosen to be run. As this approach already gives very good performance, we did not tune the parameters of the random forest machine learning algorithm. It is possible that overall performance can be improved by doing so, and we make no claims that the particular solver selection approach we use in this paper cannot be improved.

Experimental Results. We use leave-one-out cross-validation to determine the performance of our portfolio approach, as we only have 126 instances in our benchmark (which is not much for training a learning model): for each instance i, we train the selection model on all instances but i, and evaluate it on i.

The set of features is computed in 5.4 min in average, with 110 of our instances actually being under 5 min. Some instances, such as `latin_square_10`, `DSJC1000.9` or `flat1000_50_0` take a prohibitive amount of time when computing our set of features, mostly because of the two decompositions needed.

Table 3 shows us that our portfolio approach obtains results that are close to those of the Virtual Best Solver (VBS), which considers the best solver for each instance separately. It often selects either the best solver, or a solver which behaves well. The VBS (resp. our portfolio approach) uses ILP+ for 67 (resp. 78) instances, BFD 90 for 10 (resp. 13) instances, *AllDiff+Bound+Swap* for 20 (resp. 11) instances, *AllDiff+Bound+Swap+Dom* for 14 (resp. 11) instances, and *AllDiff+SumBound+Swap* for 15 (resp. 13) instances. Even when the portfolio selects the best solver, the solving time is increased by the time needed to compute features (which may be large on some instances). Note that the time needed by the model to select a solver is negligible (0.15 s on average).

Our portfolio is able to prove optimality for more than half of the 126 instances. In [12], best upper and lower bounds are reported for a set of state-of-the-art heuristic approaches, on a subset of 92 instances. Using the best of these bounds (computed with different heuristic approaches), they can prove optimality for 21 of these instances, whereas our portfolio approach is able to prove optimality for 42 of these 92 instances, *i.e.*, twice more. Finally, our portfolio

Table 3. Detailed results, for the virtual best solver and our portfolio approach. $t_{feat.}$ is the time needed to compute the features. For each method, "Algo" gives the chosen algorithm. Gec_1 denotes *AllDiff+Bound+Swap*, Gec_2 *AllDiff+Bound+Swap+Dom*, and Gec_3 *AllDiff+SumBound+Swap*. Times t_{UB} and t_{proof} for the portfolio include $t_{feat.}$.

Name	Ref. sol.	Virtual best solver				Portfolio approach				
		Algo	UB	t_{UB}	t_{proof}	$t_{feat.}$	Algo	UB	t_{UB}	t_{proof}
DSJC250.5	3210	Gec_2	3540	51076		6	Gec_1	3591	845	
DSJC1000.1	8991	Gec_3	10295	66842		126	Gec_1	10328	3466	
ash331GPIA	1432	ILP+	1432	29870		20	ILP+	1432	29890	
3-Insert._3	92	ILP+	92	0	0	0	ILP+	92	0	0
1e450_5b	1350	ILP+	1398	22554		4	ILP+	1398	22558	
qg.order60	109800	Gec_2	109800	138	139	4866	Gec_1	109800	5070	5070
r125.1	257	BFD	257	0	0	0	BFD	257	0	0
inithx.i.3	1986	ILP+	1986	9	20	8	BFD	1986	114	694
school1	2674	Gec_1	3531	75519		18	Gec_3	3648	1149	
school1_nsh	2392	BFD	2539	86400		10	Gec_3	2992	14166	
Average dist. (%)		4.25				5.51				
# Ref. sol.		83				76				
# Optim. proofs		66				65				
# Out of memory		0				0				

approach has improved (resp. reached) the best upper bounds reported in [12] for 2 (resp. 48) of the 92 instances: for DSJR500.1 the new upper bound is 2142 instead of 2156, and for 1e450_25b it is 3349 instead of 3365.

7   Conclusion and Future Work

We proposed some improvements for solving the SCP with CP and ILP, and demonstrated that they have complementary advantages: ILP is efficient on small instances, but fails to solve large instances due to its large memory needs; CP never runs out of memory but is not able to compute as good solutions as ILP on small instances. We proposed a CP/ILP combination that may be used as a compromise between CP and ILP. Besides, since it makes use of tree decomposition, this combination is better than CP or ILP alone to solve some well-structured instances. We combined those methods in a portfolio approach which obtains results close to those of the virtual best solver. It has been able to prove optimality for more than half of the considered instances. It has also been able to improve the best known upper bounds for two instances.

In the future, the use of a dedicated decomposition algorithm might be studied, in order to build a flower decomposition from scratch rather than resorting to an initial tree decomposition. The goal would be to make it more straightforward to obtain a balanced decomposition. Being able to automatically fine-tune BFD's parameters (l as well as the maximal size of separators) could also prove useful.

A major drawback of BFD is that some subproblems are solved to optimality even if they are useless due to poor assignments in the root. An interesting improvement that we shall investigate in further research would be to prevent ILP from spending more that a set amount of time on a leaf cluster, and to ask for a new assignment on the root if necessary. It could be seen as another form of restarts, as seen in [1].

Acknowledgements. This work has been supported by the ANR project SoLStiCe (ANR-13-BS02-0002-01).

References

1. Allouche, D., Givry, S., Katsirelos, G., Schiex, T., Zytnicki, M.: Anytime hybrid best-first search with tree decomposition for weighted CSP. In: Pesant, G. (ed.) CP 2015. LNCS, vol. 9255, pp. 12–29. Springer, Cham (2015). doi:10.1007/978-3-319-23219-5_2
2. Beldiceanu, N., Carlsson, M., Petit, T., Régin, J.C.: An o(nlog n) bound consistency algorithm for the conjunction of an alldifferent and an inequality between a sum of variables and a constant, and its generalization. In: ECAI, vol. 12, pp. 145–150 (2012)
3. Benlic, U., Hao, J.-K.: A study of breakout local search for the minimum sum coloring problem. In: Bui, L.T., Ong, Y.S., Hoai, N.X., Ishibuchi, H., Suganthan, P.N. (eds.) SEAL 2012. LNCS, vol. 7673, pp. 128–137. Springer, Heidelberg (2012). doi:10.1007/978-3-642-34859-4_13
4. Cplex, I.: High-performance software for mathematical programming and optimization (2005)
5. De Givry, S., Schiex, T., Verfaillie, G.: Exploiting tree decomposition and soft local consistency in weighted CSP. In: AAAI, vol. 6, pp. 1–6 (2006)
6. Gomes, C.P., Selman, B.: Algorithm portfolios. Artif. Intell. **126**(1–2), 43–62 (2001)
7. Helmar, A., Chiarandini, M.: A local search heuristic for chromatic sum. In: Proceedings of the 9th Metaheuristics International Conference, vol. 1101, pp. 161–170 (2011)
8. Huberman, B.A., Lukose, R.M., Hogg, T.: An economics approach to hard computational problems. Sci. **275**(5296), 51–54 (1997)
9. Jégou, P., Terrioux, C.: Hybrid backtracking bounded by tree-decomposition of constraint networks. Artif. Intell. **146**, 43–75 (2003)
10. Jégou, P., Kanso, H., Terrioux, C.: An algorithmic framework for decomposing constraint networks. In: 2015 IEEE 27th International Conference on Tools with Artificial Intelligence (ICTAI), pp. 1–8. IEEE (2015)
11. Jin, Y., Hao, J.K.: Hybrid evolutionary search for the minimum sum coloring problem of graphs. Inf. Sci. **352**, 15–34 (2016)
12. Jin, Y., Hamiez, J.P., Hao, J.K.: Algorithms for the minimum sum coloring problem: a review (2015). arXiv:1505.00449
13. Jin, Y., Hao, J.K., Hamiez, J.P.: A memetic algorithm for the minimum sum coloring problem. Comput. Oper. Res. **43**, 318–327 (2014)
14. Jussien, N., Rochart, G., Lorca, X.: Choco: an open source java constraint programming library. In: CPAIOR 2008 Workshop on Open-Source Software for Integer and Constraint Programming (OSSICP 2008), pp. 1–10 (2008)

15. Kadioglu, S., Malitsky, Y., Sellmann, M., Tierney, K.: Isac-instance-specific algorithm configuration. In: ECAI, vol. 215, pp. 751–756 (2010)
16. Kjaerulff, U.: Triangulation of graphs - algorithms giving small total state space. Technical report, Judex R.R. Aalborg, Denmark (1990)
17. Kotthoff, L.: LLAMA: Leveraging learning to automatically manage algorithms. Technical report, June 2013. http://arxiv.org/abs/1306.1031
18. Kotthoff, L.: Algorithm selection for combinatorial search problems: a survey. AI Mag. **35**(3), 48–60 (2014)
19. Kotthoff, L., Kerschke, P., Hoos, H., Trautmann, H.: Improving the state of the art in inexact TSP solving using per-instance algorithm selection. In: Dhaenens, C., Jourdan, L., Marmion, M.-E. (eds.) LION 2015. LNCS, vol. 8994, pp. 202–217. Springer, Cham (2015). doi:10.1007/978-3-319-19084-6_18
20. Kotthoff, L., McCreesh, C., Solnon, C.: Portfolios of subgraph isomorphism algorithms. In: Festa, P., Sellmann, M., Vanschoren, J. (eds.) LION 2016. LNCS, vol. 10079, pp. 107–122. Springer, Cham (2016). doi:10.1007/978-3-319-50349-3_8
21. Kubale, M.: Graph Colorings, vol. 352. American Mathematical Society, Canada (2004)
22. Lecat, C., Li, C.M., Lucet, C., Li, Y.: Exact methods for the minimum sum coloring problem. In: DPCP-2015, Cork, Ireland, Iran, pp. 61–69 (2015). https://hal.archives-ouvertes.fr/hal-01323741
23. Moukrim, A., Sghiouer, K., Lucet, C., Li, Y.: Upper and lower bounds for the minimum sum coloring problem (Submitted for Publication)
24. Moukrim, A., Sghiouer, K., Lucet, C., Li, Y.: Lower bounds for the minimal sum coloring problem. Electron. Notes Discrete Math. **36**, 663–670 (2010)
25. O'Mahony, E., Hebrard, E., Holland, A., Nugent, C., O'Sullivan, B.: Using case-based reasoning in an algorithm portfolio for constraint solving. In: Proceedings of the 19th Irish Conference on Artificial Intelligence and Cognitive Science, January 2008
26. Rice, J.R.: The algorithm selection problem. Adv. Comput. **15**, 65–118 (1976)
27. Robertson, N., Seymour, P.: Graph minors II: algorithmic aspects of tree-width. J. Algorithms **7**, 309–322 (1986)
28. Seipp, J., Braun, M., Garimort, J., Helmert, M.: Learning portfolios of automatically tuned planners. In: ICAPS (2012)
29. Team, G.: Gecode: generic constraint development environment (2006, 2008)
30. Thomassen, C., Erdös, P., Alavi, Y., Malde, P.J., Schwenk, A.J.: Tight bounds on the chromatic sum of a connected graph. J. Graph Theor. **13**(3), 353–357 (1989)
31. Wang, Y., Hao, J.K., Glover, F., Lü, Z.: Solving the minimum sum coloring problem via binary quadratic programming (2013). arXiv:1304.5876
32. Wu, Q., Hao, J.K.: An effective heuristic algorithm for sum coloring of graphs. Comput. Oper. Res. **39**(7), 1593–1600 (2012)
33. Wu, Q., Hao, J.K.: Improved lower bounds for sum coloring via clique decomposition. arXiv preprint (2013). arXiv:1303.6761
34. Xu, L., Hutter, F., Hoos, H.H., Leyton-Brown, K.: SATzilla: portfolio-based algorithm selection for SAT. J. Artif. Intell. Res. **32**, 565–606 (2008)

htd – A Free, Open-Source Framework for (Customized) Tree Decompositions and Beyond

Michael Abseher[✉], Nysret Musliu, and Stefan Woltran

Institute of Information Systems, TU Wien,
184/2, Favoritenstraße 9–11, 1040 Vienna, Austria
{abseher,musliu,woltran}@dbai.tuwien.ac.at

Abstract. Decompositions of graphs play a central role in the field of parameterized complexity and are the basis for many fixed-parameter tractable algorithms for problems that are NP-hard in general. Tree decompositions are the most prominent concept in this context and several tools for computing tree decompositions recently competed in the 1st Parameterized Algorithms and Computational Experiments Challenge. However, in practice the quality of a tree decomposition cannot be judged without taking concrete algorithms that make use of tree decompositions into account. In fact, practical experience has shown that generating decompositions of small width is not the only crucial ingredient towards efficiency. To this end, we present *htd*, a free and open-source software library, which includes efficient implementations of several heuristic approaches for tree decomposition and offers various features for normalization and customization of decompositions. The aim of this article is to present the specifics of *htd* together with an experimental evaluation underlining the effectiveness and efficiency of the implementation.

Keywords: Tree decompositions · Dynamic programming · Software library

1 Introduction

Graph decompositions are an important concept in the field of parameterized complexity and a wide variety of such approaches can be found in the literature including tree decompositions [11,23,37], branch decompositions [38], and hypertree decompositions [22] (of hypergraphs), to mention just a few. The concept of tree decompositions gained special attention since many NP-hard search problems become tractable when the parameter *treewidth* is bounded by some constant k [8,12,36]. A problem that exhibits tractability by bounding some problem-inherent constant is also called fixed-parameter tractable (FPT) [18].

The standard technique for solving a given problem using this concept is the computation of a tree decomposition followed by a dynamic programming (DP) algorithm that traverses the nodes of the decomposition and consecutively solves the respective sub-problems [36]. For problems that are FPT w.r.t. treewidth,

D. Salvagnin and M. Lombardi (Eds.): CPAIOR 2017, LNCS 10335, pp. 376–386, 2017.
DOI: 10.1007/978-3-319-59776-8_30

the general run-time of such algorithms for an instance of size n is $f(k) \cdot n^{\mathcal{O}(1)}$, where f is an arbitrary function over width k of the used tree decomposition. In fact, this approach has been used for several applications including the solving of inference problems in probabilistic networks [32], frequency assignment [30], computational biology [42], logic programming [33], routing problems [17], and solving of quantified boolean formulae [14].

From a theoretical point of view the actual width k is the crucial parameter towards efficiency for FPT algorithms that use tree decompositions. In the literature various approaches for optimizing width when computing decompositions have been proposed (see Sect. 2), but to the best of our knowledge no software frameworks exist which offer the feature to customize tree decompositions by other criteria than just minimizing the plain width. However, experience shows that even decompositions of the same width can lead to significant differences in the run-time of DP algorithms and recent results confirm that the width is indeed not the only important parameter that has a significant influence on the performance [27,33]. In particular, [4,6] has underlined that considering such additional criteria is highly beneficial. Nevertheless, a post-processing phase based on machine learning was needed to determine "good" tree decompositions. Therefore we see a strong need to offer a specialized decomposition framework that allows for directly constructing customized decompositions, i.e., decompositions which reflect certain preferences of the developer, in order to optimally fit to the DP algorithm in which they are used. In this paper we present a free, open-source framework (*htd*) which supports a vast amount of input graph types and different types of decompositions. *htd* includes efficient implementations of several heuristic approaches for computing tree decompositions. Furthermore, *htd* provides various built-in customization and manipulation operations in order to fit the needs of developers of DP algorithms. These include normalizations, optimization of tree decompositions, computation of induced edges and labeling operations (see Sect. 3).

Just recently, *htd* participated in the "First Parameterized Algorithms and Computational Experiments Challenge" ("PACE16")[1] where it was ranked at the third place in the heuristics track. Although *htd* provides lots of additional convenience functions, the results of *htd* with respect to the optimization of width are very close to those of the heuristic approaches ranked at the first two places. In Sect. 4, we will present some complementing experimental evaluation that also sheds light on the effect of customization when decompositions are used in a specific DP algorithm.

htd has been already used successfully in different projects, like D-FLAT [2], a framework for rapid-prototyping of dynamic programming algorithms on tree decompositions, or dynQBF [14], a DP-based solver for quantified boolean formulae. Our framework is available for download as free, open-source software at https://github.com/mabseher/htd. We consider *htd* as a potential starting point for researchers to contribute their algorithms in order to provide a new

---

[1] See https://pacechallenge.wordpress.com/track-a-treewidth/ for more details.

framework for all different types of graph decompositions. A detailed report on *htd* and all its features can be found in [5].

2 Background

The intuition underlying tree decompositions is to obtain a tree from a (potentially cyclic) graph by subsuming multiple vertices in one node and thereby isolating the parts responsible for the cyclicity. Formally, the notions of tree decomposition and treewidth are defined as follows [13,37].

Definition 1. *Given a graph $G = (V, E)$, a tree decomposition of G is a pair (T, χ) where $T = (N, F)$ is a tree and $\chi : N \to 2^V$ assigns to each node a set of vertices (called the node's bag), such that the following conditions hold:*

1. *For each vertex $v \in V$, there exists a node $i \in N$ such that $v \in \chi_i$.*
2. *For each edge $(v, w) \in E$, there exists an $i \in N$ with $v \in \chi_i$ and $w \in \chi_i$.*
3. *For each $i, j, k \in N$: If j lies on the path between i and k then $\chi_i \cap \chi_k \subseteq \chi_j$.*

The width *of a given tree decomposition is defined as $max_{i \in N} |\chi_i| - 1$ and the* treewidth *of a graph is the minimum width over all its tree decompositions.*

Note that the tree decomposition of a graph is in general not unique. In the following we consider rooted tree decompositions.

Definition 2. *A* normalized *(or* nice*) tree decomposition of a graph G is a rooted tree decomposition (T, χ) where each node $i \in N$ is of one of the following types: (1) Leaf: i has no child nodes;(2) Introduce Node: i has one child j with $\chi_j \subset \chi_i$ and $|\chi_i| = |\chi_j| + 1$; (3) Forget Node: i has one child j with $\chi_j \supset \chi_i$ and $|\chi_i| = |\chi_j| - 1$;(4) Join Node: i has two children j, k with $\chi_i = \chi_j = \chi_k$.*

Each tree decomposition can be transformed into a normalized one in linear time without increasing the width [29]. Apart from nice tree decompositions, one can find numerous other normalizations in the literature [3].

While the problem of constructing an optimal tree decomposition, i.e. a decomposition with minimal width, is intractable [7], researchers have proposed several exact methods for small graphs and efficient heuristic approaches that usually construct tree decompositions of almost optimal width for larger graphs. Examples of exact algorithms for tree decompositions are [9,21,39]; greedy heuristic algorithms include Minimum Degree heuristic [10], Maximum Cardinality Search (MCS) [40], and Min-Fill heuristic [16]. Metaheuristic techniques have been provided in terms of genetic algorithms [31,35], ant colony optimization [25], and local search based techniques [15,28,34]. Recent surveys [13,26] provide further details.

Several software frameworks that implement some of the tree decomposition algorithms mentioned above are publicly available. These libraries include QuickBB [21][2], htdecomp [19][3] and Jdrasil[4]. Very recently, also an open database for computation, storage and retrieval of tree decompositions was initiated [41].

---

[2] Available at http://www.hlt.utdallas.edu/~vgogate/quickbb.html.
[3] Available at http://www.dbai.tuwien.ac.at/proj/hypertree/downloads.html.
[4] Available at https://github.com/maxbannach/Jdrasil.

3   A Framework for (Customized) Tree Decompositions

The aforementioned tree decomposition software frameworks focus mainly on computing tree decompositions of small width; optimizing the resulting decomposition with respect to a concrete DP algorithm to be applied is therefore left to the user. Our system, *htd*, provides an efficient implementation of several algorithms that compute tree decompositions of small width and aims for much richer customizations of the resulting tree decomposition: this includes normalizations, further optimization criteria (for instance, minimizing the total number of children of join nodes), or augmenting the information in the bags (for instance, including the subgraph induced by the vertices of the bag) of the provided decomposition. In what follows we highlight the main features.

Computing Tree Decompositions: The framework provides efficient implementations of several heuristic approaches for tree decompositions including MCS, Minimum Degree heuristic, Min-Fill heuristic and their combination. The default tree decomposition strategy in *htd* is Min-Fill.

Manipulation Operations and Normalizations: Manipulations in the context of *htd* refer to all operations which alter a given decompositions, e.g., by ensuring that the maximum number of children of decomposition nodes is limited. Normalizations are manipulations which combine multiple base manipulation operations, e.g., one can use them to request only nice tree decompositions.

Optimization: In the context of *htd*, the term "optimization" is used to refer to operations which improve certain properties of a decomposition. Via optimization the obtained decomposition adheres to certain preferences. *htd* offers the ability to optimize by searching for the optimal root node and also by choosing the best tree decomposition based on a custom (potentially multi-level) criterion. Combining these two strategies might yield even better results.

Computation of Induced Edges: Any DP algorithm needs precise knowledge about the (hyper-)edges of the problem instance's graph representation induced by a bag because this is the information which actually matters when the problem at hand has to be solved. Computing this information inside the DP step is not only time-consuming, it also potentially destroys the property of fixed-parameter tractability in practice because in each bag one might need to scan the whole edge set of the input. *htd* computes and delivers this information very efficiently by extending decomposition algorithms appropriately.

Labels and Labeling Operations: Labels generalize the concept that is used for the induced edges to arbitrary information. They can be used to take care of supplemental information of the input instance and they can enhance the knowledge base represented by a decomposition. In both cases, one can define labeling functions and operations which automate the process of assigning labels.

All these customization features can be applied directly in the context of the decomposition procedure and no further code changes are necessary. The basic work-flow of how to obtain a customized decomposition of a problem instance using *htd* is the following: At first, the input specification which represents an instance of a given problem must be transformed to its corresponding instance graph. Depending on the actual need, a developer can choose from several built-in graph types, like directed and undirected ones as well as hypergraphs. In order to be able to manage additional information about the input instance, *htd* also offers labeled and so-called named counterparts for each graph type. A labeled graph type allows to add arbitrary labels to the vertices and edges of the graph.

Named graph types extend this helpful feature by providing a bi-directional one-to-one mapping between labels and the vertices or edges, respectively. This can be useful in cases where one wants to access vertices and edges not only by their identifiers but also by a custom "name" which can be of arbitrary data type. Importers for graph formats like DIMACS and GraphML are provided by *htd*, but one can also construct the graph manually using a custom importer.

After all information of the input instance is parsed, the next step for a developer is to decide for the decomposition algorithm to use and (optionally) choose from the wide range of built-in manipulation operations or to implement its own, custom manipulation. By running the decomposition algorithm with the instance graph as its input, we obtain a customized decomposition which then can be used in the dynamic programming algorithm. Alternatively, the decomposition can be exported or printed using built-in functionality.

4   Experimental Evaluation

In this section we give first results regarding the performance characteristics of our framework. The experiments in this paper are based on two investigations. At first, we have a look at the actual efficiency of *htd* in terms of the minimum width achieved within a fixed time period. Afterwards we highlight the usefulness of *htd*'s ability to optimize tree decompositions based on custom criteria.

All our experiments were performed on a single core of an Intel Xeon E5-2637@3.5 GHz processor running Debian GNU/Linux 8.3. For repeatability of the experiments we provide the complete testbed and the results under the following link: www.dbai.tuwien.ac.at/proj/dflat/evaluation_htd100b1.zip

4.1   Efficient Computation of Minimum-Width Decompositions

In order to have an indication for *htd*'s actual competitiveness, we compare *htd* 1.0.0-beta1 [1] to the participants of Track A ("Treewidth") of the Parameterized Algorithms and Computational Experiments Challenge 2016 ("PACE16").

The following list of algorithms contains *htd* and all further participants of the sequential heuristics track of the PACE treewidth challenge in the variant in which they were submitted. For each of the algorithms we provide the name of the binary, the ID in the PACE challenge (if applicable), the location of its source code and the exact identifier of the program version in the GitHub repository. Most of the algorithms use Min-Fill as basic decomposition strategy.

- htd 0.9.9: "htd_gr2td_minfill_exhaustive.sh" (PACE-ID: 5)
  Available at https://github.com/mabseher/htd
  GitHub-Commit-ID: f4f9b8907da2025c4c0c6f24a47ff4dd0bde1626

- htd 1.0.0-beta1: "htd_gr2td_exhaustive.sh" (No participant of PACE)
  Available at https://github.com/mabseher/htd
  GitHub-Commit-ID: f04bfd256e0be0eb536ef04410b541b15206255d
- "tw-heuristic" (PACE-ID: 1)
  Available at https://github.com/mrprajesh/pacechallenge.
  GitHub-Commit-ID: 6c29c143d72856f649de99846e91de185f78c15f
- "tw-heuristic" (PACE-ID: 6)
  Available at https://github.com/maxbannach/Jdrasil
  GitHub-Commit-ID: fa7855e4c9f33163606a0677485a9e51d26d7b0a
- "tw-heuristic" (PACE-ID: 9)
  Available at https://github.com/elitheeli/2016-pace-challenge
  GitHub-Commit-ID: 2f4acb30b5c48608859ff27b5f4e217ee8346ca5
- "tw-heuristic" (PACE-ID: 10) [20]
  Available at https://github.com/mfjones/pace2016
  GitHub-Commit-ID: 2b7f289e4d182799803a014d0ee1d76a4de70c1f
- "flow_cutter_pace16" (PACE-ID: 12) [24]
  Available at https://github.com/ben-strasser/flow-cutter-pace16
  GitHub-Commit-ID: 73df7b545f694922dcb873609ae2759568b36f9f

htd already proved its efficiency on the instances of the PACE challenge by achieving the third place in the competition[5]. Because these instances are relatively small – *htd* is able to decompose any of the instances in less than two seconds – we want to present here also results for larger instances.

For this purpose we consider in our experiments instances which are used in competitions for quantified boolean formulas (QBFs). In fact, we decompose here the Gaifman graph underlying the CNF matrix of the QBFs thus ignoring the actual quantifier prefix (DP-based solvers for QBFs like dynQBF [14] handle the quantifier information internally and only require a decomposition of the matrix as input). We used DataSet 1 from the QBFEVAL'16 competition[6]. The data set contains 825 instances, most of them being significantly larger than the instances of the PACE challenge. In the experiments, each test run was limited to a run-time of at most 100 s and 32 GB of main memory. For the actual evaluation we use the testbed of the PACE challenge[7].

In Fig. 1 we present the outcome of our experiments in a plot of the cumulative frequency of the obtained widths for each of the algorithms (IDs are those of the PACE challenge). A point (x, y) in the diagram indicates that y decompositions have a width of x or less. This means, an algorithm is better in terms of decomposition width when its line chart reaches the top with minimal width. We can see that already the "old" version of *htd* used in the PACE challenge outperforms its competitors in the region between width 250 and 1000. The recent version of *htd* improves upon these results and shows that only in regions with very high width, two of its competitors are able to decompose more instances.

---

[5] See https://pacechallenge.wordpress.com/2016/09/12/here-are-the-results-of-the-1st-pace-challenge/.

[6] Available at http://www.qbflib.org/TS2016/Dataset_1.tar.gz.

[7] Available at https://github.com/holgerdell/PACE-treewidth-testbed.

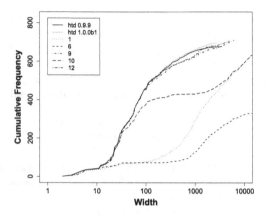

Fig. 1. Decomposition quality for instances of QBFEVAL2016 challenge

Table 1. Results for dynQBF using tree decompositions with low join node complexity

Iterations	Solved instances (Total)	Solved instances (Mean)	Total user-time
1	491	98.2	5904.54
5	512	102.4	6375.08
10	512	102.4	5908.78

4.2 Using Customized Tree Decompositions to Increase Efficiency

Next, we highlight the usefulness of *htd*'s ability to optimize tree decompositions via an application scenario using the QBF solver dynQBF [14][8]. dynQBF makes extensive use of *htd*'s features (especially concerning induced edges) and benefits from customized tree decompositions as provided by *htd*.

For the following experiments we consider the 200 instances of the 2QBF track of the QBFEVAL2014 competition[9]. These instances differ from those used in Sect. 4.1 due to the fact that after decomposing we still need to run the dynamic programming algorithm. For each of the test runs we allow a single thread, 10 min execution time and 16 GB of main memory. This time, we only consider *htd* as decomposition library as, to the best of our knowledge, currently no other framework considers custom preferences for optimization.

Table 1 shows the results of the experiments running dynQBF (using *htd* 1.0.0-beta1). For each instance, five different tree decompositions were generated using the Min-Fill heuristics with different seeds. The first column shows the number of optimization iterations, i.e., the number of iterations after which the best known decomposition is returned. The second and third column show the total and average number of instances over the five repetitions which were solved

---

[8] Available at https://github.com/gcharwat/dynqbf/releases/tag/v0.3-beta.
[9] Available at http://www.qbflib.org/TS2010/2QBF.tar.gz.

successfully using the best decomposition found and the last column shows the total amount of solving time, restricted to the instances successfully solved.

As fitness function for the optimization we aim at minimizing the complexity of join nodes given by the formula $\sum_{j \in J} \prod_{c \in C_j} |\chi_c|$ where J is the set of join nodes and C_j is the set of children of node j. That is, we want the total sum of the products of the join node children's bag sizes to be as small as possible. The intuition is that when the given measure is small, the dynamic programming algorithm is more efficient in join nodes.

We can see that with a single iteration, i.e., without optimization, dynQBF solves about 98 out of 200 instances in average. The table illustrates that the number of solved instances in our example scenario increases when we use five iterations for the optimization phase. Therefore the total solving time increases. An interesting observation is the fact that using ten optimization iterations we need almost the same amount of time as without optimization, but we still can solve more instances.[10] That means that the (on average) four additional instances come for free in our scenario.

Note that with a statistical significance of over 99.95%, the width of the obtained decompositions does not change with the number of iterations, i.e., the customized tree decompositions indeed increase the efficiency of the dynQBF algorithm. Hence, by using an optimization function of not more than ten lines of code, one can already achieve improvements using htd.

5   Conclusion

In this paper we presented a new open-source framework for tree decompositions called htd. To the best of our knowledge, htd is the first software framework which aims for optimizing tree decompositions by other criteria than just the plain width. We gave an overview over its features and provided an introduction on how to use the library (for more details, see [5]). Moreover, we evaluated our approach by comparing the performance of htd and other participants of the Parameterized Algorithms and Computational Experiments Challenge 2016. The outcome of the evaluation indicates that the performance characteristics of the new framework are indeed encouraging. Furthermore, we showed that customizing tree decompositions is a powerful feature which can improve efficiency of dynamic programming algorithms using those decompositions.

For future work we want to further improve the built-in heuristics and algorithms in order to enhance the capabilities for generation of customized decompositions of small width. Furthermore we are currently working on making some exact algorithms for tree decompositions amenable to customization. Last, but not least, we invite researchers and software developers to contribute to the

---

[10] When we use a pool of ten decompositions to choose from, the chance for obtaining an even better decomposition increases. However, no additional instance is solved when we change from five to ten iterations, but the run-time for the solved instances further decreases (compensating the time required for computing more decompositions).

library as we try to initiate a joint collaboration on a powerful framework for graph decompositions and any input is highly appreciated.

Acknowledgments. This work has been supported by the Austrian Science Fund (FWF): P25607-N23, P24814-N23, Y698-N23.

References

1. Abseher, M.: htd 1.0.0-beta1 (2016). http://github.com/mabseher/htd/tree/v1.0. 0-beta1
2. Abseher, M., Bliem, B., Charwat, G., Dusberger, F., Hecher, M., Woltran, S.: The D-FLAT system for dynamic programming on tree decompositions. In: Fermé, E., Leite, J. (eds.) JELIA 2014. LNCS, vol. 8761, pp. 558–572. Springer, Cham (2014). doi:10.1007/978-3-319-11558-0_39
3. Abseher, M., Bliem, B., Charwat, G., Dusberger, F., Hecher, M., Woltran, S.: D-FLAT: progress report. Technical report, DBAI-TR-2014-86, TU Wien (2014). http://www.dbai.tuwien.ac.at/research/report/dbai-tr-2014-86.pdf
4. Abseher, M., Dusberger, F., Musliu, N., Woltran, S.: Improving the efficiency of dynamic programming on tree decompositions via machine learning. In: Proceedings of IJCAI, pp. 275–282. AAAI Press (2015)
5. Abseher, M., Musliu, N., Woltran, S.: htd - A free, open-source framework for tree decompositions and beyond. Technical report, DBAI-TR-2016-96, TU Wien (2016). http://www.dbai.tuwien.ac.at/research/report/dbai-tr-2016-96.pdf
6. Abseher, M., Musliu, N., Woltran, S.: Improving the efficiency of dynamic programming on tree decompositions via machine learning. Technical report, DBAI-TR-2016-94, TU Wien (2016). http://www.dbai.tuwien.ac.at/research/report/dbai-tr-2016-94.pdf
7. Arnborg, S., Corneil, D.G., Proskurowski, A.: Complexity of finding embeddings in a k-tree. J. Algebraic Discrete Methods **8**(2), 277–284 (1987)
8. Arnborg, S., Proskurowski, A.: Linear time algorithms for NP-hard problems restricted to partial k-trees. Discrete Appl. Math. **23**(1), 11–24 (1989)
9. Bachoore, E.H., Bodlaender, H.L.: A branch and bound algorithm for exact, upper, and lower bounds on treewidth. In: Cheng, S.-W., Poon, C.K. (eds.) AAIM 2006. LNCS, vol. 4041, pp. 255–266. Springer, Heidelberg (2006). doi:10.1007/11775096_24
10. Berry, A., Heggernes, P., Simonet, G.: The minimum degree heuristic and the minimal triangulation process. In: Bodlaender, H.L. (ed.) WG 2003. LNCS, vol. 2880, pp. 58–70. Springer, Heidelberg (2003). doi:10.1007/978-3-540-39890-5_6
11. Bertelè, U., Brioschi, F.: On non-serial dynamic programming. J. Comb. Theor. Ser. A **14**(2), 137–148 (1973)
12. Bodlaender, H.L., Koster, A.M.C.A.: Combinatorial optimization on graphs of bounded treewidth. Comput. J. **51**(3), 255–269 (2008)
13. Bodlaender, H.L., Koster, A.M.C.A.: Treewidth computations I. Upper bounds. Inf. Comput. **208**(3), 259–275 (2010)
14. Charwat, G., Woltran, S.: Dynamic programming-based QBF solving. In: Proceedings of the 4th International Workshop on Quantified Boolean Formulas, vol. 1719, pp. 27–40. CEUR Workshop Proceedings (2016)
15. Clautiaux, F., Moukrim, A., Négre, S., Carlier, J.: Heuristic and meta-heuristic methods for computing graph treewidth. RAIRO Oper. Res. **38**, 13–26 (2004)

16. Dechter, R.: Constraint Processing. Morgan Kaufmann, USA (2003)
17. Dourisboure, Y.: Compact routing schemes for generalised chordal graphs. J. Graph Algorithms Appl. **9**(2), 277–297 (2005)
18. Downey, R.G., Fellows, M.R.: Parameterized Complexity. Monographs in Computer Science. Springer, New York (1999)
19. Ganzow, T., Gottlob, G., Musliu, N., Samer, M.: A CSP hypergraph library. Technical report, DBAI-TR-2005-50, TU Wien (2005). http://www.dbai.tuwien.ac.at/proj/hypertree/csphgl.pdf
20. Gaspers, S., Gudmundsson, J., Jones, M., Mestre, J., Rümmele, S.: Turbocharging treewidth heuristics. In: Proceedings of IPEC (2016, to appear)
21. Gogate, V., Dechter, R.: A complete anytime algorithm for treewidth. In: Proceedings of UAI, pp. 201–208. AUAI Press (2004)
22. Gottlob, G., Leone, N., Scarcello, F.: Hypertree decompositions and tractable queries. J. Comput. Syst. Sci. **64**(3), 579–627 (2002)
23. Halin, R.: S-functions for graphs. J. Geom. **8**, 171–186 (1976)
24. Hamann, M., Strasser, B.: Graph bisection with pareto-optimization. In: Proceedings of ALENEX, pp. 90–102. SIAM (2016)
25. Hammerl, T., Musliu, N.: Ant colony optimization for tree decompositions. In: Cowling, P., Merz, P. (eds.) EvoCOP 2010. LNCS, vol. 6022, pp. 95–106. Springer, Heidelberg (2010). doi:10.1007/978-3-642-12139-5_9
26. Hammerl, T., Musliu, N., Schafhauser, W.: Metaheuristic algorithms and tree decomposition. In: Kacprzyk, J., Pedrycz, W. (eds.) Springer Handbook of Computational Intelligence, pp. 1255–1270. Springer, Heidelberg (2015). doi:10.1007/978-3-662-43505-2_64
27. Jégou, P., Terrioux, C.: Bag-connected tree-width: a new parameter for graph decomposition. In: Proceedings of ISAIM, pp. 12–28 (2014)
28. Kjaerulff, U.: Optimal decomposition of probabilistic networks by simulated annealing. Stat. Comput. **2**(1), 2–17 (1992)
29. Kloks, T.: Treewidth, Computations and Approximations. LNCS, vol. 842. Springer, Heidelberg (1994)
30. Koster, A.M.C.A., van Hoesel, S.P.M., Kolen, A.W.J.: Solving frequency assignment problems via tree-decomposition 1. Electr. Notes Discrete Math. **3**, 102–105 (1999)
31. Larranaga, P., Kujipers, C.M., Poza, M., Murga, R.H.: Decomposing bayesian networks: triangulation of the moral graph with genetic algorithms. Stat. Comput. **7**(1), 19–34 (1997)
32. Lauritzen, S.L., Spiegelhalter, D.J.: Local computations with probabilities on graphical structures and their application to expert systems. J. R. Stat. Soc. Ser. B **50**, 157–224 (1988)
33. Morak, M., Musliu, N., Pichler, R., Rümmele, S., Woltran, S.: Evaluating tree-decomposition based algorithms for answer set programming. In: Hamadi, Y., Schoenauer, M. (eds.) LION 2012. LNCS, pp. 130–144. Springer, Heidelberg (2012). doi:10.1007/978-3-642-34413-8_10
34. Musliu, N.: An iterative heuristic algorithm for tree decomposition. In: Cotta, C., van Hemert, J. (eds.) Recent Advances in Evolutionary Computation for Combinatorial Optimization. Studies in Computational Intelligence, vol. 153, pp. 133–150. Springer, Heidelberg (2008)
35. Musliu, N., Schafhauser, W.: Genetic algorithms for generalized hypertree decompositions. Eur. J. Ind. Eng. **1**(3), 317–340 (2007)
36. Niedermeier, R.: Invitation to Fixed-Parameter Algorithms. Oxford Lecture Series in Mathematics and Its Applications. Oxford University Press, Oxford (2006)

37. Robertson, N., Seymour, P.: Graph minors. III. Planar tree-width. J. Comb. Theor. Ser. B **36**(1), 49–64 (1984)
38. Robertson, N., Seymour, P.: Graph minors. X. Obstructions to tree-decomposition. J. Comb. Theor. Ser. B **52**(2), 153–190 (1991)
39. Shoikhet, K., Geiger, D.: A practical algorithm for finding optimal triangulations. In: Proceedings of AAAI/IAAI, pp. 185–190. AAAI Press/The MIT Press (1997)
40. Tarjan, R.E., Yannakakis, M.: Simple linear-time algorithm to test chordality of graphs, test acyclicity of hypergraphs, and selectively reduce acyclic hypergraphs. SIAM J. Comput. **13**, 566–579 (1984)
41. van Wersch, R., Kelk, S.: Toto: an open database for computation, storage and retrieval of tree decompositions. Discrete Appl. Math. **217**, 389–393 (2017)
42. Xu, J., Jiao, F., Berger, B.: A tree-decomposition approach to protein structure prediction. In: Proceedings of CSB, pp. 247–256 (2005)

The Nemhauser-Trotter Reduction and Lifted Message Passing for the Weighted CSP

Hong Xu$^{(\boxtimes)}$ (ORCID), T.K. Satish Kumar, and Sven Koenig

University of Southern California, Los Angeles, CA 90089, USA
{hongx,skoenig}@usc.edu, tkskwork@gmail.com

Abstract. We study two important implications of the constraint composite graph (CCG) associated with the weighted constraint satisfaction problem (WCSP). First, we show that the Nemhauser-Trotter (NT) reduction popularly used for kernelization of the minimum weighted vertex cover (MWVC) problem can also be applied to the CCG of the WCSP. This leads to a polynomial-time preprocessing algorithm that fixes the optimal values of a large subset of the variables in the WCSP. Second, belief propagation (BP) is a well-known technique used for solving many combinatorial problems in probabilistic reasoning, artificial intelligence and information theory. The min-sum message passing (MSMP) algorithm is a simple variant of BP that has also been successfully employed in several research communities. Unfortunately, the MSMP algorithm has met with little success on the WCSP. We revive the MSMP algorithm for solving the WCSP by applying it on the CCG of a given WCSP instance instead of its original form. We refer to this new MSMP algorithm as the lifted MSMP algorithm for the WCSP. We demonstrate the effectiveness of our algorithms through experimental evaluations.

1 Introduction

The weighted constraint satisfaction problem (WCSP) is a combinatorial optimization problem. It is a generalization of the constraint satisfaction problem (CSP) in which the constraints are no longer "hard". Instead, each tuple in a constraint—i.e., an assignment of values to all variables in that constraint—is associated with a non-negative weight (sometimes referred to as "cost"). The goal is to find an assignment of values to all variables from their respective domains such that the total weight is minimized [1].

More formally, the WCSP is defined by a triplet $\mathcal{B} = \langle \mathcal{X}, \mathcal{D}, \mathcal{C} \rangle$, where $\mathcal{X} = \{X_1, X_2, \ldots, X_N\}$ is a set of N variables, $\mathcal{D} = \{D_1, D_2, \ldots, D_N\}$ is a set of N domains with discrete values, and $\mathcal{C} = \{C_1, C_2, \ldots, C_M\}$ is a set of M weighted constraints. Each variable $X_i \in \mathcal{X}$ can be assigned a value in its associated domain $D_i \in \mathcal{D}$. Each constraint $C_i \in \mathcal{C}$ is defined over a certain subset of the variables $S_i \subseteq \mathcal{X}$, called the scope of C_i. C_i associates a non-negative weight with each possible assignment of values to the variables in S_i. (For notational convenience, we use S_i and C_i interchangeably throughout this

© Springer International Publishing AG 2017
D. Salvagnin and M. Lombardi (Eds.): CPAIOR 2017, LNCS 10335, pp. 387–402, 2017.
DOI: 10.1007/978-3-319-59776-8_31

paper when referring to the variables participating in a weighted constraint, e.g., $X_k \in C_i \equiv X_k \in S_i$.) The goal is to find an assignment of values to all variables in $\mathcal{X}$ from their respective domains that minimizes the sum of the weights specified by each weighted constraint in $\mathcal{C}$ [1]. This combinatorial task can equivalently be characterized by having to compute

$$\arg\min_{a \in \mathcal{A}(\mathcal{X})} \sum_{C_i \in \mathcal{C}} E_{C_i}(a|C_i), \tag{1}$$

where $\mathcal{A}(\mathcal{X})$ represents the set of all $|D_1| \times |D_2| \times \ldots \times |D_N|$ complete assignments to all variables in $\mathcal{X}$. $a|C_i$ represents the projection of a complete assignment a onto the subset of variables in C_i. E_{C_i} is a function that maps each $a|C_i$ to its associated weight in C_i.

The Boolean WCSP is the WCSP in which each domain $D_i \in \mathcal{D}$ has its cardinality restricted to be 2. Despite this restriction, the Boolean WCSP is representationally as powerful as the WCSP, and it is also NP-hard to solve in general. The (Boolean) WCSP can be used to model a wide range of useful combinatorial problems arising in a large number of real-world application domains. For example, in artificial intelligence, it can be used to model user preferences [2] and combinatorial auctions. In bioinformatics, it can be used to locate RNA motifs [21]. In statistical physics, the energy minimization problem on the Potts model is equivalent to that on its corresponding pairwise Markov random field [20], which in turn can be modeled as the WCSP. In computer vision, it can be used for image restoration and panoramic image stitching [3,8].

The constraint composite graph (CCG) is a combinatorial structure associated with an optimization problem posed as the WCSP. The CCG provides a unifying framework for simultaneously exploiting the graphical structure of the variable-interactions in the WCSP as well as the numerical structure of the weighted constraints in it. The task of solving the WCSP can be reformulated as the task of finding a minimum weighted vertex cover (MWVC) on its associated CCG [9–11]. CCGs can be constructed in polynomial time and are always tripartite [9–11]. A subclass of the WCSP has instances with bipartite CCGs. This subclass is tractable since the MWVC problem can be solved in polynomial time on bipartite graphs using a staged maxflow algorithm [4].

Despite its theoretical importance, the CCG still remains largely understudied. In this paper, we study two important implications of the CCG. First, we show that the Nemhauser-Trotter (NT) reduction popularly used for kernelization of the MWVC problem [16] can also be applied to the CCG of the WCSP. This leads to a polynomial-time preprocessing algorithm that fixes the optimal values of a subset of the variables in the WCSP; and this subset is often the set of all variables. As a consequence, many WCSP instances can be solved by the polynomial-time NT reduction without search. Experimental evaluations of the NT reduction on the Boolean WCSP benchmark instances show that about $1/8^{\text{th}}$ of these benchmark instances have a kernel of size 0. In other words, about $1/8^{\text{th}}$ of these benchmark instances can be solved without search, simply by using the power of the preprocessing algorithm that encapsulates the NT reduction.

Second, belief propagation (BP) is a well-known technique used for solving many combinatorial problems in probabilistic reasoning, artificial intelligence and information theory. The min-sum message passing (MSMP) algorithm is a simple variant of BP that has also been successfully employed in several research communities. Unfortunately, the MSMP algorithm has met with little success on the WCSP. We revive the MSMP algorithm for solving the WCSP by applying it on the CCG of a given WCSP instance instead of its original form. We refer to these algorithms as the lifted MSMP algorithm (since the CCG is a lifted representation of the WCSP [9–11]) and the original MSMP algorithm, respectively. Intuitively, the lifted MSMP algorithm outperforms the original MSMP algorithm in terms of effectiveness since the CCG associated with the WCSP makes the numerical structure of its weighted constraints explicit using a tripartite graphical representation. We demonstrate the effectiveness of the lifted MSMP algorithm through experimental evaluations on the Boolean WCSP benchmark instances. We show that it outperforms the original MSMP algorithm on these benchmark instances in terms of solution quality.

2   The Constraint Composite Graph

Given an undirected graph $G = \langle V, E \rangle$, a vertex cover of G is defined as a set of vertices $S \subseteq V$ such that every edge in E has at least one of its endpoint vertices in S. A minimum vertex cover (MVC) of G is a vertex cover of minimum cardinality. When G is vertex-weighted—i.e., each vertex $v_i \in V$ has a non-negative weight w_i associated with it—its MWVC is defined as a vertex cover of minimum total weight of its vertices. The MWVC problem is to compute an MWVC on a given vertex-weighted undirected graph.

For a given graph G, the concept of the MWVC problem can be extended to the notion of projecting MWVCs onto a given independent set (IS) $U \subseteq V$. (An IS is defined as a set of vertices in which no two of them are connected by an edge.) The input to such a projection is the graph G as well as an IS $U = \{u_1, u_2, \ldots, u_k\}$. The output is a table of 2^k numbers. Each entry in this table corresponds to a k-bit vector. We say that a k-bit vector t imposes the following restrictions: (a) if the ith bit t_i is 0, the vertex u_i has to be excluded from the MWVC; and (b) if the ith bit t_i is 1, the vertex u_i has to be included in the MWVC. The projection of the MWVC problem onto the IS U is then defined to be a table with entries corresponding to each of the 2^k possible k-bit vectors $t^{(1)}, t^{(2)}, \ldots, t^{(2^k)}$. The value of the entry corresponding to $t^{(j)}$ is equal to the weight of the MWVC conditioned on the restrictions imposed by $t^{(j)}$. Figure 1 in [11] presents a simple example to illustrate this projection in a vertex-weighted undirected graph.

The table of numbers produced above can be viewed as a weighted constraint over $|U|$ Boolean variables. Conversely, given a (Boolean) weighted constraint, we design a lifted representation for it so as to be able to view it as the projection of MWVCs onto an IS in some intelligently constructed vertex-weighted undirected graph [9,10]. The benefit of constructing these representations for individual constraints lies in the fact that the lifted representation for the entire WCSP, called the CCG of the WCSP, can be obtained simply by "merging" them.

Figure 2 in [11] shows an example WCSP instance over 3 Boolean variables to illustrate the construction of the CCG. Here, there are 3 unary weighted constraints and 3 binary weighted constraints. Their lifted representations are shown next to them. The figure also illustrates how the CCG is obtained from the lifted representations of the weighted constraints: In the CCG, vertices that represent the same variable are simply "merged"—along with their edges—and every "composite" vertex is given a weight equal to the sum of the individual weights of the merged vertices. Computing the MWVC for the CCG yields a solution for the WCSP instance; namely, if X_i is in the MWVC, then it is assigned the value 1 in the WCSP instance, otherwise it is assigned the value 0 in the WCSP instance. The CCG of the WCSP can be constructed in polynomial time using the algorithms suggested in [9,10].

3   Kernelization of the WCSP: NT Reduction on CCGs

The NT reduction is a polynomial-time kernelization procedure that reduces the size of a given MWVC problem instance [16]. It can be potentially applied on CCGs as well. It is based on the observation that the MWVC problem is a half-integral problem. This means that its Integer Linear Programming (ILP) formulation exhibits the following property. Given a graph $G = \langle V, E \rangle$, let w_i be the non-negative weight associated with vertex v_i. In the ILP formulation of the MWVC problem instance on G, a Boolean decision variable Z_i is first associated with the presence of vertex v_i in the MWVC. Then, the ILP formulation is

$$\text{Minimize} \quad \sum_{i=1}^{|V|} w_i Z_i,$$
$$\forall\ v_i \in V:\quad Z_i \in \{0,1\}, \tag{2}$$
$$\forall\ (v_i, v_j) \in E:\quad Z_i + Z_j \geq 1.$$

If we relax the integrality constraints $Z_i \in \{0,1\}$ for all $i \in \{1, 2, \ldots, |V|\}$ and solve the relaxed LP, the optimal solution of the LP is guaranteed to be half-integral—i.e., $\forall i \in \{1, 2, \ldots, |V|\} : Z_i \in \{0, \frac{1}{2}, 1\}$. There then exists an MWVC on G that includes v_i if $Z_i = 1$ and excludes v_i if $Z_i = 0$. Therefore, one can kernelize the MWVC problem instance on G to an MWVC problem instance on a subgraph of G by retaining only those vertices whose Boolean variables in the optimal solution of the LP are $\frac{1}{2}$.

The half-integrality property can be further exploited to solve the LP relaxation of the MWVC problem with a maxflow algorithm instead of a general LP solver [4]. We first transform G to a vertex-weighted undirected bipartite graph $G_b = \langle V_{G_b}^L, V_{G_b}^R, E_{G_b} \rangle$ as follows. For each vertex $v_i \in V$, we create two vertices $v_i^L \in V_{G_b}^L$ and $v_i^R \in V_{G_b}^R$, both with weight w_i. For each edge $(v_i, v_j) \in E$, we create two edges $(v_i^L, v_j^R) \in E_{G_b}$ and $(v_j^L, v_i^R) \in E_{G_b}$. The MWVC problem can be solved in polynomial time on the bipartite graph G_b using a maxflow algorithm [4]; and the half-integral solution of the above LP relaxation can be retrieved as follows. If both v_i^L and v_i^R are in the MWVC of G_b, then $Z_i = 1$ and v_i can be safely included in the MWVC of G; if neither v_i^L nor v_i^R is in the MWVC of G_b, then $Z_i = 0$ and v_i can be safely excluded from the MWVC of G; if exactly one of v_i^L or v_i^R is in the MWVC of G_b, then $Z_i = \frac{1}{2}$ and v_i is retained in the kernel of the MWVC problem instance posed on G.

4 Min-Sum Message Passing on the WCSP and CCGs

BP is a well-known technique for solving many combinatorial problems across a wide range of fields such as probabilistic reasoning, artificial intelligence and information theory. It can be used to solve hard inference problems that arise in statistical physics, computer vision, error-correcting coding theory or, more generally, on graphical models such as Bayesian Networks and Markov random fields [20]. BP is an efficient algorithm that is based on local message passing. Although a complete theoretical analysis of its convergence and correctness is elusive, it works well in practice on many important combinatorial problems.

While BP performs message passing for the objective of marginalization over probabilities, the MSMP algorithm is a variant of BP that is used to find an assignment of values to all variables in X that minimizes functions of the form

$$E(X) = \sum_i E_i(X_i), \qquad (3)$$

where X is the set of all variables in the global function E; E_i is a local function constituting the ith term of E; and X_i is a subset of X containing all variables that participate in E_i.

To minimize the function $E(X)$, the MSMP algorithm first builds a factor graph, i.e., an undirected bipartite graph with one partition containing vertices that represent the variables in X and the other partition containing vertices that represent the local functions E_i for all i. An edge represents the participation of a variable in a local function. Furthermore, a message is associated with each direction of each edge. Intuitively, messages represent interactions between individual variables and local functions. The value of E_i is the potential of its corresponding vertex because it is indicative of its "potential" to affect other vertices. Messages are updated iteratively until convergence. In each iteration, the message from vertex u to vertex v is influenced by incoming messages to u as well as u's potential if it represents a local function. Upon convergence, a solution can be extracted from the messages.

The MSMP algorithm converges and produces an optimal solution if the factor graph is a tree [12]. This is, however, not necessarily the case if the factor graph is loopy [12]. Although the clique tree algorithm alleviates this problem to a certain extent by first converting loopy graphs to trees [7], the technique only scales to graphs with low treewidths. If the MSMP algorithm operates directly on loopy graphs, the theoretical underpinnings of its convergence and optimality properties still remain poorly understood. Nonetheless, it works well in practice

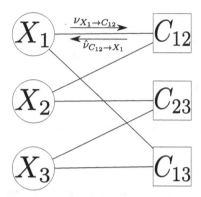

Fig. 1. Illustrates the factor graph of a Boolean WCSP instance with 3 variables $\{X_1, X_2, X_3\}$ and 3 constraints $\{C_{12}, C_{13}, C_{23}\}$. Here, $X_1, X_2 \in C_{12}$, $X_1, X_3 \in C_{13}$ and $X_2, X_3 \in C_{23}$. The circles are variable vertices, and the squares are constraint vertices. $\nu_{X_1 \to C_{12}}$ and $\hat{\nu}_{C_{12} \to X_1}$ are the messages from X_1 to C_{12} and from C_{12} to X_1, respectively. Such a pair of messages annotates each edge (not all are explicitly shown).

on a number of important combinatorial problems in artificial intelligence, statistical physics and signal processing [12,14]. Examples include the CSP [15], K-satisfiability [13] and the MVC problem [18]. Unfortunately, the MSMP algorithm has met with little success on the WCSP. In this section, we show how to revive the MSMP algorithm for the WCSP by using CCGs.

4.1   The MSMP Algorithm Applied Directly on the WCSP

We now describe how the MSMP algorithm can be applied directly to solve the Boolean WCSP defined by $\langle \mathcal{X}, \mathcal{D}, \mathcal{C} \rangle$. We refer to this as the original MSMP algorithm. As explained before, we first construct its factor graph. We create a vertex for each variable in $\mathcal{X}$ (variable vertex) and for each weighted constraint in $\mathcal{C}$ (constraint vertex). A variable vertex X_i and a constraint vertex C_j are connected by an edge if and only if C_j contains X_i. Figure 1 shows an example.

After the factor graph is constructed, a message (two real numbers) for each of the two directions along each edge is initialized, for instance, to zeros. A pair of messages $\nu_{X_1 \to C_{12}}$ and $\hat{\nu}_{C_{12} \to X_1}$ is illustrated in Fig. 1. The messages are then updated iteratively by using the min-sum update rules given by

$$\nu_{X_i \to C_j}^{(t)}(X_i = x_i) = \sum_{C_k \in \partial X_i \setminus \{C_j\}} \left[\hat{\nu}_{C_k \to X_i}^{(t-1)}(X_i = x_i) \right] + c_{X_i \to C_j}^{(t)} \tag{4}$$

$$\hat{\nu}_{C_j \to X_i}^{(t)}(X_i = x_i) = \min_{a \in \mathcal{A}(\partial C_j \setminus \{X_i\})} \left[E_{C_j}(a | \boldsymbol{X}_j) + \sum_{X_k \in \partial C_j \setminus \{X_i\}} \nu_{X_k \to C_j}^{(t)}(a | \{X_k\}) \right] \tag{5}$$
$$+ \, \hat{c}_{C_j \to X_i}^{(t)}$$

for all $X_i \in \mathcal{X}, C_j \in \mathcal{C}$ and $x_i \in \{0,1\}$ until convergence [12], where

- $\hat{\nu}_{C_j \to X_i}^{(t)}(X_i = x_i)$ for both $x_i \in \{0,1\}$ are the two real numbers of the message that is passed from the constraint vertex C_j to the variable vertex X_i in the tth iteration,
- $\nu_{X_i \to C_j}^{(t)}(X_i = x_i)$ for both $x_i \in \{0,1\}$ are the two real numbers of the message that is passed from the variable vertex X_i to the constraint vertex C_j in the tth iteration,
- ∂X_i and ∂C_j are the sets of neighboring vertices of X_i and C_j, respectively,
- $\mathbf{X}_j$ is the set of all variables in the constraint C_j, and
- $c_{X_i \to C_j}^{(t)}$ and $\hat{c}_{C_j \to X_i}^{(t)}$ are normalization constants such that

$$\min\left[\nu_{X_i \to C_j}^{(t)}(X_i = 0), \quad \nu_{X_i \to C_j}^{(t)}(X_i = 1)\right] = 0 \qquad (6)$$

$$\min\left[\hat{\nu}_{C_j \to X_i}^{(t)}(X_i = 0), \quad \hat{\nu}_{C_j \to X_i}^{(t)}(X_i = 1)\right] = 0. \qquad (7)$$

The message update rules can be understood as follows. Each message from a variable vertex X_i to a constraint vertex C_j is updated by summing up all X_i's incoming messages from its other neighboring vertices. Each message from a constraint vertex C_j to a variable vertex X_i is updated by finding the minimum of the constraint function E_{C_j} plus the sum of all C_j's incoming messages from its other neighboring vertices. The messages can be updated in various orders.

We use the superscript ∞ to indicate the values of messages upon convergence. The final assignment of values to variables in $\mathcal{X} = \{X_1, X_2, \ldots, X_N\}$ can then be found by computing

$$E_{X_i}(X_i = x_i) \equiv \sum_{C_k \in \partial X_i} \hat{\nu}_{C_k \to X_i}^{(\infty)}(X_i = x_i) \qquad (8)$$

for all $X_i \in \mathcal{X}$ and $x_i \in \{0,1\}$. Here, $E_{X_i}(X_i = 0)$ and $E_{X_i}(X_i = 1)$ can be proven to be equal to the minimum values of the total weights conditioned on $X_i = 0$ and $X_i = 1$, respectively. By selecting the value of x_i that leads to a smaller value of $E_{X_i}(X_i = x_i)$, we obtain the final assignment of values to all variables in $\mathcal{X}$.

4.2   The MSMP Algorithm Applied on CCGs

To solve a given WCSP instance, we can first transform it to an MWVC problem instance on its CCG. We can then apply the MSMP algorithm on the CCG. We refer to this procedure as the lifted MSMP algorithm.

The MWVC problem on $\langle V, E, w \rangle$—where V is the set of vertices, E is the set of edges, and w is the set of non-negative weights of the vertices—is a subclass of the Boolean WCSP. Throughout this subsection, we use the variable X_i to represent the ith vertex in V: $X_i = 1$ means the ith vertex is selected in the MWVC, and $X_i = 0$ means the ith vertex is not selected in the MWVC. The MWVC problem can therefore be rewritten as a subclass of the Boolean WCSP with only the following two types of constraints:

- Unary weighted constraints: Each of these weighted constraints corresponds to a vertex in the MWVC problem. We use C_i^V to denote the weighted constraint that corresponds to the ith vertex. C_i^V therefore only has one variable X_i. In the weighted constraint C_i^V, the tuple in which $X_i = 1$ has weight $w_i \geq 0$ and the other tuple has weight zero. This type of weighted constraints represents the minimization objective of the MWVC problem.
- Binary weighted constraints: Each of these weighted constraints corresponds to an edge in the MWVC problem. We use C_j^E to denote the weighted constraint that corresponds to the jth edge. The indices of the endpoint vertices of this edge are denoted by $j(+1)$ and $j(-1)$. C_j^E therefore has two variables $X_{j(+1)}$ and $X_{j(-1)}$. In the weighted constraint C_j^E, the tuple in which $X_{j(+1)} = X_{j(-1)} = 0$ has weight infinity and the other tuples have weight zero. This type of weighted constraints represents the requirement that at least one endpoint vertex must be selected for each edge.

Given that the MWVC problem is a subclass of the Boolean WCSP, Eqs. 4, 5 and 8 can be reused for the MSMP algorithm on it. For the MWVC problem, these equations can be further simplified. For notational convenience, we omit normalization constants in the following derivation.

For each of the unary weighted constraints C_i^V, we have

- the added weight for selecting a vertex:

$$E_{C_i^V}(X_i = x_i) = \begin{cases} w_i & x_i = 1 \\ 0 & x_i = 0 \end{cases}, \tag{9}$$

- and exactly one variable in C_i^V:

$$\partial C_i^V \setminus \{X_i\} = \emptyset. \tag{10}$$

By plugging Eqs. 9 and 10 into Eq. 5 for $t = \infty$, we have

$$\hat{\nu}_{C_i^V \to X_i}^{(\infty)}(X_i = x_i) = \begin{cases} w_i & x_i = 1 \\ 0 & x_i = 0 \end{cases} \tag{11}$$

for all C_i^V. Note that here we do not need Eq. 4 for C_i^V since it has only one variable and thus the message passed to it does not affect the final solution.

For each of the binary weighted constraints C_j^E, we have

- the requirement that at least one endpoint vertex must be selected for each edge:

$$E_{C_j^E}(X_{j(+1)} = x_{j(+1)}, X_{j(-1)} = x_{j(-1)}) = \begin{cases} +\infty & x_{j(+1)} = x_{j(-1)} = 0 \\ 0 & \text{otherwise} \end{cases}, \tag{12}$$

- and exactly two variables in C_j^E:

$$\partial C_j^E \setminus \{X_{j(\ell)}\} = \{X_{j(-\ell)}\} \quad \forall \ell \in \{+1, -1\}. \tag{13}$$

By plugging Eqs. 11, 12 and 13 into Eqs. 4 and 5 along with the fact that there exist only unary and binary weighted constraints, we have

$$
\nu^{(t)}_{X_{j(\ell)} \to C^E_j}(X_{j(\ell)} = 1) = \sum_{C_k \in \partial X_{j(\ell)} \setminus \{C^V_{j(\ell)}, C^E_j\}} \left[\hat{\nu}^{(t-1)}_{C_k \to X_{j(\ell)}}(X_{j(\ell)} = 1) \right] + w_{j(\ell)}
$$

$$(14)$$

$$
\nu^{(t)}_{X_{j(\ell)} \to C^E_j}(X_{j(\ell)} = 0) = \sum_{C_k \in \partial X_{j(\ell)} \setminus \{C^V_{j(\ell)}, C^E_j\}} \hat{\nu}^{(t-1)}_{C_k \to X_{j(\ell)}}(X_{j(\ell)} = 0) \qquad (15)
$$

$$
\hat{\nu}^{(t)}_{C^E_j \to X_{j(\ell)}}(X_{j(\ell)} = 1) = \min_{a \in \{0,1\}} \nu^{(t)}_{X_{j(-\ell)} \to C^E_j}(X_{j(-\ell)} = a) \qquad (16)
$$

$$
\hat{\nu}^{(t)}_{C^E_j \to X_{j(\ell)}}(X_{j(\ell)} = 0) = \nu^{(t)}_{X_{j(-\ell)} \to C^E_j}(X_{j(-\ell)} = 1) \qquad (17)
$$

for all C^E_j and both $\ell \in \{+1, -1\}$. By plugging Eqs. 14 and 15 into Eqs. 16 and 17, we have

$$
\hat{\nu}^{(t)}_{C^E_j \to X_{j(\ell)}}(X_{j(\ell)} = 1)
$$

$$
= \min_{a \in \{0,1\}} \left[\sum_{C_k \in \partial X_{j(-\ell)} \setminus \{C^V_{j(-\ell)}, C^E_j\}} \left[\hat{\nu}^{(t-1)}_{C_k \to X_{j(-\ell)}}(X_{j(-\ell)} = a) \right] + w_{j(-\ell)} \cdot a \right]
$$

$$(18)$$

$$
\hat{\nu}^{(t)}_{C^E_j \to X_{j(\ell)}}(X_{j(\ell)} = 0)
$$

$$
= \sum_{C_k \in \partial X_{j(-\ell)} \setminus \{C^V_{j(-\ell)}, C^E_j\}} \left[\hat{\nu}^{(t-1)}_{C_k \to X_{j(-\ell)}}(X_{j(-\ell)} = 1) \right] + w_{j(-\ell)} \qquad (19)
$$

for all C^E_j and both $\ell \in \{+1, -1\}$, where $\hat{\nu}^{(t)}_{C^E_j \to X_{j(\ell)}}(X_{j(\ell)} = b)$ for both $b \in \{0, 1\}$ are the two real numbers of the message that is passed from the jth edge to the $j(\ell)^{\text{th}}$ vertex. Since each edge has exactly two endpoint vertices, the message from an edge to one of its endpoint vertices can be viewed as a message from the other endpoint vertex to it. Formally, for the jth edge, we define the message from the $j(+1)^{\text{th}}$ vertex to the $j(-1)^{\text{th}}$ vertex in the tth iteration as

$$
\mu^{(t)}_{j(+1) \to j(-1)} \equiv \hat{\nu}^{(t)}_{C^E_j \to X_{j(-1)}}. \qquad (20)
$$

By plugging in Eq. 20 and substituting $j(\ell)$ with i and $j(-\ell)$ with j, Eqs. 18 and 19 can be rewritten (with normalization constants) in the form of messages between vertices as

$$
\mu^{(t)}_{j \to i}(X_i = 1) = \min_{a \in \{0,1\}} \left[\sum_{k \in N(j) \setminus \{i\}} \mu^{(t-1)}_{k \to j}(X_j = a) + w_j \cdot a \right] + c^{(t)}_{j \to i} \qquad (21)
$$

$$
\mu^{(t)}_{j \to i}(X_i = 0) = \sum_{k \in N(j) \setminus \{i\}} \mu^{(t-1)}_{k \to j}(X_j = 1) + w_j + c^{(t)}_{j \to i} \qquad (22)
$$

for all i and j such that the ith and j^{th} vertices are connected by an edge in E. Here, $N(j)$ is the set of neighboring vertices of the jth vertex in V and $c_{j\to i}^{(t)}$ represents the normalization constant such that $\min\left[\mu_{j\to i}^{(t)}(X_i = 1), \mu_{j\to i}^{(t)}(X_i = 0)\right] = 0$. Equations 21 and 22 are the message update rules of the MSMP algorithm adapted to the MWVC problem.

If the messages converge, by plugging Eqs. 11 and 20 into Eq. 8, the final assignment of values to variables can be found by computing

$$E_{X_i}(X_i = x_i) = \sum_{j\in N(i)} \left[\mu_{j\to i}^{(\infty)}(X_i = x_i)\right] + w_i x_i, \tag{23}$$

where the meaning of $E_{X_i}(X_i = x_i)$ is similar to that in Eq. 8.

5   Experimental Evaluation

In this section, we present experimental evaluations of the NT reduction and the lifted MSMP algorithm. We used two sets of Boolean WCSP benchmark instances for our experiments. The first set of benchmark instances is from the UAI 2014 Inference Competition[1]. Here, maximum a posteriori (MAP) inference queries with no evidence on the PR and MMAP benchmark instances can be reformulated as Boolean WCSP instances by first taking the negative logarithms of the probabilities in each factor and then normalizing them. The second set of benchmark instances is from [6][2]. This set includes the Probabilistic Inference Challenge 2011, the Computer Vision and Pattern Recognition OpenGM2 benchmark, the Weighted Partial MaxSAT Evaluation 2013, the MaxCSP 2008 Competition, the MiniZinc Challenge 2012 & 2013 and the CFLib (a library of cost function networks). The experiments were performed on those benchmark instances that have only Boolean variables.[3]

The optimal solutions of the benchmark instances in [6] were computed using toulbar2 [6]. Since toulbar2 cannot solve WCSP instances with non-integral weights, the optimal solutions of the benchmark instances from the UAI 2014 Inference Competition were computed by finding MWVCs on their CCGs. For each benchmark instance, the MWVC problem was solved by first kernelizing it using the NT reduction, then reformulating it as an ILP [19] and finally solving the ILP using the Gurobi optimizer [5] with a running time limit of 5 min.

For our experiments, we implemented the NT reduction using the Gurobi optimizer [5] as the LP solver.[4] For the MSMP algorithms, we set the initial values of all messages to zeros. If no message changed by an amount more than

---

[1] http://www.hlt.utdallas.edu/~vgogate/uai14-competition/index.html.

[2] http://genoweb.toulouse.inra.fr/~degivry/evalgm/.

[3] As shown in [10], our techniques can also be generalized to the WCSP with larger domain sizes of the variables. However, for a proof of concept, this paper focuses on the Boolean WCSP.

[4] We could have also implemented the NT reduction using a more efficient maxflow algorithm [4]; but once again, we focus only on the proof of concept in this paper.

10^{-6} in any iteration, we declared convergence. We used the synchronous message updating order, i.e., messages were updated in parallel in each iteration. This standardized the comparison between the two MSMP algorithms, factoring out the effects of different message updating orders within each iteration. In case of failure to converge within the time limit (5 min) for any benchmark instance, we reported the solution produced by the MSMP algorithm on that benchmark instance at the end of that time limit. The CCG construction algorithm and the MSMP algorithms were implemented in C++ using the Boost graph library [17] and were compiled by gcc 4.9.2 with the "–O3" option. Our experiments were performed on a GNU/Linux workstation with an Intel Xeon processor E3-1240 v3 (8MB Cache, 3.4 GHz) and 16 GB RAM.

Figure 2 shows the effectiveness of the NT reduction on the benchmark instances. The polynomial-time NT reduction solved about $1/8^{th}$ of these benchmark instances yielding empty kernels. Being able to solve this many benchmark

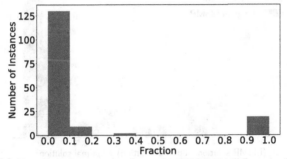

(a) Benchmark instances from the UAI 2014 Inference Competition: 19 out of 160 benchmark instances solved by the NT reduction

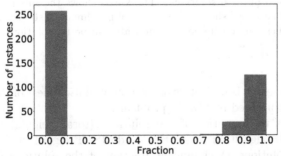

(b) Benchmark instances from [6]: 53 out of 410 benchmark instances solved by the NT reduction

Fig. 2. Shows the effectiveness of the NT reduction. The x-axis shows the fraction of variables that are eliminated by the NT reduction. The y-axis shows the number of benchmark instances on which this happens for a fraction range.

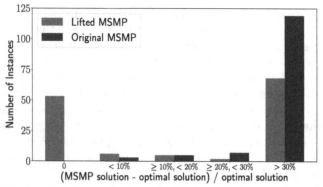

(a) Benchmark instances from the UAI 2014 Inference Competition

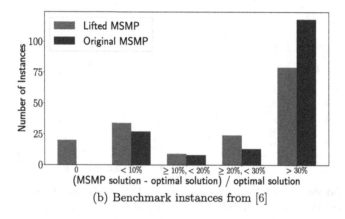

(b) Benchmark instances from [6]

Fig. 3. Shows the qualities of the solutions (total weights) produced by the original and the lifted MSMP algorithms in comparison to the optimal solutions (for benchmark instances with known optimal solutions). The x-axis shows the suboptimality of the MSMP solutions. The y-axis shows the number of benchmark instances for a range of suboptimality. Higher bars on the left are indicative of better solutions. (Color figure online)

instances without search is indicative of the potential usefulness of the NT reduction for solving structured real-world problems.

Figure 3 shows the qualities of the solutions (total weights) produced by the original MSMP algorithm versus the lifted MSMP algorithm in comparison to the optimal solutions. A significant fraction of the solutions produced by the lifted MSMP algorithm are very close to the optimal solutions. However, both MSMP algorithms produced solutions that are highly suboptimal in the > 30% suboptimality range. Therefore, Fig. 4 presents a direct comparison of the qualities of the solutions produced by the two MSMP algorithms. From this

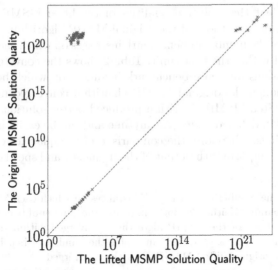

(a) Benchmark instances from the UAI 2014 Inference Competition: 126/9/18 above/below/close to the diagonal dashed line

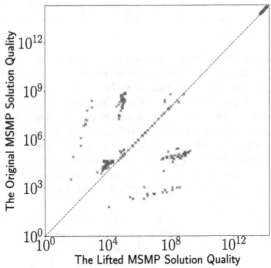

(b) Benchmark instances from [6]: 222/68/19 above/below/close to the diagonal dashed line

Fig. 4. Shows the qualities of the solutions produced by the original MSMP algorithm in direct comparison to those produced by the lifted MSMP algorithm for both sets of benchmark instances. Each point in these plots represents a benchmark instance. The x and y coordinates of a benchmark instance represent the solution qualities produced by the lifted MSMP algorithm and the original MSMP algorithm, respectively. Benchmark instances above (red)/below (blue) the diagonal dashed line have better/worse solution qualities when using the lifted MSMP algorithm instead of the original MSMP algorithm. Benchmark instances whose MSMP solution qualities differ by only 1% are considered close (green) to the diagonal dashed line. (Color figure online)

figure, it is evident that solution qualities of the lifted MSMP algorithm are significantly better than those of the original MSMP algorithm.

Table 1 shows the number of benchmark instances on which each MSMP algorithm converged within the time limit. Table 2 shows the convergence time and number of iterations for those benchmark instances on which both algorithms converged. Although the original MSMP algorithm converged more frequently and faster, the lifted MSMP algorithm produced better solutions in general. In addition, both MSMP algorithms are anytime and can be easily implemented in distributed settings. Therefore, the comparison of the qualities of the solutions produced is more important than that of the frequency and speed of convergence.

Table 1. Shows the number of benchmark instances on which each MSMP algorithm converged. The column "Neither"/"Both" indicates the number of benchmark instances on which neither/both of the MSMP algorithms converged within the time limit of 5 min. The column "Original"/"Lifted" indicates the number of benchmark instances on which only the original/lifted MSMP algorithm converged.

Benchmark instance set	Neither	Both	Original	Lifted
UAI 2014 inference competition	25	4	124	0
[6]	258	7	44	0

Table 2. Shows the number of iterations and running time for each of the benchmark instances on which both MSMP algorithms converged within the time limit of 5 min. The column "Benchmark Instance" indicates the name of each benchmark instance. The "U:" and "T:" at the beginning of the names indicate that they are from the UAI 2014 Inference Competition and [6], respectively. The columns "Iterations" and "Running Time" under "The Original MSMP" and "The Lifted MSMP" indicate the number of iterations and running time (in seconds) after which the original MSMP algorithm and the lifted MSMP algorithm converged, respectively. With a few exceptions, the number of iterations and running time for the original MSMP algorithm are in general smaller than those of the lifted MSMP algorithm.

Benchmark Instance	The Original MSMP		The Lifted MSMP	
	Iterations	Running Time	Iterations	Running Time
U:PR/relational_2	5	0.84	9	4.00
U:PR/ra.cnf	1	0.35	6	0.34
U:PR/relational_5	5	1.18	3	0.76
U:PR/Segmentation_12	9	0.04	44	0.14
T:MRF/Segmentation/4_30_s.binary	31	0.10	60	0.13
T:MRF/Segmentation/2_28_s.binary	9	0.05	44	0.11
T:MRF/Segmentation/18_10_s.binary	15	0.07	102	0.18
T:MRF/Segmentation/12_20_s.binary	31	0.13	50	0.14
T:MRF/Segmentation/11_3_s.binary	47	0.15	176	0.24
T:MRF/Segmentation/1_28_s.binary	35	0.11	60	0.14
T:MRF/Segmentation/3_20_s.binary	31	0.12	54	0.14

6   Conclusions and Future Work

We studied two important implications of the CCG associated with the WCSP. First, we showed that the NT reduction popularly used for kernelization of the MWVC problem can also be applied to the CCG of the WCSP. This leads to a polynomial-time preprocessing algorithm that fixes the optimal values of a subset of variables in a WCSP instance. This subset is often the set of all variables: We observed that the NT reduction could determine the optimal values of all variables for about $1/8^{th}$ of the benchmark instances without search.

Second, we revived the MSMP algorithm for solving the WCSP by applying it on its CCG instead of its original form. We observed not only that the lifted MSMP algorithm produced solutions that are close to optimal for a large fraction of benchmark instances, but also that, in general, it produced significantly better solutions than the original MSMP algorithm. Although the lifted MSMP algorithm requires slightly more work in each iteration since the CCG is constructed using auxiliary variables, the size of the CCG is only linear in the size of the tabular representation of the WCSP [9–11], and the lifted MSMP algorithm has the benefit of producing better solutions. Both MSMP algorithms employ local message passing techniques that avoid an exponential amount of computational effort and can be readily adapted to distributed settings as well.

There are many avenues for future work. One is to extend the lifted MSMP algorithm to handle the WCSP with variables of larger domain sizes. (This extension already exists in theory [10].) Another one is to develop a distributed version of the lifted MSMP algorithm using grid/cloud computing facilities. And a third one is to explore the usefulness of constructing the CCG recursively, i.e., constructing the CCG of the MWVC problem instance on the CCG associated with a WCSP instance, and so on.

Acknowledgement. The research at the University of Southern California was supported by the National Science Foundation (NSF) under grant numbers 1409987 and 1319966. The views and conclusions contained in this document are those of the authors and should not be interpreted as representing the official policies, either expressed or implied, of the sponsoring organizations, agencies or the U.S. government.

References

1. Bistarelli, S., Montanari, U., Rossi, F., Schiex, T., Verfaillie, G., Fargier, H.: Semiring-based CSPs and valued CSPs: frameworks, properties, and comparison. Constraints **4**(3), 199–240 (1999)
2. Boutilier, C., Brafman, R.I., Domshlak, C., Hoos, H.H., Poole, D.: CP-nets: a tool for representing and reasoning with conditional ceteris paribus preference statements. J. Artif. Intell. Res. **21**, 135–191 (2004)
3. Boykov, Y., Veksler, O., Zabih, R.: Fast approximate energy minimization via graph cuts. IEEE Trans. Pattern Anal. Mach. Intell. **23**(11), 1222–1239 (2001)
4. Goldberg, A.V., Tarjan, R.E.: A new approach to the maximum-flow problem. J. ACM **35**(4), 921–940 (1988)

5. Gurobi Optimization Inc.: Gurobi optimizer reference manual (2016). http://www.gurobi.com

6. Hurley, B., O'Sullivan, B., Allouche, D., Katsirelos, G., Schiex, T., Zytnicki, M., de Givry, S.: Multi-language evaluation of exact solvers in graphical model discrete optimization. Constraints 21(3), 413–434 (2016)

7. Koller, D., Friedman, N.: Probabilistic Graphical Models: Principles and Techniques. MIT Press, Cambridge (2009)

8. Kolmogorov, V.: Primal-dual algorithm for convex markov random fields. Technical report, MSR-TR-2005-117, Microsoft Research (2005)

9. Kumar, T.K.S.: A framework for hybrid tractability results in boolean weighted constraint satisfaction problems. In: Stuckey, P.J. (ed.) CP 2008. LNCS, vol. 5202, pp. 282–297. Springer, Heidelberg (2008). doi:10.1007/978-3-540-85958-1_19

10. Kumar, T.K.S.: Lifting techniques for weighted constraint satisfaction problems. In: the International Symposium on Artificial Intelligence and Mathematics (2008)

11. Kumar, T.K.S.: Kernelization, generation of bounds, and the scope of incremental computation for weighted constraint satisfaction problems. In: The International Symposium on Artificial Intelligence and Mathematics (2016)

12. Mézard, M., Montanari, A.: Information, Physics, and Computation. Oxford University Press, New York (2009)

13. Mézard, M., Zecchina, R.: Random k-satisfiability problem: from an analytic solution to an efficient algorithm. Phys. Rev. E 66(5), 056126 (2002)

14. Moallemi, C.C., Roy, B.V.: Convergence of min-sum message-passing for convex optimization. IEEE Trans. Inf. Theor. 56(4), 2041–2050 (2010)

15. Montanari, A., Ricci-Tersenghi, F., Semerjian, G.: Solving constraint satisfaction problems through belief propagation-guided decimation. In: The Annual Allerton Conference, pp. 352–359 (2007)

16. Nemhauser, G.L., Trotter, L.E.: Vertex packings: structural properties and algorithms. Math. Program. 8(1), 232–248 (1975)

17. Siek, J., Lee, L.Q., Lumsdain, A.: The Boost Graph Library: User Guide and Reference Manual. Addison-Wesley, Boston (2002)

18. Weigt, M., Zhou, H.: Message passing for vertex covers. Phys. Rev. E 74(4), 046110 (2006)

19. Xu, H., Kumar, T.K.S., Koenig, S.: A new solver for the minimum weighted vertex cover problem. In: Quimper, C.-G. (ed.) CPAIOR 2016. LNCS, vol. 9676, pp. 392–405. Springer, Cham (2016). doi:10.1007/978-3-319-33954-2_28

20. Yedidia, J.S., Freeman, W.T., Weiss, Y.: Understanding belief propagation and its generalizations. Exploring Artif. Intell. New Millennium 8, 236–239 (2003)

21. Zytnicki, M., Gaspin, C., Schiex, T.: DARN! A weighted constraint solver for RNA motif localization. Constraints 13(1), 91–109 (2008)

A Local Search Approach for Incomplete Soft Constraint Problems: Experimental Results on Meeting Scheduling Problems

Mirco Gelain[1], Maria Silvia Pini[1(✉)], Francesca Rossi[2],
Kristen Brent Venable[3], and Toby Walsh[4]

[1] University of Padova, Padua, Italy
mirco@gelain.it, pini@dei.unipd.it
[2] IBM T.J. Watson Research Center, Yorktown Heights, NY, USA
frossi@math.unipd.it
[3] IHMC, Tulane University, New Orleans, USA
kvenabl@tulane.edu
[4] UNSW and Data61 (formerly NICTA), Sydney, Australia
toby.walsh@nicta.com.au

Abstract. We consider soft constraint problems where some of the preferences may be unspecified. In practice, some preferences may be missing when there is, for example, a high cost for computing the preference values, or an incomplete elicitation process. Within such a setting, we study how to find an optimal solution without having to wait for all the preferences. In particular, we define a local search approach that interleaves search and preference elicitation, with the goal to find a solution which is "necessarily optimal", that is, optimal no matter the missing data, whilst asking the user to reveal as few preferences as possible. Previously, this problem has been tackled with a systematic branch & bound algorithm. We now investigate whether a local search approach can find good quality solutions to such problems with fewer resources. While the approach is general, we evaluate it experimentally on a class of meeting scheduling problems with missing preferences. The experimental results show that the local search approach returns solutions which are very close to optimality, whilst eliciting a very small percentage of missing preference values. In addition, local search is much faster than the systematic approach, especially as the number of meetings increases.

1 Introduction

Many real-life problems such as scheduling, planning, and resource allocation problems can be modeled as constraint satisfaction problems and thus solved efficiently using constraint programming techniques [4,14]. A constraint satisfaction problem (CSP) is represented by a set of variables, each with a domain of

---

F. Rossi—On leave from University of Padova

© Springer International Publishing AG 2017
D. Salvagnin and M. Lombardi (Eds.): CPAIOR 2017, LNCS 10335, pp. 403–418, 2017.
DOI: 10.1007/978-3-319-59776-8_32

values, and a set of constraints. A solution is an assignment of values to the variables which satisfies all constraints and which optionally maximizes/minimizes an objective function. For example, let us consider the meeting scheduling problem, which is the problem of scheduling meetings attended by a number of agents. This can be modeled as a CSP [15] where the variables represent the meetings, the domains represent the time slots, and constraints between two variables, that represent different meetings, model the fact that one or more agents must participate in both meetings. Every constraint is satisfied by all pairs of time slots that allow participation in both meetings according to the time needed to go between the corresponding locations. A solution is a schedule, i.e., an allocation of a time slot to each meeting, such that all agents can attend all their meetings.

Soft constraints [13] can model optimization problems by allowing for several levels of satisfiability in a constraint via the use of preference or cost values. These represent how much we like a particular instantiation of the variables of a constraint. For example, in meeting scheduling problems, each agent may give preferences to the time slots as well as to the combination of time slots for the meetings they will participate in, and the goal is to find a schedule which allows all agents to attend all their meetings and makes the agents as happy as possible in terms of their preferences.

Incomplete soft constraint problems [8–10] can model situations in which preferences are not completely known before the solving process starts. In the context of meeting scheduling problems, some agent may not specify preferences over every time slot due to computational costs, or if there is an ongoing elicitation process. In this scenario, we aim at finding a solution which is necessarily optimal, that is, optimal no matter what the missing preferences are, while asking the user to reveal as few preferences as possible. Previously this problem has been tackled with a branch and bound algorithm which was guaranteed to find a solution with this property [10]. We now show that a simple local search method can solve such problems optimally, or close to optimally, with fewer resources and can be effective in terms of both elicitation and scalability.

Our local search algorithm, defined for any kind of incomplete soft constraint problem, starts from a randomly chosen assignment to all the variables and, at each step, moves to the most promising assignment in its neighborhood, obtained by changing the value of one variable. The variable to reassign and its new value are chosen so to maximize the quality of the new assignment. To make this choice, we elicit some of the preferences missing in the neighboring assignments.

To evaluate the quality of the obtained solution and the performance of our local search approach, we compare it with the performance of the systematic method shown in [10]. We have decided to evaluate the approaches on classes of incomplete fuzzy meeting scheduling problems, and not on classes of randomly generated incomplete fuzzy problems, since structured problems are more realistic. Since in [10] tests for the systematic approach have been performed on randomly generated classes of incomplete fuzzy constraints, in this paper we first evaluate how the approach in [10] works on classes of incomplete fuzzy meeting scheduling problems. This is necessary to state which is the best systematic algorithm when we consider the class of incomplete fuzzy meeting scheduling problems. The aim of the paper is to answer all the following research questions: can we efficiently tackle incomplete meeting scheduling problems with local search? How much elicitation is needed? Do we need a sophisticated local algorithm to compete with the existing systematic approach?

Experimental results show that our local search approach returns very high quality solutions, which have a preference very close to the optimal one, whilst eliciting a small percentage of missing preferences (always smaller than 5%). Moreover, they show that a local search approach is much faster than the systematic approach when the number of meetings (i.e., the number of variables) increases (for example, 2 s vs. 325 s when there are 14 meetings). Therefore, it is not necessary to develop a complicated local search algorithm to obtain good results. Even if the local search approach is general, i.e., it holds for any kind of preference, to evaluate its performance experimentally, we consider incomplete meeting scheduling problems with fuzzy preferences.

The paper is structured as follows. In Sect. 2 we define incomplete soft constraint problems (ISCSPs) and an existing branch and bound approach to solve them, as well as meeting scheduling problems. In Sect. 3 we present our local search algorithm for ISCSPs that interleaves search and preference elicitation. In Sect. 4 we define incomplete soft meeting scheduling problems (IMSPs), that are meeting scheduling problems where constraints are replaced by preferences and some preferences may be missing, and we describe the problem generator used in the experimental studies. Next, in Sect. 5 we discuss the experimental results of the branch and bound and local search approaches on classes of fuzzy IMSPs. In Sect. 6 we compare our approach to other existing approaches to deal with incompletely specified constraint optimization problems. Finally, in Sect. 7 we summarize the results contained in this paper.

A preliminary version of some of the results of this paper and only some experiments on randomly generated fuzzy CSPs with missing preferences are contained in a short paper appeared in [11].

2 Background

We give some basic notions on incomplete soft constraint problems and meeting scheduling problems.

2.1 Incomplete Soft Constraint Problems

A soft constraint [13] involves a set of variables and it is defined by a preference function that associates a preference value from a (totally or partially ordered) set to each instantiation of its variables. This preference value is taken from a c-semiring, which is defined by $\langle A, +, \times, 0, 1 \rangle$, where A is the set of preference values, $+$ is a commutative, associative, and idempotent operator, $\times$ is used to combine preference values and is associative, commutative, and distributes over $+$, 0 is the worst element, and 1 is the best element. The c-semiring S

induces a partial or total order $\leq_S$ over the preference values, where $a \leq_S b$ iff $a + b = b$. A soft constraint satisfaction problem (SCSP) is a tuple $\langle V, D, C, S \rangle$ where V is a set of variables, D is the domain of the variables and C is a set of soft constraints over V associating values from A. An instance of the SCSP framework is obtained by choosing a particular c-semiring. For instance, choosing $\langle [0, 1], max, min, 0, 1 \rangle$ means that preferences are in $[0, 1]$ and we want to maximize the minimum preference. This defines the so-called fuzzy CSPs (FCSPs). Weighted CSPs (WCSPs) are instead defined by the c-semiring $\langle \mathcal{R}^+, min, +, +\infty, 0 \rangle$: preferences are interpreted as costs, they are in $\mathcal{R}^+$, and the goal is to minimize the sum of the costs.

Incomplete soft constraint satisfaction problems (ISCSPs) [10] generalize SCSPs by allowing incomplete soft constraints, i.e., soft constraints with missing preferences. More precisely, an incomplete soft constraint involves a set of variables and associates a value from the preference set A or ?. All tuples (i.e., partial assignments to the variables) associated with ? are called *incomplete tuples* meaning that their preference is unspecified. Given an assignment s to all the variables of an ISCSP P, the preference of s in P, denoted by $pref(P, s)$ (aka $pref(s)$), is the combination (which corresponds to the minimum in the fuzzy case and to the sum in the weighted case) of the known preferences associated to the sub-tuples of s in the constraints of P. Incomplete fuzzy constraint satisfaction problems (IFCSPs) are ISCSPs where the preferences are fuzzy. Similarly, incomplete weighted constraint satisfaction problems (IWCSPs) are weighted CSPs where some preferences may be missing.

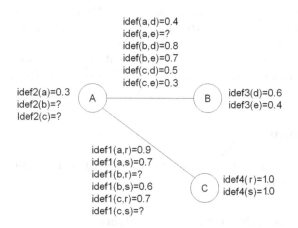

Fig. 1. An example of IFCSP.

Figure 1 shows an example of an IFCSP with three variables A, B, and C, with domains $D(A) = \{a, b, c\}$, $D(B) = \{d, e\}$, and $D(C) = \{r, s\}$. In this IFCSP there are three unary constraints on A, B, and C specified resp. by preference functions $idef2$, $idef3$, and $idef4$, a binary constraint involving A and

C specified by the preference function $idef1$, and a binary constraint involving A and B specified by the preference function $idef$. The presence of the question marks identifies the missing preference values. The preference of the assignment $s = \langle A = a, B = e, C = r \rangle$, i.e., $pref(P, s)$ is $min(0.3, 0.4, 1, 0.9) = 0.3$. Since this assignment involves some missing preferences, such as the one for the tuple $\langle A = a, B = e \rangle$, its preference should be interpreted as an upper bound of the actual preference for this assignment.

In an SCSP a complete assignment of values to all the variables is an optimal solution if its preference is the best possible. This notion is generalized to ISCSPs via the notion of *necessarily optimal solutions*, that is, complete assignments which have the best preference no matter the value of the unknown preferences.

In the example of Fig. 1, the assignment $\langle A = a, B = e, C = r \rangle$ is not necessarily optimal. In fact, if all missing preferences are 1, this assignment has preference 0.3, while the assignment $\langle A = b, B = d, C = r \rangle$ has preference 0.6. In this example, there are no necessarily optimal solutions. Consider now the same example where the preferences of both $A = b$ and $A = c$ are set to 0.2. In this new IFCSP, the assignment $\langle A = a, B = d, C = s \rangle$ has preference 0.3 and is necessarily optimal. In fact, whatever values are given to the missing preferences, all other assignments have preference at most 0.2 (if $A = b$ or c) or 0.3 (if $A = a$).

In [10] several *complete algorithms* are proposed to find a necessarily optimal solution of an ISCSP. All these algorithms follow a branch and bound schema where search is interleaved with elicitation. Elicitation is needed since the given problem may have an empty set of necessarily optimal solutions. By eliciting some preferences, this set eventually becomes non-empty. Several elicitation strategies are considered and tested on randomly generated IFCSPs. The goal was to elicit as little as possible before finding a necessarily optimal solution. In particular, the elicitation strategies depended on three parameters: *when* to elicit, *what* to elicit, and *who* chooses the value to be assigned to the next variable. For example, one may only elicit missing preferences after running branch and bound to exhaustion, or at the end of every complete branch, or at every node in the search tree. Also, one may elicit all missing preferences related to the candidate solution, or one might just ask the user for the worst preference among some missing ones. Finally, when choosing the value to assign to a variable, the system can make the choice randomly, or it can ask the user, who knows or can compute (all or some of) the missing preferences, for help. All of the details of the systematic approach can be found in [10].

2.2 Meeting Scheduling Problems

The meeting scheduling problem (MSP) is a well known benchmark for constraint satisfaction problems (CSPs) [15]. It is the problem of scheduling some meetings by allowing the participants to attend all the meetings they are involved in. More formally, a meeting scheduling problem can be described by: (i) a set of agents, (ii) a set of meetings, each with a location and a duration, (iii) a set of time slots where meetings can take place, (iv) for each meeting, a subset of agents that are supposed to attend such a meeting, (v) for each pair of locations, the time to go from one location to the other one (the so-called distance table). Some typical simplifying assumptions are making all meeting durations equal, and having the same number of meetings for each agent. Solving a MSP means allocating each meeting to a time slot in a way that all agents can participate in their meetings. An agent cannot participate in two meetings if they overlap in time, or if the time to get between locations prevents him from getting to the second meeting. The MSP can be modelled as a CSP [15] with variables representing the meetings and domains representing the time slots. We have a constraint between two variables when one or more agents participate in the two meetings. Each constraint is satisfied by all pairs of time slots that allow the participation to both meetings taking into account the time needed to get between the two locations.

3   Local Search on ISCSPs

A local search approach has been defined in [1,3] for soft constraint problems. Here we adapt it to deal with incompleteness.

Our local search algorithm for ISCSPs interleaves elicitation with search. To start, we randomly generate an assignment of all the variables. To assess the quality of such an assignment, we compute its preference. Since some preferences involved in the chosen assignment may be missing, we ask the user to reveal them. At each step, we have a current assignment $(v_1, \ldots, v_n)$ for variables $X_1, \ldots, X_n$ and we choose a variable to change its value. To do this, we compute a 'local' preference for each pair (X_i, v_i), by combining all the specified preferences of the tuples containing $X_i = v_i$ in all the constraints. If the pair (X_i, v_i) is the one with the worst local preference, we choose another value for X_i.

Consider again the example in Fig. 1 and the assignment $\langle A = a, B = e, C = r \rangle$. In this assignment, the local preference of variable A is $min(0.3, 0.4, 0.9) = 0.3$, while for B is 0.4, and for C is $min(0.9, 1) = 0.9$. Thus our algorithm would choose variable A.

To choose the new value for the selected variable, we compute the preferences of the variable assignments obtained by choosing the other values for this variable. We then choose the value which is associated to the best preference. If two values are associated to assignments with the same preference, we choose the one associated to the assignment with the smaller number of incomplete tuples. In this way, we aim at moving to a new assignment which is better than the current one and has the fewest number of missing preferences.

In the running example, from assignment $\langle A = a, B = e, C = r \rangle$, once we know that variable A will be changed, we compute $pref(\langle A = b, B = e, C = r \rangle) = 0.4$ and $pref(\langle A = c, B = e, C = r \rangle) = 0.3$. Thus, we would select the value b for A.

Since the new assignment, say s', could have incomplete tuples, we ask the user to reveal enough of this data to compute the actual preference of s'. We call ALL the elicitation strategy that elicits all the missing preferences associated to the sub-tuples of s' in the constraints.

For fuzzy constraints, we also consider another elicitation strategy, called WORST, that asks the user to reveal only the worst preference among the missing ones, if it is less than the worst known preference. In the fuzzy case, this is enough to compute the actual preference of s', since the preference of an assignment coincides with the worst preference in its constraints.

For weighted constraints, we consider the following three strategies (besides ALL):

- WW: we elicit the worst missing cost (that is, the highest) until either all the costs are elicited or the current global cost of the assignment is higher than the preference of the best assignment found so far;
- BB: we elicit the best (i.e., the minimum) cost with the same stopping condition as for WW;
- BW: we elicit the best and the worst cost in turn, with the same stopping condition as for WW.

As in many classical local search algorithms, to avoid stagnation in local minima, we employ tabu search and random moves. Our algorithm has two parameters: p, which is the probability of a random move, and t, which is the tabu tenure. When we have to choose a variable to re-assign, the variable is either randomly chosen, with probability p or, with probability $(1-p)$, we perform the procedure described above. Also, if no improving move is possible, i.e., all new assignments in the neighborhood are worse than or equal to the current one, then the chosen variable is marked as tabu and not used for t steps.

Notice that, while in classical local search scenarios the underlying problem is always the same, and we just move from one of its solutions to another one, in our scenario we also change the problem via the elicitation strategies. Since the change involves only the preference values, the solution set remains the same, although the preferences of the solutions may decrease over time.

During search, the algorithm maintains the best solution found so far, which is returned when the maximum number of allowed steps is reached. In the ideal case, the returned solution is a necessarily optimal solution of the initial problem with the preferences added by the elicitation. However, there is no formal guarantee that this is so: we can reach a problem with necessarily optimal solutions, but the algorithm may fail to find one of those. Our experimental results on classes of meeting scheduling problems with missing preferences will show that, even when the returned solutions are not necessarily optimal, their quality is very close to that of the necessarily optimal solutions.

Notice that systematic and local search solvers are completely re-usable, i.e., the algorithms are applicable to other problems as well.

4   Experimental Setting and Problem Generator

We now consider a generalization of the MSP, called IMSP, where constraints are replaced by soft constraints with preferences and each agent may decide to give only some of the preferences over the meetings he would like to attend. We recall that in these problems the variables are meetings and the variables' domains contain all the time slots in which the meeting can be allocated. Preferences associated to each variable assignment (i.e., to each time slot for each meeting) model the aggregation of the participants' preferences and they are computed as follows. For every meeting and for each time slot of this meeting the system asks the agents to reveal their preferences over this time slot. However, some agents may decide to not reveal them. The system then collects all the received preferences for that time slot and, if the majority of agents involving in this meeting express a preference over it, then it assigns to this time slot the average preference value, otherwise it marks that preference as unknown.

Similarly, in every binary soft constraint the preferences of every pairs of time slots, which are compatible with the distance table, is computed as follows: the system collects the preferences given by the agents and it sets the preference value of the pair as unknown, if one of the values has an associated incomplete preference, otherwise it takes the average preference.

Notice that other measures, beside the average, can be used to compute a collective preference, such as for example the minimum or the maximum [2], and there are also other criteria that can be considered to decide when a preference is unknown, for example when a certain number of agents have not revealed their preferences.

Solving an IMSP concerns then finding time slots for the meetings such that all agents can participate and, among all possible solutions, the solution has the best preference.

In this context, given an IMSP P, the necessarily optimal solutions are schedules that are optimal no matter how the missing preferences are revealed. Thus, if there is at least one such solution, this is certainly preferred to any other.

For eliciting a missing preference, a query is submitted to the participating community. We get a preference value by asking all those participants that didn't give their preference upfront. The user's effort shown in the paper is therefore the effort of such a community. More precisely, the user's effort is the number of preference values the users are required to consider to answer the elicitation queries.

Our generator of incomplete meeting scheduling problems has the following parameters:

- m: number of meetings;
- n: number of agents;
- k: number of meetings per agent;
- l: number of time slots;
- min and max: minimal and maximal distance in time slots between two locations;
- i: percentage of incomplete preferences.

We randomly generate an IMSP with m variables, representing the meetings, each with domain in $\{1, \ldots, l\}$ to represent the time slots from 1 to l. Such time slots are assumed to be one time unit long and to be adjacent to each other.

Given two time slots a and b, they can be used for two meetings only if the time needed to go from a location to the other one is at most $b - a - 1$.

For each of the n agents, we generate randomly k integers between 1 and m, representing the meetings in which he participates. Also, for each pair of time slots, we randomly generate an integer between min and max that represents the time needed to go from one location to the other one. These integers will form what we call the distance table. Notice that $min = 1$ means that the time distance between two meetings' locations is at least 1 time slot.

Given two meetings, if there is at least one agent who needs to participate in both, we generate a binary constraint between the corresponding variables. Such a constraint is satisfied by all pairs of time slots that are compatible according to the distance table.

We then generate the fuzzy preferences over the domain values according to parameter i. We set the preference of each compatible pair in the binary constraints by assigning the average preference that the values in the pair have in their domain when both preferences are known. Otherwise, we assign an unknown preference.

As an example, assume $m = 5$, $n = 3$, $k = 2$, $l = 5$, $min = 1$, $max = 2$, and $i = 30$. According to these parameters, we generate an IMSP with the following features:

- 5 meetings: m_1, m_2, m_3, m_4, and m_5;
- 3 agents: a_1, a_2, and a_3;
- participation in meetings: we randomly generate 2 meetings for each agent, for example: a_1 must participate in meetings m_1 and m_2, a_2 must participate in meetings m_4 and m_5, and a_3 must participate in meetings m_2 and m_3;
- 5 time slots: $t_1, \ldots, t_5$;
- distance table: we randomly generate its values;
- we randomly generate the fuzzy preferences associated to the variable values in a way that 30% of the preferences are missing. Then, we compute the preferences of the compatible pairs in the binary constraints (or we state that the preference is missing) as stated before.

In our experimental evaluation we randomly generate IMSPs by varying the number of meetings m from 5 to 14, the number of agents n from 4 to 10, the number of meetings per agent k from 2 to 5, and the percentage of incompleteness i from 10% to 100%. The distance distribution is uniform random. Moreover, we will focus on IMSPs where the preferences are fuzzy.

5   Experimental Evaluation

In this section we show how the branch and bound and local search algorithms perform on classes of fuzzy IMSPs. All our experiments have been performed on an AMD Athlon 64×2 2800+, with 1 Gb RAM, Linux operating system, and using JVM 6.0.1. Results are averaged over 100 problem instances.

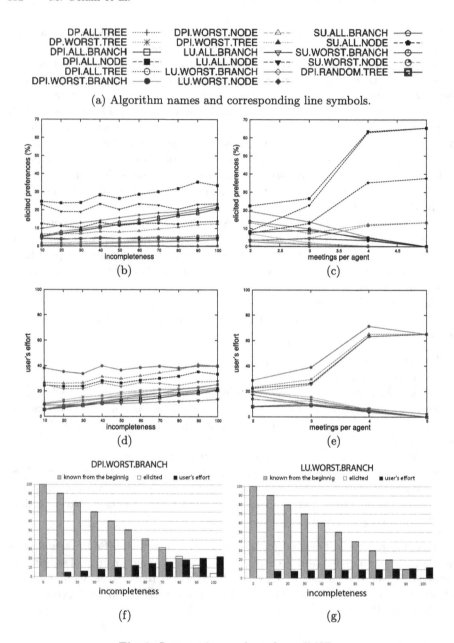

(a) Algorithm names and corresponding line symbols.

(b)   (c)

(d)   (e)

(f)   (g)

Fig. 2. Systematic search on fuzzy IMSPs.

5.1 The Systematic Approach

We implemented all the algorithms presented in [10], which have been obtained by instantiating in all the possible ways the parameters *who*, *when*, and *what*.

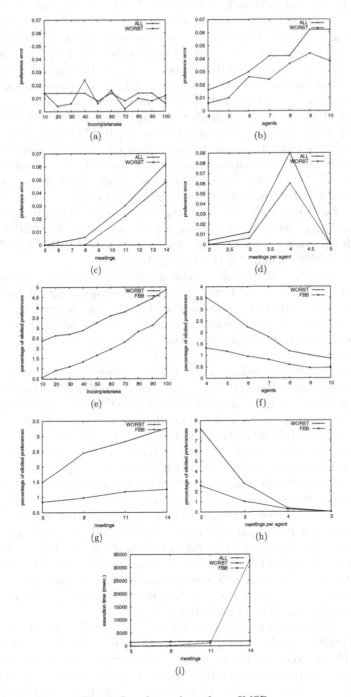

Fig. 3. Local search on fuzzy IMSPs.

While in [10] these algorithms have been tested on randomly generated IFCSPs, we now test their performance on the classes of fuzzy IMSPs defined above.

In all the experimental results for the systematic approach, the association between an algorithm name and a line symbol is shown in Fig. 2(a). The names of the algorithms coincide with those used in the literature [10]. Every acronym describes the elicitation strategy that is used, i.e., who chooses the value to be assigned to the next variable, what to elicit, and when to elicit. For example, one may only elicit missing preferences after running branch and bound to exhaustion (TREE), or at the end of every complete branch (BRANCH), or at every node in the search tree (NODE). Then, one may elicit all missing preferences (ALL) related to the candidate solution, or one might just ask the user for the worst preference (WORST) among some missing ones. Finally, when choosing the value to assign to a variable, the system can assign this value in decreasing order (DP and DPI) or one might ask the user, who knows or can compute (all or some of) the missing preferences, for help in a lazy or a smart way (LU and SU). Thus, each name has the form $X.Y.Z$, where $X \in \{$TREE, BRANCH, NODE$\}$, $Y \in \{$ALL, WORST$\}$, and $Z \in \{$DPI, DP, LU, SU$\}$.

To evaluate the performance of the algorithms, as in [10], we measured the percentage of the elicited tuples, i.e., the percentage of missing preferences revealed by the agents, and the user's effort, i.e., the percentage of missing preferences that the user has to consider to answer the queries of the system. In these experimental tests the default values are $m = 12$, $n = 5$, $k = 3$, $l = 10$, $min = 1$, and $max = 2$.

As expected, as the incompleteness in the problem rises, the percentage of elicited tuples tends to increase slightly (see Fig. 2(b)). The algorithm that elicits fewest tuples in this context is LU.WORST.BRANCH (less than 1% of incomplete tuples), among the ones that need the user's help to choose the values to instantiate, and DPI.WORST.BRANCH (less than 5% of incomplete tuples) among the others. The common features of these algorithms are that they elicit the worst preference every time they reach a complete assignment.

Moreover, these two best algorithms tend to elicit fewer tuples as the meetings per agent increases, as shown in Fig. 2(c). They reach close to 0% at 5 meetings per agent because, in that setting, the optimal preference is 0 and so they need to ask only for the missing preferences to fix the lower bound in the first branch and then they do not reach a complete assignment again. They need more or less the same amount of elicitation in percentage terms as the number of these meetings per agent increases, while eliciting almost always less than 5% of the missing tuples. Moreover, they are the best ones in terms of user's effort which is always smaller than 20% (Figs. 2(d) and (e)).

Figure 2(g) shows the results of LU.WORST.BRANCH. It elicits a very small number of preferences also in totally incomplete settings. The elicitation is always less than 1% and the user has to consider only around 10% of the incomplete tuples to give the preferences asked by the system. This behaviour is due to the fact that the search is guided by the user that selects the value to instantiate.

If we are working in settings where the system decides how to instantiate the domain values, the best algorithm is DPI.WORST.BRANCH which, as shown in Fig. 2(f), elicits a percentage of preferences which remains always under 5% so it gives very good results. Also, the effort required to the user is very small: it is at most 23% even if the incompleteness is 100% and the percentage of the elicited preferences is smaller than 5%. We will consider this algorithm, which we will call FBB (which stands for fuzzy Branch and Bound), when comparing it to our local search algorithms.

5.2 The Local Search Approach

We test the performance of our local search algorithm, instantiated with both ALL and WORST elicitation strategies, on fuzzy IMSPs by considering the quality of the returned solution, the percentage of elicited preferences, and the execution time needed to find a solution. Moreover, we compare our best local search algorithm to FBB on the same test set in terms of percentage of elicited preferences and execution time. In these experimental tests we use a step limit of 100000, a random walk probability of 0.2, and a tabu tenure of 1000. Also, we consider an average on 100 problems and the default values are $m = 10$, $n = 5$, $k = 3$, $l = 10$, $i = 30\%$, $min = 1$, $max = 2$.

First, we measure the quality of the returned solution in terms of the preference error: this is the distance, in percentage over the whole range of preference values, between the preference of the solution returned by the local search algorithm and the preference of a necessarily optimal solution. When we vary the amount of incompleteness in the problem, surprisingly, the preference error is very small, and always less than 2.5% both for ALL and WORST elicitation strategies (see Fig. 3(a)). When we vary the number of agents, meetings, and meetings per agent, the preference error rises as we increase each parameter, though it remains in the worst case below 10%. The WORST strategy is the best elicitation strategy in all settings (see Figs. 3(b), (c) and (d). Note that for 5 meetings per agent in Fig. 3(d) the problems almost all have solution preferences equal to 0 and so the error is 0 as well.

Next, we consider the percentage of elicited preferences of our best local search algorithm (i.e., with the WORST elicitation strategy) and compare it to FBB (see Figs. 3(e), (f), (g) and (h)). We omit the results of the ALL strategy as it elicits from 60% to 80% of the incomplete preferences, while the WORST strategy and FBB elicit always less than 10%. In all settings, the WORST strategy elicits more preferences than FBB. However, the local search approach is more efficient in terms of execution time when the number of meetings increases (see Fig. 3(i)). In fact, for more than 11 meetings, FBB takes much more time than the WORST and ALL local search strategies.

To better evaluate the performance of our best local search algorithm (i.e., with the WORST strategy), we consider also its runtime behavior (Fig. 4). The left part of the figure shows the percentage of elicited preferences, where one of the parameters among i, n, m, k, is not fixed, in turn. The same holds in

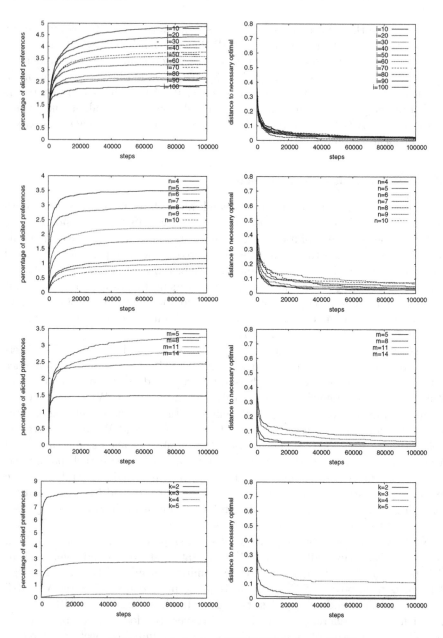

Fig. 4. Runtime behavior of WORST in fuzzy IMSPs.

the right part of the figure, which shows the distance to the preference of the necessarily optimal solution.

As for other classes of incomplete soft constraint problems, WORST elicits all the preferences needed to reach a solution during the first steps. Moreover, the distance to the necessarily optimal preference decreases very rapidly during the first 20000 steps and then decreases slightly as the search continues. This behavior is the same no matter which parameter is varying. Thus we could stop the algorithm after only 20000–30000 steps whilst still ensuring good solutions and a fast execution time.

6 Related Work

Some works have addressed issues similar to those considered in this paper by allowing for open settings in CSPs: both open CSPs [6,7] and interactive CSPs [12] work with domains that can be partially specified, and in dynamic CSPs [5] variables, domains, and constraints may change over time.

While the main goal of the work on interactive CSPs is to minimize the run time of the solver, we emphasize the minimization of the number of queries to the user and/or of the user effort. The work closest to ours is the one on open CSPs. An open CSP is a possibly unbounded, partially ordered set of constraint satisfaction problems, each of which contains at least one more domain value than its predecessor. Thus, in [7] the approach is to solve larger and larger problems until a solution is found, while minimizing the number of variable values asked to the user. To compare it to our setting, we assume all the variable values are known from the beginning, while some of the preferences may be missing, and our algorithms work on different completions of the given problem. Also, open CSPs exploit a *Monotonicity Assumption* that each agent provides values for variables in strictly non-decreasing order of preference. Even when there are no preferences, each agent gives only values for variables that are feasible. Working under this assumption means that the agent that provides new values/costs for a variable must know the bounds on the remaining possible costs, since they are provided best value first. If the bound computation is expensive or time consuming, then this is not desirable. This is not needed in our setting, where single preferences are elicited.

Algorithms that interleave elicitation and solving for incomplete problems have been considered also in [16]. However, we consider incomplete *soft* constraint problems, while in [16] they consider incomplete constraint problems and they assume to have probabilistic information regarding whether an unknown tuple satisfies a constraint. Also, we apply local search.

7 Conclusions

We defined a local search approach for solving incomplete soft constraint problems and we compared it to an existing branch and bound approach with the same goal. The comparison was done on classes of incomplete fuzzy meeting scheduling problems. Experimental results show that our local search approach returns solutions which are very close to optimality, whilst eliciting a very small percentage of missing preference values. In addition, our simple local search is much faster than the systematic approach, especially as the number of meetings increases. Therefore, it is not necessary to use a sophisticated algorithm to compete with the complete method.

References

1. Aglanda, A., Codognet, P., Zimmer, L.: An adaptive search for the NSCSPs. In: Proceedings of CSCLP 2004 (2004)
2. Arrow, K.J., Sen, A.K., Suzumura, K.: Handbook of Social Choice and Welfare. North Holland, Elsevier (2002)
3. Codognet, P., Diaz, D.: Yet another local search method for constraint solving. In: Steinhöfel, K. (ed.) SAGA 2001. LNCS, vol. 2264, pp. 73–90. Springer, Heidelberg (2001). doi:10.1007/3-540-45322-9_5
4. Dechter, R.: Constraint Processing. Morgan Kaufmann, San Francisco (2003)
5. Dechter, R., Dechter, A.: Belief maintenance in dynamic constraint networks. In: AAAI 1988, pp. 37–42 (1988)
6. Faltings, B., Macho-Gonzalez, S.: Open constraint satisfaction. In: Hentenryck, P. (ed.) CP 2002. LNCS, vol. 2470, pp. 356–371. Springer, Heidelberg (2002). doi:10.1007/3-540-46135-3_24
7. Faltings, B., Macho-Gonzalez, S.: Open constraint programming. AI J. **161**(1–2), 181–208 (2005)
8. Gelain, M., Pini, M.S., Rossi, F., Venable, K.B.: Dealing with incomplete preferences in soft constraint problems. In: Bessière, C. (ed.) CP 2007. LNCS, vol. 4741, pp. 286–300. Springer, Heidelberg (2007). doi:10.1007/978-3-540-74970-7_22
9. Gelain, M., Pini, M.S., Rossi, F., Venable, K.B., Walsh, T.: Elicitation strategies for fuzzy constraint problems with missing preferences: algorithms and experimental studies. In: Stuckey, P.J. (ed.) CP 2008. LNCS, vol. 5202, pp. 402–417. Springer, Heidelberg (2008). doi:10.1007/978-3-540-85958-1_27
10. Gelain, M., Pini, M.S., Rossi, F., Venable, K.B., Walsh, T.: Elicitation strategies for soft constraint problems with missing preferences: properties, algorithms and experimental studies. AI J. **174**(3–4), 270–294 (2010)
11. Gelain, M., Pini, M.S., Rossi, F., Venable, K.B., Walsh, T.: A local search approach to solve incomplete fuzzy CSPs. In: Proceedings of ICAART 2011, Poster Paper (2011)
12. Lamma, E., Mello, P., Milano, M., Cucchiara, R., Gavanelli, M., Piccardi, M.: Constraint propagation and value acquisition: why we should do it interactively. In: IJCAI, pp. 468–477 (1999)
13. Meseguer, P., Rossi, F., Schiex, T.: Soft constraints. In: Rossi, F., Van Beek, P., Walsh, T. (eds.) Handbook of Constraint Programming. Elsevier, Amsterdam (2006)
14. Rossi, F., Van Beek, P., Walsh, T.: Handbook of Constraint Programming. Elsevier, Amsterdam (2006)
15. Shapen, U., Zivan, R., Meisels, A.: Meeting scheduling problem (MSP). http://www.cs.st-andrews.ac.uk/~ianm/CSPLib/prob/prob046/index.html (2010)
16. Wilson, N., Grimes, D., Freuder, E.C.: Interleaving solving and elicitation of constraint satisfaction problems based on expected cost. Constraints **15**(4), 540–573 (2010)

Author Index

Printed in the United States
By Bookmasters